INTRODUCTION TO TECHNICAL MATHEMATICS
With Problem Solving

Third Edition

WALTER W. LEFFIN
Professor Emeritus, University of Wisconsin - Oshkosh

GEORGE L. HENDERSON
Professor Emeritus, University of Wisconsin - Eau Claire

MARY VAN BECK VOELKER
Wisconsin Gas Company

FRED C. JANUSEK
Northeast Wisconsin Technical Institute

WAVELAND
PRESS, INC.
Long Grove, Illinois

For information about this book, contact:
Waveland Press, Inc.
4180 IL Route 83, Suite 101
Long Grove, IL 60047-9580
(847) 634-0081
info@waveland.com
www.waveland.com

10-digit ISBN 1-57766-023-4
13-digit ISBN 978-1-57766-023-1

Printed in the United States of America

14   13   12   11   10   9

TABLE OF CONTENTS

PREFACE

In this Third Edition we have added a new chapter on using the quadratic formula to solve quadratic equations. We also have included extra problem sets at the ends of chapters 3, 5, 6, 7, 8, and 12 that emphasize technical applications. Finally, we have increased the use of metric units after chapter 5. These changes are in response to feedback from adopters of the previous editions.

During the organizing and writing of the First Edition in the early 1980s, the authors were guided by two principles: (1) students entering post-secondary education programs must be guaranteed the prerequisite capabilities necessary for the study of technical mathematics, and (2) a preliminary course should be provided to prepare students for success in a technical mathematics course, some of which at that time had failure rates as high as 75 percent. The result was *Introduction to Technical Mathematics*, designed for an introductory course and intended for students with minimal mathematics backgrounds who wished to prepare for further study in technical areas.

Later, when problem-solving ability became an emphasis in mathematics education (and other areas), our Second Edition incorporated two new chapters and several subsections designed to develop and enhance students' problem-solving abilities. The book then became *Introduction to Technical Mathematics with Problem Solving*.

As with earlier editions, the Third Edition's topical coverage is comprehensive but is presented in an easy-to-read manner so that students should be able to use the text without difficulty. A wealth of applications is contained in the nearly 800 worked out examples and some 6,000 exercises and problems. Formulas, rules of operations, and summaries of important procedures are highlighted and/or boxed when introduced.

Sets of "Review Exercises" are provided at the end of most sections to help students retain what they have learned. Calculators may be used at the instructor's option, and calculator exercises are given at the end of most sections after chapter 4. (Calculator exercises are not provided for the first three chapters so that manipulative skills can be emphasized.)

Answers for all *odd-numbered* exercises and problems are given at the end of the book, with the exception of those for chapters 4 and 11—the problem-solving chapters—which are provided within their respective chapters. Answers to *all* exercises and problems can be found in the *Instructor's Manual*.

The authors feel that material covered in the first three chapters must be mastered by students *before* beginning the study of later chapters. They also recognize that some students will already have mastered arithmetic and its applications and can begin this course with either chapter 4 or chapter 5, depending on the instructor's emphasis on problem solving. A diagnostic test over chapters 1, 2, and 3 is included in the *Instructor's Manual* for teachers to use to determine whether student's have mastered those skills.

Those instructors who wish to teach a *minimal course* from this textbook should include chapters 1, 2, 3, 5, 6, 7, 8, 9, and 10. Those who wish to teach a *comprehensive course* will include all chapters in the text.

The First Edition benefitted from reviews and comments from many people, including those we would especially like to thank: Glenn R. Boston, George W. Brewer, Reuben C. Drake, Tom O. Eller, Lanny Hendrickson, John Hutchinson, Stanley Kohli, Richard Rowe, Richard Semmler, Morris L. Shoss, Gerald Skidmore, Lawrence Trivieri, and Frank Weeks.

The Second Edition benefitted from comments made by instructors at Rose State College, Rochester Technical College, Pratt Community College, Women Unlimited, Fox Valley Technical College, and Greenville Technical College.

The Third Edition benefitted from comments by instructors from Medicine Hat College, KVTC in Fairfield, Maine, Eastern Maine Technical College, Delaware Technical and Community College, Richmond Community College, Western Piedmont Community College, Bainbridge College, Lake Michigan College, North Western Iowa Technical College, Gateway Technical College, and University of Wisconsin-Eau Claire.

The authors each appreciate the efforts of the others, blending areas of expertise into the whole that comprises this book: George L. Henderson, problem solving, arithmetic, quadratics, and trigonometry; Walter W. Leffin, geometry and measurement; Mary Van Beck Voelker, algebra, equations, and proportions; Fred C. Janusek, applications. Special thanks go to Dr. William K. Applebaugh, University of Wisconsin-Eau Claire, for his help with the revisions for the Third Edition, and also to Tracy A. Gunderson, for her work on chapter 13.

We appreciate the efforts of our editor, Laurie Prossnitz, and Neil J. Rowe, publisher of Waveland Press, in guiding the development of this edition.

Walter W. Leffin
George L. Henderson
Mary Van Beck Voelker
Fred C. Janusek

Special note to instructors: Some students' answers to problems (for example, for Section 6.5) may vary slightly from those given in the Instructor's Manual. Reasons include variances in decimal approximations used for Pi and unit conversions. For the most part, the decimal approximation for Pi used in this text is 3.14.

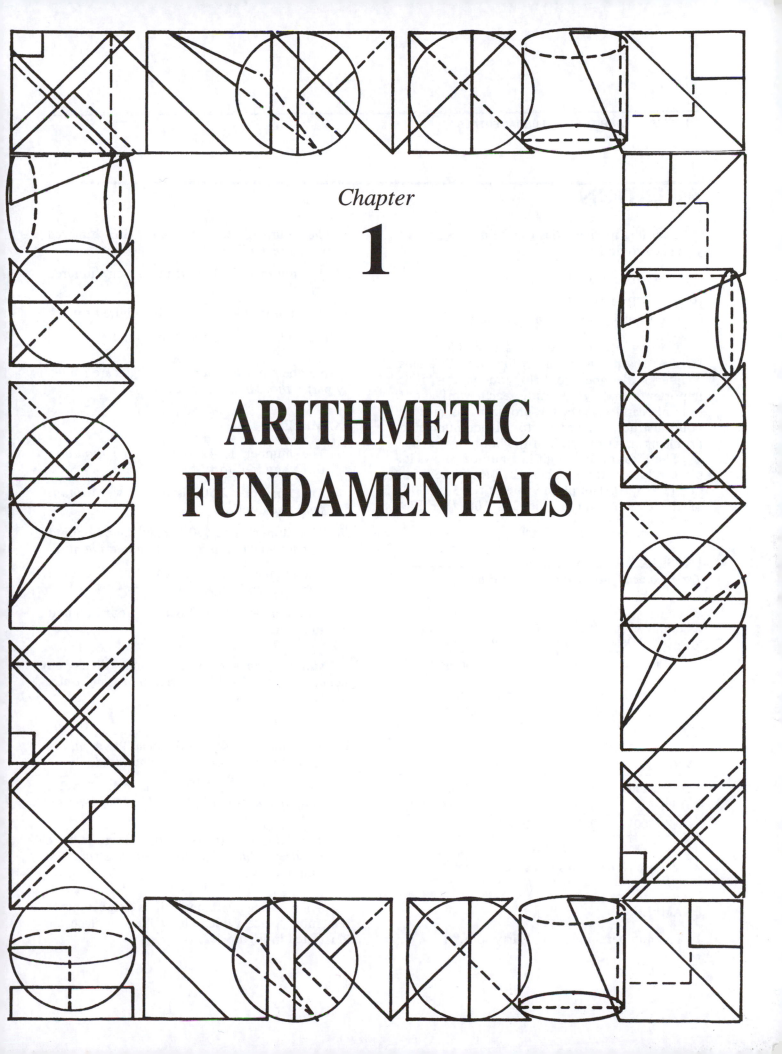

Chapter

1

ARITHMETIC
FUNDAMENTALS

1.1

ADDITION

Learning mathematics can be quite easy. All you need to do is:

(a)   get off to a good start;

(b)   go step-by-step; and

(c)   continually *review* what you learn.

Getting off to a good start in this course involves sharpening your arithmetic skills. This chapter is designed to help you do this. It is important that you study each worked-out example carefully. Do the problems assigned by your instructor. Answers to all odd-numbered problems are at the back of the book.

The location of a digit in a number shows its **place value.** In the number 3278 the place value of the 7 is 10. The place value of the 3 is 1000, of the 8 is 1, and of the 2 is 100. (See Figure 1.1.)

FIGURE 1.1

Three thousand two hundred seventy-eight

The value of each place in a number is *ten times* the value of the place to its right, as shown in Table 1.1.

TABLE 1.1

Ten thousands place	Thousands place	Hundreds place	Tens place	Ones place
10 × 1000	10 × 100	10 × 10	10 × 1	1

We use place values to name numbers.

EXAMPLES

1   The number 34 is read "thirty-four."

2   The number 227 is read "two hundred twenty-seven."

3   The number 708 is read "seven hundred eight."

4   The number 640 is read "six hundred forty."

5   The number 500 is read "five hundred."

Numbers having four to six digits are named by parts—*thousands* and *ones.*

EXAMPLES

6   The number 1,342 is read "one *thousand* three hundred forty-two."

7   The number 12,500 is read "twelve *thousand* five hundred."

8   The number 132,420 is read "one hundred thirty-two *thousand* four hundred twenty."

9   The number 305,050 is read "three hundred five *thousand* fifty."

10   The number 400,000 is read "four hundred thousand."

Numbers having seven to nine digits are named by parts—*millions, thousands,* and *ones.*

EXAMPLES

11   The number 2,400,200 is read "two *million* four hundred *thousand* two hundred."

12   The number 12,650,025 is read "twelve *million* six hundred fifty *thousand* twenty-five."

13   The number 100,500,105 is read "one hundred *million* five hundred *thousand* one hundred five."

When doing arithmetic it is important to remember place value. The values of place are illustrated in Figure 1.2.

FIGURE 1.2

Twenty-four thousand four hundred twenty-seven

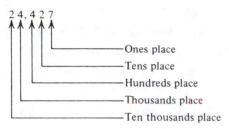

When we add columns of numbers we add to get the number in each *place* of the answer, *beginning with the ones place.*

EXAMPLES

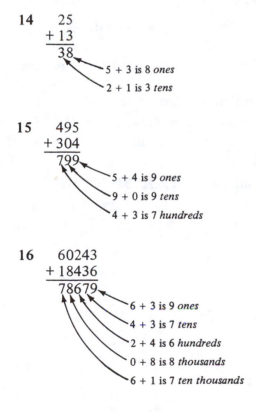

14   25
      + 13
       38

           5 + 3 is 8 *ones*
           2 + 1 is 3 *tens*

15   495
      + 304
       799

           5 + 4 is 9 *ones*
           9 + 0 is 9 *tens*
           4 + 3 is 7 *hundreds*

16   60243
      + 18436
       78679

           6 + 3 is 9 *ones*
           4 + 3 is 7 *tens*
           2 + 4 is 6 *hundreds*
           0 + 8 is 8 *thousands*
           6 + 1 is 7 *ten thousands*

Whenever the sum of the numbers for a given place is 10 or more, we have to **carry** to the next place. This is shown in Example 17.

EXAMPLE 17

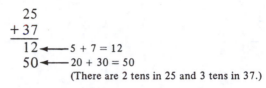

    25
  + 37
    12 ◀—— 5 + 7 = 12
    50 ◀—— 20 + 30 = 50
         (There are 2 tens in 25 and 3 tens in 37.)

This problem can be completed two different ways:

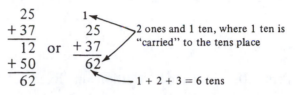

    25   1
  + 37   25   2 ones and 1 ten, where 1 ten is
    12 or + 37   "carried" to the tens place
  + 50   62
    62   1 + 2 + 3 = 6 tens

The second way is the traditional way to add numbers.

EXAMPLES

18   1 ◀———— 1 ten carried from 5 + 8 = 13
     2 5 5
  + 1 3 8
    3 9 3

19   1 1 ◀————— 1 hundred carried from
               10 + 60 + 70 = 140
    3 6 8 ◀—— 1 ten carried from 8 + 5 = 13
  + 2 7 5
    6 4 3

20   1 1
    8 7 6
  + 9 4 5
  1 8 2 1

            1 thousand carried from
            100 + 800 + 900 = 1800

21   1 1
    7 0 9
  + 4 9 6
  1 2 0 5

22   1 1 1
    8 3 2 6
  + 1 7 8 4
  1 0 1 1 0

```
23    2   2 ◄──────── 2 tens carried
        3   9   6
        2   9   9
    +   4   8   6
    ─────────────
    1   1   8   1
```

```
24    1       1   1   1   1
        9   4   3   8   9   5   6
    +   2   7   0   5   9   6   4
    ─────────────────────────────
    1   2   1   4   4   9   2   0
```

Review Exercises

Perform the indicated addition in Exercises 1–12.

1. $3 + 4$
2. $4 + 5$
3. $8 + 9$
4. $7 + 6$
5. $6 + 0$
6. $0 + 9$
7. $1 + 7$
8. $6 + 1$
9. $8 + 6$
10. $8 + 7$
11. $5 + 4$
12. $9 + 9$

EXERCISES
1.1

1. Write 76 in words.
2. Write 95 in words.
3. Write 93 in words.
4. Write 47 in words.
5. Write 354 in words.
6. Write 493 in words.
7. Write 2405 in words.
8. Write 3600 in words.
9. Write the numeral indicating "four hundred thirty-five."
10. Write the numeral indicating "five hundred six."
11. Write the numeral indicating "three hundred ninety-four."
12. Write the numeral indicating "seven hundred three."
13. Write the numeral indicating "six thousand two hundred three."
14. Write the numeral indicating "eight thousand six."
15. In which place is the 5 in 38,521?
16. In which place is the 8 in 38,521?
17. In which place is the 8 in 42,850?
18. In which place is the 2 in 42,850?
19. In which place is the 0 in 409,256?
20. In which place is the 0 in 490,256?
21. Write 700,060 in words.
22. Write 640,140 in words.
23. Write 8,000,400 in words.
24. Write 6,500,405 in words.

For Exercises 25–30 write the numerals indicating the named numbers.

25. Three million forty thousand five
26. Six million four thousand one hundred
27. Seven million five thousand three hundred four
28. Eighteen million forty-nine
29. Four thousand seventeen
30. Eight thousand six hundred.

Add the numbers in Exercises 31–54.

31.
```
   61
 + 24
```
32.
```
   43
 + 26
```
33.
```
   30
 + 25
```
34.
```
   29
 + 50
```
35.
```
  346
 +  23
```
36.
```
  255
 +  21
```
37.
```
  334
 + 125
```
38.
```
  203
 + 874
```
39.
```
 7943
 +  154
```
40.
```
 6045
 +  234
```
41.
```
   28
 + 59
```
42.
```
   64
 + 19
```
43.
```
  708
 + 188
```
44.
```
  809
 + 177
```

45.   499
      + 38

46.   289
      + 47

47. 345
      23
     109

48. 765
      24
     254

49. 5493
       25
      476
        5

50. 3864
       37
      342
        7

51. 25
     39
     64
     78
     59
     40
     24

52. 33
     69
     45
     87
     59
     30
     14

53. 2423689
     6942457
     3814126
     4033894
     5205432

54. 1345678
     7834321
     4760521
     3044069
     5020058

55. A carpenter bought 540 board feet of pine, 854 board feet of oak, and 58 board feet of birch lumber. How many board feet did she buy in all?

56. A contractor needs 12 yards of concrete for a driveway, 48 yards for a basement wall, and 33 yards for a floor. How many yards of concrete does he need in all?

57. A painter must cover 1455 square feet in white, 1275 square feet in light blue, 1248 square feet in beige, and 244 square feet in lavender. How many square feet in all must he cover?

58. An electrician needs 540 feet of heavy wire, 1035 feet of medium wire, and 763 feet of light wire. How many feet of wire in all does she need?

59. Henry's father collected $23, $2015, $419, $49, $642, and $108 from six customers. How many dollars in all did he collect?

60. A surveyor measured around a five-sided piece of property and recorded the following: side A, 549 feet; side B, 723 feet; side

C, 807 feet; side D, 439 feet; and side E, 296 feet. How many feet is it around the property?

61. John counted the items in his collections and found that he had 3476 stamps, 843 coins, 6003 bottlecaps, and 395 tabs. How many items in all did he have in his collections?

62. Susan was shown an inventory list which included the following:

Applesauce: 65 jars, 74 cans

Pears: 43 jars, 255 cans

Peaches: 75 cans, 400 jars

Green beans: 225 cans

Corn: 347 cans

Apricots: 200 cans, 47 jars.

How many cans of fruit were listed on the inventory? How many cans of vegetables? How many jars of fruit?

63. Helen Jones earns $15,475 per year, her husband earns $9,472 per year, and their son earns $3,246 per year. What is the total amount earned by the three people in one year?

64. One company produced 3,459,110 automobiles during a given time period. Two other companies produced 1,042,025 and 2,500,554 autos, respectively, during the same time period. How many autos in all were produced by the three companies during the time period?

1.2
SUBTRACTION

When we subtract numbers we do so by columns, beginning with the *ones* place.

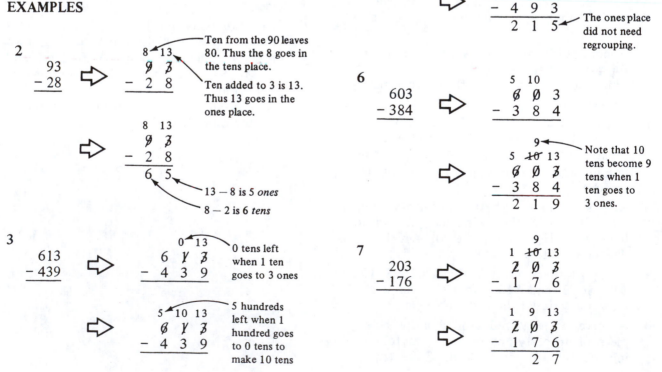

```
        5  10 13
         6̸  7̸  3̸
     -   4  3  9
     ───────────
         1  7  4
```

EXAMPLE 1

```
   9764
 - 7321
 ──────
   2443
```

4 — 1 is 3 *ones*
6 — 2 is 4 *tens*
7 — 3 is 4 *hundreds*
9 — 7 is 2 *thousands*

4
```
    534  ⇨
  - 276
```
```
        2  14
         5  3̸  4̸
     -   2  7  6
     ───────────
```
```
     4  12 14
      8̸  3̸  4̸
  -   2  7  6
  ───────────
      2  5  8
```

Regrouping (borrowing) is necessary when the top number in a column is less than the bottom number. When regrouping, the value of the place to the left of the column needing regrouping is added to the number in the column.

5
```
    708  ⇨
  - 493
```
```
     6  10
      7̸  0̸  8
  -   4  9  3
```
Regroup the 0 in the tens place first using 1 hundred from the 7 hundred.

```
     6  10
      7̸  0̸  8
  -   4  9  3
  ───────────
      2  1  5
```
The ones place did not need regrouping.

EXAMPLES

2
```
     93  ⇨
   - 28
```
```
      8  13
       9̸  3̸
   -   2  8
```
Ten from the 90 leaves 80. Thus the 8 goes in the tens place.

Ten added to 3 is 13. Thus 13 goes in the ones place.

```
      8  13
       9̸  3̸
   -   2  8
   ──────────
       6  5
```
13 — 8 is 5 *ones*
8 — 2 is 6 *tens*

6
```
    603  ⇨
  - 384
```
```
     5  10
      6̸  0̸  3
  -   3  8  4
```
```
     5  1̸0̸ 13
      6̸  0̸  3̸
  -   3  8  4
  ───────────
      2  1  9
```
Note that 10 tens become 9 tens when 1 ten goes to 3 ones.

3
```
    613  ⇨
  - 439
```
```
      0  13
       6  7̸  3̸
   -   4  3  9
```
0 tens left when 1 ten goes to 3 ones

```
      5  10 13
       6̸  7̸  3̸
   -   4  3  9
```
5 hundreds left when 1 hundred goes to 0 tens to make 10 tens

7
```
    203  ⇨
  - 176
```
```
         9
      1  1̸0̸ 13
       2̸  0̸  3̸
   -   1  7  6
```
```
      1  9  13
       2̸  0̸  3̸
   -   1  7  6
   ───────────
            2  7
```

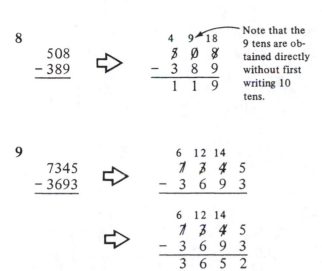

8

$$\begin{array}{r} 508 \\ -\,389 \\ \hline \end{array}$$

$$\begin{array}{r} {\scriptstyle 4\;\;9\;18} \\ \cancel{5}\;\cancel{0}\;\cancel{8} \\ -\,3\;8\;9 \\ \hline 1\;1\;9 \end{array}$$

Note that the 9 tens are obtained directly without first writing 10 tens.

9

$$\begin{array}{r} 7345 \\ -\,3693 \\ \hline \end{array}$$

$$\begin{array}{r} {\scriptstyle 6\;12\;14} \\ \cancel{7}\;\cancel{3}\;\cancel{4}\;5 \\ -\,3\;6\;9\;3 \\ \hline \end{array}$$

$$\begin{array}{r} {\scriptstyle 6\;12\;14} \\ \cancel{7}\;\cancel{3}\;\cancel{4}\;5 \\ -\,3\;6\;9\;3 \\ \hline 3\;6\;5\;2 \end{array}$$

(*Note:* When regrouping 7345 for this problem, first change 4 tens to 14 tens; then change the 2 remaining hundreds to 12 hundreds; then show the 6 thousands.)

10

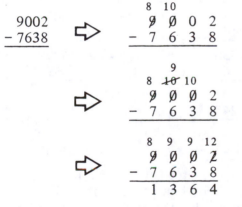

$$\begin{array}{r} 9002 \\ -\,7638 \\ \hline \end{array}$$

(*Note:* Each 0 in the top number converts to a 9 when the top number is regrouped completely.)

11

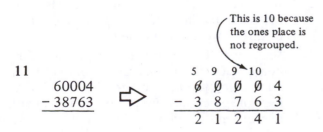

$$\begin{array}{r} 60004 \\ -\,38763 \\ \hline \end{array}$$

This is 10 because the ones place is not regrouped.

12

$$\begin{array}{r} 600403 \\ -\,487161 \\ \hline \end{array}$$

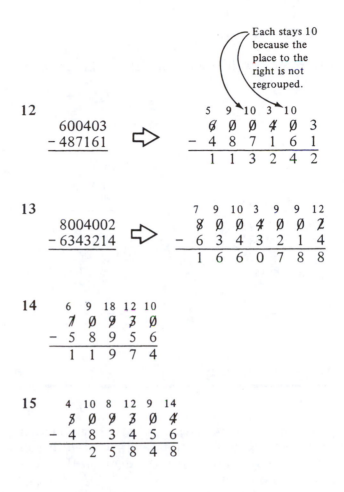

Each stays 10 because the place to the right is not regrouped.

$$\begin{array}{r} {\scriptstyle 5\;\;9\;10\;\;3\;10} \\ \cancel{6}\;\cancel{0}\;\cancel{0}\;\cancel{4}\;\cancel{0}\;3 \\ -\,4\;8\;7\;1\;6\;1 \\ \hline 1\;1\;3\;2\;4\;2 \end{array}$$

13

$$\begin{array}{r} 8004002 \\ -\,6343214 \\ \hline \end{array}$$

$$\begin{array}{r} {\scriptstyle 7\;\;9\;10\;\;3\;\;9\;\;9\;12} \\ \cancel{8}\;\cancel{0}\;\cancel{0}\;\cancel{4}\;\cancel{0}\;\cancel{0}\;\cancel{2} \\ -\,6\;3\;4\;3\;2\;1\;4 \\ \hline 1\;6\;6\;0\;7\;8\;8 \end{array}$$

14

$$\begin{array}{r} {\scriptstyle 6\;\;9\;18\;12\;10} \\ \cancel{7}\;\cancel{0}\;\cancel{9}\;\cancel{3}\;\cancel{0} \\ -\,5\;8\;9\;5\;6 \\ \hline 1\;1\;9\;7\;4 \end{array}$$

15

$$\begin{array}{r} {\scriptstyle 4\;10\;\;8\;12\;\;9\;14} \\ \cancel{5}\;\cancel{0}\;\cancel{9}\;\cancel{3}\;\cancel{0}\;\cancel{4} \\ -\,4\;8\;3\;4\;5\;6 \\ \hline 2\;5\;8\;4\;8 \end{array}$$

The answer to a subtraction problem can be checked by adding it to the "bottom" number to get the "top" number. The check for Example 15 can be shown as follows.

Subtraction   *Check*

$$\begin{array}{r} 509304 \\ -\,483456 \\ \hline 25848 \end{array} \qquad \begin{array}{r} {\scriptstyle 1\;\quad1\;\;1\;\;1} \\ 2\;5\;8\;4\;8 \\ +\,4\;8\;3\;4\;5\;6 \\ \hline 5\;0\;9\;3\;0\;4 \end{array}$$

It is a good practice to check all subtractions. It is not necessary to write an extra addition problem; the check can be made by adding "upwards," as shown here:

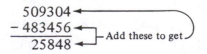

$$\begin{array}{r} 509304 \\ -\,483456 \\ \hline 25848 \end{array}$$ Add these to get

Review Exercises

Perform the indicated operations in Exercises 1–12.

1. $3 + 5$
2. $6 + 3$
3. $7 + 9$
4. $8 + 6$
5. $16 - 9$
6. $15 - 8$
7. $17 - 8$
8. $13 - 5$

9. $\begin{array}{r} 79634 \\ + 30785 \\ \hline \end{array}$

10. $\begin{array}{r} 603945 \\ + \ 62948 \\ \hline \end{array}$

11. $\begin{array}{r} 76893 \\ 4089 \\ 25 \\ 80069 \\ 284 \\ + \quad 6 \\ \hline \end{array}$

12. $\begin{array}{r} 43765 \\ 3764 \\ 90035 \\ 37 \\ 395 \\ + \quad 6 \\ \hline \end{array}$

EXERCISES
1.2

Regrouping is necessary in many subtraction problems. Regrouped notation means writing all of the regroupings above the digits in the number, as shown here:

$$37 = \overset{2\ \ 17}{\cancel{3}\ \cancel{7}} \quad \text{and} \quad 4706 = \overset{3\ \ 16\ \ 9\ \ 16}{\cancel{4}\ \cancel{7}\ \cancel{0}\ \cancel{6}}.$$

Write the numbers in Exercises 1–22 in regrouped notation, as shown in the two preceding examples.

1. 64
2. 38
3. 40
4. 70
5. 19
6. 16
7. 328
8. 576
9. 780
10. 690
11. 803
12. 706
13. 600
14. 400
15. 6397
16. 5683
17. 5307
18. 4609
19. 8003
20. 7006
21. 60000
22. 40000

Subtract as indicated in Exercises 23–48.

23. $\begin{array}{r} 74639 \\ - 32428 \\ \hline \end{array}$
24. $\begin{array}{r} 63876 \\ - 41635 \\ \hline \end{array}$

25. $\begin{array}{r} 13621 \\ - \quad 410 \\ \hline \end{array}$
26. $\begin{array}{r} 30582 \\ - \quad 371 \\ \hline \end{array}$

27. $\begin{array}{r} 64 \\ - 29 \\ \hline \end{array}$
28. $\begin{array}{r} 38 \\ - 19 \\ \hline \end{array}$

29. $\begin{array}{r} 40 \\ - 27 \\ \hline \end{array}$
30. $\begin{array}{r} 70 \\ - 56 \\ \hline \end{array}$

31. $\begin{array}{r} 19 \\ - \ 9 \\ \hline \end{array}$
32. $\begin{array}{r} 16 \\ - \ 8 \\ \hline \end{array}$

33. $\begin{array}{r} 328 \\ - \ 79 \\ \hline \end{array}$
34. $\begin{array}{r} 576 \\ - \ 98 \\ \hline \end{array}$

35. $\begin{array}{r} 780 \\ - 263 \\ \hline \end{array}$
36. $\begin{array}{r} 690 \\ - 357 \\ \hline \end{array}$

37. $\begin{array}{r} 803 \\ - \ 78 \\ \hline \end{array}$
38. $\begin{array}{r} 706 \\ - \ 39 \\ \hline \end{array}$

39. $\begin{array}{r} 600 \\ - 237 \\ \hline \end{array}$
40. $\begin{array}{r} 400 \\ - 186 \\ \hline \end{array}$

41. $\begin{array}{r} 6397 \\ - 1499 \\ \hline \end{array}$
42. $\begin{array}{r} 5683 \\ - 2996 \\ \hline \end{array}$

43. $\begin{array}{r} 5307 \\ - \ 649 \\ \hline \end{array}$
44. $\begin{array}{r} 4609 \\ - \ 839 \\ \hline \end{array}$

45. $\begin{array}{r} 8003 \\ - 2458 \\ \hline \end{array}$
46. $\begin{array}{r} 7006 \\ - 3758 \\ \hline \end{array}$

47. $\begin{array}{r} 60000 \\ - \ 4385 \\ \hline \end{array}$
48. $\begin{array}{r} 40000 \\ - \ 3465 \\ \hline \end{array}$

49. A contractor is to receive four million five hundred thousand dollars to build a road. He has been paid $2,495,500 so far. How much is he yet to be paid?

50. Harry cut 754 feet of wire from a 1000-foot coil. How many feet of wire remained in the coil?

51. Rhonda used 347 feet of thread from a spool containing 800 feet. How many feet of thread were still on the spool?

52. Jerry had 2000 board feet of pine and 500 board feet of birch. After completing a remodeling job he had a total of 795 board feet of pine and birch. How many board feet of lumber did he use?

53. The distance around a five-sided piece of land is 747 feet. One side is 79 feet long. What is the sum of the lengths of the other four sides?

54. One company makes 1,750,600 more autos in a year than does a second company. The first company made 2,200,000 autos. How many did the second company make?

1.3

MULTIPLICATION

Equal rows of dots can be used to explain the meaning of multiplication. Figure 1.3 illustrates the multiplication fact 4 X 5 = 20. All the basic facts of multiplication can be shown in this fashion. These basic facts include all possible combinations of two one-digit numbers, from 0 X 0 = 0 to 9 X 9 = 81.

FIGURE 1.3

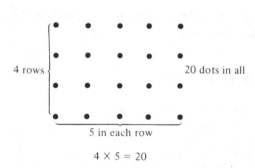

4 rows   20 dots in all

5 in each row

4 X 5 = 20

Two numbers being multiplied are called **factors**, and the result of a multiplication is called the **product**, as shown here:

$$4 \times 5 = 20$$

Factor X Factor = Product.

To multiply any whole number by 10 you simply affix a zero to the number, e.g., 7 X 10 = 70, 18 X 10 = 180, 456 X 10 = 4560, and 300 X 10 = 3000. Multiplying by 100 involves affixing two zeros, e.g., 7 X 100 = 700, 18 X 100 = 1800, 456 X 100 = 45600, and 300 X 100 = 30000. Similarly, multiplying by 1000 involves affixing three zeros, e.g., 456 X 1000 = 456000.

Zero times any number is 0, e.g., 0 X 5 = 0, 0 X 25 = 0, and 0 X 75936 = 0. One times any number is the number, e.g., 1 X 5 = 5, 1 X 25 = 25, and 1 X 75936 = 75936.

The following paragraphs, including examples 1–8, explain the process involved when you multiply by a one-digit number. Our traditional, "shortcut" method follows. Multiplying by a one-digit number can be done by multiplying the value in each *place* by the one-digit number and adding the results.

EXAMPLES

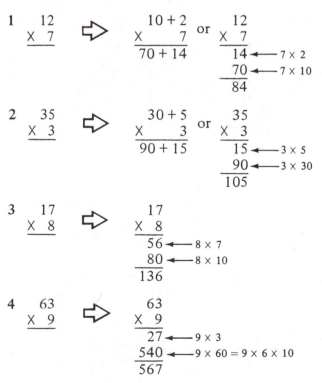

EXAMPLE 5

```
    93
 X   7
    21
   630
   651
```

In Example 5, the 21 and 630 are called **partial products** (parts of the final product). Here are some examples of partial products.

EXAMPLES

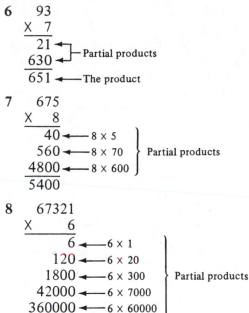

6
```
    93
 X   7
    21 ◄┐
   630 ◄┘— Partial products
   651 ◄——— The product
```

7
```
    675
 X    8
     40 ◄— 8 × 5
    560 ◄— 8 × 70      } Partial products
   4800 ◄— 8 × 600
   5400
```

8
```
    67321
 X      6
        6 ◄— 6 × 1
      120 ◄— 6 × 20
     1800 ◄— 6 × 300      } Partial products
    42000 ◄— 6 × 7000
   360000 ◄— 6 × 60000
   403926
```

Our traditional "shortcut" method for showing multiplication combines the partial products, adding them together as we proceed. Using the "shortcut" method, Example 8 can be redone as follows:

Steps 1 & 2:
```
        1 ◄——— The 1 in (6 × 2) tens
    67321        = 120 (Note: The
 X      6        place value of this 1
       26        shows that it repre-
                 sents 1 hundred.)
              ◄— 6 × 1
              — The 2 in (6 × 2) tens
                = 120
```

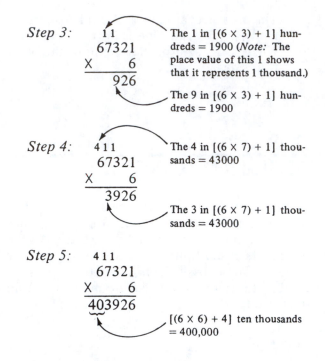

Step 3:
```
       11 ◄——— The 1 in [(6 × 3) + 1] hun-
    67321        dreds = 1900 (Note: The
 X      6        place value of this 1 shows
      926        that it represents 1 thousand.)
              — The 9 in [(6 × 3) + 1] hun-
                dreds = 1900
```

Step 4:
```
      411 ◄——— The 4 in [(6 × 7) + 1] thou-
    67321        sands = 43000
 X      6
     3926
              — The 3 in [(6 × 7) + 1] thou-
                sands = 43000
```

Step 5:
```
      411
    67321
 X      6
   403926
              [(6 × 6) + 4] ten thousands
              = 400,000
```

Now here are some examples of traditional multiplication notation.

EXAMPLES

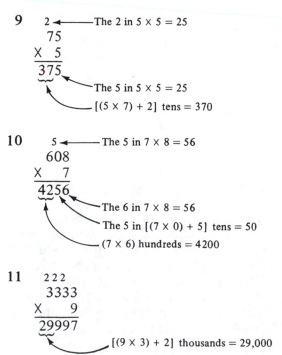

9
```
     2 ◄——— The 2 in 5 × 5 = 25
    75
 X   5
   375
       ◄— The 5 in 5 × 5 = 25
       — [(5 × 7) + 2] tens = 370
```

10
```
     5 ◄——— The 5 in 7 × 8 = 56
   608
 X   7
  4256
       — The 6 in 7 × 8 = 56
       — The 5 in [(7 × 0) + 5] tens = 50
       — (7 × 6) hundreds = 4200
```

11
```
    222
   3333
 X    9
  29997
       [(9 × 3) + 2] thousands = 29,000
```

Multiplying by a number which has two or more digits involves multiplying by each digit (and its place value) and adding the results.

EXAMPLES

12
```
    1
   23
 X 25
 ───
  115 ◄──── 5 × 23
  460 ◄──── 20 × 23  or  2 × 23 × 10
 ───
  575
```

13
```
    7 ◄──────── The 7 in 8 × 9 × 10 = 720
    6 ◄──────── The 6 in 7 × 9 = 63
   39
 X 87
 ────
  273 ◄──── 7 × 39
 3120 ◄──── 80 × 39  or  8 × 39 × 10
 ────
 3393
```

14
```
   12
   45
  257
 X 38
 ────
 2056 ◄──── 8 × 257
 7710 ◄──── 3 × 257 × 10
 ────
 9766
```

15
```
   12
   45
  257
 X 38
 ────
 2056 ◄──── 8 × 257
  771 ◄──── 3 × 257 without the 0 for ten times it
 ────
 9766
```

16
```
    361
    251
    241
   3492
 X  765
 ─────
  17460 ◄──── 5 × 3492
  20952 ◄──── 6 × 3492 (no zero for 10 times it)
  24444 ◄──── 7 × 3492 (no zero zero for 100 times it)
 2671380
```

Example 16 shows the traditional way to indicate multiplication. If it were written with "true" partial products it would look like this:

```
    361
    251
    241
   3492
 X  765
 ──────
   17460
  209520  } Note zeros.
 2444400
 ───────
 2671380
```

Normally we do not write the extra zeros, because they do not affect the sum of the partial products. However, it is good strategy, for some people, to write the zeros in to help keep the place value columns in order.

Multiplication can be expressed three different ways, using dots (raised periods), times signs, or parentheses, as shown in Examples 17–24.

EXAMPLES

17   The expression 4 X 3 means "4 times 3."

18   The expression 4 · 3 means "4 times 3."

19   The expression 4(3) means "4 times 3."

20   The expression (4)3 means "4 times 3."

21   The expression (4)(3) means "4 times 3."

22   The expression 5(3 + 4) means "5 times 7."

23   The expression (7 − 2)4 means "5 times 4."

24   The expression (8 + 9)(15 − 4) means "17 times 11."

Examples 22, 23, and 24 illustrate the fact that operations *inside* parentheses are done first when evaluating expressions. When a series of different operations (+, −, X) are included in an expression, care must be taken to do the operations in the proper order, as shown in the following steps.

Step 1:   Do the computations inside sets of parentheses first.

EXAMPLES

25   $3 - (4 - 2) = 3 - 2$
$$= 1$$

26   $(8 - 2)3 = 6 \cdot 3$
$$= 18$$

27   $(3 + 9)(4 \cdot 7) = (12)(28)$
$$= 336$$

Step 2:   Do the multiplications.

EXAMPLES

28   $7 + 3 \cdot 4 = 7 + 12$
$$= 19$$

29   $(5 - 2) + 8 \cdot 3 = 3 + 8 \cdot 3$
$$= 3 + 24$$
$$= 27$$

Step 3:   Finally, do additions and subtractions in order from left to right.

EXAMPLES

30   $12 - 8 + 3 = 4 + 3$
$$= 7$$

31   $5 + (2 \cdot 4) - 9 = 5 + 8 - 9$
$$= 13 - 9$$
$$= 4$$

32   $(3 \cdot 4) - (8 + 2) + 6 = 12 - 10 + 6$
$$= 2 + 6$$
$$= 8$$

Examples 33–35 combine these steps.

EXAMPLES

33   $7[15 - (3 \cdot 4)] = 7(15 - 12)$
$$= 7(3)$$
$$= 21$$

34   $[13 + (2 \cdot 3)][(4 \cdot 2) - 3] = (13 + 6)(8 - 3)$
$$= (19)(5)$$
$$= 95$$

35   $13 + 12(16 - 4) - 8 + 5$
$$= 13 + 12(12) - 8 + 5$$
$$= 13 + 144 - 8 + 5$$
$$= 157 - 8 + 5$$
$$= 149 + 5$$
$$= 154$$

Review Exercises

Perform the indicated operations in Exercises 1–5.

1. 7×9
2. 8×6

3. $\begin{array}{r} 4983 \\ 276 \\ 50 \\ + 1063 \end{array}$

4. $\begin{array}{r} 49836 \\ - 10967 \end{array}$

5. $(843 - 75) - 179$

6. Three shipments of 1-inch native pine are received by a contractor: 7556, 8750, and 9898 board feet, respectively. What is the total number of board feet of lumber delivered?

7. The monthly production of motors for refrigerators was as follows: January 29,220; February 32,416; March 37,240; April 39,374; May 45,666; June 52,487; July 36,458; August 35,000; September 32,250; October 51,750; November 62,475; December 50,525. Determine the total output for the year.

8. The kilowatt (kw) hours of electrical energy consumed monthly for six months were 1412; 1839; 27,000; 29,787; 32,496; and 1934. Determine the total kilowatts of energy used.

9. Determine the number of miles traveled during each of five weeks from the odometer readings shown in the table.

Week	1	2	3	4	5
Reading (Start)	32,119	32,899	33,988	35,976	37,065
Reading (End)	32,899	33,988	35,976	37,065	39,001

10. A container (drum) holds 55 gallons of turpentine. In a one-month period, the following quantities were used: 5 gallons, 10 gallons, 8 gallons, 7 gallons, 16 gallons, and 8 gallons. How much turpentine is left?

11. A customer's service bill for electricity shows that a total of 1,235 kilowatt hours were used. Of this total, 367 kilowatt hours were used for lighting service and the balance for domestic hot water. How many kilowatt hours were used for hot water?

12. An auto supply store sold $598 worth of goods on Monday; $873 on Tuesday; $724 on Wednesday; $552 on Thursday; and $359 on Saturday. What were the sales for Friday, if the total sales for the week were $3901?

EXERCISES
1.3

Multiply in Exercises 1–22.

1. 14
    6

2. 22
    7

3. 78
    9

4. 69
    8

5. 240
    8

6. 121
    7

7. 212
    6

8. 343
    5

9. 508
    9

10. 689
    6

11. 411
    14

12. 627
    26

13. 303
    97

14. 879
    78

15. 687
    90

16. 8165
    73

17. 6057
    423

18. 5007
    530

19. 7879
    896

20. 97009
    307

21. 5046
    3597

22. 6603
    4986

23. A crew of 17 men worked on a construction job 8 hours a day for 153 days. How many total hours did the crew work?

24. A bricklayer lays an average of 145 bricks an hour. At this rate, how many bricks can he lay in 37 hours?

25. A mason purchased 223 cubic yards of ready-mix concrete at $18 a cubic yard; 51 barrels of lime at $9 a barrel; and 38 cubic yards of sand at $4 per cubic yard. What was the total cost of materials?

26. The cost of a certain size brass elbow is 98 cents per pound. Determine the cost of 147 pounds of elbows.

27. If it takes 760 shingles per square (100 square feet) when laid 5 inches to weather, how many will be needed to cover 37 squares with the same weathering?

28. A bundle of white cedar shingles contains 250 shingles. How many shingles are there in 378 bundles?

29. Two electromagnets are wound. The first has 57 layers of 98 turns each; the second, 38 layers of 179 turns each. Give the total number of turns in both coils.

30. An electrical contractor purchased 196 conduit bodies at 69 cents each and 87 of another type for 86 cents each. Give the total cost of these materials.

31. A car has three 34-lumen bulbs, two 26-lumen, four 9-lumen, and three 15-lumen. Find the total lumens.

32. A truck averages 42 miles an hour for 6 hours daily for 19 days. Another averages 37 miles an hour for 6 hours daily for 23

days. A third averages 39 miles an hour for 6 hours daily for 22 days. Determine the total mileage of the three trucks.

33. The salary for each of five female executives is $50,000 per year, and the salary for each of thirteen male executives is $49,108 per year. How much money does it take to pay the salaries of the eighteen executives for one year?

Perform the indicated operations in Exercises 34–45.

34. $12 - (7 - 2)$ 35. $19 - (12 - 5)$
36. $13 + (7 \cdot 9)$ 37. $15 + (6 \cdot 12)$
38. $15 - 9 + 4$ 39. $19 - 8 + 5$
40. $5[18 - (4 \cdot 3)]$ 41. $7[12 - (2 \cdot 5)]$
42. $[7 + (2 \cdot 8)][(7 \cdot 4) - 11]$
43. $[11 + (3 \cdot 4)][(5 \cdot 5) - 12]$
44. $17 + 3(18 - 3) - 7 + 4$
45. $15 + 7(12 - 5) - 23 + 17$

1.4
FACTORING WHOLE NUMBERS

The **whole numbers** are 0, 1, 2, 3, 4, 5, 6, 7, 8, 9, 10, 11, ..., where the ... means that the pattern of numbers continues indefinitely. When multiplying, the result is called the **product**, and the numbers being multiplied are called **factors**.

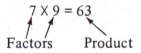

$$7 \times 9 = 63$$

When the factors 7 and 9 are multiplied, the product is 63. It is also true that $9 = 3 \times 3$. Therefore, the three factors 7, 3, and 3 can be multiplied to get 63.

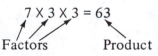

$$7 \times 3 \times 3 = 63$$

Finding the factors of a product is called **factoring**.

A **prime number** is a number greater than 1 whose only factors are *one* and *itself*. A number greater than 1 having factors other than 1 and itself is called a **composite number**. Some examples of prime and composite numbers are shown in Table 1.2.

TABLE 1.2

Number	Factored Form	Prime or Composite?
20	4 X 5 or 2 X 2 X 5	composite
6	2 X 3	composite
13	none	prime
21	3 X 7	composite
8	2 X 4 or 2 X 2 X 2	composite
54	6 X 9 or 3 X 2 X 9 or 3 X 2 X 3 X 3	composite
19	none	prime
11	none	prime

In Table 1.2 one factored form of 54 is 3 X 2 X 3 X 3, which can be written as 2 X 3 X 3 X 3. Whenever the factored form of a number is such that all the factors are primes, it is said to be the **prime factorization** of the number. Table 1.3 contains examples of prime factorization.

TABLE 1.3

Number	Prime Factorization
8	$2 \times 2 \times 2$
9	3×3
15	3×5
30	$2 \times 3 \times 5$
12	$2 \times 2 \times 3$
32	$2 \times 2 \times 2 \times 2 \times 2$

A way to find the prime factorization of a number is to factor it first, then factor all composite factors. This is shown in Examples 1–6.

EXAMPLE 1   Find the prime factorization of 18.

Solution $18 = 2 \times 9$

$\qquad = 2 \times 3 \times 3$, where the 3×3 comes from the 9

EXAMPLE 2   Find the prime factorization of 72.

Solution $72 = 9 \times 8$

$\qquad = 3 \times 3 \times 2 \times 4$

$\qquad = 3 \times 3 \times 2 \times 2 \times 2$, where the last 2×2 comes from the 4

EXAMPLES

3   $121 = 11 \times 11$

4   $35 = 5 \times 7$

5   $70 = 7 \times 10$

$\qquad = 7 \times 2 \times 5$

6   $32 = 4 \times 8$

$\qquad = 2 \times 2 \times 2 \times 4$

$\qquad = 2 \times 2 \times 2 \times 2 \times 2$

In this section we factor only products which come from basic multiplication facts. Other numbers can be factored, of course, but this involves division, which we will deal with in Section 1.5. An example is $1001 = 7 \times 11 \times 13$, where division is used to find the factors.

The order of factors in a prime factorization can vary; but the standard form lists the "smallest" factors first. For example, $42 = 7 \times 6$, which is $7 \times 3 \times 2$, which, in standard form, is $2 \times 3 \times 7$.

A **common factor** of two or more numbers is any number that is a factor of each number. For example, the common factors of 30 and 40 are 2, 5, and 10.

$$30 = 2 \times 15 \qquad 30 = 5 \times 6 \qquad 30 = 10 \times 3$$
$$40 = 2 \times 20 \qquad 40 = 5 \times 8 \qquad 40 = 10 \times 4$$

The **greatest common factor (gcf)** of two or more numbers is the greatest number that is a factor of each number.

To determine the greatest common factor of two or more numbers:

(a)   write each number in prime factor form;

(b)   list each common prime factor as many times as it is common to each prime factor form; and

(c)   compute the product of all common prime factors listed in Step (b).

The product of common prime factors is the gcf of the numbers.

EXAMPLE 7   Find the greatest common factor of 42 and 63.

Solution

(a) $42 = 2 \times 3 \times 7$

$\qquad 63 = 3 \times 3 \times 7$

(b)   The common prime factors are 3 and 7.

(c)   The gcf is 3×7, or 21.

EXAMPLE 8   Find the greatest common factor of 72 and 48.

Solution

(a) $72 = 2 \times 2 \times 2 \times 3 \times 3$

$\qquad 48 = 2 \times 2 \times 2 \times 2 \times 3$

(b) The common factors are 2, 2, 2, and 3. Note that there are three factors of 2 in each prime factor form.

(c) The gcf is 2 X 2 X 2 X 3, or 24.

EXAMPLE 9 Find the greatest common factor of 54, 90, and 36.

Solution

(a) 54 = 2 X 3 X 3 X 3
90 = 2 X 3 X 3 X 5
36 = 2 X 2 X 3 X 3

(b) The common prime factors are 2, 3, and 3.

(c) The gcf is 2 X 3 X 3, or 18.

If two or more numbers have no common prime factors, their greatest common factor is 1. This is shown in Examples 10 and 11.

EXAMPLE 10 Find the greatest common factor of 35 and 18.

Solution

(a) 35 = 5 X 7
18 = 2 X 3 X 3

(b) There are no common prime factors.

(c) The gcf is 1.

EXAMPLE 11 Find the greatest common factor of 23 and 61.

Solution

(a) 23 = 1 X 23 (23 is a prime number)
61 = 1 X 61 (61 is a prime number)

(b) There are no common prime factors.

(c) The gcf is 1.

Review Exercises

In Exercises 1–12 multiply the given numbers.

1. 7 X 9   2. 8 X 6
3. 5 X 6   4. 6 X 8
5. 27   6. 94
   X 6   X 8
7. 358   8. 495
   X 7   X 6
9. 28   10. 94
   X 46   X 29
11. 347   12. 654
   X 38   X 95

EXERCISES
1.4

Find the prime factorization of each number in Exercises 1–34.

1. 6   2. 10
3. 14   4. 9
5. 15   6. 21
7. 25   8. 35
9. 49   10. 8
11. 12   12. 16
13. 18   14. 24
15. 27   16. 32
17. 36   18. 45
19. 40   20. 20
21. 30   22. 50
23. 28   24. 42
25. 48   26. 54
27. 70   28. 56
29. 63   30. 64
31. 72   32. 80
33. 81   34. 90

Find the greatest common factor of each set of numbers in Exercises 35–48.

35. 12, 42   36. 18, 45

37. 36, 16   38. 42, 63

39. 56, 72   40. 81, 54

41. 64, 49   42. 19, 63

43. 15, 35, 70   44. 21, 36, 12

45. 48, 35, 81   46. 50, 18, 54

47. 45, 72, 27   48. 56, 81, 21

1.5
DIVISION

Division is the opposite of multiplication. "Twenty-one divided by three" means "what number times 3 equals 21?" The answer is 7, because $7 \times 3 = 21$.

The division of one number by another, the **dividend** by the **divisor**, can be written three different ways, as shown in Table 1.4. Figure 1.4 identifies the dividend, divisor, and quotient in a division problem.

TABLE 1.4

The division	3 ways to indicate it
21 divided by 3	$21 \div 3$; $3\overline{)21}$; $\dfrac{21}{3}$
14 divided by 2	$14 \div 2$; $2\overline{)14}$; $\dfrac{14}{2}$
81 divided by 9	$81 \div 9$; $9\overline{)81}$; $\dfrac{81}{9}$

FIGURE 1.4

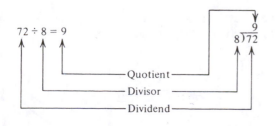

When computing, division is usually written using the symbol $\overline{)\quad}$. For example, 39 divided by 13 is written $13\overline{)39}$. The quotient is obtained by finding how many 13's there are in 39. One way to find out is to subtract 13 from 39, then subtract 13 from the result, etc., keeping a record of how many 13's were subtracted, as shown here:

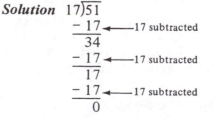

This shows that $39 \div 13 = 3$. Multiplying 13 by 3 yields $13 \times 3 = 39$, which "checks" the result.

Examples 1–3 show how subtractive division eventually leads to the usual long division method.

EXAMPLE 1 Find $51 \div 17$.

Solution $17\overline{)51}$

$$\begin{array}{r} -17 \\ \hline 34 \\ -17 \\ \hline 17 \\ -17 \\ \hline 0 \end{array}$$

17 subtracted
17 subtracted
17 subtracted

Since three 17's were subtracted, the answer is 3.

EXAMPLE 2   Find $95 \div 19$.

Solution   19)95

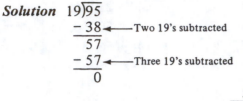

Therefore, $95 \div 19 = 2 + 3 = 5$, or 19)$\overset{5}{95}$.

EXAMPLE 3   Find $158 \div 53$.

Solution   53)158
$\quad -106 \longleftarrow$ Two 53's are subtracted.
$\quad \overline{52} \longleftarrow$ *Less than 53 remains.*

Therefore, $158 \div 53 = 2$, with a *remainder* of 52. The **remainder** is what's left over after the divisor is subtracted as many times as possible.

The traditional way of doing division involves a systematic way of determining the greatest number of times the divisor can be subtracted from the dividend. It involves leaving off zeros from multiples of 10, 100, 1000, etc., much the same way as in multiplication.

EXAMPLE 4   Find $95 \div 19$.

Solution   The problem is 19)95. *Think:* "19 is approximately 20, and 95 is approximately 100." Leaving one zero off of 20 and 100, you get 2 and 10; and you know that 2 goes into 10 five times. Try 5 as an answer. Since $5 \times 19 = 95$, 5 is correct.

EXAMPLE 5   Find 53)158.

Solution   *Think:* "53 is approximately 50, and 158 is approximately 160." Leaving a zero off of each yields 5)16, which is 3, approximately. So try 3. But $3 \times 53 = 159$. Since 159 is greater than 158, 2 must be the quotient, with a remainder of 52. (See Example 3 above.)

EXAMPLE 6   Find $279 \div 23$.

Solution   *Think:* "23 goes into 27 once" and

write a 1 above the tens place of the dividend, to get:

$$\overset{1}{23)279.}$$

Since the 1 is in the tens place, *think* "$10 \times 23 = 230$" and subtract 230 from 279 to get:

$$\begin{array}{r} 1 \\ 23)\overline{279} \\ 230 \\ \hline 49. \end{array}$$

Think: "23 goes into 49 two times" and write a 2 above the ones place of the dividend, to get:

$$\begin{array}{r} 12 \\ 23)\overline{279} \\ 230 \\ \hline 49. \end{array}$$

Think: "$2 \times 23 = 46$," and subtract 46 from 49 to get:

$$\begin{array}{r} 12 \\ 23)\overline{279} \\ 230 \\ \hline 49 \\ 46 \\ \hline 3 \end{array}$$

which shows that the quotient is 12, with a remainder of 3.

EXAMPLE 7   Find $7967 \div 48$.

Solution   *Think:* "48 goes into 79 once," and write a 1 above the hundreds place of the dividend to get:

$$\overset{1}{48)7967.}$$

Since the 1 is in the hundreds place, subtract 100×48 from 7967 to get:

$$\begin{array}{r} 1 \\ 48)\overline{7967} \\ 4800 \\ \hline 3167. \end{array}$$

Since 48 won't "go into" 31, *think* "48 into 316 is approximately 50 into 300, which is 6," and write a 6 in the tens place of the quotient to get:

$$
\begin{array}{r}
16 \\
48\overline{)7967} \\
4800 \\
\hline
3167. \\
\end{array}
$$

Since the 6 is in the tens place, subtract 60 X 48 from the 3167 to get:

$$
\begin{array}{r}
16 \\
48\overline{)7967} \\
4800 \\
\hline
3167 \\
2880 \\
\hline
287. \\
\end{array}
$$

Think: "48 into 287 is approximately 50 into 250, which is 5," and write a 5 in the ones place to get:

$$
\begin{array}{r}
165 \\
48\overline{)7967} \\
4800 \\
\hline
3167 \\
2880 \\
\hline
287. \\
\end{array}
$$

Now subtract 5 X 48 from the 287 to get:

$$
\begin{array}{r}
165 \\
48\overline{)7967} \\
4800 \\
\hline
3167 \\
2880 \\
\hline
287 \\
240 \\
\hline
47. \\
\end{array}
$$

(Notice that the 1 in 165 is above the 9, the 6 is above the 6, and the 5 is above the 7. That is, the places in both dividend and quotient correspond.) The quotient is 165, with a remainder of 47.

EXAMPLE 8   Find $90034 \div 295$.

Solution *Think:* "295 goes into 900 three

times," and write a 3 in the hundreds place in the quotient to get:

$$
\begin{array}{r}
3 \\
295\overline{)90034.} \\
\end{array}
$$

Since the 3 is in the hundreds place of the quotient, subtract 300 X 295 from 90034 to get:

$$
\begin{array}{r}
3 \\
295\overline{)90034} \\
88500 \\
\hline
1534. \\
\end{array}
$$

Since 295 won't "go into" 153, write a 0 in the tens place of the quotient to get:

$$
\begin{array}{r}
30 \\
295\overline{)90034} \\
88500 \\
\hline
1534. \\
\end{array}
$$

Think: "300 into 1500 is 5," and write a 5 in the ones place of the quotient to get:

$$
\begin{array}{r}
305 \\
295\overline{)90034} \\
88500 \\
\hline
1534. \\
\end{array}
$$

Now subtract 5 X 295 from the 1534 to get:

$$
\begin{array}{r}
305 \\
295\overline{)90034} \\
88500 \\
\hline
1534 \\
1475 \\
\hline
59. \\
\end{array}
$$

The quotient is 305 with a remainder of 59.

EXAMPLE 9   Find $11570 \div 89$.

Solution
$$
\begin{array}{r}
13 \\
89\overline{)11570} \\
8900 \\
\hline
2670 \\
2670 \\
\hline
0 \\
\end{array}
$$

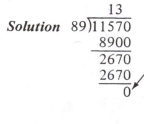

Even though this is 0, the problem is incomplete because there is no number in the ones place of the quotient. Since $0 \div 89 = 0$, a 0 must go in the ones place of the quotient to get:

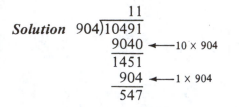

```
        130
  89)11570
      8900  ←——100 × 89
      2670
      2670  ←——30 × 89
         0
         0  ←——0 × 89
         0
```

EXAMPLE 10   Find $10491 \div 904$.

Solution
```
          11
  904)10491
      9040  ←——10 × 904
      1451
       904  ←——1 × 904
       547
```

The quotient is 11 with a remainder of 547.

Review Exercises

Multiply the given numbers in Exercises 1–12.

1. 9×7 2. 5×9
3. 6×40 4. 7×50
5. 27×70 6. 39×60
7. 27×500 8. 49×400
9. 76×8000 10. 54×7000
11.   43856   12.   92064
      X   794   X   867

EXERCISES
1.5

Divide as indicated in Exercises 1–16.

1. $6)\overline{126}$ 2. $4)\overline{120}$
3. $7)\overline{147}$ 4. $3)\overline{135}$
5. $\dfrac{182}{7}$ 6. $\dfrac{110}{11}$
7. $12)\overline{132}$ 8. $27)\overline{810}$

9. $53)\overline{742}$ 10. $92)\overline{2024}$
11. $1250 \div 25$ 12. $1638 \div 39$
13. $1782 \div 162$ 14. $14091 \div 33$
15. $9002 \div 45$ 16. $80208 \div 121$

17. A mason plastered an area of 425 square yards in 5 days. What was the average number of square yards plastered each day?

18. A contractor agreed to furnish and pour 27 cubic yards of concrete for $621. What is the cost per cubic yard?

19. A plumber sets 14 water closets and uses 42 pounds of putty. How much putty is used on an average for each closet?

20. A tank holds 4620 cubic inches of coolant. How many gallons of liquid are needed to fill the tank? (Each gallon contains 231 cubic inches.)

21. How many pieces 13 inches long can be stamped from a roll of sheet brass 325 inches long?

22. A total load of 23,256 watts is distributed equally over 18 branch circuits. Find the load per circuit in watts.

23. What is the average number of feet of wire per outlet used on a job which takes 1886 feet of wire for 82 outlets?

24. A total load of 14,004 watts is distributed equally over 36 branch circuits. Find the load per circuit in watts.

25. How many full lengths of wire 27 inches long can be cut from a roll of wire 1215 inches long?

26. The daily consumption of energy for appliances A and B and equipment C, D, and E is given in the table shown here. Find the hourly rates of consumption for A through E.

Kilowatt Hours Consumed Each 24-Hour Day

A	B	C	D	E
72	264	576	648	1176

27. Given the area of a rectangle in square feet and its length in feet, you divide the area by the length to get the width. How wide

is a rectangle 573 feet long if its area is 61,884 square feet?

28. How long is a rectangle if its width is 98 feet and its area is 56,056 square feet?

29. A jet plane averaged 595 miles each hour while traveling 16,065 miles. How many hours did it take?

30. If a car traveled 288 miles on 18 gallons of gasoline, how many miles did it travel on each gallon?

31. A rocket travels 238,800 miles to get to the moon. It did so in 15 hours. What was its average speed per hour?

32. A shipment of parts cost $514 less $10 shipping costs. How much did each part cost if there were 14 parts in the shipment?

1.6
EXPONENTS, SQUARE ROOTS, AND CUBE ROOTS

When a number can be factored into equal factors, the result can be written using an **exponent**. An exponent tells how many times its **base** is used as a factor. For example, $8 = 2 \cdot 2 \cdot 2$; thus 8 can be factored into three equal factors, each factor a 2. A small raised 3 can be used as an **exponent** of the **base** 2 to indicate this fact, i.e., $8 = 2^3$. Other examples are shown in Table 1.5.

TABLE 1.5

Number of Factors	Exponential Notation	This is read as:
$8 = 2 \cdot 2 \cdot 2$	2^3 ← exponent 3, base 2	"2 cubed"
$16 = 4 \cdot 4$	4^2 ← exponent 2, base 4	"4 squared"
$16 = 2 \cdot 2 \cdot 2 \cdot 2$	2^4 ← exponent 4, base 2	"2 to the 4th"
$25 = 5 \cdot 5$	5^2 ← exponent 2, base 5	"5 squared"
$27 = 3 \cdot 3 \cdot 3$	3^3 ← exponent 3, base 3	"3 to the 3rd" or "3 cubed"

EXAMPLES

1 The number 7^2 means $7 \cdot 7$, where the base 7 is used as a factor 2 times.

2 The number 5^3 means $5 \cdot 5 \cdot 5$, where the base 5 is used as a factor 3 times.

3 The number $(17)^2$ means $17 \cdot 17$, where the base 17 is used as a factor 2 times. It can be written as either $(17)^2$ or 17^2.

4 The number 749^3 means $749 \cdot 749 \cdot 749$.

5 The number 13^5 means $13 \cdot 13 \cdot 13 \cdot 13 \cdot 13$.

When a number is factored into *two equal factors*, each is called the **square root** of the number.

EXAMPLES

6 Since $16 = 4 \cdot 4 = 4^2$, 4 is the square root of 16.

7 Since $25 = 5 \cdot 5 = 5^2$, 5 is the square root of 25.

8 Since $64 = 8 \cdot 8 = 8^2$, 8 is the square root of 64.

When a number is multiplied by itself the result is called the **square** of the number.

EXAMPLES

9   Since $7 \cdot 7 = 49$, 49 is the square of 7.

10   Since $51 \cdot 51 = 2601$, 2601 is the square of 51.

11   Since $600 \cdot 600 = 360000$, 360000 is the square of 600.

A table of squares and square roots contains the square of each given number as well as the square root of each given number. A partial listing of squares and square roots is shown in Table 1.6.

TABLE 1.6

Number	Square	Square Root
1	1	1
4	16	2
9	81	3
25	625	5
36	1296	6
49	2401	7
64	4096	8
81	6561	9
100	10000	10

It should be noted that all square roots in this section are whole numbers. Later in this course it will be necessary to find square roots that are not whole numbers. A table of square roots is provided at the back of the book for this purpose.

Squares and cubes of fractions will be dealt with in Chapter 2. Squares and cubes of decimals will be dealt with in Chapter 3. When a number is factored into *three equal factors,* each is called the **cube root** of the number.

EXAMPLES

12   Since $8 = 2 \cdot 2 \cdot 2 = 2^3$, 2 is the cube root of 8.

13   Since $27 = 3 \cdot 3 \cdot 3 = 3^3$, 3 is the cube root of 27.

14   Since $64 = 4 \cdot 4 \cdot 4 = 4^3$, 4 is the cube root of 64.

When a number is *used as a factor three times,* the product is called the **cube** of the number.

EXAMPLES

15   Since $5 \cdot 5 \cdot 5 = 125$, 125 is the cube of 5.

16   Since $7 \cdot 7 \cdot 7 = 343$, 343 is the cube of 7.

17   Since $8 \cdot 8 \cdot 8 = 512$, 512 is the cube of 8.

A table of cubes and cube roots contains the cube of each given number as well as the cube root of each given number. A partial listing of cubes and cube roots is contained in Table 1.7.

TABLE 1.7

Number	Cube	Cube Root
1	1	1
8	512	2
27	19683	3
64	262144	4
125	1953125	5

The symbol for square root is $\sqrt{}$. The symbol for cube root is $\sqrt[3]{}$

EXAMPLES

18   The number $\sqrt{49}$ is read as "the square root of 49."

19   The number $\sqrt[3]{27}$ is read as "the cube root of 27."

20 $\sqrt{49}$ = 7 because 7 X 7 = 49

21 $\sqrt[3]{27}$ = 3 because 3 X 3 X 3 = 27

22 $\sqrt{100}$ = 10 because 10 X 10 = 100

23 $\sqrt[3]{1000}$ = 10 because 10 X 10 X 10 = 1000

Exponents and root symbols are always considered parts of numerals. For example, $\sqrt[3]{27}$ is a symbol for the number three, and $\sqrt{16}$ represents four. This fact is important when a sequence of operations is involved.

EXAMPLES

24 The expression $3 \cdot 4^2$ means 3 X 4 X 4, which is 48.

WATCH OUT! The numbers $3 \cdot 4^2$ and $(3 \cdot 4)^2$ are *not* the same.

$$3 \cdot 4^2 = 3 \text{ X } 4 \text{ X } 4 = 48$$
$$(3 \cdot 4)^2 = (3 \text{ X } 4)(3 \text{ X } 4)$$
$$= 3 \text{ X } 3 \text{ X } 4 \text{ X } 4$$
$$= 144$$

25 The expression $3\sqrt{16}$ means $3 \cdot 4$, which is 12.

26 The expression $7 + \sqrt[3]{27}$ means $7 + 3$, which is 10.

27 $6(3^2) = 6 \text{ X } 9 = 54$

28 $6(3)^2 = 6 \text{ X } 9 = 54$

29 $6^2(3) = 36 \text{ X } 3 = 108$

30 $(\sqrt{25} - 3)^2 = (5 - 3)^2 = 2^2 = 4$

31 $(\sqrt{36} - \sqrt[3]{8})(3^2 - 2^3) = (6 - 2)(9 - 8) = 4 \text{ X } 1$
    $= 4$

Review Exercises

Perform the indicated operations in Exercises 1–8.

1. 4598 + 3064   2. 9058 + 3465

3. 4598 − 3064   4. 9058 − 3465

5. 724 X 39   6. 842 X 76

7. 8693 ÷ 48   8. 9049 ÷ 76

Write the prime factorization of each number in Exercises 9–12.

9. 360   10. 600

11. 550   12. 1001

EXERCISES
1.6

1. What is the exponent in 7^3?

2. What is the exponent in 3^2?

3. What is the exponent in 16^4?

4. What is the exponent in 30^5?

Find the value of the numbers in Exercises 5–16.

5. 7^3 6. 3^2

7. $(6)^4$ 8. $(4)^5$

9. 16^2 10. 15^2

11. 24^3 12. 23^3

13. $3(4^2)$ 14. $4(3^2)$

15. $3(\sqrt{4} - 1)$ 16. $4(\sqrt[3]{8} - 1)$

Find the *square root* in each of Exercises 17–22.

17. 49   18. 64

19. 121   20. 169

21. 196   22. 144

Find the *cube root* in each of Exercises 23–28.

23. 512   24. 64

25. 125   26. 216

27. 343   28. 729

Square each number in Exercises 29–34.

29. 13   30. 15

31. 18   32. 21

33. 59   34. 67

Cube each number in Exercises 35–40.

35. 9   36. 11
37. 13   38. 14
39. 15   40. 16

Perform the indicated operations in Exercises 41–50.

41. $5(4^2)$ 42. $6(7^2)$
43. $7\sqrt[3]{64}$ 44. $5\sqrt{81}$
45. $5(7^2 - 2^3)$ 46. $6(8^2 - 3^3)$
47. $17 - 2^3$ 48. $300 - 11^2$
49. $(\sqrt{25} + \sqrt[3]{27})(3^3 - 2^2)$
50. $\sqrt{5^2 - 4^2}$

51. The distance in feet that an object falls under the influence of gravity is 16 times the square of the number of seconds it falls. How far will an object fall in 6 seconds?

52. How far will an object fall under the influence of gravity in 13 seconds? (See Exercise 51.)

53. The volume of a cube is the cube of the length of one edge. What is the volume of a cube having an edge with length 7?

54. What is the volume of a cube having an edge with length 6? (See Exercise 53.)

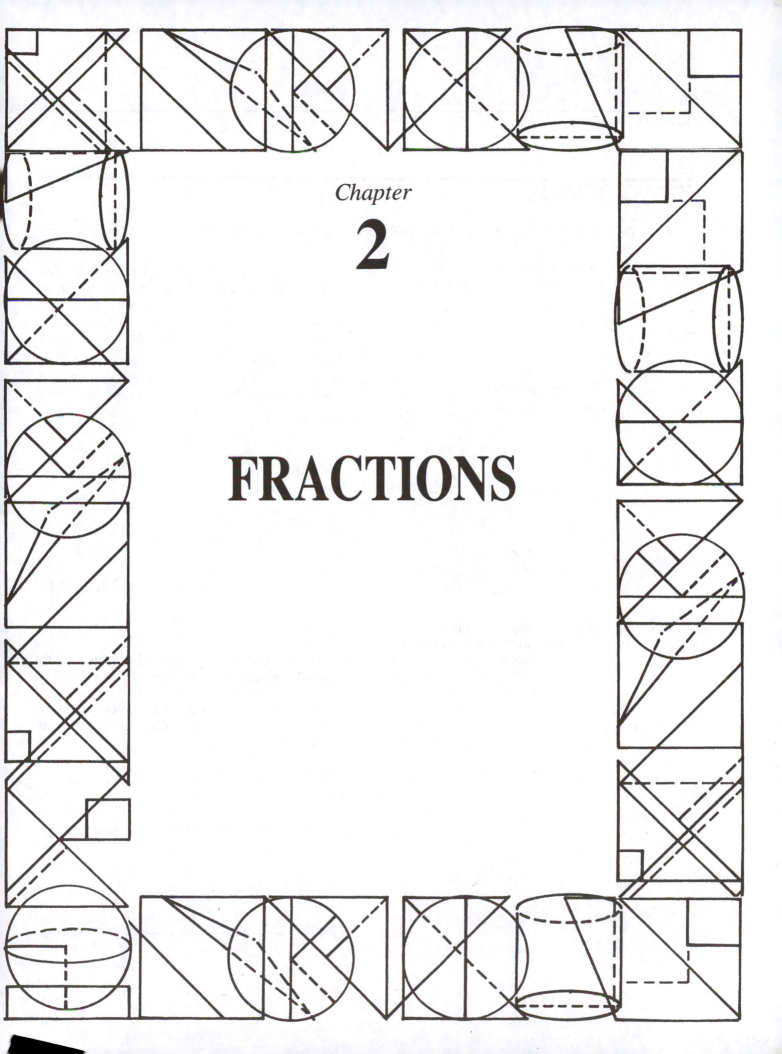

Chapter

2

FRACTIONS

2.1
FRACTIONS

The fraction $\frac{3}{4}$ (see Figure 2.1) represents *three of four equal parts*. Figure 2.2 contains two examples of using $\frac{3}{4}$. In part (i), $\frac{3}{4}$ of the circle is shaded, and in part (ii), $\frac{3}{4}$ of the dimes are shaded.

FIGURE 2.1

The fraction "three-fourths"

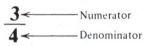

FIGURE 2.2

Two uses of the fraction $\frac{3}{4}$

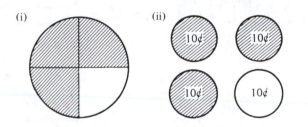

Fractions are called proper or improper. A **proper fraction's** denominator is greater than its numerator. An **improper fraction's** numerator is greater than (or equal to) its denominator. Figure 2.3 contains examples of proper and improper fractions. In part (i), $\frac{4}{5}$ of the circle is shaded, representing a *proper fraction*. Part (ii), in which $\frac{5}{4}$ of a circle has been shaded, is an example of an *improper fraction*. Some other proper fractions are $\frac{3}{4}, \frac{2}{3}, \frac{1}{2}, \frac{7}{9}, \frac{99}{100}$, and $\frac{4}{7}$. Other

FIGURE 2.3

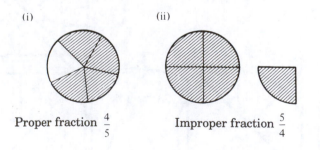

Proper fraction $\frac{4}{5}$ Improper fraction $\frac{5}{4}$

improper fractions are $\frac{4}{3}, \frac{3}{2}, \frac{2}{2}, \frac{9}{7}, \frac{3}{1}, \frac{7}{1}, \frac{100}{99}, \frac{17}{4}$, and $\frac{15}{4}$.

We can think of a fraction as an indicated division of one number by another. Zero cannot be used for the denominator of such a fraction because it is impossible to divide by zero. Some examples are $\frac{8}{4} = 8 \div 4$; $\frac{37}{5} = 37 \div 5$; and $\frac{5}{8} = 5 \div 8$. Whenever the numerator equals the denominator of a fraction, the value of the fraction is *one*. Figure 2.4 illustrates that $\frac{3}{3} = 1$. Similarly, $\frac{2}{2} = 1$, $\frac{4}{4} = 1$, $\frac{5}{5} = 1$, $\frac{6}{6} = 1$, $\frac{10}{10} = 1$, $\frac{62}{62} = 1$, etc.

FIGURE 2.4

$\frac{3}{3}$ *of a circle is 1 circle*

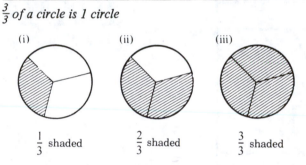

$\frac{1}{3}$ shaded $\frac{2}{3}$ shaded $\frac{3}{3}$ shaded

On a number line, fractions can be used to represent specific points. Figure 2.5 shows some points represented by fractions having 4 as a denominator. Fractions with denominators other than 4 are shown on the number line in Figure 2.6.

FIGURE 2.5

Fractions on a number line I

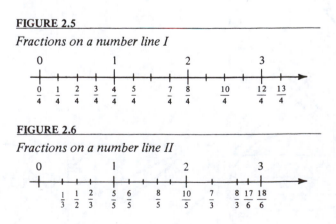

FIGURE 2.6

Fractions on a number line II

Fractions can be used to represent different types of situations. Table 2.1 contains examples of four such situations.

TABLE 2.1

Situation	Summary	Symbol
A parking lot contains 450 cars, and 75 of the cars are Buicks.	Seventy-five out of 450 cars are Buicks.	The fraction representing the portion of the total number of cars that are Buicks is $\frac{75}{450}$.
A store shelf contains 27 cans of beans. Thirteen of the cans are Van Camp's.	Thirteen out 27 cans contain Van Camp's beans.	The fraction representing the portion of the total number of cans that are Van Camp's is $\frac{13}{27}$.
John and his friend have $11 together. John has $3.	John has $3 of $11.	The fraction representing the part of the money that John has is $\frac{3}{11}$.
Susan drove 35 miles and Jim drove 65 miles. The entire trip was 100 miles.	Susan drove 35 of a total of (35 + 65) miles, or 35 of 100 miles.	The part of the total driven by Susan is represented by $\frac{35}{100}$.

Any number can be expressed as a fraction by using 1 as the denominator, e.g., $5 = \frac{5}{1}$, $7 = \frac{7}{1}$, and $39 = \frac{39}{1}$.

Review Exercises

Perform the indicated operations in Exercises 1–12.

1. $13 + 29 + 46$
2. $104 - 67$
3. 69836
    $- 10967$
4. 479056
    30089
    421
    $+ 6358$
5. $14 + 6(19 - 11) - 25 + 4$
6. 653
    $\times 94$
7. $19\overline{)643}$
8. $7005 \div 18$
9. $\sqrt{121}$
10. $\sqrt[3]{64}$
11. $(13)^2$
12. $247 - 5^3$

EXERCISES
2.1

1. What is the numerator of $\frac{7}{16}$?
2. What is the denominator of $\frac{9}{32}$?
3. The shaded parts of the figure represent _____ of _____ equal parts.

4. The shaded parts of the figure represent _____ of _____ equal parts.

5. Which of the following represent the number one?

    $$\frac{2}{3}, \frac{4}{4}, \frac{5}{6}, \frac{7}{7}, \frac{8}{9}, \frac{10}{10}$$

6. Which of the following *do not* represent the number one?

    $$\frac{2}{3}, \frac{3}{3}, \frac{4}{3}, \frac{5}{4}, \frac{5}{5}, \frac{5}{6}$$

7. Wisconsin has an area that, when multiplied by 48, gives the total area of the U.S. What fraction of the area of the U.S. is in Wisconsin?

8. Seventeen of 68 automobiles in a parking lot are Fords. What fraction of the autos in the parking lot are Fords?

9. A store display contains 31 boxes of soap. Thirteen of the boxes are Brand *X*. What fraction of the total display is made up of Brand *X*?

10. Harry drove 37 miles. His entire trip was to be 100 miles. What fraction of his trip had he completed?

11. What fractions represent points A and B in the figure below?

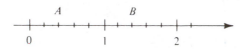

12. What fractions represent points C and D in the figure below?

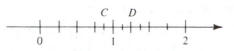

13. Which fraction has the greatest numerator?

$$\frac{39}{65}, \frac{48}{65}, \frac{4}{65}, \frac{13}{65}, \frac{48}{51}, \frac{65}{100}$$

14. Which fraction has the greatest denominator?

$$\frac{15}{16}, \frac{5}{8}, \frac{25}{26}, \frac{13}{19}, \frac{49}{21}$$

Refer to the following list of fractions for Exercises 15–26.

$$\frac{3}{4}, \frac{3}{2}, \frac{2}{3}, \frac{4}{3}, \frac{2}{2}, \frac{9}{7}, \frac{1}{2}, \frac{105}{99}, \frac{99}{100}, \frac{4}{7}, \frac{15}{4}$$

15. Which fractions in the preceding list are improper?

16. Which fractions in the list are proper?

17. Which fractions have the same denominators?

18. Which fractions have the same numerators?

19. A parking lot contains 835 cars, 49 of which are Oldsmobiles. What fraction of the cars in the lot are Oldsmobiles?

20. John and Harry have $27 together. Harry has $15. What fraction represents the part of the total money Harry has?

21. In Exercise 20, what fraction represents the part of the total money John has?

22. A store has 95 cans of tomato soup, 49 of which are Campbell's. What fraction represents the part of the tomato soup in the store that is *not* Campbell's? (Assume all cans are the same size.)

23. Gordon drove 79 miles and Jack drove 53 miles. What fraction of the entire trip did Jack drive?

24. In Exercise 23, what fraction of the entire trip did Gordon drive?

25. A water tank, when filled, holds 750 kiloliters. There are 428 kiloliters of liquid in the tank. What fraction represents the filled portion of the tank?

26. In Exercise 25, what fraction represents the unfilled portion of the tank?

2.2
EQUIVALENT FRACTIONS

Equivalent fractions are fractions that name the same number (represent the same fractional part of a unit region). Two examples of equivalent fractions are shown in Figures 2.7 and 2.8. In Figure 2.7 we see that $\frac{2}{4}$ and $\frac{3}{6}$ are equivalent fractions since both represent one-half of a unit region. Likewise, in Figure 2.8, the fractions $\frac{20}{12}$ and $\frac{15}{9}$ are equivalent since both represent one and two-thirds of a unit region.

FIGURE 2.7

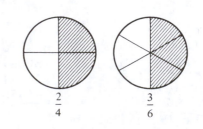

$$\frac{2}{4} \qquad \frac{3}{6}$$

FIGURE 2.8

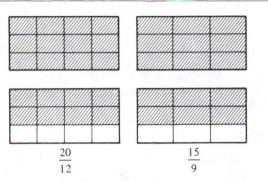

$$\frac{20}{12} \qquad \frac{15}{9}$$

Equivalent forms of a given fraction can be formed by writing the fraction in higher terms or by reducing the fraction to lower terms.

> To convert a fraction to higher terms, multiply both numerator and denominator by the same number.

EXAMPLE 1 Convert $\frac{1}{5}$ to higher terms by multiplying numerator and denominator by 2, 3, 5, and 8.

Solution

$$\frac{1}{5} = \frac{1 \times 2}{5 \times 2} = \frac{2}{10} \qquad \frac{1}{5} = \frac{1 \times 5}{5 \times 5} = \frac{5}{25}$$

$$\frac{1}{5} = \frac{1 \times 3}{5 \times 3} = \frac{3}{15} \qquad \frac{1}{5} = \frac{1 \times 8}{5 \times 8} = \frac{8}{40}$$

(*Note:* $\frac{1}{5}$, $\frac{2}{10}$, $\frac{3}{15}$, $\frac{5}{25}$, and $\frac{8}{40}$ are all equivalent fractions.)

EXAMPLE 2 Find a fraction that is equivalent to $\frac{3}{7}$ and has 28 as a denominator.

Solution We know that $28 = 7 \times 4$, so we multiply both numerator and denominator of $\frac{3}{7}$ by 4 to obtain the desired equivalent fraction.

$$\frac{3}{7} = \frac{3 \times 4}{7 \times 4} = \frac{12}{28}$$

EXAMPLE 3 Convert $\frac{11}{8}$ to an equivalent fraction with a denominator of 48.

Solution We know that $48 = 8 \times 6$. Thus,

$$\frac{11}{8} = \frac{11 \times 6}{8 \times 6} = \frac{66}{48}.$$

> To convert a fraction to lower terms, divide both numerator and denominator by the same number.

EXAMPLE 4 Convert $\frac{8}{24}$ to lower terms by dividing numerator and denominator by 2, 4, and 8.

Solution

$$\frac{8}{24} = \frac{8 \div 2}{24 \div 2} = \frac{4}{12}$$

$$\frac{8}{24} = \frac{8 \div 4}{24 \div 4} = \frac{2}{6}$$

$$\frac{8}{24} = \frac{8 \div 8}{24 \div 8} = \frac{1}{3}$$

(*Note:* $\frac{8}{24}$, $\frac{4}{12}$, $\frac{2}{6}$, and $\frac{1}{3}$ are all equivalent fractions.)

EXAMPLE 5 Find a fraction that is equivalent to $\frac{30}{54}$ and has 9 as a denominator.

Solution Since $9 = 54 \div 6$, we divide both numerator and denominator of $\frac{30}{54}$ by 6 to obtain the desired equivalent fraction.

$$\frac{30}{54} = \frac{30 \div 6}{54 \div 6} = \frac{5}{9}$$

EXAMPLE 6 Find three fractions equivalent to $\frac{18}{60}$ that are in lower terms than the given fraction.

Solution

$$\frac{18}{60} = \frac{18 \div 2}{60 \div 2} = \frac{9}{30}$$

$$\frac{18}{60} = \frac{18 \div 3}{60 \div 3} = \frac{6}{20}$$

$$\frac{18}{60} = \frac{18 \div 6}{60 \div 6} = \frac{3}{10}$$

A fraction can be reduced if the numerator and denominator have a common factor. Divisibility tests for whole numbers are useful in identifying what common factors, if any, are contained in the numerator and denominator of a given fraction. A few of the divisibility tests for whole numbers follow. A whole number, except zero, is divisible by any of its factors.

> **A whole number is divisible by two if the digit in the ones place is 0, 2, 4, 6, or 8.**

Numbers such as 18, 352, 876, 1590, and 254 are divisible by two.

$$18 \div 2 = 9 \qquad 876 \div 2 = 438$$
$$352 \div 2 = 176 \qquad 1590 \div 2 = 795$$
$$254 \div 2 = 127$$

> **A whole number is divisible by three if the sum of the digits in the number are divisible by three.**

The sum of the digits in 48 $(4 + 8 = 12)$ is divisible by 3, so 48 is divisible by 3: $48 \div 3 = 16$. The sum of the digits in 1302 $(1 + 3 + 0 + 2 = 6)$ is divisible by 3, so 1302 is divisible by 3: $1302 \div 3 = 434$.

> **A whole number is divisible by five if the digit in the ones place is zero or five.**

Numbers such as 10, 35, 470, 9325, and 6530 are divisible by five.

$$10 \div 5 = 2 \qquad 470 \div 5 = 94$$
$$35 \div 5 = 7 \qquad 9325 \div 5 = 1865$$
$$6530 \div 5 = 1306$$

> **A whole number is divisible by nine if the sum of the digits in the number is divisible by nine.**

The sum of the digits in 117 $(1 + 1 + 7 = 9)$ is divisible by 9, so 117 is divisible by 9:

$117 \div 9 = 13$. The sum of the digits in 7803 $(7 + 8 + 0 + 3 = 18)$ is divisible by 9, so 7803 is divisible by 9: $7803 \div 9 = 867$.

EXAMPLE 7 Is 4680 divisible by (a) 2? (b) 3? (c) 5? (d) 9?

Solutions

(a) The number 4680 is divisible by 2 since the digit in the ones place is 0.

$$4680 \div 2 = 2340$$

(b) The number 4680 is divisible by 3 since $4 + 6 + 8 + 0 = 18$ and 18 is divisible by 3.

$$4680 \div 3 = 1560$$

(c) The number 4680 is divisible by 5 since the digit in the ones place is 0.

$$4680 \div 5 = 936$$

(d) The number 4680 is divisible by 9 since $4 + 6 + 8 + 0 = 18$ and 18 is divisible by 9.

$$4680 \div 9 = 520$$

EXAMPLE 8 Is 195 divisible by (a) 2? (b) 3? (c) 5? (d) 9?

Solutions

(a) The number 195 is not divisible by 2 since the digit in the ones place is not 0, 2, 4, 6, or 8.

(b) It is divisible by 3 since $1 + 9 + 5 = 15$ and 15 is divisible by 3.

$$195 \div 3 = 65$$

(c) It is divisible by 5 since the digit in the ones place is 5.

$$195 \div 5 = 39$$

(d) It is not divisible by 9 since $1 + 9 + 5 = 15$ and 15 is not divisible by 9.

A fraction is said to be in **lowest terms** when the numerator and the denominator have no common factor (except the number 1). A fraction can be reduced to its lowest terms by dividing the numerator and denominator by their common factors.

EXAMPLE 9   Reduce $\frac{18}{42}$ to lowest terms.

Solution   Since 18 and 42 have a common factor 2,

$$\frac{18}{42} = \frac{18 \div 2}{42 \div 2} = \frac{9}{21}.$$

Since 9 and 21 have a common factor 3,

$$\frac{9}{21} = \frac{9 \div 3}{21 \div 3} = \frac{3}{7}.$$

Three and 7 have no common factor, other than 1. Thus, $\frac{18}{42}$ in lowest terms is the equivalent fraction $\frac{3}{7}$.

EXAMPLE 10   Reduce $\frac{210}{150}$ to lowest terms.

Solution

$$\frac{210}{150} = \frac{210 \div 2}{150 \div 2} = \frac{105}{75}$$

$$\frac{105}{75} = \frac{105 \div 5}{75 \div 5} = \frac{21}{15}$$

$$\frac{21}{15} = \frac{21 \div 3}{15 \div 3} = \frac{7}{5}$$

Thus, $\frac{210}{150}$ in lowest terms is the equivalent fraction $\frac{7}{5}$.

The process of reducing a fraction to lowest terms may be simplified by dividing the numerator and denominator by their greatest common factor (gcf). If the greatest common factor cannot be seen easily it can be determined by writing the numerator and denominator in prime factor form. Review Section 1.4 for the process

of prime factorization and the process for determining the greatest common factor of two or more numbers.

EXAMPLE 11   Reduce $\frac{84}{180}$ to lowest terms by dividing numerator and denominator by their greatest common factor.

Solution   The prime factors of 84 and 180 are:

$$84 = 2 \times 2 \times 3 \times 7$$
$$180 = 2 \times 2 \times 3 \times 3 \times 5.$$

The greatest common factor of 84 and 180 is $2 \times 2 \times 3 = 12$.

$$\frac{84}{180} = \frac{84 \div 12}{180 \div 12} = \frac{7}{15}$$

Thus, $\frac{84}{180}$ in lowest terms is the equivalent fraction $\frac{7}{15}$.

EXAMPLE 12   Reduce $\frac{32}{56}$ to lowest terms.

Solution

$$32 = 2 \times 2 \times 2 \times 2 \times 2$$
$$56 = 2 \times 2 \times 2 \times 7$$

The gcf of 32 and 56 is $2 \times 2 \times 2 = 8$.

$$\frac{32}{56} = \frac{32 \div 8}{56 \div 8} = \frac{4}{7}$$

EXAMPLE 13   Reduce $\frac{210}{252}$ to lowest terms.

Solution

$$210 = 2 \times 3 \times 5 \times 7$$
$$252 = 2 \times 2 \times 3 \times 3 \times 7$$
$$\text{gcf} = 2 \times 3 \times 7 = 42$$
$$\frac{210}{252} = \frac{210 \div 42}{252 \div 42} = \frac{5}{6}$$

Dividing the numerator and denominator of a fraction by the same number is often called

cancelling like factors. Cancellation is shown in the following example:

$$\frac{24}{36} = \frac{2 \times 2 \times 2 \times 3}{2 \times 2 \times 3 \times 3} = \frac{1 \times 1 \times 2 \times 1}{1 \times 1 \times 3 \times 1} = \frac{2}{3}.$$

The slash marks indicate each division by a common factor, so the quotient is 1 in each case. Quotients of 1 are usually not written when cancelling, but they must be understood. This is very important in examples like

$$\frac{6}{18} = \frac{2 \times 3}{2 \times 3 \times 3} = \frac{2 \times 3}{2 \times 3 \times 3} = \frac{1}{3}.$$

All of the factors in the numerator have been "cancelled out," so the numerator of the reduced form is (1 × 1), or 1.

EXAMPLE 14   Reduce $\frac{56}{84}$ to lowest terms by cancellation.

Solution

$$\frac{56}{84} = \frac{2 \times 2 \times 2 \times 7}{2 \times 2 \times 3 \times 7} = \frac{2 \times 2 \times 2 \times 7}{2 \times 2 \times 3 \times 7} = \frac{2}{3}$$

EXAMPLE 15   Reduce $\frac{210}{165}$ to lowest terms by cancellation.

Solution

$$\frac{210}{165} = \frac{2 \times 3 \times 5 \times 7}{3 \times 5 \times 11} = \frac{2 \times 7}{11} = \frac{14}{11}$$

EXAMPLE 16   Reduce $\frac{24}{72}$ to lowest terms by cancellation.

Solution

$$\frac{24}{72} = \frac{2 \times 2 \times 2 \times 3}{2 \times 2 \times 2 \times 3 \times 3} = \frac{1}{3}$$

WATCH OUT! The numerator of

$$\frac{2 \times 2 \times 2 \times 3}{2 \times 2 \times 2 \times 3 \times 3}$$

is 1, not 0.

EXAMPLE 17   Reduce $\frac{90}{18}$ to lowest terms by cancellation.

Solution

$$\frac{90}{18} = \frac{2 \times 3 \times 3 \times 5}{2 \times 3 \times 3} = \frac{5}{1} = 5$$

Review Exercises

1. In the fraction $\frac{8}{11}$ the numerator is _____ .
2. In the fraction $\frac{19}{5}$ the denominator is _____ .
3. Write $5 \div 9$ as a fraction.
4. Write $7 \div 15$ as a fraction.
5. Which fractions are equal to one?

$$\frac{5}{5}, \frac{4}{8}, \frac{1}{3}, \frac{7}{7}, \frac{13}{13}$$

6. Which fractions are proper fractions?

$$\frac{3}{7}, \frac{8}{2}, \frac{3}{3}, \frac{6}{12}, \frac{5}{15}$$

7. What is the prime factor form of 630?
8. What is the prime factor form of 264?
9. What is the greatest common factor of 63 and 42?
10. What is the greatest common factor of 75 and 60?
11. What is the greatest common factor of 268 and 252?
12. What is the greatest common factor of 630 and 840?

EXERCISES
2.2

For each fraction in Exercises 1–6 write three equivalent fractions in higher terms.

1. $\dfrac{1}{2}$ 2. $\dfrac{1}{3}$

3. $\dfrac{3}{5}$ 4. $\dfrac{2}{7}$

5. $\dfrac{8}{5}$ 6. $\dfrac{9}{4}$

For each fraction in Exercises 7–12 write three equivalent fractions in lower terms.

7. $\dfrac{12}{18}$ 8. $\dfrac{15}{60}$

9. $\dfrac{32}{56}$ 10. $\dfrac{27}{81}$

11. $\dfrac{40}{24}$ 12. $\dfrac{75}{30}$

For each fraction in Exercises 13–18 find the greatest common factor (gcf) of the numerator and denominator. (*Hint:* Use prime factorization of both numerator and denominator to help determine the gcf.)

13. $\dfrac{24}{36}$ 14. $\dfrac{18}{30}$

15. $\dfrac{42}{105}$ 16. $\dfrac{45}{75}$

17. $\dfrac{80}{48}$ 18. $\dfrac{90}{54}$

Find the gcf of numerator and denominator in Exercises 19–34; then reduce to lowest terms.

19. $\dfrac{3}{9}$ 20. $\dfrac{4}{16}$

21. $\dfrac{18}{15}$ 22. $\dfrac{24}{18}$

23. $\dfrac{27}{36}$ 24. $\dfrac{14}{28}$

25. $\dfrac{40}{64}$ 26. $\dfrac{56}{64}$

27. $\dfrac{28}{40}$ 28. $\dfrac{20}{32}$

29. $\dfrac{49}{63}$ 30. $\dfrac{30}{75}$

31. $\dfrac{120}{72}$ 32. $\dfrac{90}{36}$

33. $\dfrac{840}{300}$ 34. $\dfrac{108}{48}$

In Exercises 35–44 use cancellation to reduce to lowest terms. Show all of your work.

35. $\dfrac{60}{156}$ 36. $\dfrac{140}{84}$

37. $\dfrac{36}{132}$ 38. $\dfrac{105}{135}$

39. $\dfrac{210}{98}$ 40. $\dfrac{126}{630}$

41. $\dfrac{300}{210}$ 42. $\dfrac{195}{153}$

43. $\dfrac{546}{910}$ 44. $\dfrac{268}{402}$

2.3
MULTIPLYING FRACTIONS AND MIXED NUMBERS

The **product of two fractions** can be thought of as *a fraction of a fraction*. Figure 2.9 illustrates the product $\frac{1}{2}$ times $\frac{1}{4}$, which can be thought of as $\frac{1}{2}$ *of* $\frac{1}{4}$. The shaded part is $\frac{1}{2}$ of $\frac{1}{4}$ of the entire region, or $\frac{1}{8}$ of the entire region. Figure 2.10 illustrates the product $\frac{2}{3}$ times $\frac{4}{5}$, which can be thought of as $\frac{2}{3}$ *of* $\frac{4}{5}$. The shaded part is $\frac{2}{3}$ of $\frac{4}{5}$ of the entire region, or $\frac{8}{15}$ of the entire region.

FIGURE 2.9

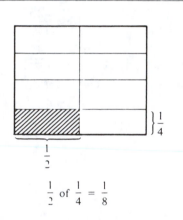

$$\frac{1}{2} \text{ of } \frac{1}{4} = \frac{1}{8}$$

FIGURE 2.10

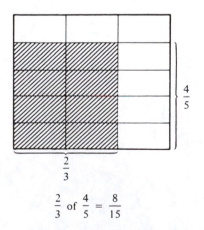

$$\frac{2}{3} \text{ of } \frac{4}{5} = \frac{8}{15}$$

Notice that the products of the fractions illustrated in Figures 2.9 and 2.10 could be obtained as follows:

$$\frac{1}{2} \cdot \frac{1}{4} = \frac{1 \cdot 1}{2 \cdot 4} = \frac{1}{8}$$

$$\frac{2}{3} \cdot \frac{4}{5} = \frac{2 \cdot 4}{3 \cdot 5} = \frac{8}{15}$$

(*Note:* The raised dot (·) means "times.") In each case the product of the two fractions is represented by a fraction whose numerator is the product of the numerators and whose denominator is the product of the denominators. This process can always be used to compute the product of two or more fractions.

> The product of two or more fractions is the product of the numerators over the product of the denominators.

EXAMPLES

1 $\frac{1}{3} \cdot \frac{2}{3} = \frac{1 \cdot 2}{3 \cdot 3} = \frac{2}{9}$

2 $\frac{3}{4} \cdot \frac{5}{2} = \frac{3 \cdot 5}{4 \cdot 2} = \frac{15}{8}$

3 $\frac{9}{16} \cdot \frac{3}{8} = \frac{9 \cdot 3}{16 \cdot 8} = \frac{27}{128}$

4 $\frac{15}{7} \cdot \frac{23}{20} = \frac{15 \cdot 23}{7 \cdot 20} = \frac{345}{140}$

5 $\frac{1}{2} \cdot \frac{3}{5} \cdot \frac{7}{8} = \frac{1 \cdot 3 \cdot 7}{2 \cdot 5 \cdot 8} = \frac{21}{80}$

6 $\frac{17}{5} \cdot \frac{6}{7} \cdot \frac{13}{11} \cdot \frac{8}{3} = \frac{17 \cdot 6 \cdot 13 \cdot 8}{5 \cdot 7 \cdot 11 \cdot 3} = \frac{10608}{1155}$

Exponents can be used on fractions to indicate multiplication.

EXAMPLES

7 $\left(\frac{1}{2}\right)^2 = \left(\frac{1}{2}\right)\left(\frac{1}{2}\right) = \frac{1 \cdot 1}{2 \cdot 2} = \frac{1}{4}$

$$8 \quad \left(\frac{1}{3}\right)^4 = \left(\frac{1}{3}\right)\left(\frac{1}{3}\right)\left(\frac{1}{3}\right)\left(\frac{1}{3}\right) = \frac{1 \cdot 1 \cdot 1 \cdot 1}{3 \cdot 3 \cdot 3 \cdot 3} = \frac{1}{81}$$

$$9 \quad \left(\frac{2}{5}\right)^3 = \left(\frac{2}{5}\right)\left(\frac{2}{5}\right)\left(\frac{2}{5}\right) = \frac{2 \cdot 2 \cdot 2}{5 \cdot 5 \cdot 5} = \frac{8}{125}$$

$$10 \quad \left(\frac{7}{8}\right)^2 = \left(\frac{7}{8}\right)\left(\frac{7}{8}\right) = \frac{7 \cdot 7}{8 \cdot 8} = \frac{49}{64}$$

Reducing the product of two or more fractions to lowest terms can be simplified by cancellation before multiplying the numerators and denominators of the fractions. If the common factors in the numerators and denominators of the fractions are not easy to see, write the prime factor form of each numerator and denominator before you try to cancel. Examples 11–15 show this process.

EXAMPLE 11   Find the product of $\frac{3}{4}$ and $\frac{8}{9}$ in lowest terms.

Solution

$$\frac{3}{4} \cdot \frac{8}{9} = \frac{3}{4} \cdot \frac{4 \cdot 2}{3 \cdot 3} = \frac{\cancel{3} \cdot \cancel{4} \cdot 2}{\cancel{4} \cdot \cancel{3} \cdot 3} = \frac{2}{3}$$

EXAMPLE 12   Find the product of $\frac{4}{15}$ and $\frac{5}{12}$ in lowest terms.

Solution

$$\frac{4}{15} \cdot \frac{5}{12} = \frac{4}{3 \cdot 5} \cdot \frac{5}{3 \cdot 4} = \frac{\cancel{4} \cdot \cancel{5}}{3 \cdot \cancel{5} \cdot 3 \cdot \cancel{4}} = \frac{1}{9}$$

Notice that the numerator of the product in reduced form is 1, not 0.

EXAMPLE 13   Find the product of $\frac{12}{25}$ and $\frac{35}{54}$ in lowest terms.

Solution

$$\frac{12}{25} \cdot \frac{35}{24} = \frac{2 \cdot 2 \cdot 3}{5 \cdot 5} \cdot \frac{5 \cdot 7}{2 \cdot 3 \cdot 3 \cdot 3}$$

$$= \frac{\cancel{2} \cdot 2 \cdot \cancel{3} \cdot \cancel{5} \cdot 7}{\cancel{5} \cdot 5 \cdot \cancel{2} \cdot \cancel{3} \cdot 3 \cdot 3} = \frac{2 \cdot 7}{5 \cdot 3 \cdot 3} = \frac{14}{45}$$

EXAMPLE 14   Find the product of $\frac{6}{35}$, $\frac{25}{16}$, and $\frac{8}{45}$ in lowest terms.

Solution

$$\frac{6}{35} \cdot \frac{25}{16} \cdot \frac{8}{45} = \frac{\cancel{2} \cdot \cancel{3}}{\cancel{5} \cdot 7} \cdot \frac{\cancel{5} \cdot \cancel{5}}{\cancel{2} \cdot \cancel{2} \cdot \cancel{2} \cdot 2} \cdot \frac{\cancel{2} \cdot \cancel{2} \cdot \cancel{2}}{\cancel{3} \cdot 3 \cdot \cancel{5}}$$

$$= \frac{1}{21}$$

EXAMPLE 15   Find the product of $\frac{105}{72}$, $\frac{48}{90}$, and $\frac{126}{21}$ in lowest terms.

Solution

$$\frac{105}{24} \cdot \frac{48}{90} \cdot \frac{126}{21}$$

$$= \frac{\cancel{3} \cdot \cancel{5} \cdot \cancel{7}}{2 \cdot 2 \cdot 2 \cdot \cancel{3}} \cdot \frac{2 \cdot 2 \cdot 2 \cdot 2 \cdot \cancel{3}}{2 \cdot \cancel{3} \cdot 3 \cdot \cancel{5}} \cdot \frac{2 \cdot 3 \cdot 3 \cdot 7}{\cancel{3} \cdot \cancel{7}}$$

$$= \frac{14}{1}$$

$$= 14$$

A **mixed number form** represents the sum of a whole number and a fraction. Table 2.2 contains several examples of mixed numbers.

TABLE 2.2

Sum	Mixed Number Form	This is read as:
$1 + \frac{3}{4}$	$1\frac{3}{4}$	one and three-fourths.
$3 + \frac{5}{16}$	$3\frac{5}{16}$	three and five-sixteenths.
$12 + \frac{4}{7}$	$12\frac{4}{7}$	twelve and four-sevenths.
$29 + \frac{15}{64}$	$29\frac{15}{64}$	twenty-nine and fifteen sixty-fourths.

An improper fraction can be converted to a whole number or mixed number form by dividing its numerator by its denominator. (See Figures 2.11 and 2.12.) In Figure 2.11 we see that $7 \div 4 = 1$ with remainder 3; thus, $\frac{7}{4} = 1\frac{3}{4}$. Similarly, in Figure 2.12, $29 \div 8 = 3$ with remainder

5; thus, $\frac{29}{8} = 3\frac{5}{8}$. Examples 16–18 illustrate the process of converting fractions to whole number form or mixed number form.

FIGURE 2.11

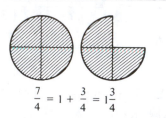

$$\frac{7}{4} = 1 + \frac{3}{4} = 1\frac{3}{4}$$

FIGURE 2.12

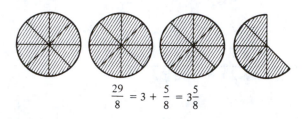

$$\frac{29}{8} = 3 + \frac{5}{8} = 3\frac{5}{8}$$

EXAMPLE 16   Convert $\frac{14}{5}$ to whole number or mixed number form.

Solution

$$14 \div 5 = 2 \text{ with remainder } 4$$

Therefore, $\frac{14}{5} = 2\frac{4}{5}$.

EXAMPLE 17   Convert $\frac{47}{7}$ to whole number or mixed number form.

Solution

$$47 \div 7 = 6 \text{ with remainder } 5$$

Therefore, $\frac{47}{7} = 6\frac{5}{7}$.

EXAMPLE 18   Convert $\frac{51}{3}$ to whole number or mixed number form.

Solution

$$51 \div 3 = 17 \text{ with remainder } 0$$

Therefore, $\frac{51}{3} = 17$.

A fraction in mixed number form can be converted to fraction form by considering its meaning. For example, $3\frac{2}{3}$ means *three wholes plus two-thirds of a whole*. Since the fractional part, $\frac{2}{3}$, has 3 as a denominator, all *wholes* are divided into thirds.

$$3\frac{2}{3} = 3 + \frac{2}{3}$$
$$= \frac{3}{1} + \frac{2}{3}$$
$$= \frac{3 \cdot 3}{1 \cdot 3} + \frac{2}{3}$$
$$= \frac{9}{3} + \frac{2}{3}$$
$$= \frac{9 + 2}{3}$$
$$= \frac{11}{3}$$

Figure 2.13 illustrates the process. The computation in this process makes use of the fact that you can add fractions having like denominators by adding their numerators and using the result as the numerator of the sum. Following are several examples of adding fractions having like denominators.

FIGURE 2.13

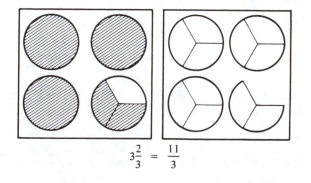

$$3\frac{2}{3} = \frac{11}{3}$$

EXAMPLES

19 $\frac{9}{3} + \frac{2}{3} = \frac{9 + 2}{3} = \frac{11}{3}$

20 $\frac{3}{4} + \frac{6}{4} = \frac{3 + 6}{4} = \frac{9}{4}$

21 $\dfrac{1}{12}+\dfrac{10}{12}=\dfrac{1+10}{12}=\dfrac{11}{12}$

22 $\dfrac{3}{7}+\dfrac{8}{7}=\dfrac{3+8}{7}=\dfrac{11}{7}$

23 $\dfrac{1}{5}+\dfrac{1}{5}=\dfrac{1+1}{5}=\dfrac{2}{5}$

The usual method for converting mixed numbers to improper fractions is as follows:

$$2\dfrac{3}{4}=2+\dfrac{3}{4}$$
$$=\dfrac{8}{4}+\dfrac{3}{4}$$
$$=\dfrac{11}{4}.$$

Now here is a "shortcut" method:

$$2\dfrac{3}{4}=\dfrac{(4\cdot 2)+3}{4}=\dfrac{11}{4}.$$

This conversion is illustrated in Figure 2.14.

FIGURE 2.14

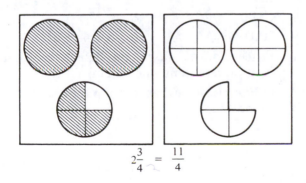

$$2\dfrac{3}{4}=\dfrac{11}{4}$$

Since $2\dfrac{3}{4}=\dfrac{8}{4}+\dfrac{3}{4}$, it is clear that $2\dfrac{3}{4}=\dfrac{2\cdot 4}{4}+\dfrac{3}{4}$, which is the same as $\dfrac{(2\cdot 4)+3}{4}$ or $\dfrac{8+3}{4}$. Thus, $2\dfrac{3}{4}=\dfrac{11}{4}$ can be obtained quickly by multiplying the 2 by the denominator 4, then adding the numerator 3 and using the result as the numerator and 4 as the denominator of the improper fraction. Figure 2.15 outlines this procedure.

FIGURE 2.15

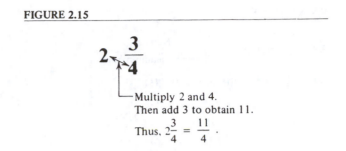

Multiply 2 and 4. Then add 3 to obtain 11. Thus, $2\dfrac{3}{4}=\dfrac{11}{4}$.

Here are some other examples of this procedure.

EXAMPLES

24 $2\dfrac{3}{5}=\dfrac{(2\cdot 5)+3}{5}=\dfrac{10+3}{5}=\dfrac{13}{5}$

25 $5\dfrac{7}{8}=\dfrac{(5\cdot 8)+7}{8}=\dfrac{40+7}{8}=\dfrac{47}{8}$

26 $4\dfrac{1}{3}=\dfrac{(4\cdot 3)+1}{3}=\dfrac{12+1}{3}=\dfrac{13}{3}$

27 $12\dfrac{7}{16}=\dfrac{(12\cdot 16)+7}{16}=\dfrac{192+7}{16}=\dfrac{199}{16}$

Multiplying mixed numbers involves converting them to improper fractions. This is illustrated in Examples 28–32.

EXAMPLES

28 $3\dfrac{2}{3}\cdot 5\dfrac{3}{4}=\dfrac{11}{3}\cdot\dfrac{23}{4}=\dfrac{11\cdot 23}{3\cdot 4}=\dfrac{253}{12}$

29 $2\dfrac{1}{3}\cdot\dfrac{11}{38}=\dfrac{7}{3}\cdot\dfrac{11}{38}=\dfrac{77}{114}$

30 $3\dfrac{3}{8}\cdot 4\dfrac{2}{3}=\dfrac{27}{8}\cdot\dfrac{14}{3}=\dfrac{3\cdot 9}{2\cdot 4}\cdot\dfrac{2\cdot 7}{3}=\dfrac{63}{4}$

31 $5\cdot 7\dfrac{11}{16}=\dfrac{5}{1}\cdot\dfrac{123}{16}=\dfrac{615}{16}$

32 $3\dfrac{2}{3}\cdot 4\dfrac{1}{4}=\dfrac{11}{3}\cdot\dfrac{17}{4}=\dfrac{187}{12}$ or $15\dfrac{7}{12}$

The following examples summarize the ideas in Sections 2.1, 2.2, and 2.3.

EXAMPLES

33 A fraction: $\dfrac{7}{16}$ ← Numerator, ← Denominator

34 Five of eight equal parts:

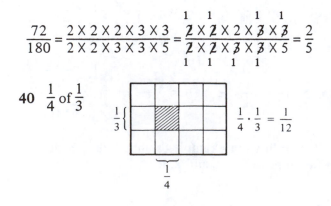

$\dfrac{5}{8}$

35

Situation	Summary	Symbol
A loaf of bread has 23 slices, and 5 of them are moldy.	Five of the 23 slices in a loaf of bread are moldy.	The fraction that shows the part of the loaf that is moldy is $\frac{5}{23}$.

36 $\dfrac{0}{5} = 0 \qquad \dfrac{5}{5} = 1 \qquad \dfrac{10}{5} = 2$

$$\begin{array}{c} 0 \qquad\qquad 1 \qquad\qquad 2 \\ \hline \frac{0}{5}\ \frac{1}{5}\ \frac{2}{5}\ \frac{3}{5}\ \frac{4}{5}\ \frac{5}{5}\ \frac{6}{5}\ \frac{7}{5}\ \frac{8}{5}\ \frac{9}{5}\ \frac{10}{5}\ \frac{11}{5}\ \frac{12}{5}\ \frac{13}{5} \end{array}$$

37 Equivalent fractions in higher terms:

$$\dfrac{3}{4} = \dfrac{3 \cdot 5}{4 \cdot 5} = \dfrac{15}{20} \qquad \dfrac{9}{5} = \dfrac{9 \cdot 7}{5 \cdot 7} = \dfrac{63}{35}$$

38 Equivalent fractions in lower terms:

$$\dfrac{12}{18} = \dfrac{12 \div 2}{18 \div 2} = \dfrac{6}{9} \qquad \dfrac{24}{72} = \dfrac{24 \div 6}{72 \div 6} = \dfrac{4}{12}$$

39 Reduce to lowest terms:

$$\dfrac{72}{180} = \dfrac{2 \times 2 \times 2 \times 3 \times 3}{2 \times 2 \times 3 \times 3 \times 5} = \dfrac{\overset{1}{\cancel{2}} \times \overset{1}{\cancel{2}} \times 2 \times \overset{1}{\cancel{3}} \times \overset{1}{\cancel{3}}}{\underset{1}{\cancel{2}} \times \underset{1}{\cancel{2}} \times \underset{1}{\cancel{3}} \times \underset{1}{\cancel{3}} \times 5} = \dfrac{2}{5}$$

40 $\dfrac{1}{4}$ of $\dfrac{1}{3}$

$\dfrac{1}{3}\Big\{ \qquad \qquad \dfrac{1}{4} \cdot \dfrac{1}{3} = \dfrac{1}{12}$

$\underbrace{\qquad}_{\frac{1}{4}}$

41 $\dfrac{13}{15} \cdot \dfrac{4}{7} = \dfrac{13 \cdot 4}{15 \cdot 7} = \dfrac{52}{105}$

42 $\dfrac{27}{32} \cdot \dfrac{40}{21} = \dfrac{\cancel{3} \cdot 9}{4 \cdot \cancel{8}} \cdot \dfrac{5 \cdot \cancel{8}}{\cancel{3} \cdot 7} = \dfrac{45}{28}$

43 Mixed number: $5 + \dfrac{2}{3} = 5\dfrac{2}{3}$

44 $13\dfrac{5}{9} = \dfrac{(13 \cdot 9) + 5}{9} = \dfrac{117 + 5}{9} = \dfrac{122}{9}$

45 $2\dfrac{1}{3} \cdot 3\dfrac{7}{8} = \dfrac{7}{3} \cdot \dfrac{31}{8} = \dfrac{7 \cdot 31}{3 \cdot 8} = \dfrac{217}{24}$ or $9\dfrac{1}{24}$

Review Exercises

Perform the indicated operations in Exercises 1–10.

1. $\begin{array}{r} 73897 \\ -\ 5968 \\ \hline \end{array}$ 2. $\begin{array}{r} 39654 \\ 8506 \\ 395 \\ 41 \\ +\ 67932 \\ \hline \end{array}$

3. $29 - 5(27 - 24) + 13 - 3(8 \div 4)$

4. $[(7 \cdot 12) - 4] - 4[13 - (2 \cdot 3)] + 15$

5. $\begin{array}{r} 495 \\ \times\ 28 \\ \hline \end{array}$ 6. $39\overline{)758}$

7. $\sqrt{100}$ 8. $\sqrt[3]{27}$

9. $300 - 4^3$ 10. $5 + 4^2 + 2^3$

11.   Fifteen of 49 fuses are defective. What fraction represents the defective fuses?

12.   In Exercise 11, what fraction represents the "good" fuses?

EXERCISES
2.3

Either multiply or find the missing numerator in Exercises 1–20.

1. $\dfrac{12}{13} \cdot \dfrac{5}{7}$ 2. $\dfrac{9}{5} \cdot \dfrac{11}{14}$

3. $\dfrac{11}{14} \cdot \dfrac{3}{8}$ 4. $\dfrac{8}{5} \cdot \dfrac{14}{13}$

5. $5 \cdot \dfrac{3}{8}$

6. $\dfrac{4}{9} \cdot 13$

7. $\dfrac{13}{5} = \dfrac{?}{15}$

8. $\dfrac{24}{8} = \dfrac{?}{4}$

9. $\dfrac{3}{7} \cdot 12$

10. $\dfrac{9}{4} \cdot \dfrac{3}{2}$

11. $\dfrac{5}{9} = \dfrac{?}{27}$

12. $\dfrac{49}{14} = \dfrac{?}{2}$

13. Convert $\dfrac{17}{3}$ to mixed number form.

14. Convert $\dfrac{58}{19}$ to mixed number form.

15. Convert $15\dfrac{4}{13}$ to an improper fraction.

16. Convert $17\dfrac{3}{19}$ to an improper fraction.

17. Convert $3\dfrac{13}{3}$ to mixed number form.

18. Convert $5\dfrac{11}{4}$ to mixed number form.

19. Convert $3\dfrac{13}{3}$ to an improper fraction.

20. Convert $5\dfrac{11}{4}$ to an improper fraction.

Multiply in Exercises 21–36.

21. $\dfrac{17}{19} \cdot \dfrac{13}{23}$

22. $\dfrac{19}{13} \cdot \dfrac{14}{3}$ (Write your answer in mixed number form.)

23. $\dfrac{17}{19} \cdot \dfrac{23}{13}$ (Write your answer in mixed number form.)

24. $\dfrac{113}{151} \cdot \dfrac{23}{87}$ 25. $3\dfrac{2}{5} \cdot 4\dfrac{3}{7}$

26. $2\dfrac{1}{3} \cdot \dfrac{11}{38}$

27. $1\dfrac{5}{8} \cdot 2\dfrac{7}{15}$ (Write your answer in mixed number form.)

28. $4\dfrac{1}{5} \cdot \dfrac{19}{35}$ (Write your answer in mixed number form.)

29. $\left(\dfrac{2}{7} \cdot 2\dfrac{2}{3}\right) \cdot 4\dfrac{5}{9}$ 30. $3\dfrac{4}{9} \cdot \left(\dfrac{4}{5} \cdot 2\dfrac{2}{3}\right)$

31. $1\dfrac{2}{3} \cdot \dfrac{1}{4} \cdot \dfrac{5}{7} \cdot 2\dfrac{5}{6}$ 32. $\dfrac{1}{3} \cdot 1\dfrac{1}{4} \cdot \dfrac{5}{9} \cdot 1\dfrac{4}{7}$

33. $\left(\dfrac{3}{4}\right)^3$

34. $\left(\dfrac{1}{2}\right)^5$

35. $\left(\dfrac{3}{8}\right)^2$

36. $\left(\dfrac{4}{5}\right)^3$

In Exercises 37–50 multiply and reduce to lowest terms.

37. $\dfrac{14}{9} \cdot \dfrac{12}{35}$

38. $\dfrac{4}{15} \cdot \dfrac{9}{20}$

39. $\dfrac{32}{40} \cdot \dfrac{25}{28}$

40. $\dfrac{12}{20} \cdot \dfrac{35}{24}$

41. $\dfrac{7}{12} \cdot \dfrac{9}{14} \cdot \dfrac{15}{21}$

42. $\dfrac{8}{15} \cdot \dfrac{21}{12} \cdot \dfrac{40}{18}$

43. $3\dfrac{2}{3} \cdot 4\dfrac{1}{5}$

44. $2\dfrac{1}{3} \cdot \dfrac{12}{37}$

45. $2\dfrac{1}{4} \cdot 2\dfrac{2}{3} \cdot 5\dfrac{1}{6}$

46. $4\dfrac{1}{2} \cdot \dfrac{2}{63} \cdot \dfrac{7}{8}$

47. $3\dfrac{3}{7} \cdot \dfrac{5}{24} \cdot \dfrac{7}{15}$

48. $8\dfrac{3}{4} \cdot 4\dfrac{2}{5} \cdot 1\dfrac{6}{7}$

49. $1\dfrac{3}{4} \cdot 2\dfrac{2}{3} \cdot 5\dfrac{1}{2}$

50. $2\dfrac{1}{4} \cdot 2\dfrac{2}{7} \cdot 4\dfrac{2}{3}$

2.4
ADDING FRACTIONS AND MIXED NUMBERS

Processes introduced in Section 2.3 that will be useful in this section are illustrated here.

$$5 = \frac{5}{1} = \frac{5 \cdot 4}{1 \cdot 4} = \frac{20}{4}$$

$$\frac{2}{3} = \frac{2 \cdot 7}{3 \cdot 7} = \frac{14}{21}$$

$$\frac{12}{16} = \frac{12 \div 4}{16 \div 4} = \frac{3}{4}$$

Adding fractions having like denominators, such as $\frac{5}{6}$ and $\frac{8}{6}$, is accomplished by simply adding the numerators and using the result as a numerator together with the common denominator. Thus,

$$\frac{5}{6} + \frac{8}{6} = \frac{5+8}{6} = \frac{13}{6}.$$

When adding fractions with unlike denominators, it is necessary to convert them to fractions having like denominators. This is illustrated in Figure 2.16.

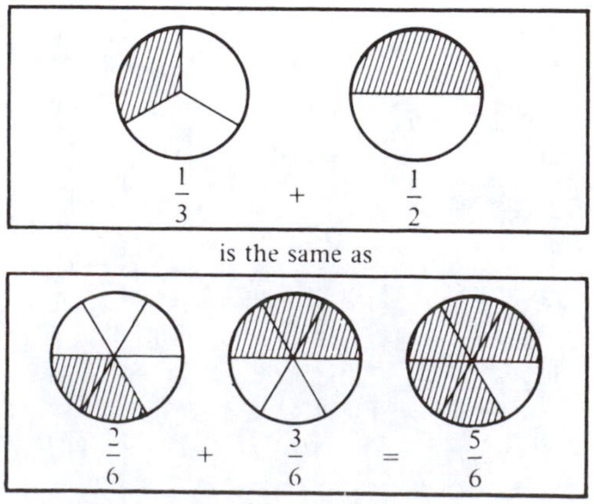

Two fractions can be converted to fractions with like denominators by multiplying or dividing both numerator and denominator by the same number. The number used for each fraction must be one that produces like denominators. If the two fractions are $\frac{1}{3}$ and $\frac{1}{2}$, you can obtain like denominators by multiplying the numerator and denominator of $\frac{1}{3}$ by 2, and multiplying the numerator and denominator of $\frac{1}{2}$ by 3. Thus,

$$\frac{1 \cdot 2}{3 \cdot 2} = \frac{2}{6} \quad \text{and} \quad \frac{1 \cdot 3}{2 \cdot 3} = \frac{3}{6}.$$

Note that 2 is the denominator of $\frac{1}{2}$ and 3 is the denominator of $\frac{1}{3}$. Often you can see by inspection that some other multiplier will produce like denominators. For example, for $\frac{1}{4}$ and $\frac{3}{8}$, like denominators can be obtained by multiplying both the numerator and denominator of $\frac{1}{4}$ by 2, i.e., $\frac{1 \cdot 2}{4 \cdot 2} = \frac{2}{8}$, which has the same denominator as $\frac{3}{8}$. Some examples of converting pairs of fractions to like denominators are shown in Table 2.3.

FIGURE 2.16

$$\frac{1}{3} + \frac{1}{2} = \frac{2}{6} + \frac{3}{6} = \frac{5}{6}$$

TABLE 2.3

Fractions	Converting	Like Denominators
$\frac{2}{3}$	$\frac{2}{3} = \frac{2 \cdot 4}{3 \cdot 4} = \frac{8}{12}$	$\frac{8}{12}$
$\frac{3}{4}$	$\frac{3}{4} = \frac{3 \cdot 3}{4 \cdot 3} = \frac{9}{12}$	$\frac{9}{12}$
$\frac{1}{2}$	$\frac{1}{2} = \frac{1 \cdot 7}{2 \cdot 7} = \frac{7}{14}$	$\frac{7}{14}$
$\frac{5}{7}$	$\frac{5}{7} = \frac{5 \cdot 2}{7 \cdot 2} = \frac{10}{14}$	$\frac{10}{14}$
$\frac{8}{16}$	$\frac{8}{16} = \frac{8}{16}$	$\frac{8}{16}$
$\frac{3}{4}$	$\frac{3}{4} = \frac{3 \cdot 4}{4 \cdot 4} = \frac{12}{16}$	$\frac{12}{16}$

When denominators are small numbers, common denominators are usually easy to determine, as in Table 2.3. When the denominators are large numbers, the following process for determining the least common denominator will be useful.

The **least common denominator** of two or more fractions is the smallest denominator, other than zero, that contains each of the denominators as a factor. The usual abbreviation for least common denominator is lcd. For example, the lcd of $\frac{2}{3}$ and $\frac{1}{4}$ is 12, since 12 is the smallest number that has 3 and 4 as factors; the lcd of $\frac{5}{12}$ and $\frac{7}{15}$ is 60, since 60 is the smallest number that has 12 and 15 as factors; and the lcd of $\frac{1}{6}$, $\frac{3}{5}$, and $\frac{2}{9}$ is 90, since 90 is the smallest number that has 6, 5, and 9 as factors.

To find the least common denominator (lcd) of two or more fractions:

(a) Write the prime factorization of each denominator.

(b) Select all of the prime factors of the largest denominator and all those additional prime factors to insure that each denominator is a factor of the product of these factors. The product of all of these factors is the lcd. (*Note:* A "shortcut" is to write each prime factorization of the denominators in exponential form. The lcd is then the product of all the different prime factors with the highest power of each factor found in the denominators. See Examples 3 and 4.)

(c) The numerator and denominator of each fraction can then be multiplied by the appropriate prime factors to obtain an equivalent fraction whose denominator is the lcd.

EXAMPLE 1 Find the lcd of $\frac{5}{6}$ and $\frac{3}{20}$.

Solution Using the 3-step process outlined above:

(a) $6 = 2 \cdot 3$
   $20 = 2 \cdot 2 \cdot 5$

(b) lcd $= 2 \cdot 2 \cdot 3 \cdot 5 = 60$

(c) $\frac{5}{6} = \frac{5}{2 \cdot 3} = \frac{(5)(2 \cdot 5)}{(2 \cdot 3)(2 \cdot 5)} = \frac{50}{60}$

   $\frac{3}{20} = \frac{3}{2 \cdot 2 \cdot 5} = \frac{(3)(3)}{(2 \cdot 2 \cdot 5)(3)} = \frac{9}{60}$

EXAMPLE 2 Find the lcd of $\frac{5}{18}$ and $\frac{7}{24}$.

Solution

(a) $18 = 2 \cdot 3 \cdot 3$
   $24 = 2 \cdot 2 \cdot 2 \cdot 3$

(b) lcd $= 2 \cdot 2 \cdot 2 \cdot 3 \cdot 3 = 72$

(c) $\frac{5}{18} = \frac{5}{2 \cdot 3 \cdot 3} = \frac{(5)(2 \cdot 2)}{(2 \cdot 3 \cdot 3)(2 \cdot 2)} = \frac{20}{72}$

   $\frac{7}{24} = \frac{7}{2 \cdot 2 \cdot 2 \cdot 3} = \frac{(7)(3)}{(2 \cdot 2 \cdot 2 \cdot 3)(3)} = \frac{21}{72}$

EXAMPLE 3 Find the lcd of $\frac{5}{24}$ and $\frac{1}{36}$.

Solution

(a) $24 = 2 \cdot 2 \cdot 2 \cdot 3 = 2^3 \cdot 3$
   $36 = 2 \cdot 2 \cdot 3 \cdot 3 = 2^2 \cdot 3^2$

(b) lcd $= 2^3 \cdot 3^2 = 72$

(c) $\frac{5}{24} = \frac{5}{2^3 \cdot 3} = \frac{(5)(3)}{(2^3 \cdot 3)(3)} = \frac{15}{72}$

   $\frac{1}{36} = \frac{1}{2^2 \cdot 3^2} = \frac{(1)(2)}{(2^2 \cdot 3^2)(2)} = \frac{2}{72}$

EXAMPLE 4 Find the lcd of $\frac{5}{48}$, $\frac{17}{30}$, and $\frac{12}{27}$.

Solution

(a) $48 = 2 \cdot 2 \cdot 2 \cdot 2 \cdot 3 = 2^4 \cdot 3$
   $30 = 2 \cdot 3 \cdot 5$
   $27 = 3 \cdot 3 \cdot 3 = 3^3$

(b) lcd $= 2^4 \cdot 3^3 \cdot 5 = 2160$

(c) $\frac{5}{48} = \frac{5(3^2 \cdot 5)}{48(3^2 \cdot 5)} = \frac{225}{2160}$

   $\frac{17}{30} = \frac{17(2^3 \cdot 3^2)}{30(2^3 \cdot 3^2)} = \frac{1224}{2160}$

   $\frac{12}{27} = \frac{12(2^4 \cdot 5)}{27(2^4 \cdot 5)} = \frac{960}{2160}$

We can use the process for determining the lcd of two or more fractions to help us find the sum of fractions with unlike denominators.

EXAMPLE 5 Find the sum of $\frac{9}{16}$ and $\frac{5}{12}$.

Solution

$$\text{lcd} = 2^4 \cdot 3 = 48$$

$$\frac{9}{16} + \frac{5}{12} = \frac{9 \cdot 3}{16 \cdot 3} + \frac{5 \cdot 4}{12 \cdot 4} = \frac{27}{48} + \frac{20}{48} = \frac{47}{48}$$

EXAMPLE 6 Find the sum of $\frac{8}{15}$ and $\frac{3}{10}$. Write the answer in lowest terms.

Solution

$$\text{lcd} = 2 \cdot 3 \cdot 5 = 30$$

$$\frac{8}{15} + \frac{3}{10} = \frac{8 \cdot 2}{15 \cdot 2} + \frac{3 \cdot 3}{10 \cdot 3} = \frac{16}{30} + \frac{9}{30} = \frac{25}{30} = \frac{5}{6}$$

EXAMPLE 7 Add the fractions and write the answer as a mixed number:

$$\frac{2}{9} + \frac{5}{8} + \frac{7}{12}.$$

Solution

$$\text{lcd} = 2^3 \cdot 3^2 = 72$$

$$\frac{2}{9} + \frac{5}{8} + \frac{7}{12} = \frac{2 \cdot 8}{9 \cdot 8} + \frac{5 \cdot 9}{8 \cdot 9} + \frac{7 \cdot 6}{12 \cdot 6}$$

$$= \frac{16}{72} + \frac{45}{72} + \frac{42}{72}$$

$$= \frac{103}{72}$$

$$= 1\frac{31}{72}$$

EXAMPLE 8 Add the fractions and reduce the answer to lowest terms:

$$\frac{7}{15} + \frac{3}{10} + \frac{7}{12} + \frac{2}{5}.$$

Solution

$$\text{lcd} = 2^2 \cdot 3 \cdot 5 = 60$$

$$\frac{7}{15} + \frac{3}{10} + \frac{7}{12} + \frac{2}{5}$$

$$= \frac{7 \cdot 4}{15 \cdot 4} + \frac{3 \cdot 6}{10 \cdot 6} + \frac{7 \cdot 5}{12 \cdot 5} + \frac{2 \cdot 12}{5 \cdot 12}$$

$$= \frac{28}{60} + \frac{18}{60} + \frac{35}{60} + \frac{24}{60}$$

$$= \frac{105}{60}$$

$$= 1\frac{45}{60}$$

$$= 1\frac{3}{4}$$

EXAMPLE 9 Find the sum of $3\frac{3}{10}$, $1\frac{4}{5}$, and $2\frac{1}{6}$ in lowest terms.

Solution

$$\text{lcd} = 2 \cdot 3 \cdot 5 = 30$$

$$3\frac{3}{10} + 1\frac{4}{5} + 2\frac{1}{6} = 3\frac{3 \cdot 3}{10 \cdot 3} + 1\frac{4 \cdot 6}{5 \cdot 6} + 2\frac{1 \cdot 5}{6 \cdot 5}$$

$$= 3\frac{9}{30} + 1\frac{24}{30} + 2\frac{5}{30}$$

$$= 3 + 1 + 2 + \frac{9}{30} + \frac{24}{30} + \frac{5}{30}$$

$$= 6 + \frac{38}{30}$$

$$= 6 + 1\frac{8}{30}$$

$$= 6 + 1\frac{4}{15}$$

$$= 7\frac{4}{15}$$

Review Exercises

Perform the indicated operations in Exercises 1–8.

1. $31 - 4(28 - 25) + 13 - 3(9 \div 3)$
2. $[(19 \cdot 14) - 59] - 3[18 - (5 \cdot 3)] + 17$

3. $\begin{array}{r} 784 \\ \times\ 79 \end{array}$

4. $27\overline{)7694}$

5. $442 - (15)^2$

6. $(2)^3 + (4)^2 - (2)^4$

7. $\dfrac{13}{17} \cdot \dfrac{3}{4}$

8. $\dfrac{4}{5} \cdot 23$

9. $\dfrac{5}{8} = \dfrac{?}{16}$

10. $\dfrac{25}{55} = \dfrac{5}{?}$

11. Convert $\frac{57}{13}$ to mixed number form.

12. Convert $15\frac{7}{8}$ to an improper fraction.

EXERCISES
2.4

Convert the fractions in Exercises 1–12 to equivalent fractions with the least common denominator.

1. $\dfrac{2}{3}, \dfrac{5}{12}$

2. $\dfrac{1}{4}, \dfrac{3}{16}$

3. $\dfrac{1}{2}, \dfrac{1}{6}$

4. $\dfrac{4}{5}, \dfrac{11}{20}$

5. $\dfrac{7}{45}, \dfrac{8}{35}$

6. $\dfrac{5}{28}, \dfrac{8}{63}$

7. $\dfrac{1}{3}, \dfrac{3}{4}, \dfrac{7}{8}$

8. $\dfrac{1}{2}, \dfrac{2}{3}, \dfrac{5}{9}$

9. $\dfrac{3}{10}, \dfrac{4}{5}, \dfrac{1}{6}$

10. $\dfrac{3}{8}, \dfrac{5}{12}, \dfrac{9}{16}$

11. $\dfrac{7}{24}, \dfrac{3}{18}, \dfrac{4}{15}$

12. $\dfrac{7}{8}, \dfrac{11}{36}, \dfrac{3}{20}$

In Exercises 13–34 add and reduce answers to lowest terms.

13. $\dfrac{1}{5} + \dfrac{3}{5}$

   WATCH OUT! The number $\frac{1}{5} + \frac{3}{5}$ does *not* equal $\frac{1+3}{5+5}$.

14. $\dfrac{3}{11} + \dfrac{5}{11}$

15. $\dfrac{5}{18} + \dfrac{7}{18}$

16. $\dfrac{9}{20} + \dfrac{3}{20}$

17. $\dfrac{3}{4} + \dfrac{5}{8}$

18. $\dfrac{7}{10} + \dfrac{2}{5}$

19. $\dfrac{1}{2} + \dfrac{2}{3}$

20. $\dfrac{1}{4} + \dfrac{5}{6}$

21. $3\dfrac{3}{4} + 4\dfrac{3}{5}$

22. $2\dfrac{4}{9} + 1\dfrac{3}{5}$

23. $2\dfrac{7}{15} + \dfrac{7}{10}$

24. $4\dfrac{5}{12} + \dfrac{4}{9}$

25. $\dfrac{1}{9} + \dfrac{3}{8} + \dfrac{5}{12}$

26. $\dfrac{1}{4} + \dfrac{3}{8} + \dfrac{5}{6}$

27. $\dfrac{3}{4} + \dfrac{5}{8} + \dfrac{9}{10}$

28. $\dfrac{2}{5} + \dfrac{1}{6} + \dfrac{4}{9}$

29. $6\dfrac{3}{16} + 8\dfrac{9}{24} + 18\dfrac{5}{8}$

30. $2\dfrac{4}{15} + 3\dfrac{7}{10} + 1\dfrac{3}{4}$

31. $\dfrac{1}{3} + 14\dfrac{5}{6} + \dfrac{5}{8}$

32. $2\dfrac{3}{4} + 5\dfrac{7}{9} + \dfrac{15}{20}$

33. $\dfrac{1}{12} + \dfrac{3}{20} + \dfrac{4}{15} + \dfrac{2}{5}$

34. $\dfrac{2}{3} + \dfrac{5}{16} + \dfrac{8}{9} + \dfrac{3}{4}$

35. Two-thirds of a building's exterior is painted white, and one-fifth of its exterior is painted green. The remainder is not painted. What fraction of the building's exterior is painted?

36. Gaylord took $\frac{2}{5}$ of a pizza and Harold took $\frac{1}{4}$ of the same pizza. What fraction of the pizza did the two boys take?

37. Sarah painted until she had used $\frac{1}{4}$ of the gallon of paint. Then Mary painted until she had used $\frac{1}{3}$ of the gallon. Finally, Susan painted until she had used $\frac{2}{5}$ of the gallon of paint. What fraction of the gallon of paint did the three girls use?

38. Three electrical resistors have resistances of $2\frac{1}{9}$ ohms, $3\frac{4}{15}$ ohms, and $1\frac{5}{18}$ ohms. What is the sum of their resistances?

39. A man drove his car $\frac{3}{8}$ of an hour; his wife drove it $1\frac{2}{3}$ hours; and their son drove it $2\frac{4}{15}$ hours, all in the same day. How long was the car driven that day?

40. Five boy scouts have $1\frac{2}{3}$ liters, $2\frac{1}{4}$ liters, $3\frac{1}{6}$ liters, $1\frac{3}{4}$ liters, and $1\frac{1}{6}$ liters of water, respectively. How many liters of water do the five boys have altogether?

41. A carpenter needs $5\frac{1}{2}$ feet of one type of trim, $27\frac{3}{4}$ feet of another type, and $125\frac{1}{3}$ feet of a third type. How many feet of trim does he need in all?

42. A cook's recipe calls for $3\frac{1}{3}$ cups of flour, $2\frac{3}{4}$ cups of sugar, and $1\frac{1}{2}$ cups of water, among other ingredients. How many cups in all of flour, sugar, and water are needed for the recipe?

2.5
SUBTRACTING FRACTIONS AND MIXED NUMBERS

The usual method of subtracting fractions involves finding common denominators. Fractions having like denominators can be subtracted directly.

EXAMPLES

1 $\dfrac{7}{8} - \dfrac{4}{8} = \dfrac{7-4}{8} = \dfrac{3}{8}$

2 $\dfrac{9}{5} - \dfrac{3}{5} = \dfrac{9-3}{5} = \dfrac{6}{5}$

3 $\dfrac{12}{16} - \dfrac{5}{16} = \dfrac{7}{16}$

Examples 4–7 illustrate subtracting fractions with unlike denominators. Notice that, before you can subtract, you must convert the fractions to equivalent fractions with like denominators.

EXAMPLES

4 $\dfrac{1}{2} - \dfrac{1}{4} = \dfrac{2}{4} - \dfrac{1}{4} = \dfrac{1}{4}$

5 $\dfrac{2}{3} - \dfrac{3}{5} = \dfrac{2 \cdot 5}{3 \cdot 5} - \dfrac{3 \cdot 3}{5 \cdot 3} = \dfrac{10}{15} - \dfrac{9}{15} = \dfrac{1}{15}$

6 $4 - \dfrac{5}{8} = \dfrac{4 \cdot 8}{1 \cdot 8} - \dfrac{5}{8} = \dfrac{32}{8} - \dfrac{5}{8} = \dfrac{27}{8} = 3\dfrac{3}{8}$

 or $4 - \dfrac{5}{8} = 3\dfrac{8}{8} - \dfrac{5}{8} = 3\dfrac{3}{8}$

7 $2\dfrac{1}{4} - \dfrac{3}{16} = \dfrac{9}{4} - \dfrac{3}{16} = \dfrac{36}{16} - \dfrac{3}{16} = \dfrac{33}{16} = 2\dfrac{1}{16}$

(Notice that $2\frac{1}{4}$ was converted to an improper fraction before subtracting.)

Examples 8 and 9 show two other ways to compute the difference shown in Example 7.

EXAMPLES

8 $2\dfrac{1}{4} - \dfrac{3}{16} = 2 + \left(\dfrac{1}{4} - \dfrac{3}{16}\right) = 2 + \left(\dfrac{4}{16} - \dfrac{3}{16}\right)$

  $= 2 + \dfrac{1}{16} = 2\dfrac{1}{16}$

9 $\begin{array}{r} 2\dfrac{1}{4} = 2\dfrac{4}{16} \\ - \quad \dfrac{3}{16} = \quad \dfrac{3}{16} \\ \hline 2\dfrac{1}{16} \end{array}$

Examples 10 through 13 illustrate a "vertical form" of subtraction like that in Example 9.

EXAMPLES

10   $\begin{array}{r} 75\dfrac{3}{4} \\ - 24\dfrac{1}{8} \\ \hline \end{array}$ ⟹ $\begin{array}{r} 75\dfrac{6}{8} \\ - 24\dfrac{1}{8} \\ \hline 51\dfrac{5}{8} \end{array}$

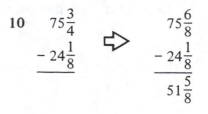

11 $13\frac{1}{5}$

$-\ 7\frac{4}{5}$

Since $\frac{1}{5}$ is less than $\frac{4}{5}$, we can't subtract fractional parts; so we change $13\frac{1}{5}$ to $12\frac{6}{5}$, as follows:

$$13\frac{1}{5} = 13 + \frac{1}{5}$$

$$= 12 + 1 + \frac{1}{5}$$

$$= 12 + \frac{5}{5} + \frac{1}{5}$$

$$= 12 + \frac{6}{5}$$

$$= 12\frac{6}{5}.$$

Usually this result can be obtained mentally.

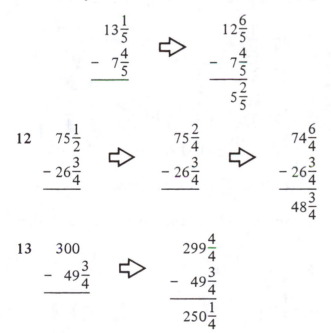

Examples 14–16 show a method for subtracting several fractions in the same problem. Parentheses can be used to show how to group the fractions. The successive subtractions are done in order from left to right.

EXAMPLES

14 $\dfrac{7}{8} - \dfrac{3}{8} - \dfrac{1}{8} = \left(\dfrac{7}{8} - \dfrac{3}{8}\right) - \dfrac{1}{8}$

$$= \dfrac{4}{8} - \dfrac{1}{8}$$

$$= \dfrac{3}{8}$$

15 $\dfrac{7}{8} - \dfrac{2}{3} - \dfrac{1}{9} = \dfrac{63}{72} - \dfrac{48}{72} - \dfrac{8}{72}$

$$= \left(\dfrac{63}{72} - \dfrac{48}{72}\right) - \dfrac{8}{72}$$

$$= \dfrac{15}{72} - \dfrac{8}{72}$$

$$= \dfrac{7}{72}$$

16 $4\dfrac{1}{2} - 1\dfrac{4}{5} - \dfrac{2}{3} = 4\dfrac{15}{30} - 1\dfrac{24}{30} - \dfrac{20}{30}$

$$= \left(3\dfrac{45}{30} - 1\dfrac{24}{30}\right) - \dfrac{20}{30}$$

$$= 2\dfrac{21}{30} - \dfrac{20}{30}$$

$$= 2\dfrac{1}{30}$$

Examples 17–20 show a method for finding an answer when addition and subtraction are combined in the same problem. Note again that the successive additions and subtractions are done in order from left to right.

EXAMPLES

17 $\dfrac{9}{16} + \dfrac{3}{16} - \dfrac{5}{16} = \left(\dfrac{9}{16} + \dfrac{3}{16}\right) - \dfrac{5}{16}$

$$= \dfrac{12}{16} - \dfrac{5}{16}$$

$$= \dfrac{7}{16}$$

EXAMPLES

18 $\dfrac{11}{32} - \dfrac{3}{32} + \dfrac{5}{32} = \left(\dfrac{11}{32} - \dfrac{3}{32}\right) + \dfrac{5}{32}$

$\qquad\qquad = \dfrac{8}{32} + \dfrac{5}{32}$

$\qquad\qquad = \dfrac{13}{32}$

19 $\dfrac{2}{3} - \dfrac{1}{4} + \dfrac{1}{5} = \dfrac{40}{60} - \dfrac{15}{60} + \dfrac{12}{60}$

$\qquad\qquad = \left(\dfrac{40}{60} - \dfrac{15}{60}\right) + \dfrac{12}{60}$

$\qquad\qquad = \dfrac{25}{60} + \dfrac{12}{60}$

$\qquad\qquad = \dfrac{37}{60}$

20 $\dfrac{3}{8} + \dfrac{13}{16} - \dfrac{1}{4} - \dfrac{5}{8} = \dfrac{6}{16} + \dfrac{13}{16} - \dfrac{4}{16} - \dfrac{10}{16}$

$\qquad\qquad = \left(\dfrac{6}{16} + \dfrac{13}{16}\right) - \dfrac{4}{16} - \dfrac{10}{16}$

$\qquad\qquad = \dfrac{19}{16} - \dfrac{4}{16} - \dfrac{10}{16}$

$\qquad\qquad = \left(\dfrac{19}{16} - \dfrac{4}{16}\right) - \dfrac{10}{16}$

$\qquad\qquad = \dfrac{15}{16} - \dfrac{10}{16}$

$\qquad\qquad = \dfrac{5}{16}$

Review Exercises

Perform the indicated multiplication or find the missing number in Exercises 1–10.

1. $3 \cdot \dfrac{5}{8}$ 2. $\dfrac{4}{9} \cdot 12$

3. $\dfrac{13}{5} = \dfrac{?}{10}$ 4. $\dfrac{24}{8} = \dfrac{?}{2}$

5. $\dfrac{3}{5} \cdot 12$ 6. $\dfrac{9}{3} \cdot \dfrac{5}{2}$

7. $\dfrac{5}{8} = \dfrac{?}{32}$ 8. $\dfrac{49}{21} = \dfrac{?}{3}$

9. $4\dfrac{3}{5} \cdot \dfrac{11}{35}$ 10. $\dfrac{2}{3} \cdot 2\dfrac{2}{3} \cdot 4\dfrac{5}{9}$

11. Convert $\dfrac{2}{3}$ and $\dfrac{12}{18}$ to equivalent fractions with a common denominator of 9.

12. Convert $\dfrac{4}{15}$ and $\dfrac{7}{10}$ to equivalent fractions with a common denominator of 30.

EXERCISES
2.5

Subtract or add as indicated in Exercises 1–12.

1. $\dfrac{3}{7} - \dfrac{1}{9}$ 2. $\dfrac{2}{3} - \dfrac{1}{16}$

3. $\dfrac{3}{5} - \dfrac{1}{8}$ 4. $\dfrac{4}{5} - \dfrac{3}{16}$

5. $3\dfrac{1}{4} - \dfrac{1}{2}$ 6. $4\dfrac{2}{3} - 1\dfrac{7}{8}$

7. $2\dfrac{3}{5} - \dfrac{2}{3}$ 8. $3\dfrac{5}{8} - 1\dfrac{3}{16}$

9. $3\dfrac{2}{9} + \dfrac{8}{15} - \dfrac{3}{10}$ 10. $\dfrac{26}{27} - \dfrac{5}{8} + \dfrac{1}{4}$

11. Reduce $\dfrac{51}{34}$ to lowest terms.

12. Reduce $\dfrac{136}{102}$ to lowest terms.

In Exercises 13–22, convert answers to lowest terms.

13. $\dfrac{7}{8} - \dfrac{2}{3} - \dfrac{1}{9}$ 14. $\dfrac{7}{8} - \dfrac{1}{4} - \dfrac{4}{13}$

15. $\dfrac{1}{4} + \dfrac{3}{4}$ 16. $\dfrac{3}{10} + \dfrac{6}{10}$

17. $\dfrac{1}{3} + 2\dfrac{2}{5}$ 18. $2\dfrac{2}{5} - \dfrac{1}{3}$

19. $\dfrac{2}{3} + \dfrac{4}{3} + \dfrac{5}{3}$ 20. $\dfrac{3}{5} - \dfrac{1}{9}$

21. $3\dfrac{4}{9} + \dfrac{7}{15} - \dfrac{3}{10}$ 22. $\dfrac{25}{26} - \dfrac{5}{9} + \dfrac{1}{3}$

23. A board $7\frac{3}{4}$ feet long has a piece $2\frac{1}{2}$ feet long cut from it. How long is the remaining board?

24. John had $3\frac{3}{4}$ pounds of nails and Harry had $2\frac{7}{8}$ pounds of nails. How many more pounds of nails did John have?

25. Yesterday it rained $\frac{9}{10}$ inches. Today it rained $\frac{3}{8}$ inches. How much more rain fell yesterday than today?

26. A man worked $6\frac{5}{8}$ hours during an eight-hour work day. The next day he worked $5\frac{9}{10}$ hours during his eight-hour shift. How many hours was he not working during the two work days?

27. How many inches are left in a 12-inch welding rod after $2\frac{1}{3}$ inches and $1\frac{3}{4}$ inches have been burned up?

28. A grocer mixed together $3\frac{4}{5}$ pounds of peanuts and $1\frac{3}{8}$ pounds of pecans. He sold two bags of the mixture, weighing $\frac{3}{4}$ lb and $1\frac{1}{3}$ lb, respectively. How many pounds of the mixture did he have left?

29. A telephone company employee cut $89\frac{3}{4}$ feet of wire off a 600-foot coil of wire. How many feet of wire remained in the coil?

2.6
DIVIDING FRACTIONS AND MIXED NUMBERS

Multiplication and division of whole numbers are related. For every **product** (multiplication) of two whole numbers there are two different related **quotients** (division). Table 2.4 illustrates this fact.

TABLE 2.4

Product (Multiplication)	Related Quotients (Division)
$4 \cdot 5 = 20$	$20 \div 5 = 4$ $20 \div 4 = 5$
$13 \cdot 11 = 143$	$143 \div 11 = 13$ $143 \div 13 = 11$

The same relationship holds true for multiplication and division of fractions. Table 2.5 illustratates this relationship for fractions.

TABLE 2.5

Product (Multiplication)	Related Quotients (Division)
$\frac{1}{4} \cdot 3 = \frac{3}{4}$	$\frac{3}{4} \div 3 = \frac{1}{4}$ $\frac{3}{4} \div \frac{1}{4} = 3$
$\frac{1}{2} \cdot \frac{2}{3} = \frac{2}{6}$	$\frac{2}{6} \div \frac{2}{3} = \frac{1}{2}$ $\frac{2}{6} \div \frac{1}{2} = \frac{2}{3}$

When one fraction is divided by another the three numbers in the problem are named as shown here:

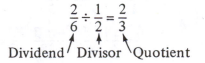

$$\frac{2}{6} \div \frac{1}{2} = \frac{2}{3}$$

Dividend Divisor Quotient

The procedure commonly used for dividing fractions is described by the following rule:

> **To divide one fraction by another, invert the divisor and multiply.**

In $\frac{5}{8} \div \frac{2}{3}$, the second fraction, $\frac{2}{3}$, is the **divisor**. Using the rule for dividing fractions to obtain an answer for $\frac{5}{8} \div \frac{2}{3}$, we have

$$\frac{5}{8} \div \frac{2}{3} = \frac{5}{8} \cdot \frac{3}{2} = \frac{15}{16}.$$

Note the inverted divisor.

Here are several additional examples which show how to apply the rule for dividing fractions.

EXAMPLES

1    $\dfrac{3}{5} \div \dfrac{2}{3} = \dfrac{3}{5} \cdot \dfrac{3}{2} = \dfrac{9}{10}$

2    $\dfrac{5}{9} \div \dfrac{6}{7} = \dfrac{5}{9} \cdot \dfrac{7}{6} = \dfrac{35}{54}$

3    $4 \div \dfrac{3}{5} = \dfrac{4}{1} \div \dfrac{3}{5} = \dfrac{4}{1} \cdot \dfrac{5}{3} = \dfrac{20}{3}$

4    $\dfrac{5}{8} \div 6 = \dfrac{5}{8} \div \dfrac{6}{1} = \dfrac{5}{8} \cdot \dfrac{1}{6} = \dfrac{5}{48}$

Cancelling can be used to reduce an answer to lowest terms when applying the rule for dividing fractions, as we see in the following examples.

EXAMPLES

5    $\dfrac{4}{15} \div \dfrac{2}{3} = \dfrac{4}{15} \cdot \dfrac{3}{2} = \dfrac{2 \cdot \cancel{2}}{\cancel{3} \cdot 5} \cdot \dfrac{\cancel{3}}{\cancel{2}} = \dfrac{2}{5}$

6    $\dfrac{9}{16} \div \dfrac{15}{32} = \dfrac{9}{16} \cdot \dfrac{32}{15} = \dfrac{3 \cdot \cancel{3}}{\cancel{16}} \cdot \dfrac{2 \cdot \cancel{16}}{\cancel{3} \cdot 5} = \dfrac{6}{5} = 1\dfrac{1}{5}$

7    $\dfrac{24}{35} \div 6 = \dfrac{24}{35} \div \dfrac{6}{1} = \dfrac{24}{35} \cdot \dfrac{1}{6} = \dfrac{4 \cdot \cancel{6}}{35} \cdot \dfrac{1}{\cancel{6}} = \dfrac{4}{35}$

> **To divide mixed numbers, write the mixed numbers as improper fractions and then divide, following the rule for dividing fractions.**

EXAMPLES

8    $1\dfrac{1}{3} \div 2\dfrac{1}{5} = \dfrac{4}{3} \div \dfrac{11}{5} = \dfrac{4}{3} \cdot \dfrac{5}{11} = \dfrac{20}{33}$

9    $5\dfrac{3}{4} \div 3\dfrac{1}{2} = \dfrac{23}{4} \div \dfrac{7}{2} = \dfrac{23}{4} \cdot \dfrac{2}{7} = \dfrac{23}{2 \cdot \cancel{2}} \cdot \dfrac{\cancel{2}}{7} = \dfrac{23}{14}$

10    $3\dfrac{1}{3} \div 6\dfrac{7}{8} = \dfrac{10}{3} \div \dfrac{55}{8} = \dfrac{10}{3} \cdot \dfrac{8}{55}$

$= \dfrac{2 \cdot \cancel{5}}{3} \cdot \dfrac{8}{\cancel{5} \cdot 11} = \dfrac{16}{33}$

Recall that a fraction bar can be used to indicate division.

EXAMPLES

11    $\dfrac{\frac{3}{4}}{\frac{1}{3}} = \dfrac{3}{4} \div \dfrac{1}{3} = \dfrac{3}{4} \cdot \dfrac{3}{1} = \dfrac{9}{4}$

12    $\dfrac{1\frac{7}{8}}{3\frac{1}{2}} = 1\dfrac{7}{8} \div 3\dfrac{1}{2}$

$= \dfrac{15}{8} \div \dfrac{7}{2}$

$= \dfrac{15}{8} \cdot \dfrac{2}{7}$

$= \dfrac{15}{4 \cdot \cancel{2}} \cdot \dfrac{\cancel{2}}{7}$

$= \dfrac{15}{28}$

$$13 \quad \frac{\frac{7}{4}}{2+\frac{5}{8}} = \frac{7}{4} \div 2\frac{5}{8}$$

$$= \frac{7}{4} \div \frac{21}{8}$$

$$= \frac{7}{4} \cdot \frac{8}{21}$$

$$= \frac{\cancel{7}}{\cancel{4}} \cdot \frac{2 \cdot \cancel{4}}{3 \cdot \cancel{7}}$$

$$= \frac{2}{3}$$

Review Exercises

Perform the indicated operations in Exercises 1–12.

1. $(349)(68)$

2. $349 \div 68$

3. $4007 - 399$

4. $\frac{3}{16} + \frac{5}{32}$

5. $\frac{3}{16} - \frac{5}{32}$

6. $\frac{3}{4} \cdot \frac{3}{8}$

7. $\frac{2}{3} \cdot \frac{1}{7} \cdot \frac{2}{5} \cdot \frac{1}{3}$

8. $\frac{14}{5} + \frac{13}{7} - 1\frac{3}{8}$

9. Reduce $\frac{38}{136}$ to lowest terms.

10. $3\frac{5}{8} + 1\frac{3}{16}$

11. $2\frac{3}{5} - \frac{3}{8}$

12. $3\frac{5}{8} + 2\frac{3}{5} - 1\frac{3}{16}$

EXERCISES
2.6

Perform the indicated operations in Exercises 1–20.

1. $\frac{7}{8} \div \frac{3}{5}$

2. $\frac{2}{5} \div 3$

3. $5 \div \frac{1}{2}$

4. $\frac{8}{7} \div \frac{5}{3}$

5. $1\frac{2}{3} \div \frac{2}{11}$

6. $1\frac{3}{4} \div \frac{5}{7}$

7. $4 \div \frac{8}{19}$

8. $3 \div \frac{2}{5}$

9. $\frac{7}{17} \div \frac{11}{35}$

10. $\frac{21}{44} \div \frac{33}{27}$

11. $\left(\frac{3}{8} \cdot \frac{27}{28}\right) \div \frac{3}{7}$

12. $\left(\frac{4}{9} \cdot \frac{28}{27}\right) \div \frac{3}{7}$

13. $\frac{17}{23} \div \left(\frac{8}{5} \div \frac{34}{7}\right)$

14. $\frac{17}{21} \div \left(\frac{7}{4} \div \frac{18}{3}\right)$

15. $3\frac{1}{4} \div \frac{3}{4}$

16. $2\frac{2}{3} \div \frac{1}{3}$

17. $\dfrac{\frac{7}{8}}{2\frac{1}{4}}$

18. $\dfrac{\frac{5}{6}}{3\frac{2}{3}}$

19. $\dfrac{7\frac{1}{8}}{2}$

20. $\dfrac{5}{3\frac{1}{2}}$

21. What is the sum of the voltages of 49 batteries, each of which is $1\frac{1}{2}$ volts? (*Hint:* Use multiplication.)

22. How many pieces of wire, each $1\frac{2}{3}$ feet long, can be cut from a piece of wire 50 feet long?

23. A factory worker can assemble 13 complete units in 4 hours. How many units can he assemble in 3 hours? Give the answer as a mixed number.

24. A farmer plowed $5\frac{2}{3}$ acres in $7\frac{1}{4}$ hours. On the average, what fraction of an acre did he plow per hour?

25. If there are approximately $3\frac{3}{4}$ bushels of corn in a full barrel, how many bushels of corn are there in $4\frac{2}{3}$ full barrels?

26. Each and every chapter in a book contains $8\frac{1}{3}$ pages of print. There are seven chapters in the book. How many pages of print are there in two copies of the book (counting only the pages in chapters)?

27. John had $1\frac{3}{4}$ pounds of candy. Harry had $1\frac{3}{8}$ pounds of the same candy. They combined the candy and divided the result into seven equal parts. How many pounds were there in each of the seven parts?

28. A cook needs $\frac{3}{4}$ of a recipe which calls for $2\frac{3}{4}$ cups of flour. How much flour does he need?

29. A mechanic has 19 bolts, each $2\frac{7}{8}$ inches long. If he placed the bolts end-to-end, how long a string of bolts would be formed?

30. If a piece of metal trim $19\frac{5}{8}$ inches long is cut into five pieces of equal length, what will be the length of each piece?

2.7
COMPARING FRACTIONS AND MIXED NUMBERS

Ordering fractions means placing them in the order in which they would appear on a number line. Figure 2.17 shows several fractions on a number line.

FIGURE 2.17

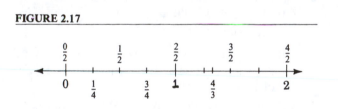

Using the symbol "$<$" to indicate "is less than," or "to the *left* of on the number line," the fractions shown in Figure 2.17 can be compared in various ways, some of which are shown here:

$\frac{0}{2} < \frac{1}{2}$ Zero halves is less than one-half.

$\frac{2}{2} < \frac{3}{2}$ Two halves is less than three halves.

$\frac{1}{4} < \frac{1}{2}$ One-fourth is less than one-half.

$\frac{3}{4} < \frac{4}{3}$ Three-fourths is less than four-thirds.

We use the symbol "$>$" to indicate "is greater than," or "to the *right* of on the number line." Some of the ways we can compare the fractions in Figure 2.17 using this symbol are shown here:

$\frac{1}{2} > \frac{1}{4}$ One-half is greater than one-fourth.

$\frac{3}{2} > \frac{1}{2}$ Three-halves is greater than one-half.

$\frac{4}{3} > \frac{3}{4}$ Four-thirds is greater than three-fourths.

$\frac{4}{2} > \frac{4}{3}$ Four-halves is greater than four-thirds.

The preceding comparisons specify some of the relationships shown in Figure 2.17. However, to *completely order* a set of fractions is to place *all* of the fractions in an increasing sequence. For example, to completely order $\frac{3}{2}, \frac{1}{2}, \frac{4}{3},$ and $\frac{1}{4},$ you must write them as follows:

$$\frac{1}{4} < \frac{1}{2} < \frac{4}{3} < \frac{3}{2}.$$

This ordering may be checked by referring to the order of the fractions on the number line in Figure 2.17.

It may be inconvenient, when ordering fractions, to use a number line. Therefore, a technique for determining which of two fractions is the greater is needed. One such technique involves conversion to equivalent fractions with like denominators and comparing numerators.

> **When fractions have a common denominator, the fractions are in the same order as the order of their numerators.**

EXAMPLE 1 Which is greater, $\frac{2}{7}$ or $\frac{3}{8}$?

Solution The lowest common denominator for $\frac{2}{7}$ and $\frac{3}{8}$ is 56.

$$\frac{2}{7} = \frac{2 \cdot 8}{7 \cdot 8} = \frac{16}{56}$$

$$\frac{3}{8} = \frac{3 \cdot 7}{8 \cdot 7} = \frac{21}{56}$$

Since $21 > 16$ we know that $\frac{21}{56} > \frac{16}{56}$. Therefore, $\frac{3}{8} > \frac{2}{7}$.

EXAMPLE 2 Which is greater, $\frac{4}{9}$ or $\frac{5}{12}$?

Solution The lcd $= 3^2 \cdot 4 = 36$.

$$\frac{4}{9} = \frac{16}{36}$$

$$\frac{5}{12} = \frac{15}{36}$$

Since $\frac{16}{36} > \frac{15}{36}$ we know that $\frac{4}{9} > \frac{5}{12}$.

EXAMPLE 3 Place the fractions $\frac{2}{3}$, $\frac{3}{4}$, and $\frac{3}{5}$ in order from smallest to largest.

Solution The lcd $= 3 \cdot 4 \cdot 5 = 60$.

$$\frac{2}{3} = \frac{40}{60}$$

$$\frac{3}{4} = \frac{45}{60}$$

$$\frac{3}{5} = \frac{36}{60}$$

Since $\frac{36}{40} < \frac{40}{60} < \frac{45}{60}$ we know that $\frac{3}{5} < \frac{2}{3} < \frac{3}{4}$.

Another method for comparing two fractions is called the **criss-cross method**. This method may be a shortcut for many comparison problems. It is particularly useful when denominators are large numbers and when it is difficult to compute the least common denominator. Criss-cross products are important when comparing fractions. As we can see from Figure 2.18, criss-

cross products are the product of the numerator of the first fraction and the denominator of the second fraction, and the product of the denominator of the first and the numerator of the second.

FIGURE 2.18

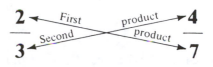

In Figure 2.18 the criss-cross products are (i) $2 \cdot 7$ and (ii) $3 \cdot 4$.

It is important that the correct criss-cross product be identified as the *first*. The **first criss-cross product is always the product of the numerator of the left fraction and the denominator of the right fraction.**

EXAMPLES

4 In $\frac{4}{5}$ and $\frac{3}{4}$, the first criss-cross product is $4 \cdot 4$.

5 In $\frac{7}{8}$ and $\frac{4}{9}$, the first criss-cross product is $7 \cdot 9$ and the second criss-cross product is $8 \cdot 4$.

6 In $\frac{13}{15}$ and $\frac{11}{14}$, the first criss-cross product is $13 \cdot 14$ and the second is $15 \cdot 11$.

The criss-cross products of $\frac{2}{7}$ and $\frac{3}{8}$ are $2 \cdot 8$ and $7 \cdot 3$, which happen to be 16 and 21. Note that 16 and 21 are the numerators when the fractions are converted to common denominator form, i.e.,

$$\frac{2}{7} = \frac{2 \cdot 8}{7 \cdot 8} = \frac{16}{56}$$

$$\frac{3}{8} = \frac{3 \cdot 7}{8 \cdot 7} = \frac{21}{56}.$$

Therefore, the criss-cross products really are the *numerators* when the two fractions are converted to fractions having like denominators.

(Like denominators are the products of the original denominators.) This fact justifies the following definition:

> One fraction is *less than* another fraction only if the first criss-cross product is less than the second.

EXAMPLE 7   Is $\frac{3}{4} < \frac{4}{5}$?

Solution   Since $3 \cdot 5$ is less than $4 \cdot 4$, it is true that $\frac{3}{4} < \frac{4}{5}$.

EXAMPLE 8   Is $\frac{7}{9} < \frac{6}{8}$?

Solution   Since $7 \cdot 8$ is *not* less than $9 \cdot 6$, $\frac{7}{9}$ is *not* less than $\frac{6}{8}$.

EXAMPLE 9   Is $\frac{6}{9} < \frac{8}{12}$?

Solution   Since $6 \cdot 12$ is *not* less than $9 \cdot 8$ (they are equal), $\frac{6}{9}$ is *not* less than $\frac{8}{12}$.

Example 9 above illustrates the fact that *if the criss-cross products are equal, the two fractions are equal*. Using criss-cross products to compare two fractions is summarized below.

> 1. If the first criss-cross product is *less than* the second, then the first fraction is less than the second. For example, for $\frac{2}{3}$ and $\frac{3}{4}$, $2 \cdot 4 < 3 \cdot 3$; therefore, $\frac{2}{3} < \frac{3}{4}$.
> 2. If the first criss-cross product is *greater than* the second, then the first fraction is greater than the second. For example, for $\frac{2}{3}$ and $\frac{3}{5}$, $2 \cdot 5 > 3 \cdot 3$; therefore, $\frac{2}{3} > \frac{3}{5}$, or $\frac{3}{5} < \frac{2}{3}$.
> 3. If the criss-cross products are equal, then the fractions are equal. For instance, for $\frac{6}{10}$ and $\frac{9}{15}$, $6 \cdot 15 = 10 \cdot 9$; therefore, $\frac{6}{10} = \frac{9}{15}$.

When ordering several fractions, compare them in pairs. For example, to order $\frac{1}{7}$, $\frac{3}{10}$, and $\frac{2}{11}$, you can establish the following relationships:

$$\frac{1}{7} < \frac{3}{10} \text{ because } 1 \cdot 10 < 3 \cdot 7$$

$$\frac{3}{10} > \frac{2}{11} \text{ because } 33 > 20$$

$$\frac{1}{7} < \frac{2}{11} \text{ because } 11 < 14.$$

Since $\frac{1}{7} < \frac{2}{11}$ and $\frac{2}{11} < \frac{3}{10}$, the ordering of the fractions is $\frac{1}{7} < \frac{2}{11} < \frac{3}{10}$. As another example, let us order the fractions $\frac{3}{5}$, $\frac{9}{16}$, $\frac{7}{13}$, and $\frac{4}{7}$.

$$\frac{3}{5} > \frac{9}{16} \text{ because } 3 \cdot 16 > 5 \cdot 9$$

$$\frac{9}{16} > \frac{7}{13} \text{ because } 9 \cdot 13 > 16 \cdot 7$$

$$\frac{3}{5} > \frac{7}{13} \text{ because } 39 > 35$$

$$\frac{9}{16} < \frac{4}{7} \text{ because } 63 < 64$$

$$\frac{4}{7} < \frac{3}{5} \text{ because } 20 < 21$$

$$\frac{4}{7} > \frac{7}{13} \text{ because } 52 > 49$$

To summarize: $\frac{3}{5}$ is *greater than* each of the other three; $\frac{7}{13}$ is *less than* each of the other three; $\frac{9}{16} < \frac{4}{7}$. Therefore,

$$\frac{7}{13} < \frac{9}{16} < \frac{4}{7} < \frac{3}{5}.$$

Ordering mixed numbers may involve converting each to a fraction, as shown in Example 10.

EXAMPLE 10   Order $2\frac{1}{3}$, $2\frac{1}{2}$, and $2\frac{3}{8}$.

Solution

$$2\frac{1}{3} = \frac{7}{3}; \ 2\frac{1}{2} = \frac{5}{2}; \ 2\frac{3}{8} = \frac{19}{8}$$

$\dfrac{7}{3} < \dfrac{5}{2}$ because $14 < 15$

$\dfrac{5}{2} > \dfrac{19}{8}$ because $40 > 38$

$\dfrac{7}{3} < \dfrac{19}{8}$ because $56 < 57$

Therefore, $\dfrac{7}{3} < \dfrac{19}{8} < \dfrac{5}{2}$ or

$$2\dfrac{1}{3} < 2\dfrac{3}{8} < 2\dfrac{1}{2}.$$

(*Note:* The numbers $2\dfrac{1}{3}$, $2\dfrac{1}{2}$, and $2\dfrac{3}{8}$ can also be ordered by ordering $\dfrac{1}{3}$, $\dfrac{1}{2}$, and $\dfrac{3}{8}$, since the whole number is 2 in each case.)

Review Exercises

In Exercises 1–4 perform the indicated multiplications and divisions, writing each answer in lowest terms.

1. $\dfrac{7}{8} \cdot \dfrac{3}{5}$ 2. $\dfrac{2}{5} \cdot 3$

3. $\dfrac{7}{8} \div \dfrac{3}{5}$ 4. $\dfrac{2}{5} \div 3$

In Exercises 5–12 perform the indicated operations, writing each answer in lowest terms.

5. $\dfrac{8}{7} \div \dfrac{5}{3}$ 6. $\dfrac{5}{2} \cdot 3$

7. $1\dfrac{2}{3} \div \dfrac{2}{11}$ 8. $\dfrac{3}{8} \cdot \dfrac{7}{5}$

9. $3 \div \dfrac{2}{5}$ 10. $\dfrac{2}{11} \cdot 1\dfrac{2}{3}$

11. $\dfrac{5}{16} \div \left(\dfrac{3}{8} \div \dfrac{1}{4} \right)$ 12. $\dfrac{\dfrac{5}{8}}{\dfrac{3}{16} + \dfrac{7}{8}}$

EXERCISES
2.7

In Exercises 1–8 compare the two fractions, using $>$ or $<$.

1. $\dfrac{2}{3}, \dfrac{8}{27}$ 2. $\dfrac{11}{5}, \dfrac{12}{6}$

3. $\dfrac{13}{15}, \dfrac{12}{14}$ 4. $\dfrac{8}{12}, \dfrac{12}{18}$

5. $\dfrac{3}{4}, \dfrac{7}{24}$ 6. $\dfrac{10}{7}, \dfrac{11}{8}$

7. $\dfrac{14}{16}, \dfrac{13}{15}$ 8. $\dfrac{9}{15}, \dfrac{12}{20}$

In Exercises 9–16 which "criss-cross" products should be used to show the relationship? (For example, $\dfrac{2}{3} > \dfrac{1}{2}$ because $2 \cdot 2 > 3 \cdot 1$.)

9. $\dfrac{4}{9} > \dfrac{3}{8}$ 10. $\dfrac{5}{9} < \dfrac{3}{5}$

11. $\dfrac{12}{15} = \dfrac{16}{20}$ 12. $\dfrac{18}{27} = \dfrac{20}{30}$

13. $\dfrac{5}{9} > \dfrac{3}{6}$ 14. $\dfrac{11}{13} > \dfrac{9}{11}$

15. $\dfrac{7}{9} = \dfrac{21}{27}$ 16. $\dfrac{16}{28} = \dfrac{20}{35}$

Compare the two fractions in Exercises 17–21.

17. $\dfrac{279}{999}, \dfrac{288}{1008}$ 18. $\dfrac{253}{572}, \dfrac{255}{574}$

19. $\dfrac{5034}{5604}, \dfrac{5031}{5601}$ 20. $\dfrac{137}{99}, \dfrac{135}{97}$

21. $7\dfrac{9}{16}, 7\dfrac{10}{17}$

22. Completely order the following fractions and locate them on a number line:

$$\dfrac{3}{4}, \dfrac{1}{3}, \dfrac{4}{7}, \dfrac{5}{8}.$$

23. Completely order the following fractions and locate them on a number line:

$$\dfrac{2}{3}, \dfrac{1}{4}, \dfrac{3}{7}, \dfrac{3}{8}.$$

24. Completely order the following fractions:

$$\dfrac{2}{7}, \dfrac{3}{8}, \dfrac{1}{6}, \dfrac{4}{9}.$$

25. Completely order the following fractions:

$$\frac{3}{7}, \frac{4}{8}, \frac{2}{6}, \frac{5}{9}.$$

26–28. A rate can be represented by a fraction. For example, the following problem can be solved using fractions.

Which is the greater cost per ounce, 12 ounces for 162 cents or 16 ounces for 216 cents?

Solution: The first cost can be represented by $\frac{162}{12}$, and the second cost by $\frac{216}{16}$. Since $(162 \cdot 16 = 2592)$

$= (12 \cdot 216 = 2592)$, the two costs are the same. Note that $162 \cdot 16$ and $12 \cdot 216$ are the criss-cross products.

Solve Exercises 26–28 using fractions, as in the preceding example.

26. Which is the greater cost per banana, 17 bananas for 93 cents or 27 bananas for 134 cents?

27. Which is the greater cost per orange, 12 oranges for 94¢ or 11 oranges for 92¢?

28. Which is the cheaper cost per pair, 6 pairs of socks for 178 cents or 9 pairs of socks for 267 cents?

2.8
USING FRACTIONS TO CONVERT UNITS

When converting a measure from one unit to another we often use a **conversion factor**. A conversion factor may be written as a fraction equal to 1, where numerator and denominator are equal measures. Some examples are shown in Table 2.6. When converting from one unit of measurement to another, a number is multiplied by a conversion factor. Multiplying by a conversion factor is like multiplying by 1.

TABLE 2.6

Equal Measures	Conversion Factors
1 foot = 12 inches	$\frac{1 \text{ ft}}{12 \text{ in.}}$ or $\frac{12 \text{ in.}}{1 \text{ ft}}$
1 yard = 3 feet	$\frac{1 \text{ yd}}{3 \text{ ft}}$ or $\frac{3 \text{ ft}}{1 \text{ yd}}$
1 meter = 100 centimeters	$\frac{1 \text{ m}}{100 \text{ cm}}$ or $\frac{100 \text{ cm}}{1 \text{ m}}$

EXAMPLE 1   Seven inches is how many feet?

Solution   We need to multiply 7 inches by a conversion factor to make the conversion from inches to feet. There are two choices: $\frac{12 \text{ in.}}{1 \text{ ft}}$ and $\frac{1 \text{ ft}}{12 \text{ in.}}$. We select the conversion factor that has *in.* as the unit in the *denominator* so we can cancel the inch units as we multiply. The computation is shown here:

$$7 \text{ in.} \left(\frac{1 \text{ ft}}{12 \text{ in.}} \right) = \frac{7 \cdot 1 \text{ (in.)(ft)}}{12 \text{ (in.)}}$$

$$= \frac{7 \cdot 1 \text{ (ft)(in.)}}{12 \text{ (in.)}} \quad \text{Can be cancelled}$$

$$= \frac{7 \text{ (ft)(in.)}}{12 \text{ (in.)}} \quad \text{Cancelling}$$

$$= \frac{7 \cdot 1 \text{ (ft)}}{12}$$

$$= \frac{7}{12} \text{ ft.}$$

Therefore, $7 \text{ in.} = \frac{7}{12}$ ft. Note that selecting $\frac{12 \text{ in.}}{1 \text{ ft}}$ as the conversion factor would result in

$7 \text{ in.} \left(\frac{12 \text{ in.}}{1 \text{ ft}} \right) = \frac{7 \cdot 12 \text{ (in.)(in.)}}{1 \text{ (ft)}}$, which yields a final fraction which does not contain units in numerator and denominator which can be cancelled.

EXAMPLE 2   Convert $1\frac{1}{2}$ feet to inches.

Solution   The choices of conversion factors are $\frac{1 \text{ ft}}{12 \text{ in.}}$ or $\frac{12 \text{ in.}}{1 \text{ ft}}$. Since the $1\frac{1}{2}$ ft will be multiplied by the conversion factor, we must select a conversion factor which has (ft) in the *denominator*. Therefore,

$$1\frac{1}{2} \text{ ft} \left(\frac{12 \text{ in.}}{1 \text{ ft}} \right) = \left(\frac{3}{2} \right) \text{ ft} \left(\frac{12 \text{ in.}}{1 \text{ ft}} \right)$$

$$= \left(\frac{3}{2} \right) \left(\frac{12 \text{ (in.)(ft)}}{1 \text{ (ft)}} \right) \longleftarrow \text{Cancelling}$$

$$= \left(\frac{3}{2} \right) \left(\frac{12}{1} \right) \text{ in.}$$

$$= \frac{36}{2} \text{ in.}$$

$$= 18 \text{ in.}$$

EXAMPLE 3   Convert 5 feet to yards.

Solution

$$5 \text{ ft} \left(\frac{1 \text{ yd}}{3 \text{ ft}} \right) = \frac{5 \cdot 1 \text{ (yd)(ft)}}{3 \text{ (ft)}} \longleftarrow \text{Cancelling}$$

$$= \frac{5}{3} \text{ yd}$$

$$= 1\frac{2}{3} \text{ yd}$$

EXAMPLE 4   Convert $5\frac{1}{2}$ yards to feet.

Solution

$$5\frac{1}{2} \text{ yd} \left(\frac{3 \text{ ft}}{1 \text{ yd}} \right) = \frac{11}{2} \left(\frac{3 \text{ ft yd}}{1 \text{ yd}} \right) \longleftarrow \text{Cancelling}$$

$$= \frac{33}{2} \text{ ft}$$

EXAMPLE 5   Convert $3\frac{3}{4}$ meters to centimeters.

Solution

$$3\frac{3}{4} \text{ m} \left(\frac{100 \text{ cm}}{1 \text{ m}} \right) = \frac{15}{4} \left(\frac{100 \text{ cm m}}{1 \text{ m}} \right)$$

$$= \frac{1500}{4} \text{ cm}$$

$$= 375 \text{ cm}$$

EXAMPLE 6   Convert 3 feet 7 inches to feet.

Solution   One way is to first convert 3 feet to inches, then *add* 7 inches and convert the result to feet. That is,

$$3 \text{ ft} \left(\frac{12 \text{ in.}}{1 \text{ ft}} \right) = 36 \text{ in.}$$

$$36 \text{ in.} + 7 \text{ in.} = 43 \text{ in.}$$

$$43 \text{ in.} \left(\frac{1 \text{ ft}}{12 \text{ in.}} \right) = \frac{43}{12} \text{ ft}$$

$$= 3\frac{7}{12} \text{ ft.}$$

Another way is to first convert 7 inches to feet, then add the results. That is,

$$7 \text{ in.} \left(\frac{1 \text{ ft}}{12 \text{ in.}} \right) = \frac{7}{12} \text{ ft}$$

$$\frac{7}{12} \text{ ft} + 3 \text{ ft} = 3\frac{7}{12} \text{ ft.}$$

This is the simpler solution because everything is converted to feet immediately.

EXAMPLE 7   Convert 5 yards 2 feet to yards.

Solution   Convert the 2 feet to yards, then add the 5 yards. That is,

$$2 \text{ ft} \left(\frac{1 \text{ yd}}{3 \text{ ft}} \right) = \frac{2}{3} \text{ yd}$$

$$\frac{2}{3} \text{ yd} + 5 \text{ yd} = 5\frac{2}{3} \text{ yd.}$$

EXAMPLE 8 Convert $8\frac{3}{4}$ yards to feet and inches.

Solution

$$8\frac{3}{4}\,\text{yd}\left(\frac{3\text{ ft}}{1\text{ yd}}\right)=\frac{35}{4}\left(\frac{3\text{ ft}}{1}\right)=\frac{105}{4}\text{ ft}$$

$$\frac{105}{4}\text{ ft}=26\frac{1}{4}\text{ ft}$$

$$26\frac{1}{4}\text{ ft}=26\text{ ft}+\frac{1}{4}\text{ ft}$$

$$\frac{1}{4}\,\text{ft}\left(\frac{12\text{ in.}}{1\text{ ft}}\right)=\frac{12}{4}\text{ in.}=3\text{ in.}$$

The answer is 26 ft 3 in.

EXAMPLE 9 Convert 5 yards 2 feet to feet.

Solution Convert the 5 yards to feet and add 2 feet:

$$5\,\text{yd}\left(\frac{3\text{ ft}}{1\text{ yd}}\right)=15\text{ ft}$$

$$15\text{ ft}+2\text{ ft}=17\text{ ft}$$

EXAMPLE 10 Convert 17 yards 2 feet 9 inches to feet.

Solution Convert everything to feet and add:

$$17\,\text{yd}\left(\frac{3\text{ ft}}{1\text{ yd}}\right)=51\text{ ft}$$

$$9\,\text{in.}\left(\frac{1\text{ ft}}{12\text{ in.}}\right)=\frac{9}{12}\text{ ft}=\frac{3}{4}\text{ ft}$$

$$51\text{ ft}+2\text{ ft}+\frac{9}{12}\text{ ft}=53\frac{3}{4}\text{ ft}$$

In Example 10, when conversion factors by which to multiply 17 yards and 9 inches were selected, the factors selected contained the *desired* unit, feet, in their numerators. This leads us to the following general rule:

> **RULE FOR SELECTING CONVERSION FACTORS**
>
> Select a conversion factor which contains the desired unit in its numerator and the given unit in its denominator.

EXAMPLE 11 Convert $3\frac{4}{5}$ yards to inches.

Solution The conversion factor is $\frac{36\text{ in.}}{1\text{ yd}}$ because 36 in. = 1 yd. Thus,

$$3\frac{4}{5}\,\text{yd}\left(\frac{36\text{ in.}}{1\text{ yd}}\right)=\frac{19}{5}\left(\frac{36\text{ in.}}{1}\right)$$

$$=\frac{684}{5}\text{ in.}$$

$$=136\frac{4}{5}\text{ in.}$$

Notice that $\frac{3\text{ ft}}{1\text{ yd}}$ and $\frac{12\text{ in.}}{1\text{ ft}}$ can be combined to get:

$$\left(\frac{3\text{ ft}}{1\text{ yd}}\right)\left(\frac{12\text{ in.}}{1\text{ ft}}\right)=\frac{36\text{ in.}}{1\text{ yd}}.$$

Review Exercises

Perform the indicated operations or find the missing number in Exercises 1–5.

1. $\frac{3}{4}\cdot 13$ 2. $\frac{4}{7}=\frac{?}{28}$

3. $4\frac{2}{3}\cdot\frac{9}{32}$ 4. $\frac{5}{9}+\frac{3}{4}-\frac{2}{5}$

5. $\left(\frac{9}{10}-\frac{2}{5}\right)-\frac{3}{32}$

6. Write $\frac{18}{5}$ as a mixed number.

7. Write $18\frac{2}{13}$ as an improper fraction.

8. Write $13\frac{5}{9}$ as an improper fraction.

Reduce to lowest terms in Exercises 9–12.

9. $\frac{54}{63}$ 10. $\frac{43}{220}$

11. $\frac{52}{169}$ 12. $\frac{132}{36}$

EXERCISES
2.8

Write the conversion factor for each equality in Exercises 1–12.

1. 12 in. = 1 ft
2. 1 yd = 3 ft
3. 1000 mm = 1 m
4. 10 dm = 1 m
5. 1 dm = 10 cm
6. 1 rd = $16\frac{1}{2}$ ft
7. $5\frac{1}{2}$ yd = 1 rd
8. 4 cups = 1 qt
9. 1 gal = 4 qt
10. 1 pt = 2 cups
11. 1 cup = $\frac{1}{2}$ pt
12. 640 a. = 1 sq mi

Use the conversion factors you wrote for Exercises 1–12 to fill in the blanks in Exercises 13–44.

13. 5 in. = _____ ft
14. 5 ft = _____ in.
15. 6 yd = _____ ft
16. 28 ft = _____ yd
17. $7\frac{1}{2}$ m = _____ cm
18. $3\frac{1}{4}$ m = _____ mm
19. 575 cm = _____ dm
20. $\frac{1}{4}$ dm = _____ cm
21. 23 rd = _____ ft
22. 100 ft = _____ rd
23. $8\frac{1}{2}$ qt = _____ cups
24. $8\frac{1}{2}$ pt = _____ cups
25. 35 qt = _____ gal
26. 40 a. = _____ sq mi
27. 75 ft = _____ rd
28. 75 yd = _____ rd
29. 500 in. = _____ yd
30. 12 yd = _____ in.
31. 1000 in. = _____ rd
32. 12 rd = _____ in.
33. 750 cups = _____ gal
34. $5\frac{1}{2}$ gal = _____ cups
35. 17 yd 2 ft = _____ ft
36. 5 yd 2 ft = _____ in.
37. 12 yd 2 ft 5 in. = _____ ft
38. 3 yd 4 ft = _____ in.
39. $7\frac{1}{3}$ ft = _____ ft _____ in.
40. $19\frac{3}{4}$ ft = _____ ft _____ in.
41. 2 yd 2 ft = _____ ft _____ in.
42. 5 yd 1 ft = _____ ft _____ in.
43. 17 yd 2 ft 5 in. = _____ ft _____ in.
44. 5 yd 1 ft 11 in. = _____ ft _____ in.
45. There are 1000 watts in 1 kilowatt. How many kilowatts are there in 5500 watts?
46. There are 27 cubic feet in 1 cubic yard. How many cubic feet in $3\frac{1}{2}$ cubic yards?

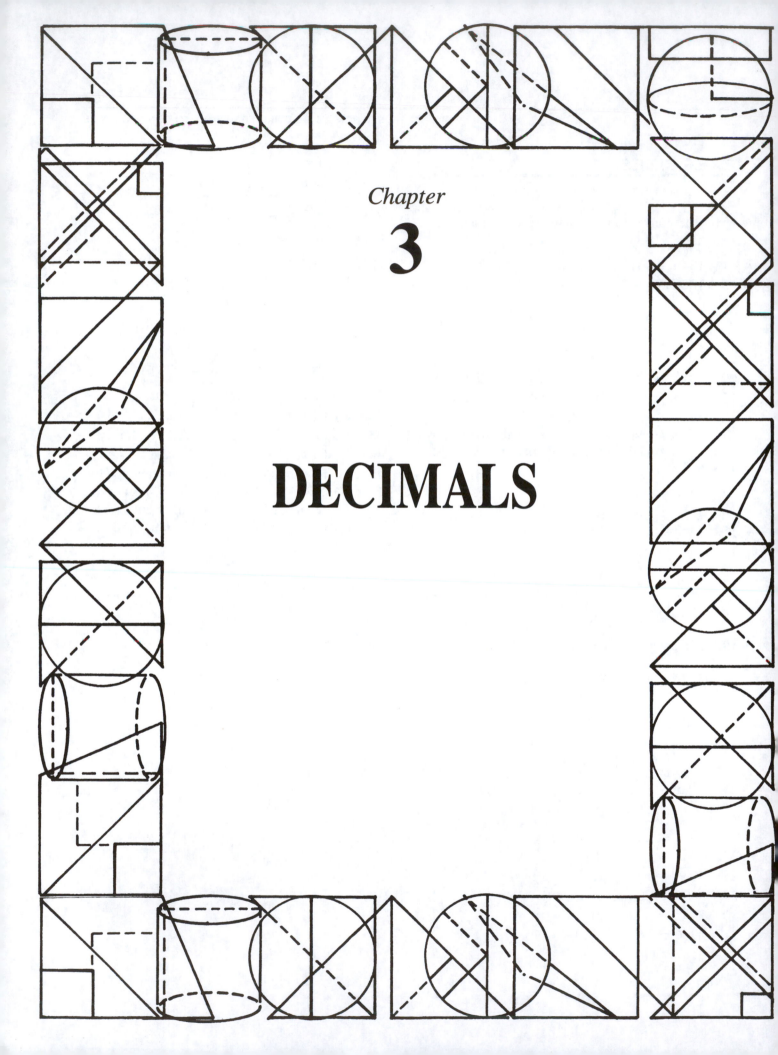

Chapter

3

DECIMALS

3.1
DECIMALS AND PLACE VALUE

Figure 3.1 illustrates **place value**. In a base ten numeral, each **place** has a value ten times the place value of its neighbor to the right and one-tenth the place value of its neighbor to the left.

FIGURE 3.1

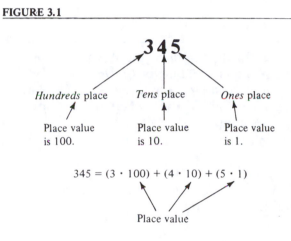

$$345 = (3 \cdot 100) + (4 \cdot 10) + (5 \cdot 1)$$

Decimal numeration extends whole number numeration. The ones place is not necessarily the last place to the right. A decimal point (.) separates the ones place from all places to its right. This is illustrated in Figure 3.2.

FIGURE 3.2

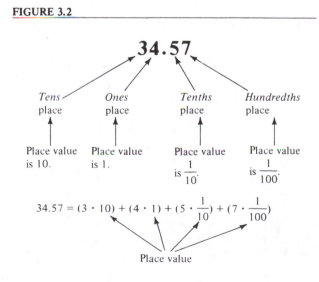

In similar fashion, $234.754 = (2 \cdot 100) + (3 \cdot 10) + (4 \cdot 1) + \left(7 \cdot \frac{1}{10}\right) + \left(5 \cdot \frac{1}{100}\right) + \left(4 \cdot \frac{1}{1000}\right)$. The following illustrates the common practice of using a zero in the ones place for decimal numbers less than 1 when results of measurements are indicated.

$$0.39 = \left(3 \cdot \frac{1}{10}\right) + \left(9 \cdot \frac{1}{100}\right) = 0.39$$

Note the placeholder "zero."

$$0.1234 = 0 + \left(1 \cdot \frac{1}{10}\right) + \left(2 \cdot \frac{1}{100}\right) + \left(3 \cdot \frac{1}{1000}\right) + \left(4 \cdot \frac{1}{10000}\right)$$

This is the *ten thousandths* place.

This is the *thousandths* place.

The number 0.39 is read "thirty-nine one-hundredths." The number 0.1234 is read "twelve hundred thirty-four ten-thousandths." The number 127.35 is read "one hundred twenty-seven *and* thirty-five one-hundredths." The number 5.003 is read "five *and* three thousandths." The word "and" is used to show the location of the decimal point. The number in Figure 3.3 is read "three thousand one hundred seven *and*

FIGURE 3.3

Place value in decimal notation

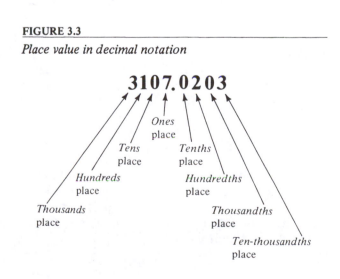

two hundred three ten-thousandths," and can be written to show the place value of each digit as follows:

$$3107.0203 = (3 \cdot 1000) + (1 \cdot 100) + (0 \cdot 10)$$
$$+ (7 \cdot 1) + \left(0 \cdot \frac{1}{10}\right) + \left(2 \cdot \frac{1}{100}\right)$$
$$+ \left(0 \cdot \frac{1}{1000}\right) + \left(3 \cdot \frac{1}{10000}\right).$$

Rounding off a decimal numeral means rewriting it in shorter form to represent an approximation of the number. For example, 3.14159 can be rounded off to 3.14 (to two decimal places), or to 3.1416 (to four decimal places), etc. As another example, 235.103846 can be rounded off to:

235.10385   (to five decimal places)

235.1038   (to four decimal places)

235.104   (to three decimal places)

235.10   (to two decimal places)

235.1   (to one decimal place).

Rounding off involves looking at the digit to the right of the round-off place and making decisions based on the following:

> 1. **If the digit to the right of the round-off place is less than five, simply eliminate all digits to the right of the round-off place.**

EXAMPLE 1   To round off 3.145265 to three places, look at the digit 2, which is to the right of the round-off place. Since $2 < 5$, simply eliminate the last three digits to get 3.145.

> 2. **If the digit to the right of the round-off place is equal to or greater than five (*and*, if it is equal to five it is *not* the last non-zero digit), increase the last desired digit by one.**

EXAMPLE 2   To round off 3.145265 to four places, look at the digit 6 (which is to the right of the round-off place). Since $6 > 5$, increase the 2 by one to get 3.1453.

EXAMPLE 3   To round off 3.145265 to two places, write 3.15, because the digit after the 4 is equal to 5.

> 3. **If the digit to the right of the round-off place is a 5 and is the *last* non-zero digit, a standard practice is to leave an *even* digit at the end of the rounded decimal.**

EXAMPLE 4   To round off 3.145 to two places, write 3.14, because 5 is the last non-zero digit and 4 is even.

EXAMPLE 5   To round off 3.135 to two places, change it to 3.14, because 5 is the last non-zero digit, and it is customary to leave an *even* digit.

Here is another example of rounding off:

$$0.918273652 = 0.9 \text{ (to one place)}$$
$$= 0.92 \text{ (to two places)}$$
$$= 0.918 \text{ (to three places)}$$
$$= 0.9183 \text{ (to four places)}$$
$$= 0.91827 \text{ (to five places)}$$
$$= 0.918274 \text{ (to six places)}$$
$$= 0.9182737 \text{ (to seven places).}$$

When rounding off a number, go to the desired round-off place and look only at the next digit to the right. If there are two or more digits to the right of the round-off place, *do not* "round back" from the last digit one digit at a time. (See Example 6.)

EXAMPLE 6   Round 3.1449 to two places.

Solution

$$3.14\underset{\longleftarrow}{4}9$$
The desired round-off place

Look at 3.144 and round off to 3.14.

WATCH OUT! Do not look at 3.1449 and change to 3.145, then to 3.15.

Rounding off to the nearest 10 yields a whole number whose right-hand digit is zero. Rounding off to the nearest 100 yields a whole number whose last two right-hand digits are zeros. For example, $353 = 350$ to the nearest *ten*, and $3549 = 3500$ to the nearest *hundred*. Here are some additional examples of rounding off.

EXAMPLES

7   0.089 becomes 0.09 (to the nearest hundredth *or* to two decimal places)

8   5.0943 becomes 5.09 (to the nearest hundredth)

9   13.45 becomes 13.4 (to the nearest tenth)

10   13.55 becomes 13.6 (to the nearest tenth)

11   5.3009 becomes 5.301 (to the nearest thousandth)

12   5.3009 becomes 5.30 (to two decimal places)

13   1432.58 becomes 1400 (to the nearest *hundred*)

14   29.53 becomes 30 (to the nearest *ten*)

Review Exercises

1.   Fourteen of 76 automobiles in a parking lot are Fords. What fraction of the autos in the lot are Fords?

2.   A pole vaulter clears 17 feet, and a high jumper clears $6\frac{1}{2}$ feet. What fraction of the pole vaulter's cleared height is the high jumper's cleared height?

3.   Nineteen of 95 cars passing a given point are Chevrolets. What fraction of the cars are *not* Chevrolets?

4.   Twelve students in a class of 56 are A students, and seventeen are B students. What fraction of the students in the class have less than a B average?

Perform the indicated operations in Exercises 5–9.

5. $13\frac{3}{8} + 4\frac{5}{9}$ 6. $13\frac{3}{8} - 3\frac{9}{16}$

7. $\frac{3}{7} + \frac{4}{9} - \frac{1}{3}$ 8. $\left(2\frac{4}{5}\right)\left(1\frac{3}{7}\right)$

9. $2\frac{4}{5} \div \frac{7}{10}$

10.   In a certain truck yard, one-third of the trucks have six wheels, two-fifths of the trucks have ten wheels, and the rest of the trucks have four wheels. What fraction of the trucks have four wheels?

11.   Completely order the following:

$$\frac{5}{9}, \frac{4}{8}, \frac{6}{10}, \frac{7}{11}.$$

12.   John's heart beats five times every six seconds, and Susan's heart beats nine times every eleven seconds. Whose heart beats faster?

EXERCISES
3.1

Write each decimal in Exercises 1–10 in expanded notation. Use the following example as a guide.

$$34.56 = (3 \cdot 10) + (4 \cdot 1) + \left(5 \cdot \frac{1}{10}\right)$$

$$+ \left(6 \cdot \frac{1}{100}\right)$$

1. 478.394	2. 0.394	29. 5.389	30. 5.383
3. 20.14	4. 304.209	31. 54.564	32. 54.565
5. 1000.003	6. 666.666	33. 54.566	34. 13.009
7. 394.784	8. 0.285	35. 1.495432	36. 0.55555
9. 509.708	10. 3000.001	37. 3.5465	38. 0.09534

11–14. Write out in words the decimal numbers listed in Exercises 1–4 above.

15–20. Write out in words the decimal numbers listed in Exercises 5–10 above.

Round off the numbers in Exercises 21–28 to the nearest *ten*.

21. 2458	22. 3794
23. 25.89	24. 347.95
25. 3594	26. 4976
27. 34.05	28. 295.04

Round off the numbers in Exercises 29–48 to two decimal places.

39. 6.279	40. 6.273
41. 55.563	42. 36.385
43. 2.556	44. 11.007
45. 1.39532	46. 0.4545
47. 4.5465	48. 0.095443

Round off the numbers in Exercises 49–56 to the nearest *hundred*.

49. 950004	50. 359.79
51. 666	52. 0.394
53. 340025	54. 793.04
55. 897	56. 0.959

3.2
ADDING AND SUBTRACTING DECIMALS

Adding decimals is similar to adding whole numbers. The digits in each place are added. In column addition it is important to align all decimal points in a vertical line. For example, to add 13.34, 112.506, 0.4934, and 5.3, you write:

$$\begin{array}{l} 13.34 \\ 112.506 \\ 0.4934 \\ \underline{5.3} \end{array} \quad \text{or} \quad \begin{array}{l} 13.3400 \\ 112.5060 \\ 0.4934 \\ \underline{5.3000} \end{array}$$

The decimal point in the answer is placed just below the decimal points of the addends. The completed addition is shown here:

$$\begin{array}{r} 13.34 \\ 112.506 \\ 0.4934 \\ + \quad 5.3 \\ \hline 131.6394 \end{array}$$

Subtracting decimals is similar to subtracting whole numbers. In vertical notation, all decimal points must be aligned, as in addition. For example, 394.056 − 99.26 is written:

$$\begin{array}{r} 394.056 \\ - \quad 99.26 \\ \hline \end{array}$$

The completed subtraction is shown below:

$$\begin{array}{r} 394.056 \\ - \quad 99.26 \\ \hline 294.796 \end{array}$$

Note placement of decimal point in answer.

It is often necessary to use zeros as placeholders when subtracting decimals. Table 3.1 contains examples of this.

TABLE 3.1. USING ZEROS AS PLACEHOLDERS IN SUBTRACTION

Without Zeros	With Zeros	Subtraction
$5.43 - 3.843$	$\begin{array}{r} 5.430 \\ -3.843 \\ \hline \end{array}$	$\begin{array}{r} 5.430 \\ -3.843 \\ \hline 1.587 \end{array}$
$100.4 - 35.555$	$\begin{array}{r} 100.400 \\ -\ 35.555 \\ \hline \end{array}$	$\begin{array}{r} 100.400 \\ -\ 35.555 \\ \hline 64.845 \end{array}$
$13.51 - 8.64385$	$\begin{array}{r} 13.51000 \\ -\ 8.64385 \\ \hline \end{array}$	$\begin{array}{r} 13.51000 \\ -\ 8.64385 \\ \hline 4.86615 \end{array}$

Review Exercises

Add or subtract in Exercises 1–4.

1. $\begin{array}{r} 5039 \\ 4865 \\ 7298 \\ +\ 4813 \\ \hline \end{array}$

2. $\begin{array}{r} 7643 \\ 821 \\ 5005 \\ 16 \\ +\ \ \ 25 \\ \hline \end{array}$

3. $\begin{array}{r} 69765 \\ -\ 29987 \\ \hline \end{array}$

4. $\begin{array}{r} 900045 \\ -\ 321508 \\ \hline \end{array}$

5. The circumference of a circle is four hundred and fifteen thousandths centimeters. Indicate the circumference as a decimal.

Perform the indicated operations in Exercises 6–11.

6. $\dfrac{3}{8} + \dfrac{4}{7}$

7. $\dfrac{5}{8} - \dfrac{4}{7}$

8. $\dfrac{2}{3}\left(\dfrac{3}{8} - \dfrac{1}{6}\right)$

9. $\left(\dfrac{1}{2} - \dfrac{1}{3}\right)\left(\dfrac{2}{3} - \dfrac{1}{2}\right)$

10. $\left(\dfrac{2}{3}\right)^5$

11. $\left(\dfrac{1}{4}\right)^2 + \left(\dfrac{1}{3}\right)^3$

12. Round 789.405564 to the nearest thousandth.

EXERCISES 3.2

Perform the indicated operations in Exercises 1–16.

1. $2.34 + 13.56$ 2. $12.304 + 0.493$

3. $12.003 + 257.04 + 9.679 + 4578.0001$

4. $143.49 + 725.05 + 34.93 + 5.95 + 200.09$

5. $3.24 + 13.57$ 6. $11.409 + 0.032$

7. $9.004 + 349.03 + 43.39 + 9.55 + 900.02$

8. $9.3 + 6.05 + 241.201 + 200.039$

9. $\begin{array}{r} 13.56 \\ -\ \ 2.34 \\ \hline \end{array}$ 10. $\begin{array}{r} 285.45 \\ -\ \ 34.88 \\ \hline \end{array}$

11. $27.89 - 3.24$ 12. $42.75 - 13.83$

13. $800.56 - 79.989$ 14. $500.06 - 25.189$

15. $49.0035 - 1.549$ 16. $70.0029 - 0.934$

17. Jack told Harold that the length of his pencil was seven and fourteen one-hundredths inches. Write the length as a decimal.

18. Three men have $394.87, $109.98, and $5,432.27, respectively. How much more money do the three have together than Sam, who has $4,055.39?

19. Susan spent $5.95 on records, $8.49 on tapes, $3.58 on a phono needle, and $49.55 on clothes. How much did she spend?

20. Harry bought skis for $210.39, boots for $95.43, gloves for $6.59, and goggles for $4.49. He also paid $1.24 in sales tax. What was the total cost, including tax?

21. A machined part is 5.139 inches long. Three other parts are 14.005 inches, 12.001 inches, and 0.495 inches long, respectively. If the parts are all laid end-to-end tightly with no gaps between them, what is the overall length of the four parts?

22. Three car engines develop 104.35 horsepower, 121.05 horsepower, and 148.95 horsepower, respectively. What is the sum of their horsepowers?

23. What is the difference in power between two engines with 148.95 h.p. and 104.38 h.p., respectively?

24. How much longer is a 14.003-inch part than a 9.495-inch part?

3.3
MULTIPLYING DECIMALS

Multiplying decimal numbers by 10, 100, and 1000 is an important skill. It can be shown that $10(32.4) = 324$ and $10(25.34) = 253.4$. In both cases, the decimal point in the result is *one place to the right* of the decimal point in the original factor.

$$\begin{array}{r} 25.34 \\ \times \quad 10 \\ \hline 253.4 \end{array}$$

Decimal point moved one place to the right.

$$\begin{array}{r} 345.734 \\ \times \quad 10 \\ \hline 3457.34 \end{array}$$

Decimal point moved one place to the right.

Multiplying decimals by 100 results in the decimal point being moved two places to the right of its location in the factor. Multiplying decimals by 1000 "moves the decimal point" three places to the right. The following examples illustrate multiplying decimals by 10, 100, and 1000.

EXAMPLES

1   $10(42.35) = 423.5$
    Moved one place to right

2   $100(42.356) = 4235.6$
    Moved two places to right

3   $1000(4.2356) = 4235.6$
    Moved three places to right

4   $(0.1204)(10) = 1.204$

5   $(0.1204)(1000) = 120.4$

6   $(5.3)(100) = (5.30)(100) = 530.$
    Moved two places to right

7   $(5.3)(1000) = 5300.$
    Moved three places to right

8   $(0.000389)(100) = 0.0389$

9   $(0.000389)(1000) = 0.389$

10   $(0.1)(1000) = 100.$
    Moved three places to right

Multiplying decimal numbers by 0.1, 0.01, and 0.001 is also an important skill. It can be shown that $0.1(3.45) = 0.345$ and $0.01(3.98) = 0.0398$. Multiplying by 0.1 *moves the decimal point one place to the left* and multiplying by 0.01 *moves the decimal point two places to the left*.

$$\begin{array}{r} 3.45 \\ \times \quad 0.1 \\ \hline 0.345 \end{array}$$

Decimal point moved one place to the left

$$\begin{array}{r} 3.98 \\ \times \quad 0.01 \\ \hline 0.0398 \end{array}$$

Decimal point moved two places to the left

Similarly, multiplying by 0.001 *moves the decimal point three places to the left*. The following examples illustrate multiplying by 0.1, 0.01, and 0.001.

EXAMPLES

11   $0.1(42.35) = 4.235$
    Moved one place to left

12   $0.01(456.3) = 4.563$
      Moved two places to left

13   $(0.001)(42345.79) = 42.34579$
      Moved three places to left

14   $(0.34)(0.1) = 0.034$
      Moved one place to left

15   $(0.57)(0.01) = 0.0057$
      Moved two places to left

16   $(0.4)(0.001) = 0.0004$
      Moved three places to left

17   $(4.93)(0.001) = 0.00493$

Multiplying by 10, 100, 1000, 10000, etc. is the same as multiplying by a power of ten, i.e., 10^1, 10^2, 10^3, 10^4, etc. The result in each case is obtained by moving the decimal point *to the right as many places as the power indicates.*

EXAMPLES

18   $(3.445)(10^2) = 344.5$
      Moved two places to right

19   $(0.395789)(10^4) = 3957.89$
      Moved four places to right

In exponential notation, 0.1 can be written as 10^{-1}, 0.01 can be written as 10^{-2}, 0.001 can be written as 10^{-3}, etc. The negative exponents (see Chapter 6 for negative numbers) here can be thought of as an indication of the number of decimal places.

EXAMPLES

20   The expression $10^{-1} = 0.1$ means that there is one decimal place. That is, the decimal point is one place to the left of the right end of the numeral.

21   The expression $10^{-2} = 0.01$ means that there are two decimal places. That is, the

decimal point is two places to the left of the right end of the numeral.

22   The expression $10^{-3} = 0.001$ shows three decimal places.

Examples 23–28 illustrate the fact that, whenever you multiply a number by a power of ten, you simply move the decimal point the number of places indicated by the power—*to the right for positive powers,* and *to the left for negative powers.*

EXAMPLES

23   $(75)(100) = (75)(10^2) = 7500$
      The decimal point for 75 was moved two places to the right to get 7500.

24   $(75)(0.01) = (75)(10^{-2}) = 0.75$
      Moved two places to the left

25   $(9.832)(10^2) = 983.2$
      Moved two places to the right

26   $(9.832)(10^{-1}) = 0.9832$
      Moved one place to the left

27   $(9.832)(10^{-3}) = 0.009832$
      Moved three places to the left

28   $(9.832)(10^4) = 98320.$
      Moved four places to the right

Any decimal number can be written as a whole number *times* a power of 10.

EXAMPLES

29   $4.3 = 43(10^{-1})$

30   $0.034 = 34(10^{-3})$

31 $17.28 = 1728(10^{-2})$

32 $756 = 756(10^0)$ (where 10^0 is a name for *one*)

Multiplying decimal numbers is like multiplying whole numbers, except that proper location of the decimal point must be considered. A rule for multiplying decimal numbers is:

> When multiplying two decimal numbers, multiply as if they were whole numbers and locate the decimal point in the *answer* so that it *has as many decimal places as the sum of the number of decimal places in the factors.*

(*Note:* "Number of decimal places" means the number of digits to the right of the decimal point.)

EXAMPLE 33

$$\begin{array}{r} 34.03 \\ \times\quad 3.1 \\ \hline 3403 \\ 10209 \\ \hline 105.493 \end{array}$$

Note that the right-hand digits are in the same column.

The decimal point indicates three decimal places because the factors contain two decimal places and one decimal place, respectively, and $2 + 1 = 3$.

EXAMPLE 34

$$\begin{array}{r} 3.04012 \\ \times\quad 3.02 \\ \hline 608024 \\ 00000 \\ 912036 \\ \hline 9.1811624 \end{array}$$

5 decimal places
2 decimal places
Note the zeros.
7 decimal places because $5 + 2 = 7$

EXAMPLE 35

$$\begin{array}{r} 0.00045 \\ \times\quad 0.053 \\ \hline 135 \\ 225 \\ \hline 0.00002385 \end{array}$$

5 decimal places
3 decimal places
Zeros needed to make 8 decimal places.

EXAMPLE 36

$(0.0001)(0.000003) = 0.0000000003$

Review Exercises

Perform the indicated operations in Exercises 1–8.

1. $\left(\dfrac{4}{3} \div \dfrac{1}{2}\right)\left(\dfrac{2}{3} + \dfrac{7}{5}\right)$ 2. $34\overline{)7893}$

3. $\left(\dfrac{2}{3} \cdot \dfrac{11}{4}\right) - \left(\dfrac{3}{4} \div \dfrac{4}{9}\right) + \dfrac{1}{2}$

4. $\begin{array}{r} 9643 \\ \times\ 796 \\ \hline \end{array}$

5. $\left(\dfrac{2}{3}\right)^4 + 5^2 - 2^4$ 6. $3.45 - 1.769$

7. $25.405 + 134.69 + 3478.921 + 1.008$

8. $(25.405 + 3478.921) - (134.69 + 10.8)$

9. Round 34.50954 to the nearest hundredth.

10. Round 345.0954 to the nearest hundredth.

11. Round 345.0954 to the nearest ten.

12. Write seven hundred and forty-four ten thousandths in decimal form.

EXERCISES
3.3

In Exercises 1–6 multiply each decimal number by 10.

1. 24.35   2. 13.49
3. 0.579   4. 0.795
5. 0.0492   6. 0.0942

Multiply as indicated in Exercises 7–32.

7. $(13.245)(100)$ 8. $(42.389)(100)$

9. $(0.34945)(1000)$

10. $(0.734285)(1000)$

11. $(0.0049204)(10000)$

12. $(0.0094402)(10000)$

13. $(0.000056)(100)$

14. (0.000065)(100)

15. (34.21)(0.1)   16. (23.42)(0.1)

17. (0.043)(0.01)   18. (0.034)(0.01)

19. (49325)(0.0001)

20. (93425)(0.0001)

21. (0.0004)(0.001)   22. (0.0008)(0.001)

23. (796)(0.123)   24. (8679)(0.324)

25. (0.043)^2 26. (72.158)^2

27. (19.3)^3 28. (17.4)^3

29. (0.021)(0.593)(19.2)

30. (0.034)(0.422)(17.1)

31. (1728.594)(0.00912)

32. (7128.493)(0.00613)

33. For each piece of material manufactured, 0.038 in. is wasted. What is the total amount, in inches, wasted when one million parts are manufactured?

34. Five one-hundredths of the price of an automobile is added for tax purposes. How much is added to the price of an automobile listed at $5034.65?

35. A mason purchased 223 cubic yards of ready-mix concrete at $18.47 per cubic yard. How much did the concrete cost him?

36. An electrical contractor purchased 196 conduit bodies at $1.08 each. How much did they cost him?

37. A car averaged 51.93 miles per hour during a 2455-hour trip. How many miles long was the trip?

38. Forty-eight executives averaged $39,850 net pay per year. What was the total amount of net pay for the 48 executives during the year?

39. The area of a rectangle is found by multiplying the length by the width. What is the area of a rectangle 743.28 meters long and 323.49 meters wide? (The answer will be in square meters.)

3.4
DIVIDING DECIMALS

Dividing decimal numbers is the same as dividing whole numbers, except that proper location of the decimal point in the quotient must be taken into consideration. To locate the decimal point in a "long division" problem, simply move the decimal points in both divisor and dividend the same number of places to the right and locate the quotient decimal point immediately above the dividend's new decimal point. The number of places the decimal point is to be moved is the number it takes to change the divisor into a whole number.

EXAMPLES

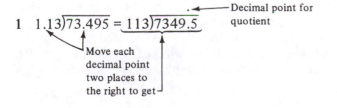

Once the quotient's decimal point location is determined, carry out the indicated division as if both dividend and quotient were whole numbers. Extra places in the dividend can be indicated by affixing zeros, making it possible to complete the division process and obtain a quotient having any desired number of decimal places. Here is an example.

EXAMPLE 4

$$1.03\overline{)5.0342} = 103\overline{)503.42}$$

$$= 103\overline{\smash)\overset{4.}{503.42}}$$
$$\underline{412}$$
$$91$$

$$= 103\overline{\smash)\overset{4.8}{503.42}}$$
$$\underline{412}$$
$$914$$
$$\underline{824}$$
$$90$$

$$= 103\overline{\smash)\overset{4.88}{503.42}} \longleftarrow \text{Answer (quotient)}$$
$$\underline{412}$$
$$914$$
$$\underline{824}$$
$$902$$
$$\underline{824}$$
$$78 \leftarrow \text{Remainder}$$

Extending the quotient in Example 4 to more decimal places involves using zeros as placeholders in the dividend. This is shown in the next example.

EXAMPLE 5

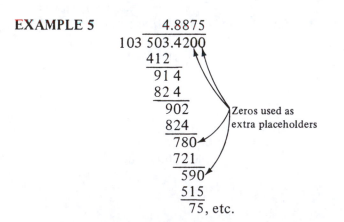

$$
\begin{array}{r}
4.8875 \\
103\overline{\smash)503.4200} \\
\underline{412} \\
91\ 4 \\
\underline{82\ 4} \\
902 \\
\underline{824} \\
780 \\
\underline{721} \\
590 \\
\underline{515} \\
75, \text{ etc.}
\end{array}
$$

Zeros used as extra placeholders

The quotient can be indicated to any specified number of decimal places. For example, to the nearest hundredth the quotient in Example 5 is 4.89 (note the rounding off of the quotient); and to the nearest thousandth it is 4.888. Here is another example of division of decimals.

EXAMPLE 6

$$0.03\overline{)0.00149} = 3\overline{)0.149}$$

$$= 3\overline{\smash)\overset{.049}{0.149}}$$
$$\underline{12}$$
$$29$$
$$\underline{27}$$
$$2$$

$$= 3\overline{\smash)\overset{.049}{0.1490}}$$
$$\underline{12}$$
$$29$$
$$\underline{27}$$
$$20$$

$$= 3\overline{\smash)\overset{.0496}{0.1490}}$$
$$\underline{12}$$
$$29$$
$$\underline{27}$$
$$20$$
$$\underline{18}$$
$$2$$

$$= 3\overline{\smash)\overset{.049666}{0.149000}}$$
$$\underline{12}$$
$$29$$
$$\underline{27}$$
$$20$$
$$\underline{18}$$
$$20$$
$$\underline{18}$$
$$20$$
$$\underline{18}$$
$$2$$

The answer to Example 6, 0.049666..., continues to repeat 6's as far as you want to carry out the division past 0.049. Thus, 0.049666... is known as a **repeating decimal**. Special notation to indicate this decimal is $0.049\overline{6}$, where the 6 with a bar over it at the right end, $\overline{6}$, means that 6 continues to repeat indefinitely.

When one number is divided by another, the quotient may be a **repeating decimal**, which contains either a repeating single digit or a repeating sequence of digits. Several examples of repeating decimals obtained by dividing whole numbers are shown in Table 3.2. When one number is divided by another and the repeating digit in the quotient is a zero, the decimal is called a **terminating decimal**. Several examples of terminating decimals are shown in Table 3.3.

TABLE 3.2

Quotient of Whole Numbers	Repeating Decimal
$1 \div 3$ or $\frac{1}{3}$	$1 \div 3 = 0.3\overline{3}$
$2 \div 3$ or $\frac{2}{3}$	$2 \div 3 = 0.6\overline{6}$
$5 \div 9$ or $\frac{5}{9}$	$5 \div 9 = 0.5\overline{5}$
$1 \div 12$ or $\frac{1}{12}$	$1 \div 12 = 0.083\overline{3}$
$7 \div 36$ or $\frac{7}{36}$	$7 \div 36 = 0.194\overline{4}$
$23 \div 99$ or $\frac{23}{99}$	$23 \div 99 = 0.23\overline{23}$
$1 \div 7$ or $\frac{1}{7}$	$1 \div 7 = 0.142857\overline{142857}$

TABLE 3.3

Quotient of Whole Numbers	Terminating Decimal
$1 \div 4$ or $\frac{1}{4}$	$1 \div 4 = 0.250\overline{0} = 0.25$
$2 \div 5$ or $\frac{2}{5}$	$2 \div 5 = 0.400\overline{0} = 0.4$
$1 \div 8$ or $\frac{1}{8}$	$1 \div 8 = 0.1250\overline{0} = 0.125$
$3 \div 25$ or $\frac{3}{25}$	$3 \div 25 = 0.1200\overline{0} = 0.12$
$1 \div 16$ or $\frac{1}{16}$	$1 \div 16 = 0.06250\overline{0} = 0.0625$
$9 \div 200$ or $\frac{9}{200}$	$9 \div 200 = 0.04500\overline{0} = 0.045$

A fraction can be converted to decimal form by dividing its numerator by its denominator. The decimal quotient will be either a terminating decimal or a repeating decimal.

EXAMPLE 7 $\frac{3}{8} = 8\overline{)3}$

$$= 8\overline{)3.0} \quad \begin{array}{r} .3 \\ \underline{2\,4} \\ 6 \end{array}$$

$$= 8\overline{)3.00000} \quad \begin{array}{r} .3750\overline{0} \\ \underline{2\,4} \\ 60 \\ \underline{56} \\ 40 \\ \underline{40} \end{array}$$

Therefore, $\frac{3}{8} = 0.3750\overline{0}$, or simply 0.375.

EXAMPLE 8 $2\frac{4}{5} = \frac{14}{5} = 5\overline{)14}$

$$= 5\overline{)14.0} \quad \begin{array}{r} 2.8 \\ \underline{10} \\ 4\,0 \\ \underline{4\,0} \end{array}$$

EXAMPLE 9 $\frac{5}{6} = 6\overline{)5}$

$$= 6\overline{)5.0000} \quad \begin{array}{r} .8333 \\ \underline{4\,8} \\ 20 \\ \underline{18} \\ 20 \\ \underline{18} \\ 20 \\ \underline{18} \end{array}$$

Therefore, $\frac{5}{6} = 0.83\overline{3}$.

EXAMPLE 10 $\frac{1}{4} = 4\overline{)1} = 0.25$

EXAMPLE 11 $\frac{2}{7} = 7\overline{)2}$

$$= 7\overline{)2.000000000000}^{.285714\overline{285714}}$$

Therefore, $\frac{2}{7} = 0.285714\overline{285714}$.

Sometimes a decimal is written with a fraction in the last place, for example, $3.44\frac{1}{3}$. The fraction simply represents its decimal value **beginning with that place**. Thus, $3.44\frac{1}{3} = 3.44\overline{3}$; $1.32\frac{1}{8} = 1.32125$; and $1.3\frac{1}{25} = 1.304$.

Often decimal numbers are indicated with terminology like "five and one-half thousandths," or "one plus three and one-half thousandths." "Five and one-half thousandths" means $\frac{5\frac{1}{2}}{1000}$. Writing $5\frac{1}{2}$ as 5.5 and then dividing by 1000 (which is the same as multiplying by 0.001) we get $\frac{5.5}{1000} = 5.5(0.001) = 0.0055$.

In similar fashion, "one plus three and one-half thousandths" means $1 + \frac{3\frac{1}{2}}{1000}$, which can be converted to decimal form as follows:

$$1 + \frac{3\frac{1}{2}}{1000} = 1 + \frac{3.5}{1000} = 1 + 0.0035 = 1.0035.$$

EXAMPLES

12 "Eight and one-quarter thousandths" is:

$$\frac{8\frac{1}{4}}{1000} = \frac{8.25}{1000} = 0.00825.$$

13 "Seven plus five and one-eighth thousandths" is:

$$7 + \frac{5\frac{1}{8}}{1000} = 7 + \frac{5.125}{1000} = 7 + 0.005125$$
$$= 7.005125.$$

Review Exercises

Find the value of each of the numbers in Exercises 1–12.

1. $\left(\frac{4}{5}\right)^3$ 2. $(395)^2$

3. $\left(\frac{1}{2}\right)^5 + \left(\frac{1}{3}\right)^3$ 4. $\left(\frac{1}{3}\right)^3 - \left(\frac{1}{2}\right)^6$

5. $(4.3)^2$ 6. $(7.05)^2$

7. $(0.004)^3$ 8. $(0.011)^3$

9. $(1.003)^2$ 10. $(2.0101)^2$

11. $(17.28)(0.0035)$ 12. $(0.01082)(0.055)$

EXERCISES
3.4

Express the quotients in Exercises 1–8.

1. $1.13\overline{).0345}$ (to the nearest tenth)

2. $30.012\overline{).0345}$ (to the nearest hundredth)

3. $13.3\overline{)55.355}$ (to the nearest thousandth)

4. $24.8\overline{)17.42}$ (to the nearest thousandth)

5. $5.0842 \div 1.02$ (to the nearest hundredth)

6. $9.043 \div 2.04$ (to the nearest hundredth)

7. $0.55555 \div 342.03$ (to the nearest hundredth)

8. $0.4384 \div 605.05$ (to the nearest hundredth)

Convert to decimal form in Exercises 9–24.

9. $\frac{7}{8}$ 10. $\frac{4}{9}$

11. $\frac{1}{3}$ 12. $\frac{5}{11}$

13. $\frac{5}{16}$ 14. $\frac{34}{6}$

15. $\frac{23}{4}$ 16. $\frac{39}{9}$

17. $0.34\frac{1}{2}$ 18. $0.43\frac{3}{4}$

19. $4.3\frac{3}{8}$ 20. $4.4\frac{5}{8}$

21. $155.0\frac{1}{5}$ 22. $104.0\frac{1}{2}$

23. $0.000\frac{2}{3}$ 24. $0.00\frac{1}{3}$

In Exercises 25–30, write the expressions as decimals.

25. Four and one-fourth thousandths
26. Six and three-fourths thousandths
27. Two and four-fifths hundredths
28. Twelve and five-eighths ten-thousandths
29. Nineteen plus four and one-half ten-thousandths
30. Seven plus seven and three-eighths ten-thousandths
31. A baseball player gets 13 hits in 42 times at bat. His batting average is the number of hits divided by the number of times at bat, to the nearest thousandth. What is his batting average?

32. A baseball team gets 126 hits in 478 times at bat. What is the team's batting average? (See Exercise 31.)

33. How many meters are there in $8\frac{3}{4}$ feet if one meter equals 39.37 inches? Round the result to the nearest hundredth.

34. How many inches are there in 457 centimeters if one inch equals 2.54 centimeters? Round the result to the nearest hundredth.

Indicate each fraction in Exercises 35–40 as a repeating decimal.

35. $\frac{2}{3}$ 36. $\frac{5}{3}$

37. $\frac{1}{4}$ 38. $\frac{1}{8}$

39. $\frac{3}{7}$ 40. $\frac{8}{7}$

3.5
DECIMALS TO FRACTIONS

Every decimal number can be converted to a fraction or a mixed number. The procedures for converting terminating decimals and repeating decimals to fractions are explained in the following discussion and examples.

Recall that terminating decimals are decimals such as 0.25, 0.018, 0.324, and 0.00057.

Mixed decimal number forms may have a terminating decimal to the right of the decimal point, such as 4.35, 2.082, 5.0075, and 45.0625. The following steps describe how to convert a terminating decimal to a fraction or a mixed number.

To convert a terminating decimal to a fraction:

(a) Use the digits to the right of the decimal point as the numerator of the fraction.
(b) Count the number of digits to the right of the decimal point. Use this number as the exponent for a power of 10 (10^1, 10^2, 10^3, 10^4, etc.). Use this power of 10 as the denominator of the fraction.
(c) Reduce the fraction to lowest terms.
(d) If the original number was greater than one, add the whole number (the digits to the left of the decimal point in the original number) to the fraction to form a mixed number.

To check your answer: Divide the numerator by the denominator. You should obtain the terminating decimal in the original decimal as your quotient.

EXAMPLE 1 Convert 0.024 to a fraction.

Solution Using steps (a)–(c):

(a) The numerator of the fraction is 24.

(b) There are three digits to the right of the decimal point, so the denominator of the fraction is $10^3 = 1000$.

(c) $\dfrac{24}{1000} = \dfrac{24 \div 8}{1000 \div 8} = \dfrac{3}{125}$

Therefore, $0.024 = \dfrac{3}{125}$.

EXAMPLE 2 Convert 4.0625 to a mixed number.

Solution Using steps (a)–(d):

(a) The numerator of the fraction is 625.

(b) The denominator of the fraction is $10^4 = 10,000$

(c) $\dfrac{625}{10,000} = \dfrac{625 \div 625}{10,000 \div 625} = \dfrac{1}{16}$

(d) $4.0625 = 4\dfrac{1}{16}$

EXAMPLE 3 Convert 0.25, 1.037, 0.0045, and 9.5325 to fractions or mixed numbers.

Solutions

$0.25 \quad = \dfrac{25}{100} = \dfrac{1}{4}$

$1.037 \quad = 1\dfrac{37}{1000}$

$0.0045 = \dfrac{45}{10,000} = \dfrac{9}{2000}$

$9.5325 = 9\dfrac{5325}{10,000} = 9\dfrac{213}{400}$

Recall that decimals in which a single digit or a group of digits repeats endlessly are called repeating decimals. Examples of repeating decimals are $0.3\overline{3}$, $0.23\overline{23}$, $0.083\overline{3}$, and $0.142\overline{142}$. The following steps describe how to convert a repeating decimal to a fraction.

To convert a repeating decimal to a fraction:

(a) Write an equality, $X = $ repeating decimal.
(b) Count the number of repeating digits and write a power of 10 with this number as exponent (10^1, 10^2, 10^3, etc.).
(c) Multiply both sides of the equality in (a) by the power of 10 from (b).
(d) Subtract the equality in (a) from the result in (c); subtract left side from left side and subtract right side from right side.
(e) Write a fraction with the right side of the result from (d) as the numerator and the number from the left side of the result from (d) as the denomina-

tor (the number to the left of X after subtracting).

(f) Reduce the fraction to lowest terms.

To check your result: Divide the numerator by the denominator. You should obtain the original repeating decimal as your quotient.

EXAMPLE 4 Convert $0.6\overline{6}$ to a fraction.

Solution

(a) $X = 0.6\overline{6}$

(b) There is one repeating digit; $10^1 = 10$.

(c) $10X = 6.6\overline{6}$ (*Note:* The right side is written so that the repeating digit symbol is lined up with the same symbol on the right side of the equality in step (a).)

(d) $\begin{array}{r} 10X = 6.6\overline{6} \\ - \quad X = 0.6\overline{6} \\ \hline 9X = 6.00 \end{array}$ (*Note:* Ten X minus one X is equal to nine X.)

(e) $\dfrac{6 \longleftarrow \text{Right side of the result in (d)}}{9 \longleftarrow \text{Number to the left of } X \text{ in the result in (d)}}$

(f) $\dfrac{6}{9} = \dfrac{2}{3}$. Therefore, $0.6\overline{6} = \dfrac{2}{3}$.

EXAMPLE 5 Convert $0.27\overline{27}$ to a fraction.

Solution

(a) $X = 0.27\overline{27}$

(b) There are two repeating digits; $10^2 = 100$.

(c) $100X = 27.27\overline{27}$

(d) $\begin{array}{r} 100X = 27.27\overline{27} \\ - \quad X = 0.27\overline{27} \\ \hline 99X = 27.0000 \end{array}$

(e) $\dfrac{27 \longleftarrow \text{Right side of the result in (d)}}{99 \longleftarrow \text{Number to the left of } X \text{ in the result in (d)}}$

(f) $\dfrac{27}{99} = \dfrac{3}{11}$. Therefore, $0.27\overline{27} = \dfrac{3}{11}$.

EXAMPLE 6 Convert $0.083\overline{3}$ to a fraction.

Solution

(a) $X = 0.083\overline{3}$

(b) There is one repeating digit; $10^1 = 10$.

(c) $10X = 0.83\overline{3}$

(d) $\begin{array}{r} 10X = 0.833\overline{3} \\ - \quad X = 0.083\overline{3} \\ \hline 9X = 0.7500 \end{array}$

(e) $\dfrac{0.75}{9}$

(f) $\dfrac{0.75}{9} = \dfrac{0.75 \times 100}{9 \times 100} = \dfrac{75}{900} = \dfrac{1}{12}$

Therefore, $0.083\overline{3} = \dfrac{1}{12}$.

EXAMPLE 7 Convert $5.0425\overline{425}$ to a mixed number.

Solution The result will be a mixed number with a whole number part of 5. We consider the fraction part here.

(a) $X = 0.0425\overline{425}$

(b) There are three repeating digits; $10^3 = 1000$.

(c) $1000X = 42.5425\overline{425}$

(d) $\begin{array}{r} 1000X = 42.5425\overline{425} \\ - \quad X = 0.0425\overline{425} \\ \hline 999X = 42.5000000 \end{array}$

(e) $\dfrac{42.5}{999}$

(f) $\dfrac{42.5}{999} = \dfrac{425}{9990} = \dfrac{85}{1998}$

Therefore, $5.0425\overline{425} = 5\dfrac{85}{1998}$.

Review Exercises

1. Divide: $35.4 \div 0.72$

2. Divide: $0.0256 \div 4.5$

3. Convert $\dfrac{7}{20}$ to a decimal.

4. Convert $\frac{5}{16}$ to a decimal.

5. Convert $\frac{25}{27}$ to a decimal.

6. Convert $\frac{14}{45}$ to a decimal.

7. Multiply: 25.7×0.35

8. Multiply: 58.6×4.028

9. Add: $25.04 + 8.369 + 47.9$

10. Add: $0.3492 + 86.57 + 432.65$

11. Write the following fractions in order, from least to greatest: $\frac{5}{11}, \frac{2}{5}, \frac{14}{34}, \frac{3}{7}$.

12. Write the following fractions in order, from greatest to least: $\frac{11}{14}, \frac{4}{5}, \frac{13}{16}, \frac{7}{9}$.

EXERCISES 3.5

Write each decimal in Exercises 1–22 as a fraction or mixed number in lowest terms.

1. 0.45	2. 0.64
3. 0.025	4. 0.0125
5. 3.375	6. 4.825
7. 9.0055	8. 8.0072
9. 0.5625	10. 0.1875
11. 14.8025	12. 19.0484
13. $0.3\overline{3}$	14. $0.5\overline{5}$
15. $0.16\overline{6}$	16. $0.06\overline{6}$
17. $0.34\overline{34}$	18. $0.19\overline{19}$
19. $0.124\overline{124}$	20. $0.427\overline{427}$
21. $2.063\overline{63}$	22. $5.027\overline{27}$

23. The thickness of gauge 00 sheet steel is 0.34375 inch. What is the thickness of this gauge steel as a fraction of an inch?

24. If the thickness of a brass plate is 0.015625 inch, what is the thickness expressed as a fraction?

25. A length on a blueprint is given as $2.416\overline{6}$ feet. What is this length as a mixed number?

26. The length of a bolt is $2.46\overline{6}$ inches. What is the length of this bolt expressed as a mixed number?

3.6
PERCENT AS A DECIMAL AND FRACTION

Percent means "per hundred." The percent symbol, %, when used with a number, implies that the number is divided by one hundred. A percent can be written as a fraction and a decimal. (In some cases the decimal equivalent of a percent will, in fact, be a whole number, e.g., Examples 5 and 6.)

EXAMPLES

1 $5\% = \frac{5}{100} = 0.05$

2 $37\% = \frac{37}{100} = 0.37$

3 $125\% = \frac{125}{100} = 1.25$

4 $695\% = \frac{695}{100} = 6.95$

5 $100\% = \frac{100}{100} = \frac{1}{1} = 1$

6 $500\% = \frac{500}{100} = \frac{5}{1} = 5$

7 $\frac{1}{2}\% = \frac{\frac{1}{2}}{100} = \frac{0.5}{100} = 0.005$

8 $\frac{3}{4}\% = \frac{\frac{3}{4}}{100} = \frac{0.75}{100} = 0.0075$

9   $7.9\% = \dfrac{7.9}{100} = 0.079$

10 $123.65\% = \dfrac{123.65}{100} = 1.2365$

11 $0.8\% = \dfrac{0.8}{100} = 0.008$

12 $33\frac{1}{3}\% = \dfrac{33.3\overline{3}}{100} = 0.3\overline{3}$

Finding a given percent *of* a number involves *multiplying* the number by the decimal value of the given percent. For example, to find 7% *of* 45, you multiply 45 by 0.07 to get 3.15. Here are some other examples.

EXAMPLES

13 $15\% \text{ of } 123 = \dfrac{15}{100}(123) = 0.15(123)$
$$= 18.45$$

14 $39\% \text{ of } 65 = 0.39(65) = 25.35$

15 $107\% \text{ of } 345 = 1.07(345) = 369.15$

16 $0.75\% \text{ of } 83 = 0.0075(83) = 0.6225$

17 $0.04\% \text{ of } 1563 = 0.0004(1563) = 0.6252$

18 $254.33\% \text{ of } 69 = 2.5433(69) = 175.4877$

To "add on" a percent of a number to the number involves finding the given percent of the number and adding it to the number. Here are some examples.

EXAMPLE 19   Add 7% of 47 to 47.

Solution

$$7\% \text{ of } 47 = 0.07(47) = 3.29$$
$$47 + 3.29 = 50.29$$

This is the same as taking 107% of 47, which is $1.07(47) = 50.29$.

EXAMPLE 20   "Add on" 13% to 75.

Solution

$$13\% \text{ of } 75 = 0.13(75) = 9.75$$
$$75 + 9.75 = 84.75$$

This is the same as taking 113% of 75, which is $1.13(75) = 84.75$.

EXAMPLE 21   "Add on" 25% to $795.50.

Solution This is the same as 100% of 795.50 plus 25% of 795.50, which is 125% of 795.50.

$$125\% \text{ of } 795.50 = 1.25(795.50) = 994.375,$$

which is $994.38.

EXAMPLE 22   "Add on" 135% to 224.

Solution This is the same as 100% of 224 *plus* 135% of 224, which is 235% of 224 or $2.35(224) = 526.4$.

To "discount" a percent of a number from the number involves subtracting the given percent of the number from the number. This yields the same result as taking (100 less the given percent) of the number.

EXAMPLE 23   Discount 75 by 12%.

Solution

$$75 - (12\% \text{ of } 75) = 75 - (0.12 \times 75)$$
$$= 75 - (9) = 66$$

This is the same as taking 88% of 75, which is $0.88 \times 75 = 66$, where $88\% = 100\% - 12\%$.

EXAMPLE 24   Discount 695 by 47%.

Solution

$$695 - (0.47 \times 695) = 695 - 326.65 = 368.35$$

This is the same as taking 53% of 695, which is $0.53(695) = 368.35$, where $53\% = 100\% - 47\%$.

EXAMPLE 25 What is the sale price of a $75 item discounted 25%?

Solution

$$100\% - 25\% = 75\%$$
$$75\% \text{ of } 75 = 0.75 \times 75 = 56.25$$

Therefore, the sale price is $56.25.

Review Exercises

Perform the indicated operations in Exercises 1–8.

1. $25.3 + 1.75 + 0.63$
2. $28.04 + 9.567 + 0.793$
3. $19.0639 - 5.8769$
4. $98.03542 - 71.96653$
5. $(1.4)(0.59)$
6. $(0.0056)(1.7)$
7. $(42.3)(7.0063)$
8. $(5.014)(0.00069)$

In Exercises 9–12 calculate to the nearest hundredth.

9. $42.3 \div 5.94$
10. $5.94 \div 42.3$
11. $(5.45)(0.08) \div 2.06$
12. $2.06 \div (5.45)(0.08)$

11. $66\frac{2}{3}\%$
12. 39.3%
13. 128.8%
14. $133\frac{1}{3}\%$

Do the indicated computations in Exercises 15–38.

15. 14% of 104
16. 12% of 119
17. 35% of 68
18. 49% of 49
19. 112% of 8
20. 235% of 6
21. 0.4% of 69
22. 0.8% of 84
23. 0.08% of 25
24. 0.09% of 32
25. 28.3% of 9
26. 14.4% of 8
27. 125.03% of 8
28. 254.04% of 5
29. Add 15% of 36 to 36.
30. Add 19% of 22 to 22.
31. Add 4% of 17.32 to 17.32.
32. Add 5% of 1.95 to 1.95.
33. Add 128% of 12 to 12.
34. Add 234% of 15 to 15.
35. Discount 25 by 13%.
36. Discount 36 by 12%.
37. Discount 254 by 83%.
38. Discount 500 by 77%.
39. What is the sale price of a $49.95 calculator discounted 25%?
40. What is the sale price of a $55 coat discounted $33\frac{1}{3}\%$?

EXERCISES
3.6

In Exercises 1–14 write each percent as a fraction having denominator 100 and as a decimal.

1. 44%
2. 29%
3. 241%
4. 314%
5. 1142%
6. 2413%
7. $\frac{1}{4}\%$
8. $\frac{3}{8}\%$
9. 6.4%
10. 5.8%

3.7
APPLICATIONS OF PERCENT (I)

Computations involving percent are done routinely when figuring various taxes. The taxes affecting almost every American include sales tax, social security tax, and income tax. Most people are also affected by gasoline tax, property tax, taxes on various commodities such as tobacco and liquor, and excise tax. Examples 1–8 illustrate how percent is involved in calculating taxes.

EXAMPLE 1   The sales tax on retail purchases of goods in a certain state is five percent. How much sales tax must be paid on the purchase of a piece of furniture which sells for $499.50?

Solution   The sales tax is computed by finding 5% of the purchase price. Since 5% = 0.05, (0.05)(499.50) = 24.975, then 24.975 rounded off to the nearest hundredth is 24.98, and the amount of tax is $24.98.

EXAMPLE 2   Mrs. Johnson purchased three items costing $27.95, $48.50, and $13.54, respectively. The rate of sales tax on these purchases was four percent. How much sales tax did she have to pay?

Solution

$$0.04(27.95 + 48.50 + 13.54) = 0.04(89.99)$$
$$= 3.5996$$

Therefore, the sales tax was $3.60.

EXAMPLE 3   The amount of social security tax paid by self-employed U.S. citizens in 1992 was 15.3 percent of all income to $55,500. If Margaret Wilson earned $22,542.74 from self-employment that year, how much social security tax did she pay?

Solution

$$15.3\% = 0.153$$
$$0.153\,(22,542.74) = 3,449.0392$$

Therefore, she paid $3,449.04.

EXAMPLE 4   A single taxpayer in the United States in 1992 paid a 15 percent income tax on all taxable income up to and including $20,350. How much income tax did a single college student earning $4,055 taxable income have to pay?

Solution

$$0.15\,(4,055) = 608.25$$

Therefore, the income tax was $608.25.

EXAMPLE 5   The income tax for single taxpayers in 1992 was 15 percent on the first $20,350; 28 percent on the next $29,050; and 31 percent on all taxable income over $49,400. Sam Jackson received $55,482 in taxable income that year. How much income tax did he pay?

Solution

$$0.15\,(20,350) = 3,052.50$$
$$0.28\,(29,050) = 8,134.00$$
$$0.31\,(55,482 - 49,400) =$$
$$0.31\,(6,082) = 1,885.42$$

Therefore, his income tax was $3,052.50 + $8,134.00 + $1,885.42; which is a total of $13,071.92.

EXAMPLE 6 The Federal U.S. Excise Tax on gasoline in 1992 was 14.1 cents per gallon. Jennifer Swartz bought 13.4 gallons of gasoline for her automobile. How much Federal Excise Tax did she pay?

Solution

$$14.1\ (13.4)\ =\ 188.94\ \text{(cents)}$$

Therefore, she paid $1.89 Federal Excise Tax.

EXAMPLE 7 Federal beer tax in 1992 was $18 per barrel; and federal cigarette tax was $21 per 1,000 cigarettes. Harold Party bought one-half barrel of beer and 80 cigarettes. How much federal tax did he pay on the purchases?

Solution

$$18\ (1/2)\ =\ 9.00$$
$$21\ (80/1000)\ =\ 21\ (0.080)$$
$$=\ 1.68$$

Therefore, he paid $9.00 + $1.68, which totals $10.68.

EXAMPLE 8 Property taxes are often computed in mills per unit of value. A mill is 0.001 of a dollar, which makes it 0.1% of a dollar. The number of mills divided by 10 produces the percent of tax rate. A typical property tax rate is 51.9 mills per $1 of assessed valuation, which is 5.19% of the assessed valuation. If a home is assessed at $69,100 and the tax rate is 5.19%, what is the total amount of property tax for the year?

Solution

$$5.19\% = 0.0519$$
$$0.0519(69100) = 3586.29$$

The tax is $3586.29.

Review Exercises

1.   Add: $4.035 + 0.0459$
2.   Subtract: $40.35 - 0.0459$
3.   Multiply: $(4.035)(0.0459)$
4.   Divide: $4.035 \div 0.0459$
5.   What is the value of $\left(\dfrac{3}{4}\right)^2$?
6.   What is the value of $(0.03)^3$?
7.   Write $\dfrac{5}{4}\%$ as a decimal.
8.   Write 3.24% as a decimal.
9.   Find 13% of 475.
10. Find 12% of 0.0034.
11. What is the result when 27% of 482 is added to 482?
12. What is the sale price on a $110.50 bicycle which has been discounted 20%?

EXERCISES
3.7

Use the examples in Section 3.7 as a guide in answering Exercises 1–11.

1. How much sales tax at 5% is charged on a $72.45 item?
2. A man purchased a car for $493. His sales tax was at the rate of 4%. How much sales tax did he pay?
3. Donald Spudlik traded in his old car for a new one. The price of the new car was $14,308.80. He was allowed $3850 on his old car. He paid sales tax on the difference at the rate of 5%. How much sales tax did he pay?

4. Mrs. Holloway purchased seven items in a department store, priced $2.98, $7.65, $4.99, $24.50, $117.95, $0.79, and $2.09, respectively. Sales tax was charged on the total purchase at 6%. How much sales tax did she pay?

5. How much social security tax did Harry Jones pay in 1992 if his total taxable income was from self-employment and amounted to $31,048.96? (See Example 3 in Section 3.7.)

6. A single taxpayer in the U.S. had a taxable income of $14,249.50 in 1992. How much income tax did she pay?

7. How much income tax was paid by single taxpayer Henry Todd in 1992 if his taxable income was $62,500?

8. Gini Flowers bought 2,450 gallons of gasoline during 1992. How much Federal Excise Tax did she pay on her year's gasoline purchases?

9. Shirley Smoker bought 7,300 cigarettes during 1992. How much federal tax did she pay on those purchases?

10. How much property tax is due for a year on a home valued at $32,450 if the tax rate is 5.17%?

11. A family's home has an assessed valuation of $41,750. The property tax mill rate is 49.3 mills per dollar of assessed valuation. What is the family's tax bill for a year?

3.8
APPLICATIONS OF PERCENT (II)

Many computations involving percent are known as "add ons." An "add on" is made when a given percent of a number is added to the number.

EXAMPLE 1 How much must be paid if the amount of a purchase is $24.33 and the sales tax is 5%?

Solution The amount of sales tax is 5% of 24.33, which is 1.2165, or $1.22. Adding $1.22 to the purchase price $24.33, you get $25.55, which is the amount to be paid.

EXAMPLE 2 A worker receives a 9 percent raise. Before the raise his hourly rate of pay was $6.05. What is his new hourly rate?

Solution The raise per hour is 9% of $6.05, which is $0.5445, or $0.54. His new hourly rate is $6.05 plus $0.54, or $6.59.

EXAMPLE 3 A distributor adds 95% to her cost in order to determine the selling price for an item. If the item cost her $2.48, what is her selling price?

Solution Two dollars and forty-eight cents plus 95% of $2.48 is $2.48 plus $2.36, which is $4.84.

EXAMPLE 4 A lender charges add-on interest at the rate of 8% per year. How much will a borrower repay if $500 is borrowed for one year?

Solution Five hundred dollars plus 8% of $500 is $500 plus $40, which is $540.

EXAMPLE 5 How much will a borrower repay on $4000 borrowed for 3 years at 7.5% add-on interest per year?

Solution The interest per year is 7.5% of $4000, or $300. For three years that's three times $300, or $900. The borrower will repay $4000 plus $900, or $4900.

EXAMPLE 6 A businessman prices his merchandise at a 30% markup. What is his selling price for an item costing $4.45?

Solution Four dollars and forty-five cents plus 30% of $4.45 is $4.45 plus $1.34, which is $5.79.

EXAMPLE 7 A broker adds 7% onto the desired price for a home so that the seller can, in turn, pay her 6% commission. If a seller wants $41,500 for a home, at what price will the broker offer it?

Solution The add-on must be 7% of $41,500, which is $2905. Therefore, the broker will charge $41,500 plus $2905, or $44,405.

EXAMPLE 8 What will be the broker's commission on the sale described in Example 7? How much will the seller receive?

Solution Her commission will be 6% of $44,405, or $2664.30. The seller will receive $44,405 less $2664.30, which is $41,740.70.

As was shown in Section 3.6, add-ons can also be calculated by multiplying the original number by 100% *plus* the add-on percent.

EXAMPLE 9 How much must be paid if the amount of purchase is $1.79 and the sales tax is 6%?

Solution The sales tax add-on is 6%. The amount paid is 106% of $1.79, which is 1.8974, or $1.90.

EXAMPLE 10 An employee receives an 8½% raise. Her annual salary was $31,500. What is her new annual salary?

Solution The add-on rate is 8.5 percent. Therefore, the new annual salary is 108.5% of $31,500, which is $34,177.50.

EXAMPLE 11 A distributor adds $66\frac{2}{3}$% to his cost in order to determine his selling price. What will he sell an item for if it cost him $44.75?

Solution

$$166\frac{2}{3}\% \text{ of } \$44.75 = 1.67 (44.75) = \$74.73$$

(*Note:* If $166\frac{2}{3}$% is written as 1.66667, the selling price is $74.58. Therefore, it is to the distributor's advantage to round off $66\frac{2}{3}$% to 67%.)

Review Exercises

In Exercises 1–8 change the percents to equivalent decimals.

1. 8%
2. 78%
3. 236%
4. 0.082%
5. 21.6%
6. 0.0064%
7. $8\frac{1}{3}$%
8. $41\frac{2}{3}$%
9. Find 2.6% of 230.
10. What is 126% of 300?
11. What is 0.52% of 1020?
12. What is $3\frac{3}{4}$% of 75?

EXERCISES 3.8

1. A customer purchased $44.83 worth of goods and had to pay 5% sales tax. What was the total amount paid by the customer?
2. A woman bought a refrigerator for $379.95

and paid sales tax on the purchase at a 4% rate. What was her total payment?

3. Harold Johnson was making $5.88 per hour when he received a 6% raise in pay. What was his new hourly rate of pay?

4. A school principal received an 11 percent raise from her $56,500 annual salary. What was her new salary?

5. A book distributor prices the materials in his catalog at 93% more than he pays for them. What is his catalog price for a book costing him $3.44?

6. A music store prices instruments at 55% more than the wholesale price. What is the retail price on an organ which cost the store owner $2055?

7. Banks used to charge "add-on" interest for automobile secured loans. If Carla Stone bought a new car and obtained a $9300 loan at 6% add-on interest for 3 years, how much interest did she pay?

8. Hank Greener obtained an $11,178 automobile loan from the same bank as Carla Stone. (See Exercise 7.) How much interest did he pay for a two-and-one-half-year loan?

9. A store owner prices her merchandise at a 33% markup. What price does she charge for a shaver costing her $28.50?

10. An implement dealer receives 46% more than he pays for garden tractors. What does he charge for a garden tractor costing him $1055?

11. A real estate broker sells property for $6\frac{1}{2}$% more than the owners expect, so that his 6% commission is covered. A homeowner wants $85,950 for his home. At what price will the broker sell the home?

12. What will the broker's selling price be on a home for which the owner wants $112,700? (See Exercise 11.)

3.9
APPLICATIONS OF PERCENT (III)

> To convert a decimal or a whole number to a percent you simply multiply by 100, which means moving the decimal point two places to the right, and affixing the percent symbol.

EXAMPLES

1 $0.35 = 35\%$

2 $2.3 = 230\%$

3 $0.075 = 7.5\%$ or $7\frac{1}{2}\%$

4 $0.25\frac{1}{2} = 0.255 = 25.5\% = 25\frac{1}{2}\%$

5 $4 = 400\%$

6 $0.0002 = 0.02\%$

7 $7\frac{3}{4} = 7.75 = 775\%$

> To convert a fraction to a percent, first convert the fraction to a decimal, then change the decimal to a percent.

EXAMPLES

8 $\frac{1}{4} = 0.25 = 25\%$

9 $\frac{4}{5} = 0.8 = 80\%$

10 $\frac{2}{3} = 0.66\overline{6} = 66.\overline{6}\%$ (Note that three 6's are needed.)

11 $\frac{3}{8} = 0.375 = 37.5\%$

12 $\frac{3}{2} = 1.5 = 150\%$

13 $4\frac{3}{5} = 4.6 = 460\%$

14 $\frac{4}{3} = 1.33\overline{3} = 133.\overline{3}\%$ (Note that three 3's are needed.)

Discounts often involve percent. Discounting a given amount by a specified percent involves calculating the percent of the amount and then subtracting the result from the original amount.

EXAMPLE 15   Discount 743 by 18%.

Solution

$$18\% \text{ of } 743 = 133.74$$
$$743 - 133.74 = 609.26$$

EXAMPLE 16   A merchant advertised a 25% discount on an item priced at $79.50. What was the discounted price?

Solution

$$25\% \text{ of } \$79.50 = \$19.88$$
$$\$79.50 - \$19.88 = \$59.62$$

EXAMPLE 17   A man was offered "one-third off" if he bought an $895 item immediately. How much did he pay for the item if he bought it immediately?

Solution   Since $\frac{1}{3} = 0.33\overline{3} = 33.\overline{3}\% = 33\frac{1}{3}\%$, he would pay $895 - 0.33(\$895)$, where $\frac{1}{3}$ is rounded off to 0.33. The answer is $599.65.

As was shown in Section 3.6, discounted amounts can be calculated directly by taking 100% *minus* the discount percent of the original amount.

EXAMPLE 18   Discount $846 by 15%.

Solution $100\% - 15\% = 85\%$
$$85\% \text{ of } \$846 = \$719.10$$

EXAMPLE 19   A merchant allowed a 7% discount on an item priced at $112.50 if the customer paid cash. What was the discounted cash price of the item?

Solution $100\% - 7\% = 93\%$
$$93\% \text{ of } \$112.50 = \$104.625$$

The discount cash price was $104.63.

EXAMPLE 20   A set of tools was advertised at 45% off the original price of $218.60. What was the advertised price?

Solution $100\% - 45\% = 55\%$
$$55\% \text{ of } \$218.60 = \$120.23$$

EXAMPLE 21   Discount 444 by 23%, and discount the result by 20%.

Solution

$$444 - 0.23(444) = 341.88$$
$$341.88 - 0.20(341.88) = 273.504$$

Thus, the answer is 273.504.

A second solution: $0.80[0.77(444)] = 273.504$

Review Exercises

Perform the indicated operations in Exercises 1–12.

1. $\frac{3}{4} + \frac{7}{8}$ 2. $\frac{7}{8} - \frac{3}{4}$

3. $\left(\frac{3}{4}\right)^3$ 4. $\frac{3}{4} \div \frac{7}{8}$

5. $(0.032)^2$ 6. $\frac{(0.14)^2}{1.3}$

7. $0.89 + 1.78 + 0.067 + 75.3$

8. $9.0005 - 7.86596$

9. 38% of 453   10. 348% of 600

11. $\frac{7}{8}$% of 500   12. 1.08% of 250

EXERCISES
3.9

Convert the numbers in Exercises 1–18 to percents.

1. 0.45   2. 0.56

3. 0.055   4. 0.065

5. $4\frac{3}{4}$ 6. $16\frac{2}{5}$

7. 0.0005   8. 0.0026

9. 3.8   10. 13.75

11. $\frac{2}{5}$ 12. $\frac{3}{5}$

13. $\frac{1}{6}$ 14. $\frac{5}{6}$

15. $\frac{7}{3}$ 16. $\frac{8}{3}$

17. $5\frac{3}{8}$ 18. $7\frac{5}{8}$

19. Discount 749 by 33%.

20. Discount 497 by 44%.

21. Discount 250 by $\frac{3}{4}$%.

22. Discount 390 by $\frac{5}{8}$%.

23. Discount 1000 by 0.014%.

24. Discount 500 by 0.104%.

25. A television set was advertised at 28% off the original price of $489. What was the advertised price?

26. A sofa was advertised at 19% off the original price of $395. What was the advertised price?

27. How much would you pay for an $88.88 item marked "one-fourth off"?

28. A bicycle was marked "one-fifth off." Its original price was $129.95. What was the sale price?

29. A dress originally marked 20% off was then sold at a discount of 15%. If the retail price of the dress was $89, what was the final sale price?

30. A $455 stove was discounted 10%, and then the sale price was discounted an additional 15%. What was the final selling price?

31. A banquet manager quoted a meal price at $7.45 plus 15% gratuity and 4% sales tax. Which is the correct total price, $8.92 or $8.87? Why?

32. A meal costs a total of $7.15 plus 15% and 4% for gratuity and sales tax, respectively. What is the total cost of the meal?

33. A man was earning $6.85 per hour. He was given a raise of 7%, and a year later an additional raise of 5%. How much per hour was he then paid?

34. A woman took a job at an annual salary of $11,500. In six months she was given a 5% raise in salary. A year later she received an 8% increase. What was her annual salary after the two raises?

35. Harry Phillips was receiving $8.05 per hour before he was given a 3% cost of living increase. Later he was assessed a 2% cost of living decrease. What was his final amount of pay per hour?

36. During three years of working at the same job, Marion Sorenson received pay raises of 4% and 7%, respectively, and then a decrease of 3%. If her beginning salary was $9000 per year, what was her salary after the three years?

3.10
USING DECIMALS TO CONVERT UNITS

Decimals can be used when converting units. Section 2.8 contains the conversion factors needed here.

EXAMPLE 1   How many feet are $8\frac{5}{16}$ inches?

Solution   The mixed number $8\frac{5}{16} = 8.3125$. The conversion factor is $\frac{1 \text{ ft}}{12 \text{ in.}}$ (see Section 2.8).

$$8.3125 \text{ in.} \left(\frac{1 \text{ ft}}{12 \text{ in.}} \right) = \frac{8.3125}{12} \text{ ft} = 0.6927 \text{ ft}$$

(to four decimal places)

EXAMPLE 2   How many feet equal 19.3 inches?

Solution

$$19.3 \text{ in.} \left(\frac{1 \text{ ft}}{12 \text{ in.}} \right) = \frac{19.3}{12} \text{ ft} = 1.608 \text{ ft}$$

(to three decimal places)

EXAMPLE 3   Convert 4.7 feet to inches.

Solution

$$4.7 \text{ ft} \left(\frac{12 \text{ in.}}{1 \text{ ft}} \right) = \frac{(4.7)(12)}{1} \text{ in.} = 56.4 \text{ in.}$$

EXAMPLE 4   Convert 29.35 feet to yards.

Solution

$$29.35 \text{ ft} \left(\frac{1 \text{ yd}}{3 \text{ ft}} \right) = \frac{29.35}{3} \text{ yd} = 9.78 \text{ yd}$$

(to the nearest hundredth)

EXAMPLE 5   Convert 8.9 yards to feet.

Solution

$$8.9 \text{ yd} \left(\frac{3 \text{ ft}}{1 \text{ yd}} \right) = (8.9)(3) \text{ ft} = 26.7 \text{ ft}$$

EXAMPLE 6   Convert 8 feet 5 inches to feet (to the nearest hundredth of a foot).

Solution

$$5 \text{ in.} \left(\frac{1 \text{ ft}}{12 \text{ in.}} \right) = \frac{5}{12} \text{ ft} = 0.42 \text{ ft}$$

Thus, the answer is 8.42 ft.

EXAMPLE 7   Convert $8\frac{5}{16}$ yards to feet and inches.

Solution

$$8\frac{5}{16} = 8.3125$$

$$8.3125 \text{ yd} \left(\frac{3 \text{ ft}}{1 \text{ yd}} \right) = 24.9375 \text{ ft}$$

$$0.9375 \text{ ft} \left(\frac{12 \text{ in.}}{1 \text{ ft}} \right) = 11.25 \text{ in.}$$

24 ft + 11.25 in. = 24 ft 11.25 in.

EXAMPLE 8   Convert 17 yards 2 feet 9 inches to feet.

Solution

$$17 \text{ yd} \left(\frac{3 \text{ ft}}{1 \text{ yd}} \right) = 51 \text{ ft}$$

$$9 \text{ in.} \left(\frac{1 \text{ ft}}{12 \text{ in.}} \right) = \frac{9}{12} \text{ ft} = 0.75 \text{ ft}$$

51 ft + 2 ft + 0.75 ft = 53.75 ft

EXAMPLE 9   Convert 84.7 feet to yards, feet, and inches.

Solution

$$84.7 \text{ ft} \left(\frac{1 \text{ yd}}{3 \text{ ft}} \right) = \frac{84.7}{3} \text{ yd} = 28.233 \text{ yd}$$

(to three decimal places)

28.233 yd = 28 yd + 0.233 yd

$$0.233 \text{ yd} \left(\frac{3 \text{ ft}}{1 \text{ yd}} \right) = 0.699 \text{ ft}$$

$$0.699 \text{ ft} \left(\frac{12 \text{ in.}}{\text{ft}} \right) = 8.388 \text{ in.}$$

Therefore, the answer is 28 yd 0 ft 8.388 in.

EXAMPLE 10   Convert 8 yards 2.45 feet to feet and inches.

Solution

$$8 \text{ yd} \left(\frac{3 \text{ ft}}{1 \text{ yd}} \right) = 24 \text{ ft}$$

$$0.45 \text{ ft} \left(\frac{12 \text{ in.}}{1 \text{ ft}} \right) = 5.4 \text{ in.}$$

Therefore, 24 ft + 2 ft + 5.4 in. = 26 ft 5.4 in.

EXAMPLE 11   Convert 0.2 inch to the nearest $\frac{1}{8}$ inch.

Solution   We know that 1 in. = $\frac{8}{8}$ in. Thus, 0.2 in. = (0.2)(1 in.), and

$$0.2 \text{ in.} = 0.2 \left(\frac{8}{8} \right) \text{ in.} = \frac{1.6}{8} \text{ in.}$$

Since 1.6 to the nearest unit is 2, the answer is 0.2 in. = $\frac{2}{8}$ in., to the nearest $\frac{1}{8}$ in.

EXAMPLE 12   Convert 4.4 inches to the nearest $\frac{1}{16}$ inch.

Solution   First we find that $0.4 \left(\frac{16}{16} \right) = \frac{6.4}{16}$, which is $\frac{6}{16}$, to the nearest $\frac{1}{16}$. Then the answer is 4.4 in. = $4\frac{6}{16}$ in., to the nearest $\frac{1}{16}$ in.

EXAMPLE 13   Convert 3 yards 2 feet $5\frac{5}{8}$ inches to the nearest 0.001 inch.

Solution   The answer can be expressed in inches, or in yards, feet, and inches. The latter is simpler because you need only convert $\frac{5}{8}$ in.

to decimal form. Thus, $\frac{5}{8}$ in. = 0.625 in., so the answer is 3 yd 2 ft 5.625 in.

If the answer is to be expressed in inches, you get

$$3 \text{ yd} \left(\frac{36 \text{ in.}}{1 \text{ yd}} \right) + 2 \text{ ft} \left(\frac{12 \text{ in.}}{1 \text{ ft}} \right) + 5.625 \text{ in.,}$$

which is

108 in. + 24 in. + 5.625 in.,

or 137.625 in.

Review Exercises

Perform the indicated operations in Exercises 1–8.

1. 17(756)
2. 70.89 + 505.763
3. $\frac{2}{3} - \frac{1}{16}$
4. $3\frac{3}{4} + 5\frac{2}{15}$
5. $4\frac{7}{8} \div 3\frac{1}{4}$
6. 7.003 − 5.69485
7. (7.39)(0.0048)
8. 7.39 ÷ 0.0048 (to nearest hundredth)
9. "Add-on" 0.045% to 74.83.
10. "Discount" 0.03% from 5.85.
11. Convert $8\frac{4}{16}$ feet to inches.
12. Convert $8\frac{5}{8}$ yards to feet and inches.

EXERCISES
3.10

1. How many feet (to the nearest hundredth) is $8\frac{5}{8}$ inches?

2. How many feet (to the nearest tenth) is $6\frac{7}{8}$ inches?

3. 14.4 in. = _____ ft

4. 19.04 in. = _____ ft

5. 9.45 ft = _____ in.

6. 10.04 ft = _____ in.

7. 45.9 ft = _____ yd

8.   70.08 ft = _____ yd

9.   13.06 yd = _____ ft

10.   15.54 yd = _____ ft

Give answers to Exercises 11–24 to the nearest hundredth.

11.   7 ft 7 in. = _____ ft

12.   3 ft 11 in. = _____ ft

13.   83 yd 23 ft = _____ yd

14.   74 yd 13 ft = _____ yd

15.   0.3 yd 2.3 ft = _____ yd

16.   4.3 yd 0.6 ft = _____ yd

17. $5\frac{3}{8}$ yd = _____ ft _____ in.

18. $3\frac{5}{7}$ yd = _____ ft _____ in.

19.   16 yd 2 ft 8 in. = _____ ft

20.   12 yd 1 ft 7 in. = _____ ft

21.   100 ft = _____ yd _____ ft _____ in.

22.   200 ft = _____ yd _____ ft _____ in.

23.   7 yd 3.7 ft = _____ ft _____ in.

24.   5 yd 1.09 ft = _____ ft _____ in.

25.   Convert 0.5 in. to the nearest $\frac{1}{8}$ in.

26.   Convert 0.7 in. to the nearest $\frac{1}{8}$ in.

27.   Convert 3.8 in. to the nearest $\frac{1}{16}$ in.

28.   Convert 13.05 in. to the nearest $\frac{1}{16}$ in.

29.   Convert 2 yd 2 ft $2\frac{1}{8}$ in. to inches (to the nearest hundredth).

30.   Convert 5 yd 7 ft $1\frac{1}{4}$ in. to feet (to the nearest hundredth).

EXTRA PROBLEMS
CHAPTER 3

1.   Mary spent three-fourths of $25, then she spent seven-eighths of what remained. How much did she spend?

2.   Craig gave away three-fourths of $25, then he gave away three-sevenths of what remained. How much did he give away?

3.   A plumber bought three garbage disposal units for $150 each, and sold them for $225 each. How much profit did he make on the transactions?

4.   A publishing company invested $13 per copy to print 2,000 copies of a book, then sold the books for $27 each. What was the gross profit on the 2,000 copies?

5.   A carpenter sold 17 windows, installed, for $55 each. He had paid $19.50 for each window. What was his gross profit (without considering the cost of labor to install or delivery cost)?

6.   Harry subscribed to a computer "on-line" service, paying $19.95 per month PLUS $2.95 per hour after the first five hours. Every time he goes "on-line" his telephone charge is 20 cents. During January he was on-line for a total of 600 minutes, and made 60 calls. Not counting taxes, what was Harry's January cost for the on-line service?

7.   A certain on-line service for e-mail is FREE. However, each time a user sends a message, or set of messages, he or she must pay 35 cents for the long distance telephone call. Ignoring taxes, what will it cost Nancy to send 48 separate e-mail messages?

8.   A computer software package costs $79 plus 5 percent sales tax; and a mail-in rebate form refunds the buyer $10. It costs 34 cents for the stamp and envelope. What is the total net cost of the software package?

9.   A 900-number phone call adds $2.95 per minute to your telephone bill. Your average regular monthly telephone bill is $44.50 without taxes. During a given month you make two 900-number calls of five minutes each. What is your expected total telephone bill for that month (ignoring taxes)?

10.   An item of computer software is on sale in a catalog for $39.95. When a person orders it, that person must add 6 percent for sales tax and use a 32-cent stamp on the order. What will be the total cost for the software?

11.   An automobile repair shop charges each customer the retail cost of parts used plus $55

per hour for labor, plus 6 percent of the total for sales tax. Harriet's car needed $28.70 worth of parts and the mechanic took one and one-half hours for the repairs. What was her total cost?

12. An automobile repair shop charges each customer the retail cost for parts used plus $52.50 per hour for labor plus five percent sales tax on the total. Jerry's car needed $94 worth of parts that took two hours to replace. What was his total cost?

13. Sandra needed $500 cash. She used two credit cards in an ATM (automated teller machine) and drew $250 from each account the first day of the monthly billing period for each. One of her cards had been issued by a bank that charged a three percent fee for each withdrawal plus 21/60 of a percent for each day of the 30-day billing period. The other card had been issued by her local credit union that charged no withdrawal fees and only 15/360 of a percent for each day of the 30-day billing period. How much did she owe on each account after 30 days for the $250 withdrawal?

14. Dennis needed $600 cash. He took his two credit cards to an ATM and drew $300 from each account. His bank-issued card required an ATM fee of two and one-half percent of the amount withdrawn plus interest at 19/360 percent per day. His credit union issued card required only interest payment at the rate of 14.5/360 percent per day. There were 21 days left in the billing period for each card. How much did he owe on each account at the end of the billing period?

15. A clothing store advertised a 25 percent discount on all items for a week. During the last two days of the week the store offered an additional 15 percent off. How much did Jean pay during the last day of the sale for a blouse that was originally priced at $85?

16. A furniture store advertised a three-day sale on certain items limited to those in stock, 10 percent off the first day, an additional 10 percent off the second day, and an additional 10 percent off from the second day sale price for purchases made the third day. Mr. and Mrs. Johnson bought a chair during the third day of the sale that had been originally priced at $288. How much did they pay for the chair (ignoring taxes)?

17. Sally invested $300 for three years, earning six percent compound interest each year on her balance. How much was in her account at the end of the third year?

18. Steve borrowed $1,000 for one year, with interest accruing quarterly on his balance at the rate of thirteen percent per annum. How much did he owe at the end of the year?

19. The Rogers' heating bill was computed at 22 cents per therm for 75 therms and $.2241 per therm for the remaining 106 therms; plus a PGA commodity charge of $.4183 per therm for 75 therms and $.5243 per therm for the remaining 106 therms. Extra customer charges were added in the amounts $2.06 and $2.94. What was the total bill?

20. The Rogers' electricity bill was for 327 kwh at $.0667 per kwh, plus 468 kwh at $.06514 per kwh; plus customer charges of $1.44 and $2.94. What was the total bill?

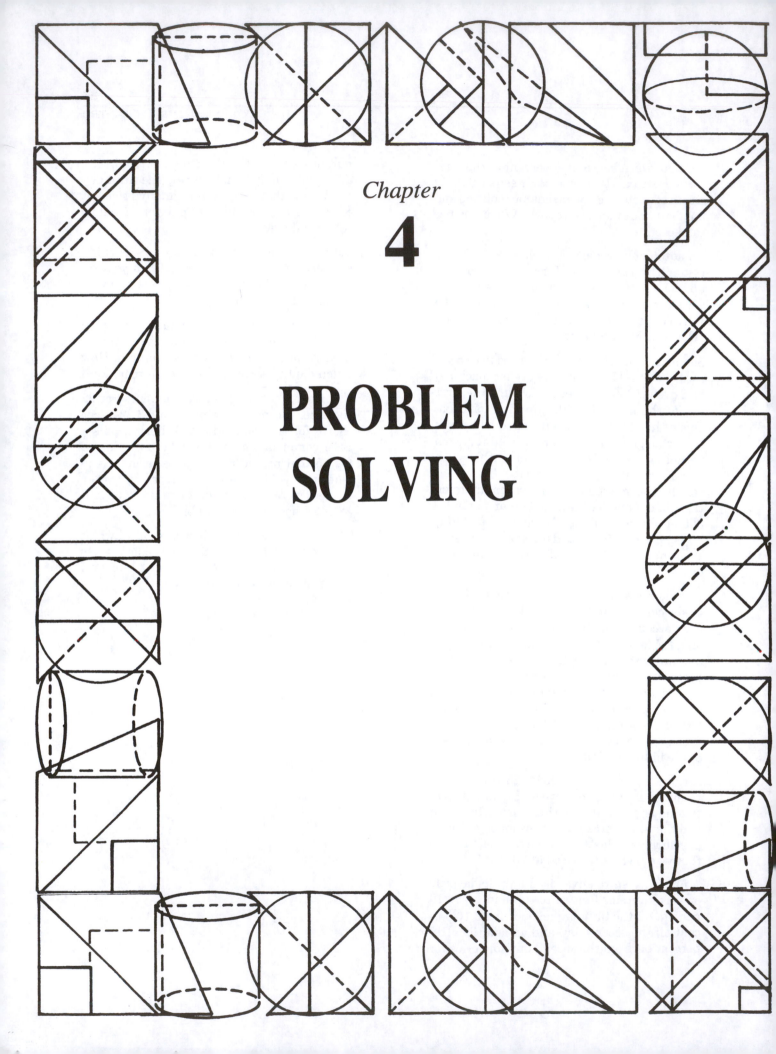

Chapter

4

PROBLEM SOLVING

4.1
INTRODUCTION TO PROBLEM SOLVING

Successfully solving a problem is truly intrinsically rewarding. Your self-image is enhanced, you feel a sense of accomplishment, you feel self-satisfied, and you have added to your knowledge base.

Problem solving is not a spectator sport. You get good at it only by solving problems, which is time consuming. Tackling the challenge of a "tough" problem often means dedicating hours, sometimes days, to concentrated mental activity without interruption. Yet it's fun, and the time flies!

Most people can learn to become better problem solvers by using certain techniques, learning to concentrate fully during mental processes, and by getting lots of practice. The problem-solving sections of this book are designed to help you improve your ability to solve problems.

The sets of materials for each problem-solving strategy incorporate the prerequisite capabilities (knowledge and skills) that are necessary if a person is to make good use of the tactic. A sample problem that can be solved using each tactic is explained in detail, and a complete step-by-step solution is shown. Other similar problems are included for you to solve. Answers for these problems are provided at the end of the section.

Study each sample problem thoroughly, making sure that you understand the tactic(s) used. Then try the extra problems, which are listed in approximate order of increasing difficulty. Refer to answers only when you have a solution, or are completely baffled. If your solution does not agree with the answer given, **forget your answer** and try again.

4.2
ESTIMATING AND CHECKING

A tactic that is often successful for many people when solving problems—where other tactics do not seem appropriate—is the process of guessing an answer, then checking to see if it is correct.

EXAMPLE 1 Sue threw three darts at a board numbered like below. She hit three different

1	2	6
8	5	3
7	4	9

numbers whose sum was 23. Which three numbers did she hit?

Solution It is quite easy to make reasonable guesses and check their sums, thinking along lines like, "To get 23 I must use some larger numbers, so I'll try 7 + 8 + 9, which gives me 24. So I'll try 6 + 7 + 8, which gives me 21, which is too few. Next I'll try 6 + 7 + 9, which is 22. So I'll try 6 + 8 + 9, which gives me 23, the correct answer!"

Practice at this type of estimating makes it possible to come up with correct answers with fewer tries.

EXERCISES
4.2

1. Four numbers, two different even numbers and two different odd numbers, add up to 34. Find one such set of numbers.

2. Consider the counting numbers, 1 to 1,000. Which one of these numbers, when multiplied by itself, is closest to 1992?

3. Within the eight nines, insert *plus signs*

so that the sum will be 1,125.

9   9   9   9   9   9   9   9

Solutions   Do not look at these answers until you have tried to solve all the problems. If any one of your answers is incorrect, **forget the answer given here** and try again to solve the problem.

1. 5, 9, 8, 12; or 3, 11, 8, 12; or 7, 11, 6, 10; or any other combination that works.

2. The number is 45.

3. 999 + 99 + 9 + 9 + 9.

4.3
A VENN DIAGRAM

Mathematicians have a special name for a diagram consisting of a set of intersecting circles surrounded by a rectangle. Named after the logician Robert Venn, this type of diagram is called a *Venn diagram*.

Venn diagrams are useful for arriving at solutions to problems that involve sets of information that possibly intersect each other. There are no particular mathematical prerequisites for using Venn diagrams.

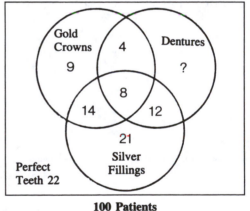

100 Patients

EXAMPLE 1   A dentist's records show that out of 100 patients, 35 have gold crowns, 55 have silver fillings, 34 have dentures, and 22 have perfect teeth. However, 8 of the patients have silver fillings and gold crowns and dentures; 12 have gold crowns and dentures; 22 have both gold crowns and silver fillings; 20 have both silver fillings and dentures; 21 have only silver fillings; and 9 have only gold crowns. How many of his patients have only dentures?

Solution   The following Venn diagram, consisting of three intersecting circles surrounded by a rectangle, can be labeled as shown to reflect the data in the problem.

Please refer to the diagram as you read the following explanation of how information from the problem has been used to fill in segments of the diagram.

When rereading the problem you should notice that only 8 of the patients have all three: silver fillings, gold crowns, and dentures. Since one circle—the upper left—represents people having gold crowns, and one circle—the upper right—represents people having dentures, and the third circle—bottom—represents people having silver fillings; where all three circles *intersect* is the region representing people who have all three. Therefore, the 8 is written in that region.

The region in the rectangle that is outside the three circles represents the people who have perfect teeth; therefore, the 22 is written there.

The problem says that 12 people have both gold crowns and dentures. The region representing these people is in the *intersection* of the two top circles. Since 8 are already represented, the remaining 4 are indicated in the top part of the region. Eight plus four is twelve.

Because 22 people have both gold crowns and silver fillings, represented in the *intersection* of the left two circles; and 8 are already represented; the 14 represents the others. Since 20 people have dentures and silver fillings, and 8 are already indicated, the 12 represents the others.

Eight plus four plus fourteen can be found in the gold crown circle. The problem says that 35 people have gold crowns; so the remaining 9 must be indicated in the upper left hand portion of the gold crown circle. Similarly, a 21 is placed in the bottom part of the silver filling circle.

There were 100 patients in all, which is indicated below the rectangle.

The question mark represents the number of people who only have dentures. Its value can be calculated by adding all the numbers inside the Venn diagram and subtracting from 100. Thus, 10 patients only have dentures.

Remember, you fill in the Venn diagram from inside out, indicating what is represented by the *intersection* of all the circles. A Venn diagram can contain as few as one circle, and as many as you can draw that all intersect clearly.

This type of diagram is useful for solving problems that contain information about intersecting sets of data.

EXERCISES
4.3

1. At a convention of butchers (B), advertisers (A), and checkout clerks (C), there were a total of 350 people. There were 50 who were both B and A but not C; 70 who were B but neither A nor C; 60 who were A but neither B nor C; 40 who were both A and C but not B; 50 who were both B and C but not A; and 80 who were C but neither A nor B. How many at the convention were all three, A, B, and C?

2. In a group of 30 men and women, every woman has at least one of the following characteristics: blue eyes, blonde hair, tall, beautiful. There are 8 beautiful women in the group, 9 blue-eyed women, 9 tall women, and 11 blonde women. Two of the women are blonde, blue-eyed, beautiful, and tall; 4 are blue-eyed and blonde; 5 are blue-eyed and beautiful; 5 are blonde and beautiful; 5 are blonde and tall; 5 are blue-eyed and tall; 6 are beautiful and tall; 4 are beautiful and blue-eyed and tall; 3 are tall, blue-eyed blondes; 3 are beautiful, blue-eyed blondes; and 4 are tall, blonde and beautiful. How many of the group are men?

3. A reporter interviewed the coaches of the football, basketball, and track teams; and the principal who said, "We have a total of 89 boys in school. Fifteen are in the marching band and do not go out for sports. All the others are out for one or more sports each." The football coach said, "Forty-three boys came out for football. Fourteen of them also went out for track, and 15 went out for basketball." The basketball coach said, "Thirty-one boys came out for basketball. Fifteen of them were also out for football; 11 of them are out for track." The track coach said, "I have 34 boys out for track. Fourteen of them went out for football; 11 of them went out for basketball." How many boys went out for all 3 sports?

Solutions   Do not look at these answers until you have tried to solve the first two problems. If any of your answers disagree with these, try again. The third problem is the most difficult, but the answer here is the correct one.

1. None!
2. 11 men
3. 6 boys

4.4
ACTING OUT, MAKING MODELS

Oftentimes a problem is better understood if you can "act it out." Making physical or pictorial models is a more advanced technique.

EXAMPLE 1   Place 20 pennies on the table in a row. Replace every fourth coin with a nickel. Then replace every third coin with a dime. Next, replace every sixth coin with a quarter. After doing this, what is the value of the 20 coins on the table?

Solution   The "acting out" can be represented by the following diagram.

```
P P P P P P P P P P P P P P P P P P P P
    N       N       N       N       N
    D   D   D   D   D   D
        Q       Q       Q
```

Which results in the following set of 20 coins:

P P D N P Q P N D P P Q P P D N P Q P N

The total value of 10 pennies, 4 nickels, 3 dimes, and 3 quarters is .10 plus .20 plus .30 plus .75, or $1.35.

A "tree diagram" serves as a model for certain types of problems. Prerequisite knowledge needed for solving the following problem includes understanding prime factors and prime factorization. A factor is one element of a multiplication. Prime factors are factors divisible *only* by themselves *and* the number 1, e.g., 2, 5, 7, 11, 13, 17, 19, 23, 29, and so on. The *Fundamental Theorem of Arithmetic* states that all non-primes greater than 3 can be uniquely factored into products of primes. For example, 24 = 2 x 2 x 2 x 3, or 2^3 x 3.

EXAMPLE 2   A tree diagram can be used when solving this problem.

Find the prime factorization of 1,170.

Solution   Begin the tree diagram by writing two factors of 1170:

Then add additional branches by factoring each of the factors:

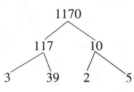

Then bring down the prime factors and factor 39, to get:

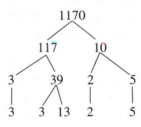

which shows the prime factorization of 1,170 to be 3 x 3 x 13 x 2 x 5, which can be simplified to

$$2 \times 3^2 \times 5 \times 13$$

EXAMPLE 3   Janet has 4 sweaters, 3 skirts, and 3 blouses. How many different outfits of one each can she wear?

Solution   Representing the sweaters by A, B, C, and D; and the skirts by F, G, and H; and the blouses by J, K, and L; you can create the following diagram:

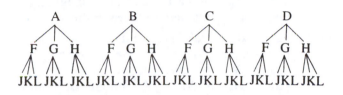

Each downward branch of the tree, 3 steps, represents a different outfit. For example, branch A, F, J represents an outfit made up of a sweater, skirt, and blouse. So does branch D, G, K. By counting the number of entries in the bottom line of the diagram you can determine the total number of different outfits she can wear, which is 36.

Drawing any type of picture to represent the action of a problem is considered *constructing a model* of the problem.

EXERCISES

4.4

1. In a line of 24 pennies, replace every third one with a quarter, then every other one with a dime, and, finally, every fourth one with a nickel. What is the value of the 24 coins now on the table?

2. Find the prime factorization of 12,012.

3. A man has four pairs of shoes, six suits, ten ties, and seven shirts. How many different outfits consisting of a shirt, a tie, a suit, and a pair of shoes is he able to wear?

Solutions   Do not look at these answers until you have tried to solve all the problems. If any one of your answers is incorrect, **forget the answer given here** and try again to solve the problem.

1. $1.98
2. 2^2 x 3 x 7 x 11 x 13
3. 1,680

4.5
WORKING BACKWARDS

If you are having difficulty sorting out the information in a problem, try reading it backwards in sentences, or from the middle both backwards and forwards in sentences. This tactic often helps make the given information fall into place in comprehensive fashion.

EXAMPLE 1   Carl ended his day with $5. He spent $20 during the day. He earned $15 during the day. How many dollars did he have at the beginning of the day?

Solution   Reading the problem backwards in sentences yields:

How many dollars did he have at the beginning of the day? Carl earned $15 during the day. He spent $20 during the day. He ended his day with $5.

Letting x represent the number of dollars Carl had in the morning, the action of the problem now can be symbolized,

$$x + 15 - 20 = 5$$

which solves to $x = 10$; thus the answer is $10.

EXAMPLE 2   In a survey of 100 students, the numbers studying various languages were: Spanish, 28; German, 30; French, 42; all three languages 3; Spanish and German, 8; Spanish and French, 10; and German and French, 5. How many of the 100 students were not studying a language?

Solution   Reading from the inside forwards, then backwards, the information becomes:

All three languages, 3;
Spanish and German, 8;
Spanish and French, 10;
German and French, 5;
French, 42; German, 30;
Spanish, 28.

And the following Venn diagram can be filled in from inside out to show that the answer to the problem is 20.

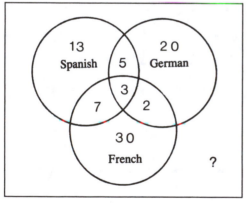

100 Students

2. In a school 12 students take English and French, 11 take English and German, 8 take French and German, 5 take all three languages, 28 take English, 23 take French, and 23 take German. If every student in school takes at least one of these languages, how many students are there in the school?

3. Allen, Bob, and Carl decide to play a game of cards. They agree on the following procedure: When a player loses a game, he will pay enough to double the amount of money that each other player has. They played three times. Carl was the third one to lose, Bob was the second one to lose, and Allen was the first one to lose. When it was all over, each player had exactly $8. How much did each of them start with?

Solutions   Do not look at these answers until you have tried to solve all the problems. If any one of your answers is incorrect, **forget the answer given here** and try again to solve the problem.

1. $50
2. 48
3. Allen, 13; Bob, 7; Carl, 4

EXERCISES
4.5

1. Marianne ended a day of charge card shopping with a credit balance of $18. During the day she charged $47 worth of merchandise and received a credit voucher for $15. What was her credit balance at the beginning of the day?

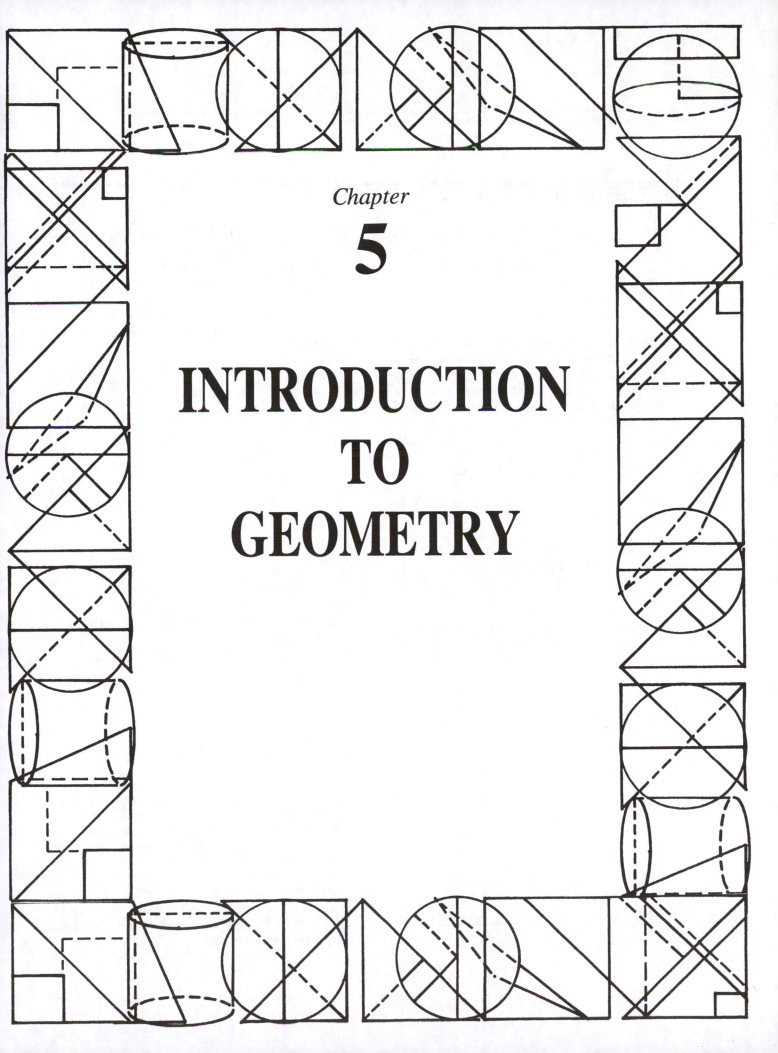

Chapter

5

INTRODUCTION TO GEOMETRY

5.1
SIMPLE GEOMETRIC FIGURES

Geometry is the study of the properties and measurements of figures, regions, and solids formed by points, lines, and planes. The terms **point, line,** and **plane** are undefined. These geometric ideas do have some important properties and may be represented by drawings or physical objects.

A **point** has no length, width, or depth. It has position only. The end of a sharpened lead pencil and the tip of a pin are physical objects which represent points. In drawings, a point is represented by a dot. Points are usually named by capital letters.

A **line** has no width or depth. It has length only. The term **line** refers to a straight, continuous set of points that has no beginning and no end.

A part of a line between two given points is called a **line segment.** The edge of a ruler, a taut piece of wire, and a sharp crease in a folded sheet of paper are physical objects which represent line segments. In drawings, line segments are usually named with two capital letters representing the endpoints, or by a single lower case letter (see Figure 5.1 (i) and (ii)).

A part of a line extending from a point on the line continuously in one direction is called a **ray.** Any point on a line divides the line into exactly two rays, and is the endpoint of each ray. A ray is named by its endpoint and any other point in the ray, as shown in parts (iii) and (iv) of Figure 5.1. Note the order of the capital

letters in part (iv) of the figure. Ray *BC is not* the same as ray *CB.*

A **plane** has no depth. It has length and width only. The surface of a mirror, a sheet of paper, and the flat surface of a desk top are physical objects which represent planes. In drawings, a plane is usually represented by a four-sided figure.

EXAMPLE 1 Identify all of the line segments represented in the figure.

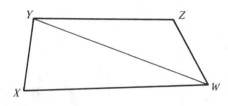

Solution There are five line segments: *XY, YZ, ZW, WX,* and *WY.*

An **angle** is formed by two rays with a common endpoint. The rays are the **sides** of the angle. The common endpoint is the **vertex** of the angle (see Figure 5.2).

FIGURE 5.2

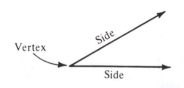

Angles can be named in several ways. When naming angles, the symbol "∠" may be used in place of the word "angle." As we can see from Figure 5.3, an angle may be named using three capital letters or a single capital letter. If three capital letters are used, the capital letter at the vertex must be *between* the other two letters (see part (i) of figure). The single letter at the vertex may be used if there is only one angle with this vertex (part (ii)).

FIGURE 5.1

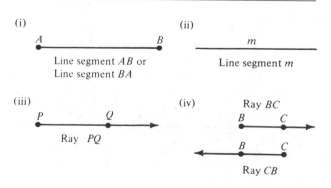

(i)

Line segment *AB* or
Line segment *BA*

(ii)

Line segment *m*

(iii)

Ray *PQ*

(iv)

Ray *BC*

Ray *CB*

FIGURE 5.3

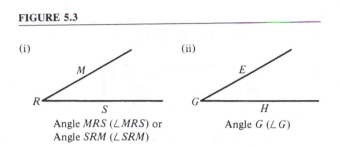

(i)

Angle *MRS* (∠*MRS*) or
Angle *SRM* (∠*SRM*)

(ii)

Angle *G* (∠*G*)

Another way of designating an angle is with a numeral, a lower case letter, or a Greek alphabet letter inside the angle. This is shown in Figure 5.4.

FIGURE 5.4

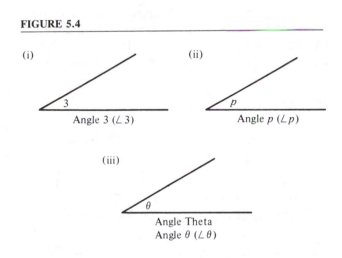

(i)

Angle 3 (∠3)

(ii)

Angle *p* (∠*p*)

(iii)

Angle Theta
Angle θ (∠θ)

EXAMPLE 2 Name eight different angles in the following figure.

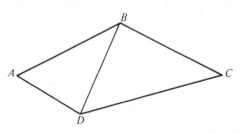

Solution ∠*A*, ∠*ABD*, ∠*ABC*, ∠*DBC*, ∠*C*, ∠*CDB*, ∠*CDA*, and ∠*BDA*.

EXAMPLE 3 Name the angle in four different ways.

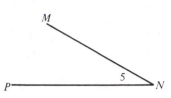

Solution ∠*MNP*, ∠*PNM*, ∠*N*, and ∠5.

The size of an angle is determined by the amount of rotation between the sides, and measured by a unit called a **degree**. One complete rotation of a side about a vertex is defined as *360 degrees*, which may also be written as 360°. This is shown in Figure 5.5.

FIGURE 5.5

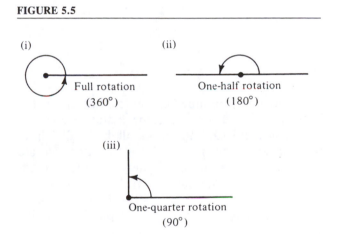

(i)

Full rotation
(360°)

(ii)

One-half rotation
(180°)

(iii)

One-quarter rotation
(90°)

Two lines, or line segments, or rays, that intersect to form a 90-degree angle are said to be **perpendicular**. The symbol "⊥" may be used in place of the words "is perpendicular to." The symbol "ㄴ" may be used to represent a 90-degree angle, or two perpendicular lines in a geometric figure. In Figure 5.6(i), line segment *AB* is perpendicular to line segment *CD*. In part (ii) of the figure, ∠*XYZ* is a 90-degree angle, and *XY* is perpendicular to *YZ*. In part (iii), *FH* is perpendicular to *EG*, and ∠*FHE* and ∠*FHG* are right angles.

FIGURE 5.6

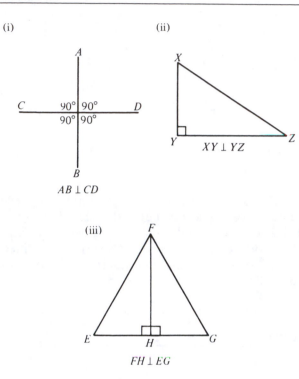

(i)

$AB \perp CD$

(ii)

$XY \perp YZ$

(iii)

$FH \perp EG$

Two lines, or line segments, that are in the same plane and are the same distance apart at any point are said to be **parallel**. Parallel lines will never meet, no matter how far they are extended. The symbol "∥" may be used in place of the words "is parallel to." In part (i) of Figure 5.7, line *RS* is parallel to line *GH;* and in part (ii), line segment *MN* is parallel to line segment *PQ*.

FIGURE 5.7

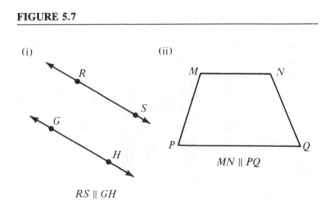

(i)

$RS \parallel GH$

(ii)

$MN \parallel PQ$

EXAMPLE 4 Identify the perpendicular and parallel line segments in the figure.

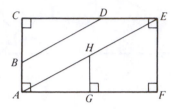

Solution Parallel line segments: *ABC* ∥ *GH*, *ABC* ∥ *EF*, *GH* ∥ *EF*, *BD* ∥ *AE*, *CDE* ∥ *AGF*.

Perpendicular line segments: *CBA* ⊥ *AGF*, *ABC* ⊥ *CDE*, *CDE* ⊥ *EF*, *EF* ⊥ *FGA*, *HG* ⊥ *AGF*.

A **polygon** is a closed figure formed by three or more line segments. Polygons are identified by the number of sides in the figure. Six different types of polygons are shown in Figure 5.8.

FIGURE 5.8

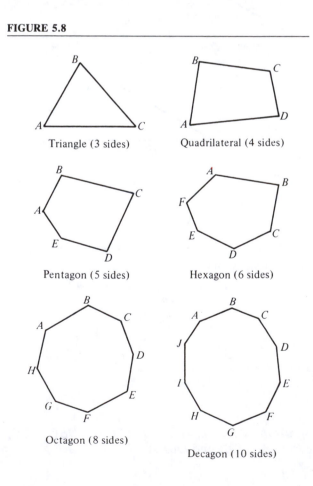

Triangle (3 sides)

Quadrilateral (4 sides)

Pentagon (5 sides)

Hexagon (6 sides)

Octagon (8 sides)

Decagon (10 sides)

Quadrilaterals can be divided into several subclasses according to their sides and angles. For example, a **parallelogram** (Figure 5.9(i)) is a quadrilateral with both pairs of opposite sides parallel. (The opposite sides are also equal in length.) A **trapezoid** (Figure 5.9(ii)) is a quadrilateral with only one pair of opposite sides parallel.

FIGURE 5.9

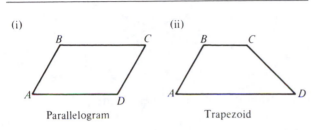

(i) Parallelogram   (ii) Trapezoid

There are also several subclasses of parallelograms. A **rectangle** (Figure 5.10(i)) is a parallelogram with all angles equal to 90°. A **square** (Figure 5.10(ii)) is a rectangle with all sides equal in length. A **rhombus** (Figure 5.10(iii)) is a parallelogram with all sides equal in length.

FIGURE 5.10

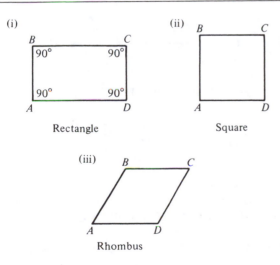

(i) Rectangle   (ii) Square   (iii) Rhombus

In order to understand the terms *circle, radius, diameter, chord, tangent, arc, sector,* and *circumference,* let's look at Figure 5.11. A **circle** is a set of points, all of which are the same

FIGURE 5.11

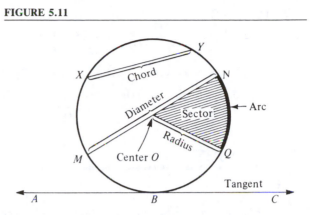

distance from a fixed point called the **center**. A circle is named by a capital letter at the center point. The circle in the figure is named circle *O*.

A **radius** is a line segment from the center to any point on the circle. The plural of radius is **radii**. Radii of circle *O* are *OQ*, *OM*, and *ON*.

A **diameter** is a line segment with endpoints on the circle, and passes through the center of the circle. Segment *MON* is a diameter of circle *O*.

A **chord** is a line segment that connects two points on the circle. Segment *XY* is a chord of circle *O*. Diameter *MON* is also a chord of circle *O*.

A **tangent** is a line that touches the circle in only one point. Line *ABC* is tangent to circle *O* at point *B*.

An **arc** is a part of the circle between any two points on the circle. One arc of circle *O* is arc *NQ* (symbol: $\overset{\frown}{NQ}$). Some of the other arcs of circle *O* are $\overset{\frown}{XY}$, $\overset{\frown}{XM}$, $\overset{\frown}{MBQ}$, and $\overset{\frown}{XYNQ}$.

A **sector** is a region inside of the circle bounded by two radii and an arc. In circle *O* the shaded region is called sector *ONQ*. The region bounded by radii *OM* and *OQ* and arc *MBQ* is called sector *OMQ*. The **circumference** of a circle is the distance around the circle.

Review Exercises

Perform the indicated operations in Exercises 1–10.

1. 0.35 X 0.0264   2. 3.92 X 0.803

3. $10.23 - 0.008$

4. $4.0382 - 0.357$

5. $3.75 \div 0.15$

6. $0.028 \div 1.6$

7. $1.9 + 0.37 + 43.9$

8. $4.05 + 0.8 + 37.946$

9. 43% of 569

10. 8% of 24.5

11. 128% of 34

12. 185% of 93

EXERCISES

5.1

1. How many lines can contain one given point? How many lines can contain two given points?

2. How many planes can contain two given points? If three points do not all lie on the same line, how many different planes can contain all three points?

3. Given five points, no three of which lie on the same line, how many line segments, each containing two of the given points, can be drawn?

4. Given four points, no three of which lie on the same line, how many line segments, each containing two of the given points, can be drawn?

5. Identify all of the line segments in the figure.

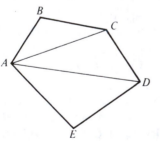

6. Identify all of the line segments in the figure.

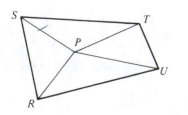

7. Identify four line segments in the figure.

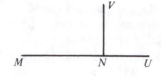

8. Identify six line segments in the figure.

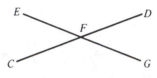

9. Name all of the angles of the polygon.

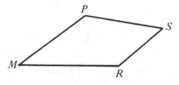

10. Name all of the angles of the polygon.

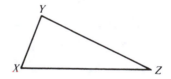

11. Name angle 1 in another way.

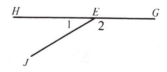

12. Name angle 2 in another way.

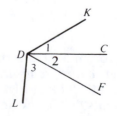

13. Name two angles that have segment QW as a side.

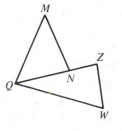

14. Name two angles that have segment BC as a side.

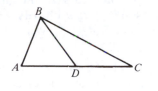

15. Name an angle that has point Y as a vertex.

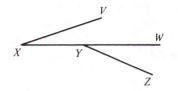

16. Name an angle that has point B as a vertex.

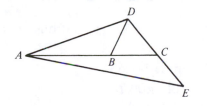

17. Name eight different angles in the figure.

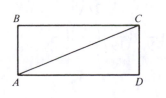

18. Name eight different angles in the figure.

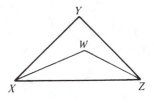

19. Name nine different angles in the figure.

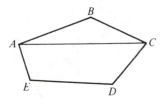

20. Name eight different angles in the figure.

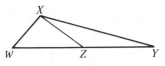

Identify the line segments that appear to be parallel or perpendicular in the figures in Exercises 21–24.

21.

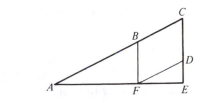

22.

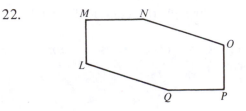

23.

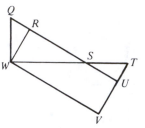

24.

25. A polygon with four sides is called a/an _____ .

26. A polygon with six sides is called a/an _____ .

27. Draw a pentagon with two of its sides parallel.

28. Draw a triangle with two of its sides perpendicular.

Are the statements in Exercises 29–40 true or false?

29. Every trapezoid is a parallelogram.

30. Every square is a rhombus.

31. A square is a parallelogram.

32. A square is a rectangle.

33. A rectangle is a quadrilateral.

34. The opposite sides of a parallelogram are equal in length.

35. A chord through the center of a circle is a diameter.

36. A tangent to a circle is also a chord of the circle.

37. The length of a radius of a circle is equal to one-half the length of a diameter of the circle.

38. A radius of a circle is also a chord of the circle.

39. If a trapezoid has one right angle, then it has two right angles.

40. Some triangles are quadrilaterals.

41. Draw a circle P. Draw a chord XY that is not a diameter. Draw diameters XR and YS, and then draw polygon $XSRY$. What kind of polygon is $XSRY$?

42. Draw a circle Q. Draw a chord AB that is not a diameter. Draw another chord CD that is parallel to AB but not the same length as AB. Draw chords AC and BD to form polygon $ABDC$. What kind of polygon is $ABDC$?

43. Draw a circle O. Draw a tangent to circle O at a point A on the circle. Draw a perpendicular to the tangent at point A. Does the perpendicular to the tangent pass through the center of circle O?

44. Draw a circle M. Draw a chord EF that is not a diameter. Draw another chord GH perpendicular to EF. What kind of polygon is $EGFH$?

Calculator Exercises

Perform the indicated operations in Exercises 45–50.

45. 4.3% of $2789.53

46. 12.8% of $409.78

47. $(83.542 \div 0.093)(5.94)$

48. $(490.62 \div 7.35)(0.028)$

49. $\left(\dfrac{3}{4}\right)\left(\dfrac{7}{5}\right)\left(\dfrac{24}{13}\right)\left(\dfrac{8}{17}\right)\left(\dfrac{29}{37}\right)$

50. $\left(\dfrac{6}{11}\right)\left(\dfrac{14}{9}\right)\left(\dfrac{26}{19}\right)\left(\dfrac{4}{13}\right)\left(\dfrac{17}{32}\right)$

5.2
ANGLE MEASURE AND TRIANGLES

One standard unit for measuring the size of an angle is the degree. A **degree** may be defined as $\frac{1}{360}$ of one complete rotation of a ray about a point. A **protractor** is an instrument that can be used to measure the approximate size of an angle. Figure 5.12 shows how to use a semicircular protractor.

FIGURE 5.12

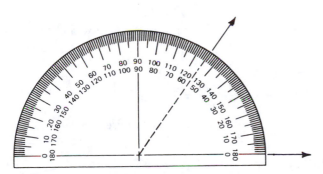

Step 1: Place the base of the protractor along one side of the angle. Line up the indicator on the protractor with the vertex of the angle.

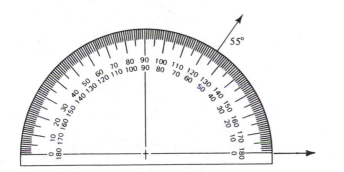

Step 2: Read the measure of the angle where the other side of the angle crosses the arc of the protractor. In this example, the measure of the angle is approximately 55°.

EXAMPLE 1 Use a protractor to check the measures of the following angles.

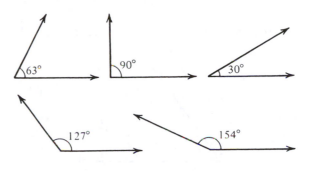

EXAMPLE 2 Use a protractor to check the measures of the following angles.

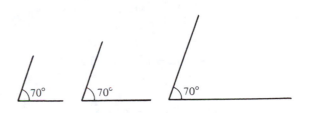

Notice that the lengths of the sides of an angle do not affect the measure (number of degrees) of the angle.

For more precise measurements, each degree is divided into 60 parts, called **minutes** ($'$):

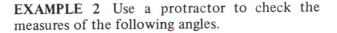

$$1° = 60' \text{ (60 minutes)}.$$

Each minute, in turn, is divided into 60 parts, called **seconds** ($''$):

$$1' = 60'' \text{ (60 seconds)}.$$

An angle measure such as $23°42'15''$ is read, "twenty-three degrees forty-two minutes fifteen seconds."

WATCH OUT! The symbols ′ and ″ are used for minutes and seconds when measuring angles, and are also used for feet and inches when measuring length. Do not confuse the meaning of these symbols.

EXAMPLE 3   What is the measure of $\angle ABC$?

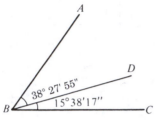

Solution

$$38°27'55''$$
$$+\ 15°38'17''$$
$$\overline{53°65'72''}$$

$$72'' = 60'' + 12'' \text{ or } 1'12'',$$

so we can say

$$53°65'72'' = 53°66'12''.$$

$$66' = 60' + 6' \text{ or } 1°6',$$

so we can say

$$53°66'12'' = 54°6'12''.$$

Thus, the sum of the two angles is $54°6'12''$.

Angle measures given in degrees, minutes, and seconds can be expressed as decimal numerals in degrees only.

EXAMPLE 4   Change 17 degrees 48 minutes to degrees.

Solution

$$48 \text{ min} = 48 \text{ min} \times \frac{1 \text{ degree}}{60 \text{ min}}$$

$$= \frac{48}{60} \text{ degree}$$

$$= 0.8 \text{ degree}$$

Therefore, 17 degrees 48 minutes = 17.8 degrees. (*Note:* We can use the symbols for degrees, minutes, and seconds and "cancel" units, as above, when converting units.)

$$48' = 48' \times \frac{1°}{60'} = 48' \times \frac{1°}{60'} = \frac{48°}{60} = 0.8°.$$

EXAMPLE 5   Change 108 degrees 37 minutes to degrees.

Solution

$$37 \text{ min} \times \frac{1 \text{ degree}}{60 \text{ min}} = \frac{37}{60} \text{ degree}$$

$$\doteq 0.617 \text{ degree} \text{ (to the nearest thousandth)}$$

Therefore, 108 degrees 37 minutes $\doteq$ 108.617 degrees. (*Note:* The symbol "$\doteq$" means "approximately equal to.")

EXAMPLE 6   Change $48°13'48''$ to degrees.

Solution

$$13'48'' = \left(13' \times \frac{60''}{1'}\right) + 48''$$

$$= 780'' + 48'' = 828''$$

$$1° = 60' \times \frac{60''}{1'} = 3600''$$

$$828'' \times \frac{1°}{3600''} = \frac{828°}{3600} = 0.23°$$

Therefore, $48°13'48'' = 48.23°$.

EXAMPLE 7 Change $26°43'17''$ to degrees.

Solution

$$43'17'' = \left(43' \times \frac{60''}{1'}\right) + 17''$$

$$= 2580'' + 17'' = 2597''$$

$$2597'' \times \frac{1°}{3600''} = \frac{2597°}{3600}$$

$$\doteq 0.721° \text{ (to the nearest thousandth)}$$

Therefore, $26°43'17'' \doteq 26.721°$.

Angle measures represented as decimal parts of a degree can be expressed as degrees, minutes, and seconds.

EXAMPLE 8 Change $35.7°$ to degrees, minutes, and seconds.

Solution

$$0.7° = \left(0.7° \times \frac{60'}{1°}\right)$$

$$= 42' \leftarrow \text{Change } 0.7° \text{ to minutes.}$$

$$35.7° = 35°42'0''$$

EXAMPLE 9 Change $15.38°$ to degrees, minutes, and seconds.

Solution

$$0.38° = \left(0.38° \times \frac{60'}{1°}\right)$$

$$= 22.8' \leftarrow \text{Change } 0.38° \text{ to minutes.}$$

$$0.8' = \left(0.8' \times \frac{60''}{1'}\right)$$

$$= 48'' \leftarrow \text{Change } 0.8' \text{ to seconds.}$$

$$15.38° = 15°22'48''$$

EXAMPLE 10 Change $29.894°$ to degrees, minutes, and seconds.

Solution

$$0.894° \times \frac{60'}{1°} = 53.64' \leftarrow \text{Change } 0.894° \text{ to minutes.}$$

$$0.64' \times \frac{60''}{1'} = 38.4'' \leftarrow \text{Change } 0.64' \text{ to seconds.}$$

$$29.894° \doteq 29°53'38'' \text{ (to the nearest second)}$$

Angles are classified according to size, as shown in Figure 5.13. A **right angle** (part (i) of the figure) is an angle with a measure of $90°$. A **straight angle** (ii) is an angle with a measure of $180°$. An **acute angle** (iii) is an angle with a measure between $0°$ and $90°$. An **obtuse angle** (iv) is an angle with a measure between $90°$ and $180°$. **Complementary angles** (v) are two angles with measures totalling $90°$. **Supplementary angles** (vi) are two angles with measures totalling $180°$.

FIGURE 5.13

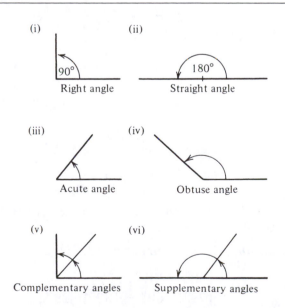

(i) Right angle
(ii) Straight angle
(iii) Acute angle
(iv) Obtuse angle
(v) Complementary angles
(vi) Supplementary angles

EXAMPLE 11 Classify each angle in polygon *ABCDE*.

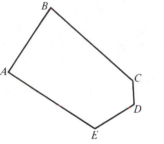

Solution Angle *A* is a right angle. Angle *B* is an acute angle. Angles *C*, *D*, and *E* are obtuse angles.

EXAMPLE 12 What is the complement and supplement of an angle of 35°?

Solution The complement of $35° = (90° - 35°) = 55°$. The supplement of $35° = (180° - 35°) = 145°$.

EXAMPLE 13 What is the complement and supplement of an angle of 53°37'?

Solution The complement of 53°37'

$$= 90° - 53°37'$$
$$= 89°60' - 53°37'$$
$$= 36°23'.$$

The supplement is 126°23'.

EXAMPLE 14 What is the complement of an angle of 37°25'46"?

Solution The complement of 37°25'46"

$$= 90° - 37°25'46''$$
$$= 89°59'60'' - 37°25'46''$$
$$= 52°34'14''.$$

EXAMPLE 15 What is the supplement of an angle of 132°16'29"?

Solution The supplement of 132°16'29"

$$= 180° - 132°16'29''$$
$$= 179°59'60'' - 132°16'29''$$
$$= 47°43'31''.$$

Two angles are **adjacent** if they have the same vertex and a common side between them. For instance, in Figure 5.14, $\angle ABC$ is adjacent to $\angle CBD$ since *B* is the vertex of both angles and *BC* is the common side between the angles.

FIGURE 5.14

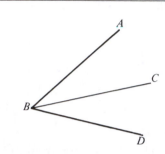

EXAMPLE 16 Name the pair(s) of adjacent angles in the figure.

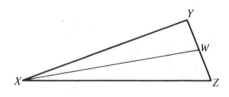

Solution $\angle YXW$ is adjacent to $\angle WXZ$.
$\angle YWX$ is adjacent to $\angle XWZ$.

A triangle is a polygon with three sides. A triangle also has three angles, and:

**The sum of
the angles of every triangle is 180°.**

EXAMPLE 17 In triangle *ABC*, $\angle A = 69°30'$ and $\angle B = 76°48'$. What is the measure of $\angle C$?

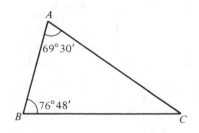

Solution

$$\angle A + \angle B + \angle C = 180°$$
$$\angle A + \angle B = 146°18'$$
$$\angle C = 180° - 146°18'$$
$$= 33°42'$$

The fact that line segments or angles in a geometric figure, or in several figures, are equal can be represented with symbols. The same symbol is used to identify the equal parts. For example, look at Figure 5.15. In part (i) of the figure, segment *AB* is equal to segment *CD;* $\angle A$ is equal to $\angle D$; and $\angle B$ is equal to $\angle C$. In part (ii), segment *EF* is equal to segment *RS;* segment *EG* is equal to segment *RT;* and $\angle E$ is equal to $\angle R$.

FIGURE 5.15

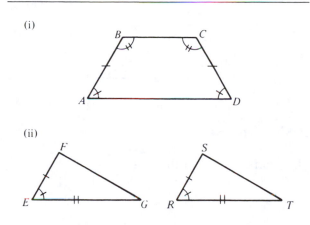

Triangles are classified by their sides and angles, as we see from Figure 5.16. A **right triangle** (part (i) of the figure) is a triangle with one right angle. The other two angles are acute angles and complementary. An **equilateral triangle** (ii) is a triangle with three equal sides. All angles of an equilateral triangle are equal. Each angle is equal to 60°. An **isosceles triangle** (iii) is a triangle with two equal sides. The angles opposite the equal sides are equal. A **scalene triangle** (iv) is a triangle with three unequal sides and three unequal angles. An **obtuse triangle** (v) is a triangle with one obtuse angle. An **acute triangle** (vi) is one having three acute angles.

FIGURE 5.16

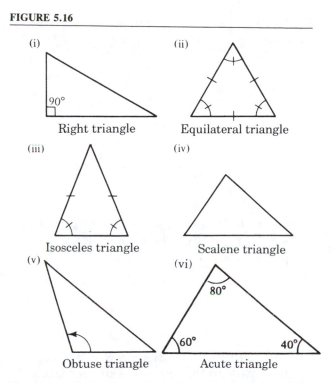

Many problems in technical work involve a right triangle. It is important that you learn the common terms associated with right triangles. As we see from part (i) in Figure 5.17, the two sides that form the right angle are called **legs**. The side opposite the right angle is called the **hypotenuse**. Now look at part (ii) of the figure. With reference to $\angle A$, *BC* is called the **side opposite** $\angle A$, and *AC* is called the **side adjacent** to $\angle A$. With reference to $\angle B$, *AC* is called the **side opposite** $\angle B$, and *BC* is called the **side adjacent** to $\angle B$.

FIGURE 5.17

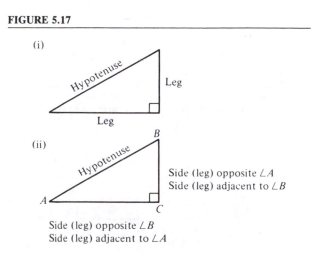

EXAMPLE 18 In right triangle *MNP*, what is the side adjacent to ∠*M*? In right triangle *MRS*, what is the side opposite ∠*MRS*?

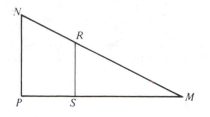

Solution In right triangle *MNP*, side *MP* is adjacent to ∠*M*. In right triangle *MRS*, side *MS* is opposite ∠*MRS*.

An **altitude** of a triangle is a line segment from a vertex to the opposite side, perpendicular to the side. Every triangle has three altitudes, as illustrated in Figure 5.18. In the case of an obtuse triangle, two of the altitudes are perpendicular to the extended sides of the triangle, as shown in Figure 5.19.

FIGURE 5.18

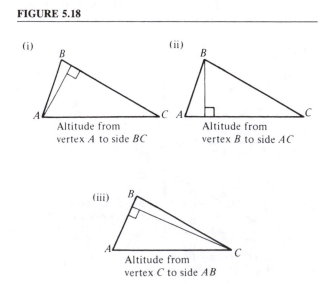

(i) Altitude from vertex *A* to side *BC*

(ii) Altitude from vertex *B* to side *AC*

(iii) Altitude from vertex *C* to side *AB*

FIGURE 5.19

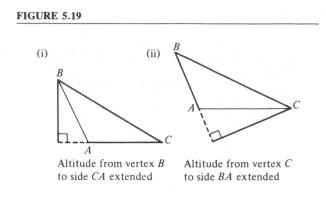

(i) Altitude from vertex *B* to side *CA* extended

(ii) Altitude from vertex *C* to side *BA* extended

In a right triangle the two sides that form the right angle are also altitudes of the triangle. For example, in Figure 5.20, side *AC* is an altitude from vertex *A* to side *CB*, and side *BC* is an altitude from vertex *B* to side *AC*.

FIGURE 5.20

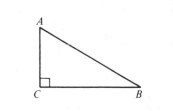

Review Exercises

1. Name ∠1 in another way.

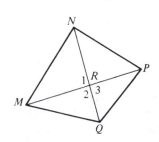

2. Name ∠3 in another way.

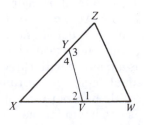

3. Change $\dfrac{7}{16}$ to a decimal.

4. Change $\dfrac{786}{1200}$ to a decimal.

5. $\dfrac{3}{4} + \dfrac{5}{6} =$ _____

6. $\dfrac{2}{3} + \dfrac{5}{8} =$ _____

7. $2\frac{3}{4} \times 1\frac{1}{6} =$ _____

8. $3\frac{1}{4} \times 2\frac{1}{3} =$ _____

9. Change 3.7 feet to inches (to the nearest eighth of an inch).

10. Change 5.9 feet to inches (to the nearest eighth of an inch).

11. Change 2.8 feet to inches (to the nearest sixteenth of an inch).

12. Change 3.3 feet to inches (to the nearest sixteenth of an inch).

EXERCISES
5.2

Use a protractor to measure the angles in Exercises 1–4.

1.

2.

3.

4.

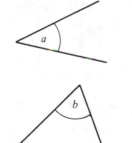

In Exercises 5–8 find the sum of the angles in each polygon. Use a protractor to measure each angle.

5.

6.

7.

8.

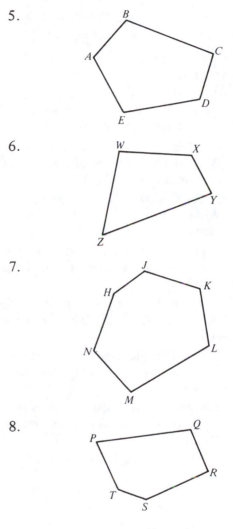

In Exercises 9–14 change each angle measure to degrees (to the nearest thousandth).

9. 25°39'   10. 46°51'

11. 4°16'24"   12. 20°43'15"

13. 135°8'30"   14. 147°57'7"

In Exercises 15–20 change each angle measure to degrees, minutes, and seconds.

15. 42.6°   16. 83.26°

17. 89.18°   18. 30.375°

19. 120.35°   20. 159.09°

Find the complement and supplement of each angle in Exercises 21–26.

21. 19°   22. 65°
23. 54°37′   24. 5°29′
25. 29°6′34″   26. 34°20′18″

27. What is the measure of each angle of an equilateral triangle?

28. One acute angle of a right triangle has a measure of 22°45′. What is the measure of the other acute angle of the triangle?

29. In triangle *PQR*, $\angle P = 27°15'$ and $\angle Q = 86°40'$. What is the measure of $\angle R$? What kind of triangle is triangle *PQR*?

30. In triangle *MNS*, $\angle M = 103°30'$ and $\angle N = 29°45'$. What is the measure of $\angle S$? What kind of triangle is triangle *MNS*?

31. Is an equilateral triangle also an isosceles triangle? Can an obtuse triangle also be a right triangle?

32. Can a right triangle also be a scalene triangle? Can an isosceles triangle also be a right triangle?

33. Can a scalene triangle also be an obtuse triangle? Can an isosceles triangle also be a scalene triangle?

34. Can an isosceles triangle also be an obtuse triangle? Can an equilateral triangle also be a right triangle?

35. Name an angle that is adjacent to $\angle HGJ$.

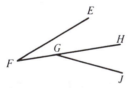

36. Name an angle that is adjacent to $\angle 1$.

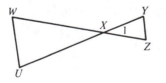

37. In right triangle *MNT*, which side is the hypotenuse? Which side is opposite $\angle NMT$?

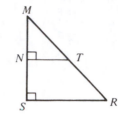

38. In right triangle *ACD*, which side is adjacent to $\angle A$? In right triangle *ABC*, which side is the hypotenuse?

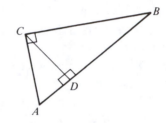

39. Draw a scalene triangle *MNP* with the altitude from *N* to side *MP*.

40. Draw an obtuse isosceles triangle *EFG* with *E* the vertex of the obtuse angle. Draw the altitude from *F* to the opposite side.

41. In an isosceles triangle an angle opposite one of the equal sides has a measure of 46°. What are the measures of the other two angles of the triangle?

42. In scalene triangle *ABC*, with altitude *BD*, the measure of angle *DBC* is 63° and the measure of angle *DBA* is 42°. What are the measures of angles *A* and *C*?

Calculator Exercises

In Exercises 43–46 perform the indicated operations.

43. $(24.38)(4.07) + (457.9)(4.7)$
44. $(83)(6.8)(0.79) \div (889.373)$
45. $(15.7)^2 + (832)^2$
46. $(903)^2 - (532)^2$
47. Change $\dfrac{3524}{8075}$ to a decimal.
48. Change $\dfrac{1862}{3525}$ to a decimal.

5.3
PERIMETER

The **perimeter** of a plane figure is the distance around the figure. If the figure is a polygon, the perimeter is computed by adding the lengths of the sides of the polygon. Compute the perimeter of each polygon in Examples 1–3.

EXAMPLE 1

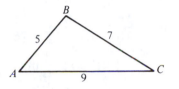

Solution   Perimeter = 5 + 7 + 9 = 21

EXAMPLE 2

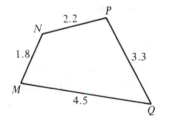

Solution   Perimeter = 1.8 + 2.2 + 3.3 + 4.5 = 11.8.

EXAMPLE 3

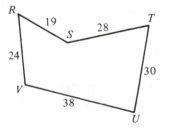

Solution   Perimeter = 24 + 19 + 28 + 30 + 38 = 139.

To compute the perimeter of a polygon, the lengths of all of the sides must be expressed in the same unit. This is illustrated in the next two examples.

EXAMPLE 4   Compute the perimeter of triangle XYZ.

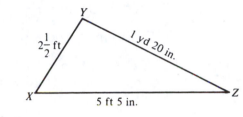

Solution

$$\text{Side } XY = 2\frac{1}{2} \text{ ft} = \frac{5}{2} \text{ ft}$$

$$= \frac{5}{2} \text{ ft} \times \frac{12 \text{ in.}}{1 \text{ ft}}$$

$$= \frac{60}{2} \text{ in.} = 30 \text{ in.}$$

$$\text{Side } YZ = 1 \text{ yd } 20 \text{ in.}$$

$$= \left(1 \text{ yd} \times \frac{36 \text{ in.}}{1 \text{ yd}}\right) + 20 \text{ in.}$$

$$= 56 \text{ in.}$$

$$\text{Side } XZ = 5 \text{ ft } 5 \text{ in.}$$

$$= \left(5 \text{ ft} \times \frac{12 \text{ in.}}{1 \text{ ft}}\right) + 5 \text{ in.}$$

$$= 65 \text{ in.}$$

The perimeter of triangle XYZ = 30 in. + 56 in. + 65 in. = 151 in.

EXAMPLE 5   Compute the perimeter of $ABCD$. Give the answer to the nearest inch.

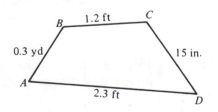

Solution

$$\text{Side } AB = 0.3 \text{ yd} = 0.3 \text{ yd} \times \frac{36 \text{ in.}}{1 \text{ yd}}$$

$$= (0.3)(36) \text{ in.} = 10.8 \text{ in.}$$

$$\text{Side } BC = 1.2 \text{ ft} = 1.2 \text{ ft} \times \frac{12 \text{ in.}}{1 \text{ ft}}$$

$$= (1.2)(12) \text{ in.} = 14.4 \text{ in.}$$

$$\text{Side } CD = 15 \text{ in.}$$

$$\text{Side } AD = 2.3 \text{ ft} = 2.3 \text{ ft} \times \frac{12 \text{ in.}}{1 \text{ ft}}$$

$$= (2.3)(12) \text{ in.} = 27.6 \text{ in.}$$

The perimeter of $ABCD$ = 10.8 in + 14.4 in. + 15 in. + 27.6 in. = 67.8 in. $\doteq$ 68 in. (to the nearest inch).

EXAMPLE 6 Compute the perimeter of triangle ABC. Give the answer in inches, to the nearest fourth of an inch.

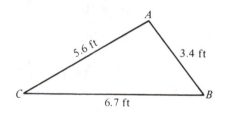

Solution

$$\text{Perimeter} = 5.6 \text{ ft} + 3.4 \text{ ft} + 6.7 \text{ ft}$$

$$= 15.7 \text{ ft}$$

$$15.7 \text{ ft} \times \frac{12 \text{ in.}}{1 \text{ ft}} = (15.7)(12) \text{ in.} = 188.4 \text{ in.}$$

$$0.4 \text{ in.} \times \frac{4}{4} = \frac{1.6}{4} \text{ in.} \doteq \frac{2}{4} \text{ in.}$$

$$188.4 \text{ in.} \doteq 188\frac{2}{4} \text{ in.} \quad \text{\small (to the nearest fourth of an inch)}$$

EXAMPLE 7 Compute the perimeter of $PQRS$. Give your answer to the nearest sixteenth of an inch.

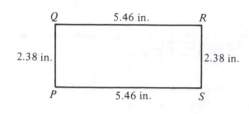

Solution

$$\text{Perimeter} = 2.38 \text{ in.} + 5.46 \text{ in.} +$$

$$2.38 \text{ in.} + 5.46 \text{ in.}$$

$$= 15.68 \text{ in.}$$

$$0.68 \text{ in.} \times \frac{16}{16} = \frac{10.88}{16} \text{ in.} \doteq \frac{11}{16} \text{ in.}$$

The perimeter of $PQRS \doteq 15\frac{11}{16}$ in.

The perimeters of parallelograms, and many other plane figures, can be expressed in ways that may simplify computation. In Table 5.1 we let p represent the perimeter of a figure and the lower case letters a and b represent the lengths of the sides of the figure.

EXAMPLE 8 Find the perimeter of a rectangle with a length of 17 inches and a width of 9 inches.

Solution

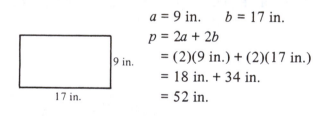

$$a = 9 \text{ in.} \qquad b = 17 \text{ in.}$$

$$p = 2a + 2b$$

$$= (2)(9 \text{ in.}) + (2)(17 \text{ in.})$$

$$= 18 \text{ in.} + 34 \text{ in.}$$

$$= 52 \text{ in.}$$

EXAMPLE 9 Find the perimeter of a rhombus where the length of one side is 3.5 inches.

Solution

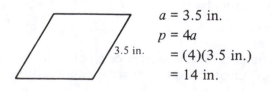

$$a = 3.5 \text{ in.}$$

$$p = 4a$$

$$= (4)(3.5 \text{ in.})$$

$$= 14 \text{ in.}$$

TABLE 5.1

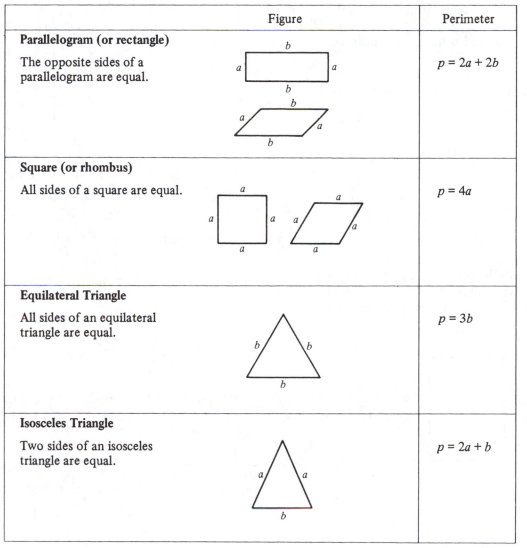

	Figure	Perimeter
Parallelogram (or rectangle) The opposite sides of a parallelogram are equal.		$p = 2a + 2b$
Square (or rhombus) All sides of a square are equal.		$p = 4a$
Equilateral Triangle All sides of an equilateral triangle are equal.		$p = 3b$
Isosceles Triangle Two sides of an isosceles triangle are equal.		$p = 2a + b$

EXAMPLE 10 Find the perimeter of an isosceles triangle in which the base is 6 units long and the length of each of the other sides is 9 units.

Solution

$$a = 9 \quad b = 6$$
$$p = 2a + b$$
$$= (2)(9) + 6$$
$$= 18 + 6$$
$$= 24 \text{ units}$$

EXAMPLE 11 Find the perimeter of an equilateral triangle where the length of one side is 1 foot 5 inches.

Solution

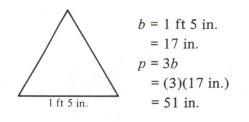

$$b = 1 \text{ ft } 5 \text{ in.}$$
$$= 17 \text{ in.}$$
$$p = 3b$$
$$= (3)(17 \text{ in.})$$
$$= 51 \text{ in.}$$

The perimeter of a complex figure may be computed, even though some measures are not given, by subdividing the figure into parallelograms, triangles, and other simple polygons.

EXAMPLE 12 Compute the perimeter of the figure.

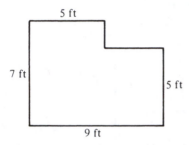

Solution Subdividing the figure into two rectangles makes it possible to see the lengths of the two sides that were not given in the original figure.

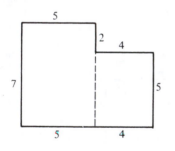

Perimeter = 7 + (5 + 4) + 5 + 4 + 2 + 5
= 32 ft

EXAMPLE 13 Compute the perimeter of the figure. Give the answer in feet and inches, to the nearest eighth of an inch.

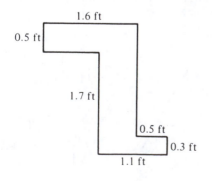

Solution

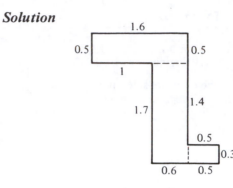

Perimeter = 0.5 + 1 + 1.7 + (0.6 + 0.5)
+ 0.3 + 0.5 + (1.4 + 0.5) + 1.6
= 8.6 ft

$$0.6 \text{ ft} = 0.6 \text{ ft} \times \frac{12 \text{ in.}}{1 \text{ ft}} \times \frac{8}{8}$$

$$= \frac{57.6}{8} \text{ in.}$$

$$= 7\frac{1.6}{8} \text{ in.} \doteq 7\frac{2}{8} \text{ in.}$$

Thus, the perimeter is approximately 8 ft $7\frac{2}{8}$ in.

EXAMPLE 14 Compute the perimeter of the figure. Give the answer to the nearest sixteenth of an inch.

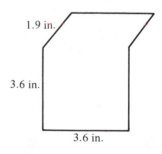

Solution

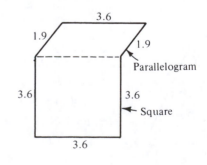

Perimeter = (4)(3.6) + (2)(1.9)

= 18.2 in.

$$0.2 \text{ in.} \times \frac{16}{16} = \frac{3.2}{16} \text{ in.}$$

$$\doteq \frac{3}{16} \text{ in.}$$

The perimeter is approximately $18\frac{3}{16}$ in.

EXAMPLE 15 Compute the perimeter of the figure.

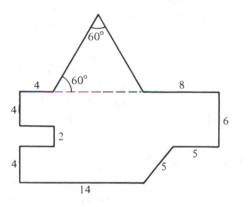

Solution

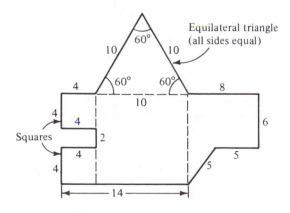

Perimeter = 4 + 10 + 10 + 8

+ 6 + 5 + 5 + 14

+ 4 + 4 + 2 + 4 + 4

= 80 units

The perimeter, or distance around a circle, is called the **circumference** of the circle. The circumference of a circle is related to its diameter. The ratio $\frac{\text{circumference}}{\text{diameter}}$ for any circle is equal to a constant called **pi**. The number pi is approximately 3.14 or $3\frac{1}{7}$. Because of this relationship between the number pi, the circumference, and the diameter of any circle, we can compute the circumference of a circle by multiplying the length of the diameter times pi.

$$\boxed{\textbf{Circumference} = \textbf{pi} \times \textbf{Diameter}}$$

This can be symbolized as:

$$\boxed{c = \pi d}$$

where c represents the circumference of a circle, d represents the length of the diameter of the circle, and π represents the number pi.

EXAMPLE 16 Find the circumference of a fly wheel with a diameter of 12 inches.

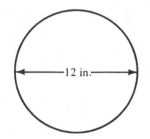

Solution

$$c = \pi d$$

$$\doteq (3.14)(12 \text{ in.})$$

$$\doteq 37.68 \text{ in.}$$

EXAMPLE 17 The radius of a circle is $2\frac{1}{3}$ inches. What is the circumference of the circle?

Solution The diameter of the circle is 2 times $2\frac{1}{3}$ in., or $\frac{14}{3}$ in.

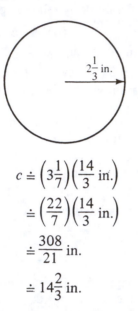

$$c \doteq \left(3\frac{1}{7}\right)\left(\frac{14}{3}\text{ in.}\right)$$

$$\doteq \left(\frac{22}{7}\right)\left(\frac{14}{3}\text{ in.}\right)$$

$$\doteq \frac{308}{21}\text{ in.}$$

$$\doteq 14\frac{2}{3}\text{ in.}$$

EXAMPLE 18 Find the length of arc *ABC* of circle *O*. The radius of circle *O* is 9.7 units. Round off your answer to the nearest tenth of a unit.

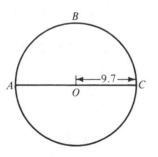

Solution

$$\text{Circumference of circle } O = \pi d$$

$$\doteq (3.14)(19.4)$$

$$\doteq 60.916\text{ units}$$

The arc *ABC* is a *semicircle*, so its length is $\frac{180}{360}$, or $\frac{1}{2}$ of the circumference of circle *O*.

$$\text{Length of arc } ABC = \left(\frac{1}{2}\right)(60.916\text{ units})$$

$$= 30.458\text{ units}$$

$$\doteq 30.5\text{ units}$$

EXAMPLE 19 Find the length of arc *QRS*. Give the answer to the nearest eighth of an inch.

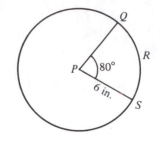

Solution

$$\text{Circumference of circle } P \doteq \left(3\frac{1}{7}\right)(12\text{ in.})$$

$$\doteq \frac{264}{7}\text{ in.}$$

Arc *QRS* is the part of the circle from point *Q* to point *S*, moving in a clockwise direction. Angle *QPS* has a measure of 80°, so the length of arc *QRS* is $\frac{80}{360}$ of the circumference of the circle.

$$\text{Length of arc } QRS = \frac{80}{360} \times \text{circumference}$$

$$\doteq \left(\frac{2}{9}\right)\left(\frac{264}{7}\text{ in.}\right)$$

$$\doteq \frac{528}{63}\text{ in.}$$

$$\doteq 8\frac{24}{63}\text{ in.}$$

$$\doteq 8\frac{3}{8}\text{ in.}$$

EXAMPLE 20 Find the length of arc *ABC*. Give the answer to the nearest sixteenth of an inch.

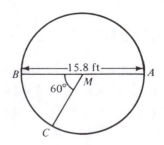

Solution

Circumference of circle $M \doteq (3.14)(15.8 \text{ ft})$

$$\doteq 49.612 \text{ ft}$$

Arc ABC is the part of the circle from point A to point C moving in a counter-clockwise direction. Arc ABC contains a semicircle plus arc BC, so the length of arc ABC is $\frac{180 + 60}{360}$, or $\frac{240}{360}$ of the circumference of the circle.

$$\text{Length of arc } ABC \doteq \left(\frac{240}{360}\right)(49.612 \text{ ft})$$

$$\doteq \frac{99.224}{3} \text{ ft}$$

$$\doteq 33.075 \text{ ft}$$

$$0.075 \text{ ft} \times \frac{12 \text{ in.}}{1 \text{ ft}} \times \frac{16}{16} = \frac{14.4}{16} \text{ in.} \doteq \frac{14}{16} \text{ in.}$$

Therefore, the length of arc $ABC \doteq 33 \text{ ft } \frac{14}{16}$ in.

By using the ideas of perimeters of polygons and circumferences of circles, we can find the perimeter of geometric figures that are combinations of basic figures and parts of figures.

EXAMPLE 21 An arched doorway is a rectangle topped by a semicircle. Find the perimeter of the entire arched doorway shown in the figure. Give the answer to the nearest inch.

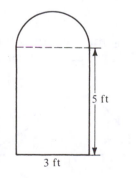

3 ft

Solution

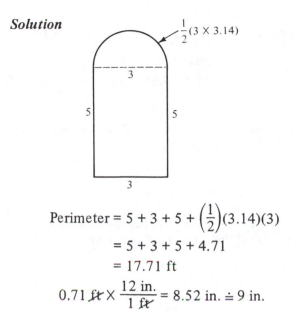

$$\text{Perimeter} = 5 + 3 + 5 + \left(\frac{1}{2}\right)(3.14)(3)$$

$$= 5 + 3 + 5 + 4.71$$

$$= 17.71 \text{ ft}$$

$$0.71 \text{ ft} \times \frac{12 \text{ in.}}{1 \text{ ft}} = 8.52 \text{ in.} \doteq 9 \text{ in.}$$

The perimeter is approximately 17 ft 9 in.

EXAMPLE 22 Metal edging is to be put around the entire outer edge of a service counter. How much edging is needed for the counter shown in the figure?

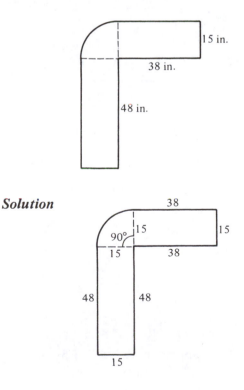

Solution

Perimeter = $(48 + 15 + 48) + \frac{1}{4}(30 \times 3.14)$

$+ (38 + 15 + 38)$

$= 111 + 23.55 + 91$

$= 225.55$

Approximately 226 inches of metal edging are needed for the counter.

EXAMPLE 23 What is the perimeter of the figure? Give the answer to the nearest fourth of an inch.

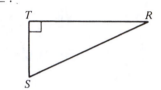

Solution

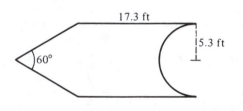

Perimeter = $17.3 + 10.6 + 10.6$

$+ 17.3 + \left(\frac{1}{2}\right)(3.14)(10.6)$

$= 55.8 + 16.642$

$= 72.442$ ft

$0.442 \, \cancel{ft} \times \frac{12 \text{ in.}}{1 \, \cancel{ft}} \times \frac{4}{4} = \frac{21.216}{4}$ in.

$= 5\frac{1.216}{4}$ in.

$\doteq 5\frac{1}{4}$ in.

The perimeter is approximately 72 ft $5\frac{1}{4}$ in.

Review Exercises

Perform the indicated operations in Exercises 1–6.

1. $\frac{5}{8} - \frac{1}{3}$ 2. $\frac{5}{6} - \frac{1}{4}$

3. $\left(\frac{1}{3}\right)(18.42)$ 4. $\left(\frac{1}{4}\right)(23.5)$

5. $189.645 \div 23.5$ 6. $3189.012 \div 453.5$

7. The hypotenuse in triangle *TRS* is side
   ———— .

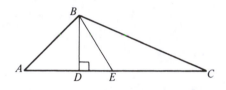

8. Line segment ———— is an altitude of triangle *ABC*.

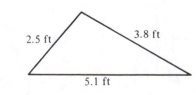

9. Change $4.37°$ to degrees, minutes, and seconds.

10. Change $27.82°$ to degrees, minutes, and seconds.

11. A polygon with four sides is called a/an
    —————— .

12. A polygon with five sides is called a/an
    —————— .

EXERCISES
5.3

Compute the perimeter of each polygon in Exercises 1–4. Give each answer to the nearest inch.

1.

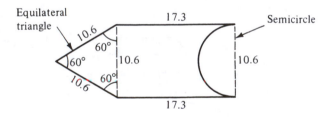

2.

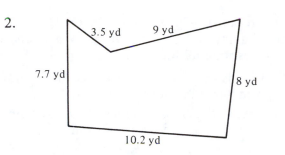

7.

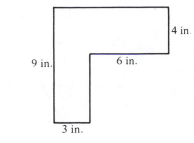

3.

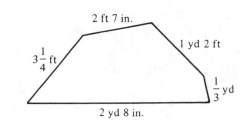

8.

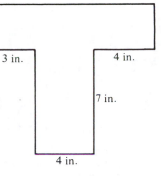

4.

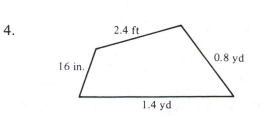

9.

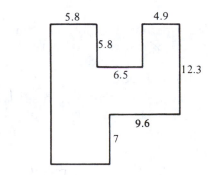

Compute the perimeter of each polygon in Exercises 5–12.

5.

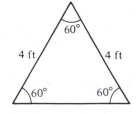

10.

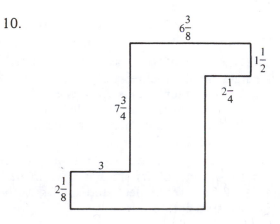

6.

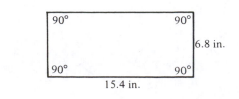

11.

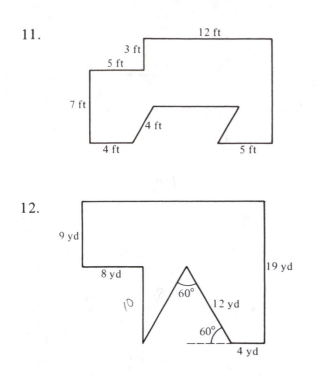

12.

13. What is the perimeter of a square where the measure of one side is equal to 2 feet 2 inches? Give the answer to the nearest thousandth of a foot.

14. What is the perimeter of a rectangle with a length of 27 feet and a width of 13 feet $3\frac{1}{2}$ inches? Give the answer to the nearest thousandth of a foot.

15. What is the perimeter of an equilateral triangle where the length of one side is $3\frac{2}{3}$ feet?

16. What is the perimeter of a rhombus if the measure of one side is $7\frac{1}{3}$ inches?

17. The measure of the base of an isosceles triangle is 12.5 inches, and the measure of one of the other sides is 9.6 inches. What is the perimeter of the triangle? Give the answer in feet and inches to the nearest eighth of an inch.

18. What is the perimeter of a rhombus if the length of one of its sides is 18.7 inches? Give the answer in feet and inches to the nearest sixteenth of an inch.

In Exercises 19–26 use $\pi = 3.14$ or $3\frac{1}{7}$. If not otherwise specified, round off your answers to the nearest hundredth.

19. What is the circumference of a circle whose diameter is 25 inches?

20. What is the circumference of a circle whose diameter has a measure of 17 units?

21. The measure of the radius of a circle is 5.3 inches. What is the circumference? Give the answer to the nearest hundredth of a foot.

22. What is the circumference of a circle with a radius of $4\frac{1}{3}$ feet? Give the answer in feet and inches to the nearest fourth of an inch.

23. What is the length of arc ABC?

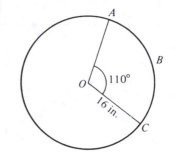

24. What is the length of arc XYZ?

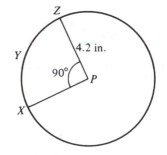

25. What is the length of arc EFG?

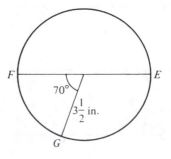

26. What is the length of arc *RST*?

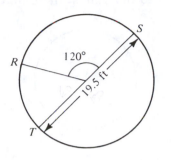

Find the perimeter of the figures in Exercises 27–30.

27.

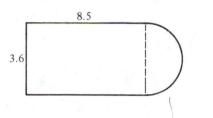

28.

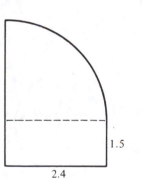

29.

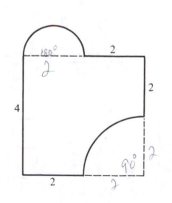

30.

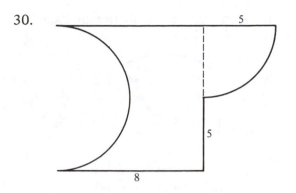

31. The perimeter of a rectangle is 42 feet. If the measure of one side of the rectangle is 8 feet, what are the measures of the other three sides?

32. The perimeter of an equilateral triangle is $19\frac{2}{3}$ inches. What is the length of each side of the triangle?

33. The circumference of a circle is 300 inches. What is the approximate measure of the diameter of the circle?

34. The circumference of a circle is 48.7 inches. What is the approximate length of the radius of the circle?

35. If the cogwheel *B* makes one revolution, how long is the path traveled by point *A* ?

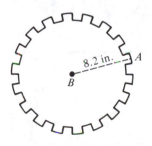

36. Two 7-inch pulleys are placed 34 inches apart at their center points. How long is the belt that is placed around the pulleys?

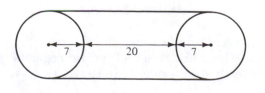

Find the perimeter of the figures in Exercises 37–40.

37.

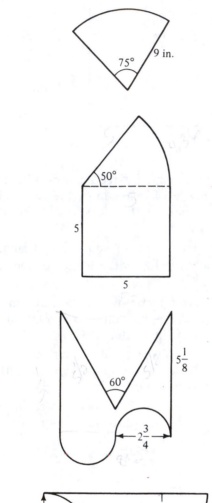

38.

39.

40.

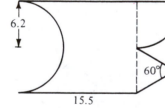

Use a calculator to do the computations. Give your answers to the nearest thousandth.

41.

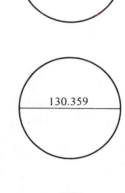

42.

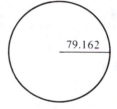

43.

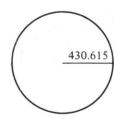

44.

45.

Calculator Exercises

A more precise approximation of the number pi is 3.1416. Use this value of pi to compute the circumference or perimeter in Exercises 41–45.

5.4
AREAS OF POLYGONS

The **area** of a region enclosed by a geometric figure is the number of units needed to completely cover the region. Standard units of area measure are square regions, such as a square inch, a square foot, or some other convenient square unit. If the region has an irregular shape it is necessary to *estimate* the number of square units needed to cover the surface.

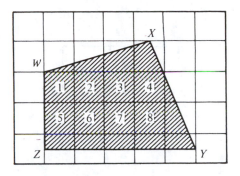

EXAMPLE 1 The region enclosed by polygon *ABCD* is completely covered by 15 square units. Thus, the area of polygon *ABCD* is 15 square units. (*Note:* When we say "the area of the polygon," or "the area of a geometric figure," we mean the area of the region enclosed by the figure.)

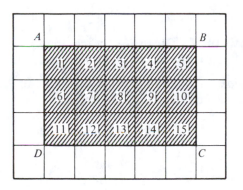

EXAMPLE 2 There are 8 square units completely inside polygon *WXYZ*. There are 11 square units partially inside of the polygon. These 11 partial units combine to form *approximately* $5\frac{1}{2}$ complete units. Thus, the area of polygon *WXYZ* is approximately $8 + 5\frac{1}{2}$, or $13\frac{1}{2}$ square units.

EXAMPLE 3 There are 9 units completely inside the figure. There are 8 units partially inside the figure, which form *approximately* 4 complete units. Thus, the area of the figure is approximately $9 + 4$, or 13 square units.

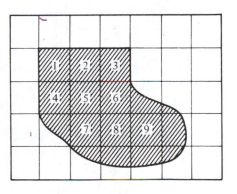

The area of a rectangle can be computed by multiplying the measure of its length times the measure of its width.

Area of a rectangle = Length × Width

This is symbolized as:

$A = lw.$

Of course, both length and width must be measured in the same linear unit to compute area in this way.

EXAMPLE 4

Area = Length × Width

= (7)(4)

= 28 square units

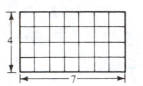

EXAMPLE 5

Area = (3.4)(5.25)

= 17.85 square units

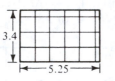

EXAMPLE 6

Area = (16 in.)(28 in.)

= (16)(28)(in.)(in.)

= 448 in.^{2}

= 448 square inches

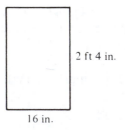

The facts in Figure 5.21 are useful in converting from one unit of area measure to another.

FIGURE 5.21

(i)

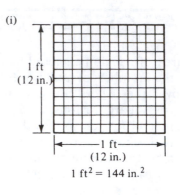

1 ft² = 144 in.²

(ii)

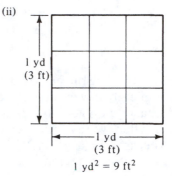

1 yd² = 9 ft²

EXAMPLE 7   The area of a rectangle is 3 square feet. What is the area measure in square inches?

Solution   To change square feet to square inches, we use the conversion fact:

$$\frac{144 \text{ in.}^2}{1 \text{ ft}^2} = 1.$$

$$3 \text{ ft}^2 = 3 \text{ ft}^2 \times \frac{144 \text{ in.}^2}{1 \text{ ft}^2} = 432 \text{ in.}^2$$

EXAMPLE 8   Find the area of the rectangle. Give the answer in square feet.

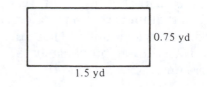

Solution

$$\text{Area} = (0.75 \text{ yd})(1.5 \text{ yd})$$
$$= (0.75)(1.5)(\text{yd})(\text{yd})$$
$$= 1.125 \text{ yd}^2$$

$$1.125 \ \cancel{\text{yd}^2} \times \frac{9 \text{ ft}^2}{1 \ \cancel{\text{yd}^2}} = 10.125 \text{ ft}^2$$

The height or width of a parallelogram is the length of a perpendicular line segment from a vertex opposite the base to the base. This is shown in Figure 5.22.

FIGURE 5.22

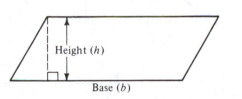

Height (*h*)

Base (*b*)

The area of a parallelogram can be computed by multiplying the measure of the base times the measure of the height.

Area of a parallelogram = Base × Height

This is symbolized as:

$A = bh.$

EXAMPLE 9

$$\text{Area} = \text{Base} \times \text{Height}$$
$$= (5)(3)$$
$$= 15 \text{ square units}$$

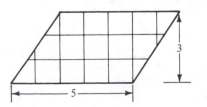

5

3

EXAMPLE 10

$$\text{Area} = (12.7)(9.8)$$
$$= 124.46 \text{ square units}$$

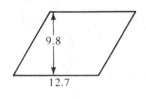

9.8

12.7

EXAMPLE 11

$$\text{Area} = (29 \text{ in.})(13 \text{ in.})$$
$$= 377 \text{ in.}^2$$

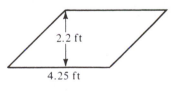

13 in.   1 ft 5 in.

2 ft 5 in.

EXAMPLE 12 The measure of the base of a parallelogram is 4.25 feet, and the height is 2.2 feet. Find the area of the parallelogram and give the answer in square inches.

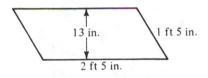

2.2 ft

4.25 ft

Solution

$$\text{Area} = (4.25 \text{ ft})(2.2 \text{ ft})$$
$$= (4.25)(2.2)(\text{ft})(\text{ft})$$
$$= 9.35 \text{ ft}^2$$

$$9.35 \ \cancel{\text{ft}^2} \times \frac{144 \text{ in.}^2}{1 \ \cancel{\text{ft}^2}} = 1346.4 \text{ in.}^2$$

EXAMPLE 13 The height of a parallelogram is 0.4 yard, and the base is 1.3 yards. What is the area of the parallelogram in square inches?

Solution

$$\text{Area} = (1.3 \text{ yd})(0.4 \text{ yd})$$
$$= 0.52 \text{ yd}^2$$

$$0.52 \text{ yd}^2 \times \frac{9 \text{ ft}^2}{1 \text{ yd}^2} \times \frac{144 \text{ in.}^2}{1 \text{ ft}^2} = 673.92 \text{ in.}^2$$

An altitude to a given side of a triangle, or the side extended, is the height of the triangle with the given side as the base of the triangle. Two examples of this are shown in Figure 5.23.

FIGURE 5.23

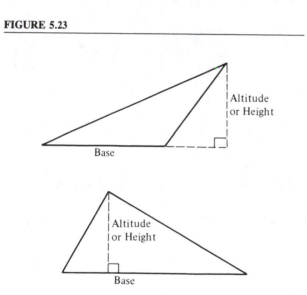

The area of a triangle is equal to one-half the area of a parallelogram with the same base and the same height (altitude). Figure 5.24 illustrates this principle. Therefore, we can compute the area of a triangle by multiplying one-half times the measure of the base times the measure of the height.

FIGURE 5.24

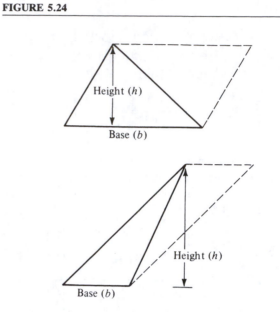

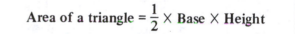

> **Area of a triangle $= \dfrac{1}{2} \times$ Base $\times$ Height**

This is symbolized as:

> $$A = \frac{1}{2}bh.$$

EXAMPLE 14

$$\text{Area} = \frac{1}{2} \times \text{Base} \times \text{Height}$$

$$= \left(\frac{1}{2}\right)(8)(4)$$

$$= 16 \text{ square units}$$

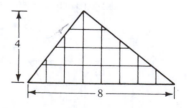

EXAMPLE 15

$$\text{Area} = \frac{1}{2} \times \text{Base} \times \text{Height}$$

$$= \left(\frac{1}{2}\right)(12.9 \text{ in.})(6.3 \text{ in.})$$

$$= 40.635 \text{ in.}^2$$

$$\doteq 40.6 \text{ in.}^2$$

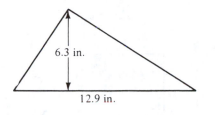

Solution

$$\text{Area} = \left(\frac{1}{2}\right)\left(9\frac{3}{4} \text{ ft}\right)\left(6\frac{1}{2} \text{ ft}\right)$$

$$= \left(\frac{1}{2}\right)\left(\frac{39}{4} \text{ ft}\right)\left(\frac{13}{2} \text{ ft}\right)$$

$$= \frac{507}{16} \text{ ft}^2$$

$$\frac{507}{16} \text{ ft}^2 \times \frac{1 \text{ yd}^2}{9 \text{ ft}^2} = \frac{507}{144} \text{ yd}^2$$

$$= 3\frac{75}{144} \text{ yd}^2$$

$$\doteq 3.52 \text{ yd}^2$$

EXAMPLE 16

$$A = \frac{1}{2}bh$$

$$= \left(\frac{1}{2}\right)(16 \text{ ft})(14.8 \text{ ft})$$

$$= 118.4 \text{ ft}^2$$

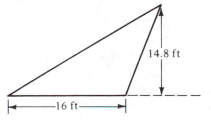

EXAMPLE 17 Find the area of the triangle. Give the answer to the nearest hundredth of a square yard.

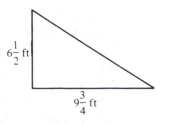

EXAMPLE 18 Find the area of the triangle. Give the answer to the nearest hundredth of a square foot.

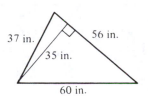

Solution Note that any side of a triangle can be used as a base.

$$\text{Area} = \left(\frac{1}{2}\right)(56 \text{ in.})(35 \text{ in.})$$

$$= 980 \text{ in.}^2$$

$$980 \text{ in.}^2 \times \frac{1 \text{ ft}^2}{144 \text{ in.}^2} = \frac{980}{144} \text{ ft}^2$$

$$\doteq 6.81 \text{ ft}^2$$

The two parallel sides of a trapezoid are called **bases** of the trapezoid, and the distance between the parallel sides is the **height** of the trapezoid. See Figure 5.25.

FIGURE 5.25

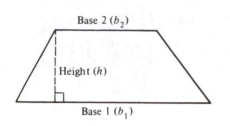

The area of a trapezoid can be computed by multiplying one-half times the measure of the height times the sum of the measures of the bases.

> **Area of a trapezoid**
> $= \frac{1}{2} \times$ **Height** $\times$ **(Sum of the bases)**

This is symbolized as:

> $A = \frac{1}{2}h(b_1 + b_2)$.

EXAMPLE 19

Area $= \frac{1}{2} \times$ Height $\times$ (Sum of the bases)

$= \left(\frac{1}{2}\right)(4)(8 + 3)$

$= \left(\frac{1}{2}\right)(4)(11)$

$= 22$ square units

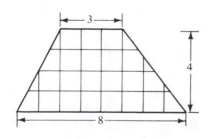

EXAMPLE 20

Area $= \frac{1}{2}h(b_1 + b_2)$

$= \frac{1}{2}(10.5)(15.3 + 8.2)$

$= \frac{1}{2}(10.5)(23.5)$

$= 123.375$ square units

$\doteq 123.4$ square units

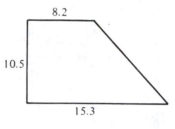

EXAMPLE 21

Area $= \left(\frac{1}{2}\right)\left(5\frac{3}{4} \text{ in.}\right)\left(9\frac{1}{2} \text{ in.} + 4\frac{1}{4} \text{ in.}\right)$

$= \left(\frac{1}{2}\right)\left(5\frac{3}{4} \text{ in.}\right)\left(13\frac{3}{4} \text{ in.}\right)$

$= \left(\frac{1}{2}\right)\left(\frac{23}{4} \text{ in.}\right)\left(\frac{55}{4} \text{ in.}\right)$

$= \frac{1265}{32} \text{ in.}^2$

$= 39\frac{17}{32} \text{ in.}^2$

$\doteq 39.53 \text{ in.}^2$

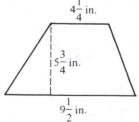

EXAMPLE 22 Find the area of the trapezoid. Give the answer to the nearest tenth of a square foot.

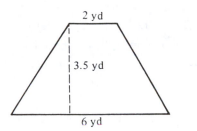

Solution

$$\text{Area} = \left(\frac{1}{2}\right)(3.5 \text{ yd})(6 \text{ yd} + 2 \text{ yd})$$

$$= 14 \text{ yd}^2$$

$$14 \text{ yd}^2 \times \frac{9 \text{ ft}^2}{1 \text{ yd}^2} = 126.0 \text{ ft}^2$$

EXAMPLE 23 Find the area of the trapezoid. Give the answer to the nearest hundredth of a square yard.

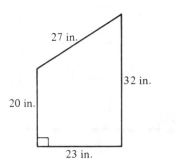

Solution

$$\text{Area} = \left(\frac{1}{2}\right)(23 \text{ in.})(32 \text{ in.} + 20 \text{ in.})$$

$$= 598 \text{ in.}^2$$

$$598 \text{ in.}^2 \times \frac{1 \text{ ft}^2}{144 \text{ in.}^2} \times \frac{1 \text{ yd}^2}{9 \text{ ft}^2} = \frac{598}{1296} \text{ yd}^2$$

$$\doteq 0.46 \text{ yd}^2$$

The area of a complex figure may be computed, even though some measures are not given, by subdividing the figure into rectangles, triangles, parallelograms, and other simple polygons.

EXAMPLE 24 Compute the area of the figure.

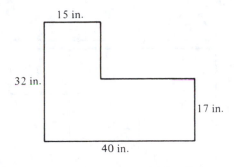

Solution Subdividing the figure into two rectangles makes it possible to determine the lengths of all of the sides of the figure.

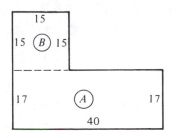

Area of rectangle A = 17 in. $\times$ 40 in.

$$= 680 \text{ in.}^2$$

Area of rectangle B = 15 in. $\times$ 15 in.

$$= 225 \text{ in.}^2$$

Total area of the figure = 680 in.^{2} + 225 in.^{2}

$$= 905 \text{ in.}^2$$

EXAMPLE 25 Compute the area of the figure.

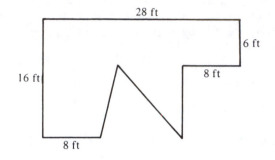

Solution

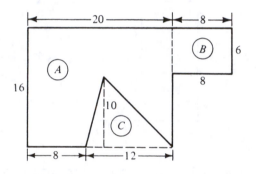

Area of rectangle A = 16 ft × 20 ft

$$= 320 \text{ ft}^2$$

Area of rectangle B = 8 ft × 6 ft

$$= 48 \text{ ft}^2$$

Area of triangle $C = \frac{1}{2} \times 12 \text{ ft} \times 10 \text{ ft}$

$$= 60 \text{ ft}^2$$

Total area of the figure $= A + B - C$

$$= 320 \text{ ft}^2 + 48 \text{ ft}^2$$
$$- 60 \text{ ft}^2$$
$$= 308 \text{ ft}^2$$

Review Exercises

Perform the indicated operations in Exercises 1–8.

1. 0.35 + 43.7 + 0.024 + 93.08

2. 74.3 + 9.038 + 46 + 0.09

3. $\frac{1}{4} + \frac{3}{5} + \frac{1}{2}$

4. $\frac{2}{3} + \frac{4}{5} + \frac{1}{6}$

5. $\frac{4}{9} \div \frac{3}{8}$

6. $\frac{6}{5} \div \frac{2}{3}$

7. 47.3 × 9.02

8. 846.2 × 0.04

9. A triangle with three equal sides is called a/an _____ triangle.

10. A triangle with no equal sides is called a/an _____ triangle.

11. Find the perimeter of the figure.

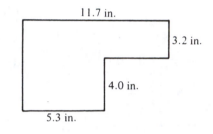

12. Find the perimeter of the figure.

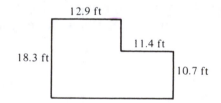

EXERCISES
5.4

Estimate the area of the shaded regions in Exercises 1–4.

1.

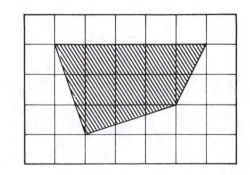

2.

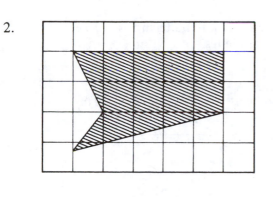

6.

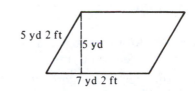

3.

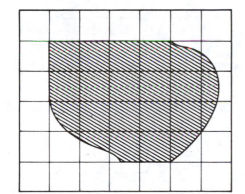

7.

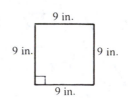

8.

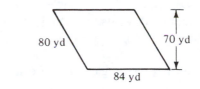

For Exercises 9–12 give the area of the parallelograms in square feet.

9.

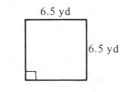

4.

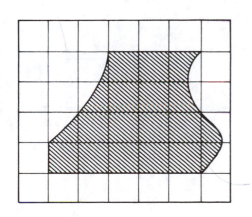

10.

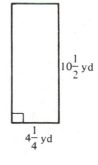

Find the area of the parallelograms in Exercises 5–8.

5.

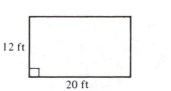

11.

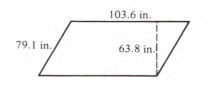

12.

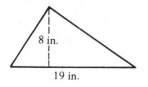

13 in.   15.8 in.

8.3 in.

13. What is the area of a rectangular metal plate that measures 4.5 inches by 6.25 inches?

14. What is the floor area in a square room that measures $12\frac{1}{2}$ feet on a side?

15. What is the area of a square where the measure of a side is 2 feet 3 inches? Give the answer in square yards.

16. What is the area of a rectangle whose length is $1\frac{1}{2}$ feet and whose width is 11 inches? Give the answer in square feet.

Find the area of the triangles in Exercises 17–20.

17.

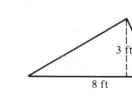

8 in.

19 in.

18.

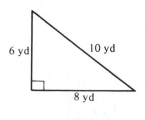

3 ft

8 ft

19.

6 yd   10 yd

8 yd

20.

13.9 in.   12 in.

7 in.

Give the area of the triangles in Exercises 21–22 in square inches.

21.

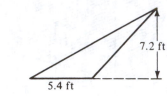

7.2 ft

5.4 ft

22.

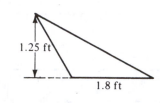

1.25 ft

1.8 ft

For Exercises 23–24 give the area of the triangles in square yards.

23.

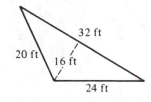

32 ft

20 ft   16 ft

24 ft

24.

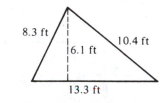

8.3 ft

6.1 ft   10.4 ft

13.3 ft

Find the area of the trapezoids in Exercises 25–30.

25.

11 in.

15 in.

18 in.

26.

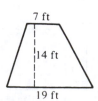

7 ft

14 ft

19 ft

27.

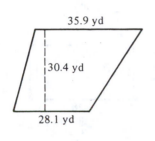

43.6 in.

31.5 in.

19.4 in.

28.

35.9 yd

30.4 yd

28.1 yd

29.

1 ft 9 in.

14 in.

10 in.

1 ft 3 in.

30.

4 yd 1 ft

5 yd 2 ft

3 yd 1 ft

4 yd

31. The measure of the base of a triangle is 17 feet, and the measure of the altitude to the base is 7 feet. What is the area of the triangle in square yards?

32. The measure of the hypotenuse of a right triangle is 35 inches, and the measures of the two legs of the triangle are 21 inches and 28 inches. What is the area of the triangle in square feet?

33. In parallelogram $ABCD$ the measure of side AB is 43.8 inches, and the measure of side

DA is 21.4 inches. If the distance between sides AB and DC is 18.7 inches, what is the area of the parallelogram?

34. In triangle EFG the measure of side FG is 14.8 units, and the altitude from E to side FG is 6.4 units. What is the area of triangle EFG?

35. What is the area of trapezoid $WXYZ$?

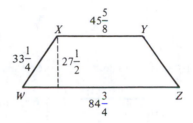

$45\frac{5}{8}$

X Y

$33\frac{1}{4}$ $27\frac{1}{2}$

W Z

$84\frac{3}{4}$

36. What is the area of triangle QNP?

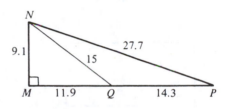

N

9.1   27.7

15

M 11.9 Q 14.3 P

37. The area of rectangle $WXYZ$ is 108 square units. The measure of side WX is 16 units. What is the measure of side XY?

38. The area of parallelogram $ABCD$ is 360 square units. What is the height of the parallelogram?

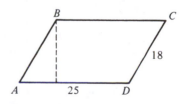

B C

18

A 25 D

39. The area of rectangle $RXYS$ is 45 square units. What is the area of trapezoid $WXYZ$?

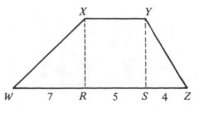

X Y

W 7 R 5 S 4 Z

40. Triangle *RST* is a right triangle. What is the length of segment *RQ*, the altitude from *R* to the hypotenuse *ST*?

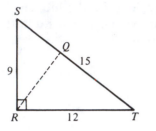

Find the area of the figures in Exercises 41–46.

41.

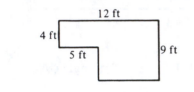

42.

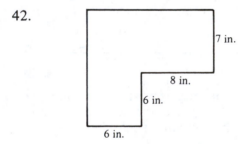

43.

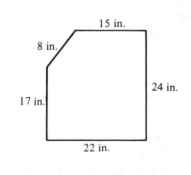

44.

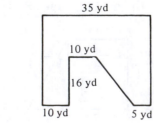

45.

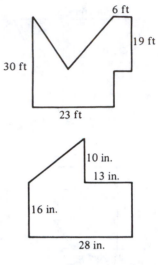

46.

Calculator Exercises

Use a calculator to compute the areas of the polygons in Exercises 47–50. Give the answers to the nearest thousandth of a square yard.

47.

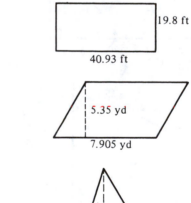

40.93 ft, 19.8 ft

48.

5.35 yd, 7.905 yd

49.

4.84 ft, 4.67 ft

50.

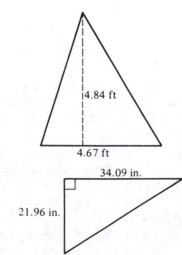

34.09 in., 21.96 in.

Give the answers for Exercises 51–52 to the nearest thousandth of a square foot.

51.

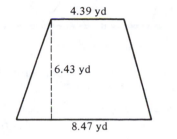

52.

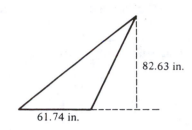

5.5
GATHERING AND ORGANIZING DATA

A problem that can easily be solved using this strategy is:

How many different diagonals can be drawn in a polygon having nineteen sides?

Prerequisite knowledge needed before attempting to solve this problem includes knowing that a polygon is a closed plane figure having line segments as sides; a diagonal of a polygon is a line segment joining two non-adjacent vertices (corners) of the polygon; and that convex polygons (where all diagonals are *inside* the figure) and concave polygons (where at least one diagonal lies *outside* the figure) having the same number of sides also have the same number of different diagonals.

A way to gather data on this problem is to look at several simple polygons and count the number of diagonals in each. The following are illustrations of four such polygons.

A triangle (3-sided polygon) has no diagonals.

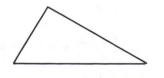

A quadrilateral (4-sided polygon) has two diagonals.

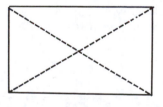

A pentagon (5-sided polygon) has five diagonals.

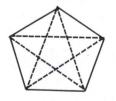

A hexagon (6-sided polygon) has nine diagonals.

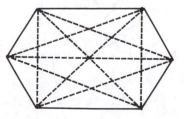

Combining this data into a table yields:

Number of sides	Number of diagonals
3	0
4	2
5	5
6	9

Do you see a pattern in the number sequence 0, 2, 5, 9? Hopefully you noticed that the difference of 2 and 0 is 2, the difference of 5 and 2 is 3, and the difference of 9 and 5 is 4. If so, you might make the conjecture that the next *difference* will be 5. Five more than nine is fourteen. Does a 7-sided polygon (septagon) have 14 different diagonals? Draw a septagon, then draw all its different diagonals, and count them. You should have 14.

Assuming that this pattern continues, an 8-sided polygon (octagon) has 6 more than 14 which is 20, diagonals; a 9-sided polygon (nonogon) has 7 more than 20, which is 27, diagonals; and a 10-sided polygon has 8 more than 27, which is 35, diagonals; and so on.

You should now be able to solve the original problem, how many different diagonals in a 19-sided polygon? Your answer should be 152 different diagonals.

EXERCISES
5.5

1. How many different diagonals in a 100-sided polygon?

2. What is the *sum* of the number of different diagonals in a 20-sided polygon and a 15-sided polygon?

3. If a straight line separates a plane into two distinct regions, how many distinct regions will eight lines, each intersecting all the others at distinct points, separate a plane?

Solutions

Do not look at these answers until you have tried to solve all three problems. If any one of your answers is incorrect, **forget the answer here** and try again to solve the problem.

1. 4,850
2. 430
3. 38

5.6
AREA OF A CIRCLE

The area of a circle can be computed by multiplying π (pi) times the square of the radius.

Area of a circle = π × Radius × Radius

This is symbolized as:

$A = \pi r^2$.

You will recall from Section 5.3 that the number pi (π) is approximately 3.14, or $3\frac{1}{7}$.

EXAMPLE 1 What is the area of a circle whose radius is 9 inches?

Solution

$$\begin{aligned}
\text{Area} &= \pi \times 9 \text{ in.} \times 9 \text{ in.} \\
&= (3.14)(9 \text{ in.})(9 \text{ in.}) \\
&= 254.34 \text{ in.}^2 \\
&\doteq 254 \text{ in.}^2
\end{aligned}$$

EXAMPLE 2 What is the area of a circle whose diameter is 1 foot 5 inches?

Solution

$$\begin{aligned}
\text{Radius} &= \frac{1}{2} \text{ diameter} \\
&= \left(\frac{1}{2}\right)(17 \text{ in.}) \\
&= 8.5 \text{ in.}
\end{aligned}$$

$$A = \pi r^2$$
$$= (3.14)(8.5 \text{ in.})^2$$
$$= (3.14)(72.25 \text{ in.}^2)$$
$$= 226.865 \text{ in.}^2$$
$$\doteq 227 \text{ in.}^2$$

EXAMPLE 3 What is the area of the circle shown? Give the answer in square feet. Round off your answer to the nearest hundredth of a square foot.

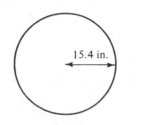

Solution

$$\text{Area} = (3.14)(15.4 \text{ in.})^2$$
$$= (3.14)(237.16 \text{ in.}^2)$$
$$= 744.6824 \text{ in.}^2$$

$$744.6824 \text{ in.}^2 \times \frac{1 \text{ ft}^2}{144 \text{ in.}^2} \doteq 5.17 \text{ ft}^2$$

EXAMPLE 4 What is the area of a "cross section" of a $\frac{1}{8}$-inch wire?

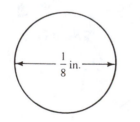

Solution

$$\text{Area} = \left(3\frac{1}{7}\right)\left(\frac{1}{16} \text{ in.}\right)^2$$
$$= \left(\frac{22}{7}\right)\left(\frac{1}{256} \text{ in.}^2\right)$$
$$= \frac{22}{1792} \text{ in.}^2$$
$$\doteq 0.01228 \text{ in.}^2$$

A **circular ring (annulus)** is defined as the area between a large circle and a smaller circle that is within it. An example of a ring is the cross section of a water pipe. The area of a ring

can be found by subtracting the area of the smaller circle from the area of the larger circle.

EXAMPLE 5 Find the area of the ring (shaded portion) shown.

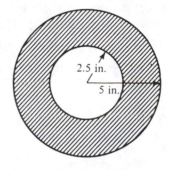

Solution

$$\text{Area of large circle} = (3.14)(5 \text{ in.})^2$$
$$= 78.5 \text{ in.}^2$$
$$\text{Area of small circle} = (3.14)(2.5 \text{ in.})^2$$
$$= 19.625 \text{ in.}^2$$
$$\text{Area of ring} = 78.5 \text{ in.}^2 - 19.625 \text{ in.}^2$$
$$= 58.875 \text{ in.}^2$$
$$\doteq 58.9 \text{ in.}^2$$

EXAMPLE 6 Find the area of the shaded portion of the figure shown. Give your answer to the nearest hundredth of a square foot.

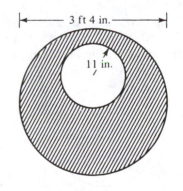

Solution

$$\text{Area of large circle} = (3.14)(20 \text{ in.})^2$$
$$= 1256 \text{ in.}^2$$
$$\text{Area of small circle} = (3.14)(11 \text{ in.})^2$$
$$= 379.94 \text{ in.}^2$$
$$\text{Area of shaded portion} = 1256 \text{ in.}^2$$
$$- 379.94 \text{ in.}^2$$

= 876.06 in.²

$$876.06 \text{ in.}^2 \times \frac{1 \text{ ft}^2}{144 \text{ in.}^2} \doteq 6.08 \text{ ft}^2$$

The **area of a sector** is found by multiplying the area of the circle by the fraction $\frac{\text{angle of the sector}}{360}$. This fraction represents the part of the whole circle occupied by the sector of the circle.

EXAMPLE 7 Find the area of a sector of 45° in a circle with a radius of 6 inches.

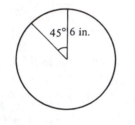

Solution

$$\text{Area of circle} = (3.14)(6 \text{ in.})^2$$
$$= 113.04 \text{ in.}^2$$
$$\text{Area of sector} = \frac{45}{360}(113.04 \text{ in.}^2)$$
$$= 14.13 \text{ in.}^2$$
$$\doteq 14 \text{ in.}^2$$

EXAMPLE 8 To form a sheet metal cone, a sector of 38°40′ is cut from a sheet metal circle with a diameter of 15 inches. What is the area of the metal used for the cone?

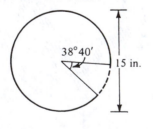

Solution

$$\text{Area of circle} = (3.14)(7.5 \text{ in.})^2$$
$$= 176.625 \text{ in.}^2$$
$$38°40' = 38.67°$$
$$\text{Area of sector} = \left(\frac{38.67}{360}\right)(176.625 \text{ in.}^2)$$
$$\doteq 18.972 \text{ in.}^2$$

Area of metal used
for the cone $\doteq 176.625 \text{ in.}^2$
$$- 18.972 \text{ in.}^2$$
$$\doteq 157.653 \text{ in.}^2$$

If a chord is drawn between the endpoints of the arc of a sector, the sector is divided into two figures, a triangle and a **segment**. The shaded portion of Figure 5.26 is a segment of the circle. The area of a segment can be found by subtracting the area of the formed triangle from the area of the sector containing the segment.

FIGURE 5.26

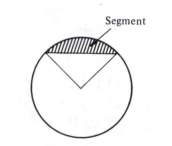

EXAMPLE 9 Find the area of the segment in the figure.

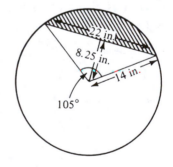

Solution

$$\text{Area of circle} = (3.14)(14 \text{ in.})^2$$
$$= 615.44 \text{ in.}^2$$
$$\text{Area of sector} = \left(\frac{105}{360}\right)(615.44 \text{ in.}^2)$$
$$\doteq 179.5 \text{ in.}^2$$
$$\text{Area of triangle} = \left(\frac{1}{2}\right)(8.25 \text{ in.})(22 \text{ in.})$$
$$= 90.75 \text{ in.}^2$$
$$\text{Area of segment} \doteq 179.5 \text{ in.}^2 - 90.75 \text{ in.}^2$$
$$\doteq 88.75 \text{ in.}^2$$

Usually, the most convenient way to find the area of a compound figure is to find the areas of the individual parts, add them together, and then subtract the areas of any holes and cut-out sections.

EXAMPLE 10 Find the area of the metal stamping shown in the figure. Give your answer to the nearest hundredth of a square inch.

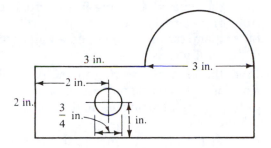

Solution You may find it convenient to label the parts of a compound figure to keep better track of your computations.

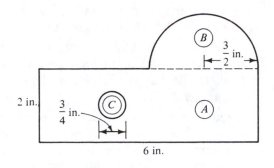

Area of part A = (2 in.)(6 in.)
$$= 12 \text{ in.}^2$$

Area of part B = $\left(\frac{1}{2}\right)\left(3\frac{1}{7}\right)\left(\frac{3}{2} \text{ in.}\right)^2$

$$= \frac{198}{56} \text{ in.}^2$$

$$\doteq 3.54 \text{ in.}^2$$

Area of part C = $\left(3\frac{1}{7}\right)\left(\frac{3}{8} \text{ in.}\right)^2$

$$= \frac{198}{448} \text{ in.}^2$$

$$\doteq 0.44 \text{ in.}^2$$

Total area $\doteq$ 12 in.^{2} + 3.54 in.^{2}
$$- 0.44 \text{ in.}^2$$
$$\doteq 15.1 \text{ in.}^2$$

EXAMPLE 11 Find the area of the machine part shown in the figure. Give your answer to the nearest hundredth of a square inch.

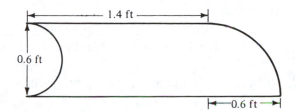

Solution

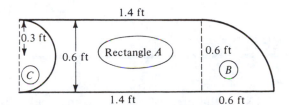

Area of part A = (0.6 ft)(1.4 ft)
$$= 0.84 \text{ ft}^2$$

Area of part B = $\left(\frac{1}{4}\right)$(3.14)(0.6 ft)^2

$$= 0.2826 \text{ ft}^2$$

Area of part C = $\left(\frac{1}{2}\right)$(3.14)(0.3 ft)^2

$$= 0.1413 \text{ ft}^2$$

Total area = 0.84 ft^2 + 0.2826 ft^2
$$- 0.1413 \text{ ft}^2$$
$$= 0.9813 \text{ ft}^2$$

$$0.9813 \text{ ft}^2 \times \frac{144 \text{ in.}^2}{1 \text{ ft}^2} \doteq 141.31 \text{ in.}^2$$

Review Exercises

Find the products in Exercises 1–6.

1. (1.3)(84.4)(0.05)
2. (2.5)(0.73)(9.426)
3. $(15.3)^2$ 4. $(0.34)^2$

5. $\frac{3}{16} \times \frac{7}{5}$

6. $\frac{4}{9} \times \frac{8}{15}$

7. Find the circumference of a circle with a diameter of 7 inches.

8. Find the circumference of a circle with a radius of 4.36 feet.

9. Change 27°34′ to degrees. Give your answer to the nearest thousandth of a degree.

10. Change 104°17′ to degrees. Give your answer to the nearest hundredth of a degree.

11. Change 14.7 square feet to square yards.

12. Change 243.57 square inches to square feet.

EXERCISES

5.6

1. What is the area of a circle whose radius is 5.7 inches?

2. What is the area of a circle whose radius is 2.3 feet?

3. What is the area of a circle whose diameter is 25.6 feet?

4. What is the area of a circle whose diameter is 1.4 yards?

5. Find the area of a circle whose radius is 6.8 inches. Give your answer to the nearest thousandth of a square foot.

6. Find the area of a circle whose diameter is 1.21 feet. Give your answer to the nearest square inch.

7. Find the area of the shaded ring.

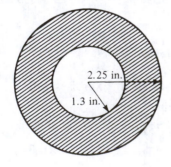

8. Find the area of the shaded ring.

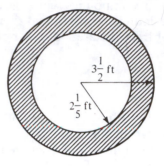

9. Find the area of the ring in the cross section of a circular concrete pipe with an outer diameter of 7 inches and an inner diameter of 4 inches.

10. Find the area of the ring in the cross section of a circular concrete pipe with an outer radius of 8 inches if the concrete is 2.5 inches thick.

11. A water pipe with a 4-inch radius can carry how many times as much water as a pipe with a 2-inch radius? (*Hint:* Compare the cross-sectional areas of the two pipes.)

12. A water main with an 18-inch diameter can carry how many times as much water as a water main with a 16-inch diameter?

13. A round steel rod will support a load of 8000 pounds per square inch of its cross section. If the bar is $\frac{15}{16}$ inch in diameter, how much will it support?

14. A circular ventilation opening is to be covered by heavy wire screen. The radius of the opening is 1 foot 10 inches. If the circular screen cover is cut from a piece four feet square how much screen is wasted? Allow a 1-inch overlap for fastening the screen to the wall.

15. What is the area of the sector shown in the figure?

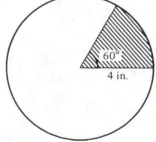

16. What is the area of the sector shown in the figure?

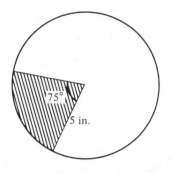

17. The water from a certain sprinkler nozzle will reach the shaded area shown in the figure. How many square yards will the water reach?

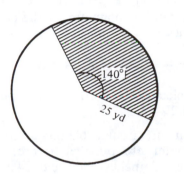

18. Determine the area of the shaded sector. Give your answer to the nearest thousandth of a square foot.

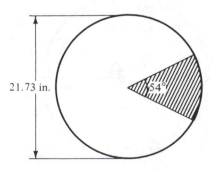

19. A tank bottom is cut from a steel plate as shown. What is the area of the tank bottom? What percent of the plate is wasted?

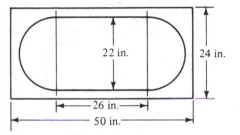

20. A metal bracket is cut from $3'' \times 6''$ stock. What is the area of the bracket shown? What percent of stock is wasted for each bracket cut?

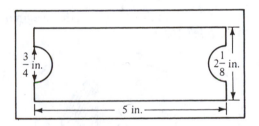

21. How many square feet of ground area are covered by the curved driveway shown?

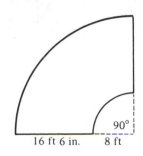

22. What is the area of the metal in the key section shown?

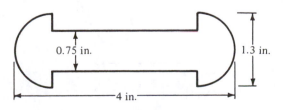

23. A segment of a circle is cut off to form a template for a cam. What is the area of the template?

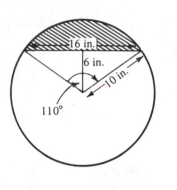

24. What is the area of the shaded segment in the circle?

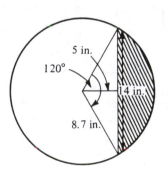

25. What is the area of the metal plate after the two holes have been drilled and tapped as shown?

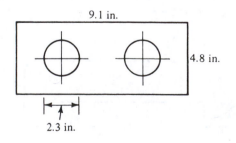

26. What is the area of the face of the pipe flange shown in the figure?

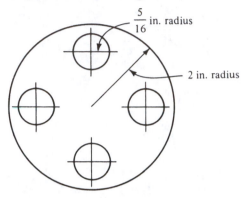

27. What is the inside area of the cross section of a piece of $1\frac{3}{4}$-inch pipe if the pipe is 0.28 inch thick?

28. What is the area of a circular plate with a diameter of 10 inches after it has been drilled and tapped for a 1-inch conduit? The outside diameter of the conduit is $1\frac{1}{4}$ inches.

Calculator Exercises

Use a calculator to do the computations in Exercises 29–34.

29. Eighteen circular blanks 3.465 inches in diameter are punched out of a sheet of metal 12 inches wide and 24 inches long. What is the area of one of the blanks? What percent of the metal sheet is wasted?

30. A regular hexagon is to be cut from a circle with a radius of 5.32 inches. How much of the metal circle is wasted?

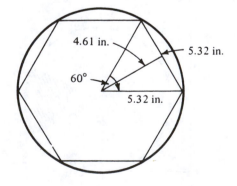

31. What is the area of the shaded region?

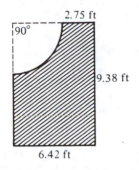

32. What is the area of the surface of the plate shown?

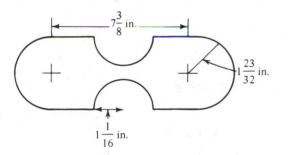

33. The brake lining of a disk brake has an outside diameter of 9.573 inches and is 2.275 inches wide. What is the total area of the braking surface to the nearest thousandth of a square inch?

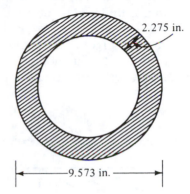

34. A circular plate with a radius of $3\frac{7}{32}$ inches is drilled and tapped for three cables with diameters of $\frac{7}{8}$ inch, $\frac{9}{32}$ inch, and $\frac{5}{16}$ inch. What is the area of the plate after the three holes have been drilled?

5.7
VOLUME

The **volume** of a three-dimensional object is the amount of space it occupies. Standard units of volume measure include a cubic inch, a cubic foot, and a cubic yard. A **cubic inch** is the amount of space occupied by a cube one inch long, one inch wide, and one inch high. This is shown in Figure 5.27. Notice that the corners are all at right angles to each other.

Likewise, a **cubic foot** is the amount of space occupied by a cube one foot long, one foot wide, and one foot high. A **cubic yard** is the amount of space occupied by a cube one

FIGURE 5.27

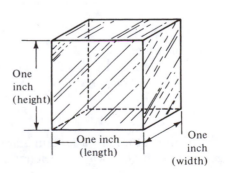

yard long, one yard wide, and one yard high. Other standard units of volume measure can be defined in a similar way.

The volume of a rectangular box can be determined by counting the number of 1-unit cubes it will take to fill the box. For example, a box 4 units long, 2 units wide, and 3 units high has a volume of 4 × 2 × 3, or 24, cubic units. As can be seen from Figure 5.28, it takes three layers of 1-unit cubes to fill the box with eight cubes in each layer, for a total of 24 cubes.

FIGURE 5.28

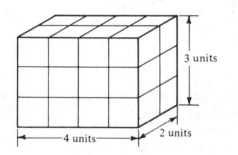

It is not always practical to fill a rectangular box with cubes to determine its volume, nor is it possible to fill the object if it is a solid. A simple way to compute the volume of a rectangular box or a rectangular solid is to multiply the measure of the length times the measure of the width times the measure of the height.

> **Volume of a rectangular solid (box) =**
> **Length × Width × Height**

This is symbolized as:

> $V = lwh.$

To compute the volume of a rectangular solid or rectangular box in this way, the length, width, and height must all be measured with the same linear unit.

EXAMPLE 1 What is the volume of a rectangular room with the dimensions 12 feet 7 inches by 10 feet 3 inches by 8 feet 2 inches? Give the answer in cubic feet to the nearest hundredth of a cubic foot.

Solution First change each measure to feet.

$$12 \text{ ft } 7 \text{ in.} = 12\frac{7}{12} \text{ ft} \doteq 12.583 \text{ ft}$$

$$10 \text{ ft } 3 \text{ in.} = 10\frac{3}{12} \text{ ft} \doteq 10.25 \text{ ft}$$

$$8 \text{ ft } 2 \text{ in.} = 8\frac{2}{12} \text{ ft} \doteq 8.167 \text{ ft}$$

$$
\begin{aligned}
\text{Volume} &= \text{Length} \times \text{Width} \times \text{Height} \\
&\doteq (12.583 \text{ ft})(10.25 \text{ ft})(8.167 \text{ ft}) \\
&\doteq (12.583)(10.25)(8.167)(\text{ft})(\text{ft})(\text{ft}) \\
&\doteq 1053.34 \text{ ft}^3 \\
&\doteq 1053.34 \text{ cubic feet}
\end{aligned}
$$

EXAMPLE 2 One cubic inch of brass weighs 0.302 pound. What is the weight of a solid rectangular brass bar 2 inches by 6 inches by 1 foot?

Solution

$$
\begin{aligned}
\text{Volume of the bar} &= 2 \text{ in.} \times 6 \text{ in.} \times 12 \text{ in.} \\
&= 144 \text{ cubic inches}
\end{aligned}
$$

$$\text{Weight of the bar} = 144 \text{ in.}^3 \times \frac{0.302 \text{ lb}}{1 \text{ in.}^3}$$

$$= 43.488 \text{ lb}$$

The facts in Figure 5.29 are useful in converting from one unit of volume measure to another.

FIGURE 5.29

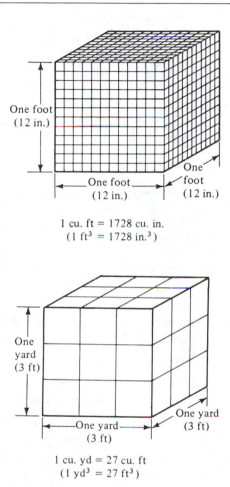

1 cu. ft = 1728 cu. in.
($1 ft^3 = 1728 in.^3$)

1 cu. yd = 27 cu. ft
($1 yd^3 = 27 ft^3$)

A solid (or hollow) three-dimensional object that has two parallel bases that are polygons with the same size and shape, and whose sides are parallelograms, is called a **prism**. A prism-like solid whose parallel bases are equal circles is called a **cylinder**. Several examples of prisms and a cylinder are shown in Figure 5.30. The volume of a prism or cylinder can be found by multiplying the area of one of the bases by the height of the object.

FIGURE 5.30

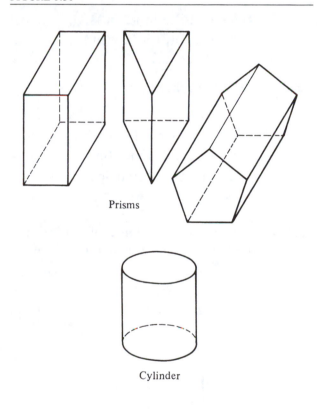

Prisms

Cylinder

EXAMPLE 3 The volume of an iron casting is 2.3 cubic feet. What is the volume of the casting in cubic inches?

Solution

$$2.3 \ ft^3 \times \frac{1728 \ in.^3}{1 \ ft^3} = 3974.4 \ in.^3$$

EXAMPLE 4 The specifications for a roadbed call for a base of 675 cubic feet of crushed rock. How many cubic yards of rock are needed?

Solution

$$675 \ ft^3 \times \frac{1 \ yd^3}{27 \ ft^3} = 25 \ yd^3$$

> **Volume of prism or cylinder =**
> **Area of base × Height**

This is symbolized as:

> $$V = Bh.$$

EXAMPLE 5 Find the volume of a cylinder with radius 3 inches and height 7 inches.

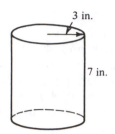

Solution The area of the base of the cylinder is:

$$B = (3.14)(3 \text{ in.})(3 \text{ in.})$$
$$= (3.14)(3)(3)(\text{in.})(\text{in.})$$
$$= 28.26 \text{ in.}^2$$

The volume of the cylinder is:

$$V = Bh$$
$$= (28.26 \text{ in.}^2)(7 \text{ in.})$$
$$= (28.26)(7)(\text{in.}^2)(\text{in.})$$
$$= 197.82 \text{ in.}^3 \text{ (cubic inches)}.$$

EXAMPLE 6 Find the volume of a triangular wedge with dimensions as shown in the figure.

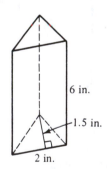

Solution The area of the triangular base of the wedge is:

$$B = \left(\frac{1}{2}\right)(2 \text{ in.})(1.5 \text{ in.})$$
$$= \left(\frac{1}{2}\right)(2)(1.5)(\text{in.})(\text{in.})$$
$$= 1.5 \text{ in.}^2$$

Volume = Area of base × Height

$$= (1.5 \text{ in.}^2)(6 \text{ in.})$$
$$= (1.5)(6)(\text{in.}^2)(\text{in.})$$
$$= 9 \text{ in.}^3$$

EXAMPLE 7 Find the volume of concrete in the section of sewer pipe shown in the figure. Give your answer to the nearest tenth of a cubic foot.

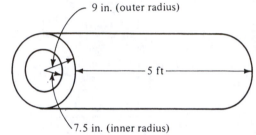

Solution The area of the base is the difference between the areas of the two circles in the cross section of the pipe.

Area of large circle = $(3.14)(9 \text{ in.})^2$
$$= 254.34 \text{ in.}^2$$
Area of small circle = $(3.14)(7.5 \text{ in.})^2$
$$= 176.625 \text{ in.}^2$$
$$B = 254.34 \text{ in.}^2$$
$$- 176.625 \text{ in.}^2$$
$$= 77.715 \text{ in.}^2$$

The height in inches is $5 \text{ ft} \times \dfrac{12 \text{ in.}}{1 \text{ ft}} = 60 \text{ in.}$

$$V = Bh$$
$$= (77.715 \text{ in.}^2)(60 \text{ in.})$$
$$= 4662.9 \text{ in.}^3$$

$$4662.9 \text{ in.}^3 \times \frac{1 \text{ ft}^3}{1728 \text{ in.}^3} \doteq 2.7 \text{ ft}^3$$

Basketballs, golf balls, and ping pong balls are examples of three-dimensional objects called **spheres**. The volume of a sphere can be computed by multiplying $\frac{4}{3}$ times π times the radius cubed.

$$\boxed{\textbf{Volume of a sphere} = \frac{4}{3} \times \pi \times \textbf{(radius)}^3}$$

This is symbolized:

$$\boxed{V = \frac{4}{3}\pi r^3 .}$$

EXAMPLE 8 Find the volume of a sphere with a radius of 4 inches.

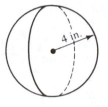

Solution

$$\text{Volume} = \frac{4}{3} \times \pi \times \text{(radius)}^3$$

$$= \left(\frac{4}{3}\right)(3.14)(4 \text{ in.})^3$$

$$= \left(\frac{4}{3}\right)(3.14)(64 \text{ in.}^3)$$

$$\doteq 267.95 \text{ in.}^3$$

The volume of a compound, or irregular, solid can be computed by subdividing the object into regular solids or parts of regular solids, computing the volume of each part, and then combining the volumes of the parts.

EXAMPLE 9 Determine the volume of the metal casting shown in the figure.

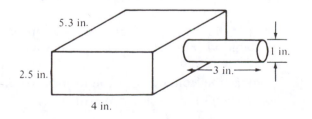

Solution The casting can be subdivided into a rectangular solid and a cylinder.

$$\begin{aligned} \text{Volume of} \\ \text{rectangular solid} &= lwh \\ &= (4 \text{ in.})(2.5 \text{ in.})(5.3 \text{ in.}) \\ &= 53 \text{ in.}^3 \end{aligned}$$

$$\begin{aligned} \text{Volume of} \\ \text{cylinder} &= Bh \\ &= [(3.14)(0.5 \text{ in.})^2](3 \text{ in.}) \\ &= [(3.14)(0.25 \text{ in.}^2)](3 \text{ in.}) \\ &= 2.355 \text{ in.}^3 \end{aligned}$$

$$\begin{aligned} \text{Total area} &= 53 \text{ in.}^3 + 2.355 \text{ in.}^3 \\ &= 55.355 \text{ in.}^3 \end{aligned}$$

EXAMPLE 10 Find the volume of the iron block illustrated in the figure. (The core hole runs the entire length of the block.)

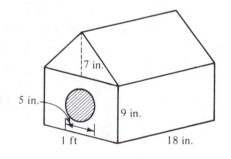

Solution

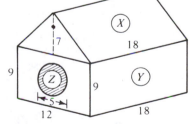

$$\text{Volume of } X = Bh$$

$$= \left[\left(\frac{1}{2}\right)(12 \text{ in.})(7 \text{ in.})\right](18 \text{ in.})$$

$$= 756 \text{ in.}^3$$

Volume of $Y = lwh$

$\qquad = (18 \text{ in.})(12 \text{ in.})(9 \text{ in.})$

$\qquad = 1944 \text{ in.}^3$

Volume of $Z = Bh$

$\qquad = [(3.14)(2.5 \text{ in.})^2](18 \text{ in.})$

$\qquad = 353.25 \text{ in.}^3$

Total volume $= X + Y - Z$

$\qquad = 756 \text{ in.}^3 + 1944 \text{ in.}^3$

$\qquad \quad - 353.25 \text{ in.}^3$

$\qquad = 2346.75 \text{ in.}^3$

Review Exercises

1. Thirty-eight percent of 450 is _____ .

2. Seventy-six percent of 25 is _____ .

3. Change 4 feet 5 inches to feet. Give your answer to the nearest hundredth of a foot.

4. Change 2 feet 11 inches to feet. Give your answer to the nearest thousandth of a foot.

5. Change 0.7 inch to the nearest sixteenth of an inch.

6. Change 0.4 inch to the nearest eighth of an inch.

7. Write the numeral three hundred and fifty-two thousandths.

8. Write the numeral six thousand seven and eight tenths.

9. Compute the area of triangle ABC.

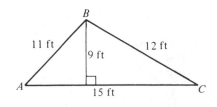

10. Compute the area of triangle XYZ.

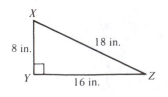

11. An eight-sided polygon is called a/an _____ .

12. A six-sided polygon is called a/an _____ .

EXERCISES

5.7

1. How many cubic feet of space are there in a storage area 9 feet long, 7 feet wide, and 8 feet high?

2. How many cubic feet of concrete are needed for a retaining wall 50 feet long, 5.5 feet high, and 1.25 feet wide?

3. How many cubic yards of concrete are needed for a driveway 20 feet wide, 70 feet long, and 6 inches thick?

4. How many cubic yards of earth must be removed for an excavation 35.4 feet long, 17 feet 8 inches wide, and 8 feet 6 inches deep?

5. One cubic inch of cast aluminum weighs 0.093 pound. How much does an aluminum bar 5 inches by 3 inches by 12 inches weigh?

6. A rectangular water tank is 5 feet long, 2.5 feet wide, and 21 inches high. If the tank is filled to the top, how many gallons will it hold? One cubic foot will hold about 7.5 gallons.

7. What is the weight of a piece of steel $\frac{1}{4}$ inch thick, 2 feet 3 inches long, and 8 inches wide? The weight of steel is 0.28 pound per cubic inch.

8. What is the weight of a brass cube 3.2 inches on a side? Brass weighs 0.302 pound per cubic inch.

9. What is the volume of a cylindrical propane gas tank with a diameter of 4 inches and a length of 13 inches?

10. What is the volume of a round iron bar 3 feet long with a diameter of $\frac{7}{8}$ inch?

11. A storage tank in the shape of a cylinder has a radius of 2 feet and is 5 feet high. What is the volume of the tank to the nearest hundredth of a cubic yard?

12. A cylindrical gasoline tank is 10 inches in diameter and 30 inches long. What is the volume of the tank to the nearest thousandth of a cubic foot?

13. What is the total piston displacement of a 6-cylinder engine if each piston has a diameter of $3\frac{3}{16}$ inches and a $4\frac{1}{2}$-inch stroke?

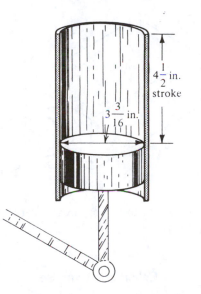

14. Two storage tanks are welded together as shown. What is the volume of the two tanks to the nearest hundredth of a cubic inch?

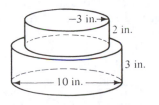

15. What is the volume of a sphere with a radius of 5 inches?

16. What is the volume of a rubber ball that has a diameter of 12 inches?

17. A spherical water tank has a radius of 8.2 feet. How many gallons of water will the tank hold if one gallon of water has a volume of approximately 231 cubic inches?

18. A solid iron sphere has a diameter of 1.14 feet. What is the weight of the sphere if iron weighs 0.26 pound per cubic inch?

19. A cross section of a concrete cistern is shown in the figure. How many cubic yards of concrete are needed to build the cistern?

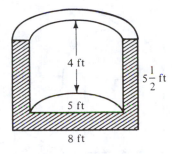

20. A loading platform is supported by four concrete pillars. Each pillar is $2\frac{1}{2}$ feet in diameter and 7 feet high. How many cubic yards of concrete are needed to build the support pillars?

Find the volume of the figures in Exercises 21–28.

21.

22.

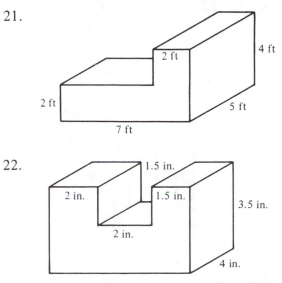

23.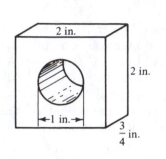

24.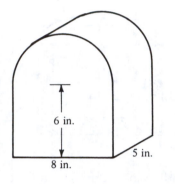

25.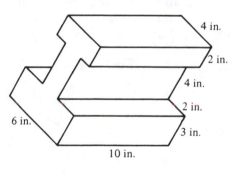

26.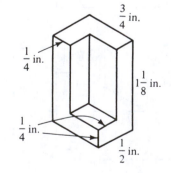

27.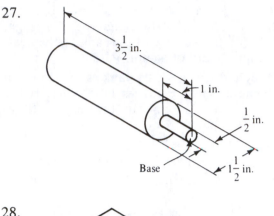

28.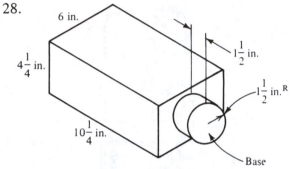

(*Note:* The symbol $1\frac{1}{2}''^{R}$ means a radius length of $1\frac{1}{2}$ inches.)

Calculator Exercises

Use a calculator to do the computations for Exercises 29–34.

29. A brass bushing has an outer diameter of 2.375 inches and an inner diameter of 1.084 inches. If brass weighs 521.7 pounds per cubic foot, what is the weight of one bushing?

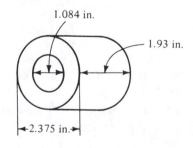

30. A hollow, wrought iron sphere has an inner diameter of 2.75 inches. The iron is 0.16 inch thick. If wrought iron weighs 0.2834 pound per cubic inch, what is the weight of the iron ball?

31. What is the weight of 300 aluminum parts shown? (Aluminum weighs about 0.093 pound per cubic inch.)

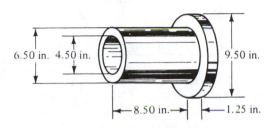

32. What is the weight of 450 steel parts shown? (Steel weighs about 0.283 pound per cubic inch.)

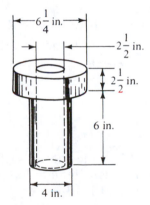

33. Find the volume of the object shown to the nearest hundredth of a cubic inch.

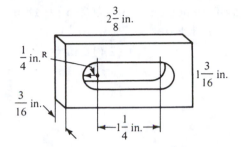

34. Find the volume of the object shown to the nearest thousandth of a cubic foot.

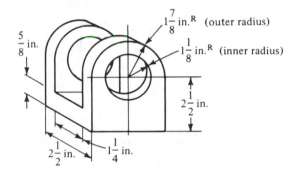

5.8
SURFACE AREA

There are two types of surface area of three-dimensional objects.

1. **Lateral surface area** is the sum of the areas of each of the sides of the object.

2. **Total surface area** is the sum of the lateral surface area and the areas of the bases of the object.

EXAMPLE 1 Find the lateral surface area of a rectangular solid with length 10 inches, width 6 inches, and height 3 inches.

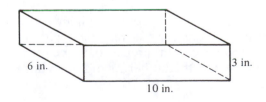

Solution

Area of large side = (3 in.)(10 in.) = 30 in.²
Area of small side = (3 in.)(6 in.) = 18 in.²
Lateral area = (2)(30 in.²)
                    + (2)(18 in.²)
                    = 60 in.² + 36 in.²
                    = 96 in.²

EXAMPLE 2 What is the total area of a metal junction box that is $7\frac{1}{2}$ inches by $5\frac{3}{4}$ inches by 12 inches?

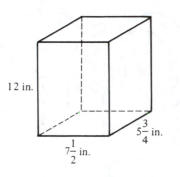

Solution

	Area	Total
2 sides	$\left(7\frac{1}{2} \text{ in.}\right)(12 \text{ in.}) = 90 \text{ in.}^2$	180 in.²
2 sides	$\left(5\frac{3}{4} \text{ in.}\right)(12 \text{ in.}) = 69 \text{ in.}^2$	138 in.²
2 bases	$\left(7\frac{1}{2} \text{ in.}\right)\left(5\frac{3}{4} \text{ in.}\right)$	
	= 43.125 in.²	86.25 in.²

Total area = 180 in.² + 138 in.² + 86.25 in.²
            = 404.25 in.²

The lateral surface of a cylinder, if it were unrolled, would form a rectangle with a length equal to the circumference of the cylinder and a width equal to the height of the cylinder (see Figure 5.31).

FIGURE 5.31

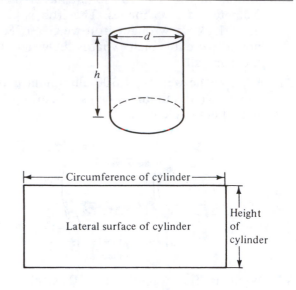

Lateral area
of a cylinder = Circumference × Height
             = π × Diameter × Height
             = πdh

EXAMPLE 3 Find the lateral surface area of a cylinder if the diameter of the cylinder is 8 inches and the height is 14 inches.

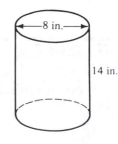

Solution

Lateral area = π × Diameter × Height
             = (3.14)(8 in.)(14 in.)
             = 351.68 in.²

EXAMPLE 4 What is the total surface area of a cylinder with a radius of 10 inches and a height of 15 inches. Compute the answer to the nearest hundredth of a square foot.

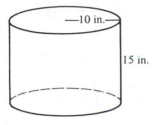

Solution

$$\text{Lateral area} = \pi dh$$
$$= (3.14)(20 \text{ in.})(15 \text{ in.})$$
$$= 942 \text{ in.}^2$$

$$\text{Area of base} = \pi r^2$$
$$= (3.14)(10 \text{ in.})^2$$
$$= 314 \text{ in.}^2$$

$$\text{Total area} = 942 \text{ in.}^2 + (2)(314 \text{ in.}^2)$$
$$= 1570 \text{ in.}^2 \quad \text{Note that there are two bases—the top and the bottom of the cylinder.}$$

$$1570 \text{ in.}^2 \times \frac{1 \text{ ft}^2}{144 \text{ in.}^2} \doteq 10.90 \text{ ft}^2$$

EXAMPLE 5 What is the area of a label used to wrap the lateral surface of a cylinder with a radius of 4.2 inches and a height of 9.7 inches? Allow 0.3 inch for overlap for gluing the label.

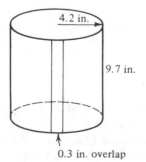

0.3 in. overlap

Solution

$$\text{Circumference}$$
$$\text{of cylinder} = \pi d$$
$$= (3.14)(8.4 \text{ in.})$$
$$= 26.376 \text{ in.}$$
$$\text{Length of label} = 26.376 \text{ in.} + 0.3 \text{ in.}$$
$$= 26.676 \text{ in.}$$
$$\text{Area of label} = (26.676 \text{ in.})(9.7 \text{ in.})$$
$$= 258.7572 \text{ in.}^2$$

The surface area of a sphere can be computed by multiplying 4 times pi times the radius squared.

> **Surface area of a sphere = $4 \times \pi \times (\text{radius})^2$**

This is symbolized as:

> $$S = 4\pi r^2.$$

Spheres are often used as storage containers since they have the greatest volume for the amount of the surface area of the container.

EXAMPLE 6 What is the surface area of a spherical water tank that has a radius of 20 feet? Compute your answer to the nearest thousandth of a square yard.

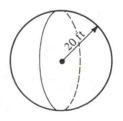

Solution

$$\text{Surface area} = 4 \times \pi \times (\text{radius})^2$$
$$= (4)(3.14)(20 \text{ ft})^2$$
$$= 5024 \text{ ft}^2$$

$$5024 \text{ ft}^2 \times \frac{1 \text{ yd}^2}{9 \text{ ft}^2} \doteq 558.222 \text{ yd}^2$$

EXAMPLE 7   What is the lateral area of a hemi-sphere (one-half of a sphere) with a diameter of $7\frac{2}{3}$ inches?

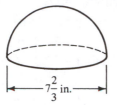

Solution

$$\text{Surface area} = \left(\frac{1}{2}\right)(4\pi r^2)$$

$$= \left(\frac{1}{2}\right)(4)(3.14)\left(\frac{23}{6}\text{ in.}\right)\left(\frac{23}{6}\text{ in.}\right)$$

$$= \frac{830.53}{9}\text{ in.}^2$$

$$\doteq 92\frac{2}{9}\text{ in.}^2$$

It is often necessary to compute the surface area of irregularly shaped three-dimensional objects. Examples 8 and 9 deal with two such cases.

EXAMPLE 8   Find the total surface area of the object shown in the figure.

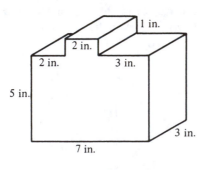

Solution

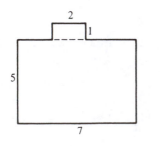

	Area	Total
2 sides	(2 in.)(1 in.) + (5 in.)(7 in.) = 37 in.²	74 in.²
1 base	(3 in.)(7 in.) = 21 in.²	21 in.²
2 sides	(3 in.)(5 in.) = 15 in.²	30 in.²
1 side	(3 in.)(3 in.) = 9 in.²	9 in.²
2 sides	(1 in.)(3 in.) = 3 in.²	6 in.²
2 sides	(2 in.)(3 in.) = 6 in.²	12 in.²
	Total surface area =	152 in.²

EXAMPLE 9   Find the total surface area of the metal casting shown in the figure.

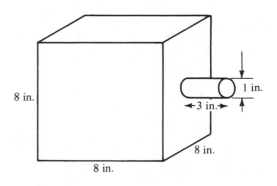

Solution   There are 6 sides of the cube. Each side has an area of (8 in.)(8 in.), or 64 square inches. The total area of the sides is 384 square inches. Note that the end of the cylinder has the same area as the hole in the side where the cylinder is attached.

Lateral area of cylinder = (3.14)(1 in.)(3 in.)

$$= 9.42\text{ in.}^2$$

Total surface area = 393.42 in.²

EXAMPLE 10   The cost of rust-proofing is 0.08 cents per square inch. What is the cost of rust-proofing the entire exterior surface of the metal storage shed shown in the figure?

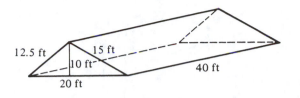

Solution

	Area		Total
1 side	(15 ft)(40 ft) = 600 ft²		600 ft²
1 side	(12.5 ft)(40 ft) = 500 ft²		500 ft²
1 base	(20 ft)(40 ft) = 800 ft²		800 ft²
2 ends	$\left(\frac{1}{2}\right)$(10 ft)(20 ft) = 100 ft²		200 ft²
		Total surface area =	2100 ft²

$$2100 \text{ ft}^2 \times \frac{144 \text{ in.}^2}{1 \text{ ft}^2} = 302{,}400 \text{ in.}^2$$

Cost of rust-proofing

$$= 302{,}400 \text{ in.}^2 \times \frac{0.08 \text{ cents}}{1 \text{ in.}^2} \times \frac{\$1}{100 \text{ cents}}$$

$$= \$241.92$$

Review Exercises

1. What is the circumference of a circle with a diameter of $15\frac{1}{4}$ inches?

2. What is the circumference of a circle with a radius of 0.78 feet?

3. What is the area of a circle with a radius of 0.35 inch?

4. What is the area of a circle with a diameter of $3\frac{1}{2}$ yards?

5. What is the area of triangle *XYZ*?

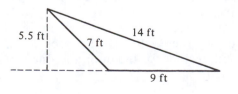

6. What is the area of triangle *MRS*?

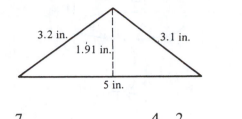

7. $\frac{3}{8} \div \frac{7}{9} =$ _____

8. $\frac{4}{5} \div \frac{2}{3} =$ _____

9. 35% of 84 is _____

10. 17% of 92 is _____

11. Write $\frac{7}{16}$ as a decimal. Round off your answer to the nearest ten-thousandth.

12. Write $\frac{5}{12}$ as a decimal. Round off your answer to the nearest thousandth.

EXERCISES
5.8

Find the lateral surface area of the objects in Exercises 1–6.

1.

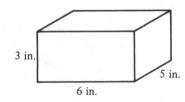

2.

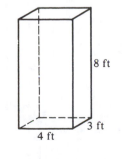

3.

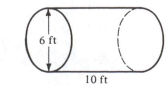

4.

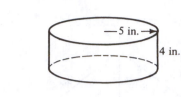

— 5 in. —

4 in.

5.

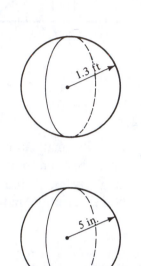

1.3 ft

6.

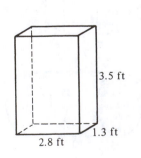

5 in.

Find the total surface area of the objects in Exercises 7–12.

7.

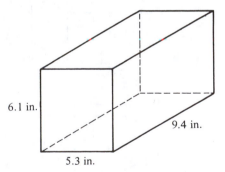

6.1 in.

9.4 in.

5.3 in.

8.

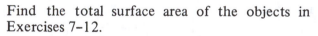

3.5 ft

1.3 ft

2.8 ft

9.

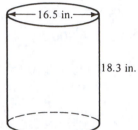

3.2 ft

8.7 ft

10.

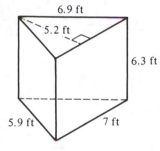

— 16.5 in. —

18.3 in.

11.

6.9 ft

5.2 ft

6.3 ft

5.9 ft   7 ft

12.

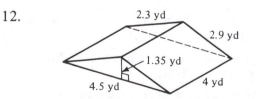

2.3 yd

2.9 yd

1.35 yd

4.5 yd   4 yd

13. What is the minimum amount of paper needed to cover a shoe box that is 13.5 inches by 6.3 inches by 5.2 inches?

14. A room is 11 feet 8 inches by 18 feet 3 inches by 7 feet 10 inches. How many square feet of wallpaper are needed to cover the walls of the room? Allow 30 square feet for the door and one window in the room.

15. A spherical storage tank with a diameter of 18 feet is to be painted. If one gallon of paint will cover approximately 200 square feet, how many gallons of paint are needed?

16. How many square inches of leather are needed to cover a basketball with a diameter of $12\frac{1}{2}$ inches, excluding waste?

17. A welded steel bin, with a top cover, is constructed as shown. How many square feet of metal plate are needed to construct the bin and cover, excluding waste?

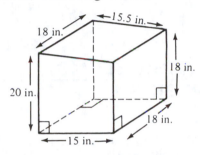

18. The cost of sheet copper is 55 cents per pound. What is the cost of the sheet copper needed to cover the sides and bottom of the container shown in the figure? The sheet copper weighs approximately 12 ounces per square foot.

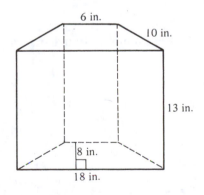

19. How many square feet of sheet metal are needed to construct a cylindrical air duct with a radius of 8 inches and a length of 12.8 feet?

20. What is the lateral surface area of a pipe with a 4-inch outside diameter and a length of 24 feet? Compute the answer to the nearest hundredth of a square foot.

21. Lead sheeting is used to line the sides and bottom of a cylindrical tank for acid storage. The tank has a diameter of 4 feet and a height of 2.5 feet. If the lead sheeting costs 0.82 cents per square inch, what is the cost of lining the tank?

22. An iron rod 8 inches long with a diameter of $1\frac{1}{4}$ inches is to be plated with a silver-copper compound. What is the cost of plating the rod if the processing cost is 78.5 cents per square inch?

23. Find the total surface area of the object in the figure.

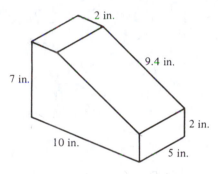

24. Find the total surface area of the object in the figure.

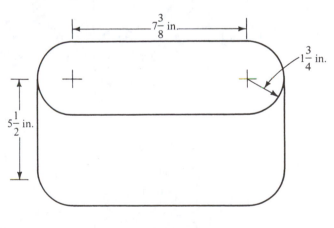

Calculator Exercises

Use a calculator to compute the total surface area of the objects in Exercises 25–29. Give each answer to the nearest thousandth of a unit.

25.

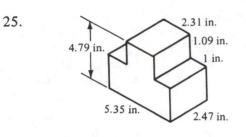

26.

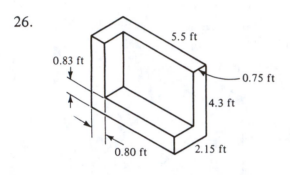

27.

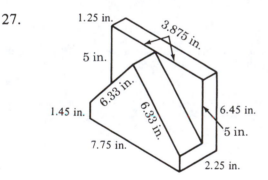

28.

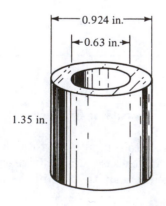

29.

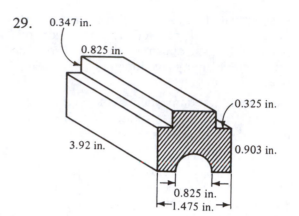

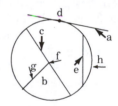

EXTRA PROBLEMS
CHAPTER 5

1. Identify the parts of the circle that are lettered in the figure.

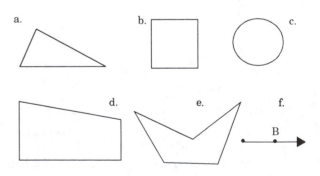

2. Identify each geometric figure in as many ways as you can.

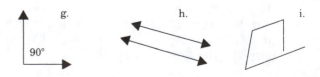

3. a. Name a line segment parallel to $\overline{AJ}$.
    b. Name a line segment parallel to $\overline{CD}$.
    c. Name a line segment parallel to $\overline{CBI}$.
    d. Name a line segment perpendicular to $\overline{FG}$.
    e. Name a line segment perpendicular to $\overline{CB}$.
    f. Name a line segment parallel to $\overline{IH}$.

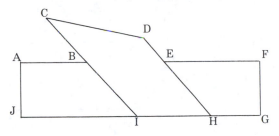

4. a. Name the lines shown in the figure.

    b. Name the line segments shown in the figure.

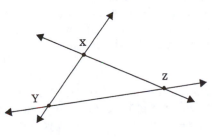

5. a. Name the lines shown in the figure.

    b. Name the rays shown in the figure.

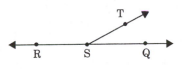

6. How long is the diameter of a circle if the radius is 5.3 inches?

7. Name the triangles in the figure.

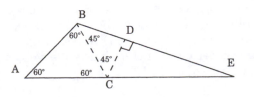

a. acute triangle

b. isosceles triangle

c. right triangle

d. equilateral triangle

e. obtuse triangle

8.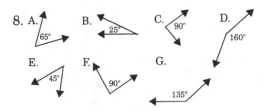

    a. Which angles are acute angles?

    b. Which angles are right angles?

    c. Which angles are obtuse angles?

    d. Which angles are complementary?

    e. Which angles are supplementary?

9. A stair is to be covered with carpet both on the rise and on the run of the stairs. Each rise is 7.5 inches and each run is 8.75 inches. Allowing for a 10 percent error margin, how long should a piece of carpet be to cover the stairs from point A to point B? Give your answer to the nearest foot.

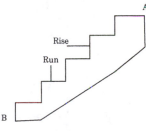

10. Find the perimeter of the hole in the plate. Round off your answer to the nearest tenth of an inch.

11. A milling machine spindle has a diameter of 3 inches. If the spindle rotates at a speed of 250 rotations per minute, what is the cutting speed of the spindle in feet per minute?

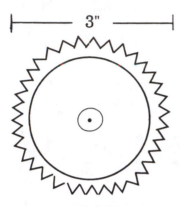

12. What is the length of arc ACB? Give your answer to the nearest tenth of an inch.

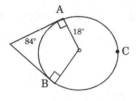

13. Find the area of the hexagonal bolt head shown in the diagram. Give your answer to the nearest hundredth of a square inch.

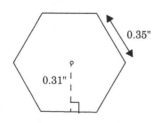

14. Triangular shaped pieces with a base 6 3/16 inches and an altitude 8 3/8 inches are blanked out of a metal strip 8 1/2 inches wide. If eight pieces are blanked out of a strip that is 50 inches long: a) find the total area of the eight triangular pieces, and b) find the area of the strip that is wasted.

15. The land parcel shown in the figure is to be graded and paved. The cost estimate is $13.65 per square yard. What is the cost estimate, to the nearest dollar, for grading and paving the entire section?

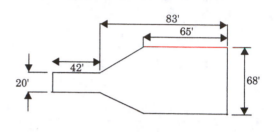

16. At the center of one side of a garage, measuring 80 feet on each side of the building, is located an air hose 60 feet in length. Find the area in which the air hose may be used outside of the building, in square yards. The hose is on the outside of the building and the maximum area is to be found.

17. What is the area of the fabric cutting pattern shown in the diagram? Give your answer to the nearest tenth of a square foot.

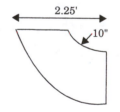

18. Find the lateral area of a cylinder with a height of 8.850 inches and the radius of the base is 3.420 inches. Round off your answer to the nearest thousandth of a square inch.

19. A hollow glass sphere has an outside circumference of 22.4 inches. The wall thickness of the sphere is 0.20 inch. Glass weighs 1.6 ounces per cubic inch. What is the weight of the glass sphere?

20. A storage container for liquids is shaped like a cylinder. The container has a radius of 2.4 feet and a height of 6.8 feet. Find the volume of the container to the nearest hundredth of a cubic foot.

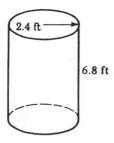

21. Find the surface of a volleyball with a diameter of 14 inches.

22. Find the total surface area of a closed cylindrical container that has a height of 5 feet, and the radius of the top and the bottom is 1.2 feet. Round off your answer to two decimal places.

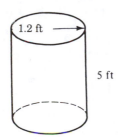

Chapter

6

THE METRIC SYSTEM

6.1
SI BASE UNITS, PREFIXES, AND LENGTH

A metric system of measure is a decimal-based system of weights and measures. The base units and the larger and smaller units derived from those base units are related to each other by **powers of ten**. The most widely accepted metric system is the **International System of Units (SI)**. The SI is a uniform language of measurement in which the symbols, prefixes, and the system itself, are exactly the same in every country that uses the system.

The SI is constructed from seven base units and two supplementary units, as we see from Table 6.1. SI units for all other quantities are derived from the nine units listed in Table 6.1. This means that all other units are expressed as products and quotients of the base and supplementary units without numerical factors. For example, the **coulomb (C)** is a derived unit representing the quantity of electricity or electric charge. A coulomb is expressed in terms of electric current and time as **A · s (ampere-second)**. Other SI derived units will be presented later in this chapter.

Multiples and submultiples of all SI units are expressed by using a set of sixteen prefixes

TABLE 6.1

Unit Name		Symbol
SI Base Units		
meter	(length)	m
kilogram	(mass)	kg
second	(time)	s
ampere	(electric current)	A
kelvin	(thermodynamic temperature)	K
candela	(luminous intensity)	cd
mole	(amount of substance)	mol
SI Supplementary Units		
radian	(plane angle)	rad
steradian	(solid angle)	sr

which represent powers of ten. These are shown in Table 6.2.

The base SI unit for length is the **meter** (also spelled metre). The prefixes presented in Table 6.2 are used to represent lengths greater or less than a meter. Table 6.3 lists most commonly used metric units of length. The comparisons

TABLE 6.2

	Prefix	Symbol	Power of Ten	
	exa	E	1 000 000 000 000 000 000	(10^{18})
	peta	P	1 000 000 000 000 000	(10^{15})
	tera	T	1 000 000 000 000	(10^{12})
	giga	G	1 000 000 000	(10^{9})
	mega	M	1 000 000	(10^{6})
Most commonly used prefixes	kilo	k	1000	(10^{3})
	hecto	h	100	(10^{2})
	deka	da	10	(10)
	deci	d	0.1	(10^{-1})
	centi	c	0.01	(10^{-2})
	milli	m	0.001	(10^{-3})
	micro	μ	0.000 001	(10^{-6})
	nano	n	0.000 000 001	(10^{-9})
	pico	p	0.000 000 000 001	(10^{-12})
	femto	f	0.000 000 000 000 001	(10^{-15})
	atto	a	0.000 000 000 000 000 001	(10^{-18})

with other familiar objects or customary units are given to help you begin thinking in terms of metric units.

TABLE 6.3

Metric Unit	Equivalent in Meters	Non-Metric Equivalent
kilometer (km)	1000 meters	About 0.6 of a mile
hectometer (hm)	100 meters	A little (about 10%) longer than a football field
dekameter (dam)	10 meters	About the length of two full-sized station wagons parked end-to-end
meter (m)	1 meter	A little (about 10%) longer than a yard
decimeter (dm)	0.1 meter	About the length of two jumbo paper-clips placed end-to-end
centimeter (cm)	0.01 meter	About the width of your little fingernail
millimeter (mm)	0.001 meter	About the thickness of a dime

To convert one metric unit to another you multiply or divide by a power of ten. For example, to convert meters to centimeters you multiply by 100; to convert meters to dekameters you divide by 10; to convert hectometers to decimeters you multiply by 1000; and so on.

The stairstep chart shown in Figure 6.1 is a useful aid when converting metric units to other metric units. The examples are all units of length, but the chart and the three-step procedure for using it can be used for all SI base units. The conversion process is based on the following principles of place value:

> Moving the decimal point to the right in a number is the same as multiplying the number by a power of ten.
>
> Moving the decimal point to the left in a number is the same as dividing the number by a power of ten.

> **PROCEDURE FOR USING THE STAIRSTEP CONVERTER**
>
> 1. Locate the original unit prefix on one of the steps.
> 2. Count the number of steps to the new unit prefix.
> 3. Move the decimal point in the original unit as many places in the same right (left) direction as you moved on the stairs. This will give the correct decimal point placement for the new unit.

FIGURE 6.1

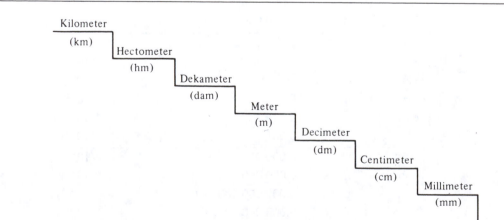

EXAMPLE 1 Change 5 dekameters to centimeters.

Solution We count three steps to the right, moving from dekameter to centimeter. Thus, we move the decimal point in 5 dam three places to the right to change to cm (multiply by 1000). Then,

$$5 \text{ dam} = 5000 \text{ cm}.$$

EXAMPLE 2 Change 4.5 meters to hectometers.

Solution We count two steps to the left moving from meter to hectometer. Thus, we move the decimal point in 4.5 m two places to the left to change to hm (divide by 100). Then,

$$4.5 \text{ m} = 0.045 \text{ hm}.$$

EXAMPLE 3 Use the stairstep chart to check the following examples.

$$6.4 \text{ m} = 6400 \text{ mm} \qquad 6.4 \text{ m} = 0.0064 \text{ km}$$
$$35.8 \text{ dam} = \quad 358 \text{ m} \qquad 35.8 \text{ dam} = \quad 3.58 \text{ hm}$$
$$0.24 \text{ km} = \quad 240 \text{ m} \qquad 47 \text{ mm} = \quad 0.047 \text{ m}$$

When adding or subtracting metric measurements, all measures must be expressed in the same unit.

EXAMPLE 4 Three metal pipes are welded together as shown in the figure. What is the overall length of the welded pipe?

Solution

$$\begin{aligned} \text{Total length} &= 115 \text{ mm} + 20 \text{ cm} + 1.5 \text{ dm} \\ &= 115 \text{ mm} + 200 \text{ mm} + 150 \text{ mm} \\ &= 465 \text{ mm} \end{aligned}$$

EXAMPLE 5 What is the perimeter of a city lot in the shape of polygon *ABCD*?

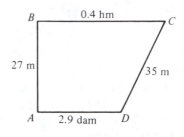

Solution

$$\begin{aligned} \text{Perimeter} &= 27 \text{ m} + 0.4 \text{ hm} + 35 \text{ m} + 2.9 \text{ dam} \\ &= 27 \text{ m} + 40 \text{ m} + 35 \text{ m} + 29 \text{ m} \\ &= 131 \text{ m} \end{aligned}$$

Review Exercises

Perform the indicated operations in Exercises 1–12.

1. 4.78×1000
2. $37.54 \times 10{,}000$
3. $6.3 \div 1000$
4. $0.84 \div 100$
5. $75.46 - 0.058$
6. $830 - 5.64$
7. $(2.8)^2$
8. $(0.35)^2$
9. $32 \div 0.8$
10. $4.2 \div 12$
11. $\sqrt{64}$
12. $\sqrt{49}$

EXERCISES

6.1

Perform the metric conversions in Exercises 1–14.

1. 4 m = _____ cm
2. 2 km = _____ m
3. 53 mm = _____ m
4. _____ hm = 16 m
5. _____ mm = 49.3 dm
6. 0.34 dam = _____ m
7. 0.059 km = _____ m
8. _____ cm = 14 hm

9. 17 cm = _____ dam

10. _____ km = 83 m

11. 16.5 dm = 1650 _____

12. 8.4 hm = 840 _____

13. 0.38 _____ = 38 cm

14. 43.8 m = 0.438 _____

15. Two pieces of metal rod, 1 dm long and 17 cm long, are cut from a piece of stock that is 1 m long. How long is the piece of stock that is left over?

16. A piece of angle iron 38 mm long is cut from a piece of stock 1 m 20 cm long. If a second piece 10.3 cm long is cut from the same piece of stock, what is the length of the piece that is left over?

17. What is the total length of the steel shaft shown in the figure in centimeters? What is the length in decimeters?

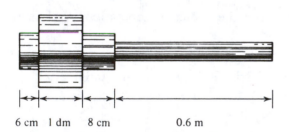

6 cm   1 dm   8 cm   0.6 m

18. What is the total length in millimeters of the welded rod shown here? What is the length in centimeters?

3 cm   54 mm   0.4 dm

19. Seventeen holes are to be drilled in a piece of plywood. Allowing 3 cm distance from each end to the center of the first and last holes, what is the distance between the centers of the holes if they are to be an equal distance apart?.

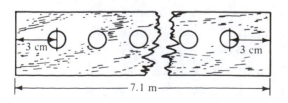

20. Thirty-one holes are to be drilled in a metal plate. Allowing 7 cm distance from each end to the center of the first and last holes, what is the distance between the centers of the holes if they are to be an equal distance apart?

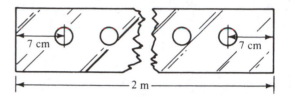

21. What is the circumference of the circle in centimeters?

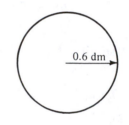

0.6 dm

22. What is the circumference of the circle in meters?

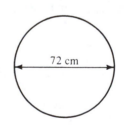

72 cm

23. What is the circumference of the circle in millimeters?

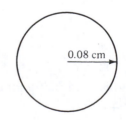

0.08 cm

24. What is the circumference of the circle in decimeters?

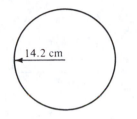

14.2 cm

Compute the perimeter of each figure in Exercises 25–32. Give each answer in meters and in centimeters.

25.

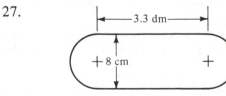

60 cm

7 dm

35 cm

1.4 m

26.

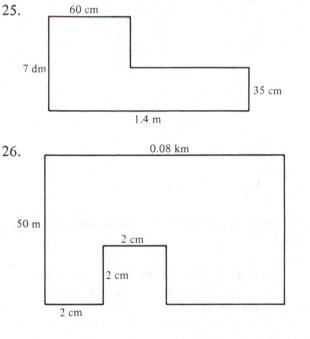

0.08 km

50 m

2 cm

2 cm

2 cm

27.

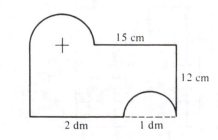

3.3 dm

8 cm

28.

15 cm

12 cm

2 dm   1 dm

29.

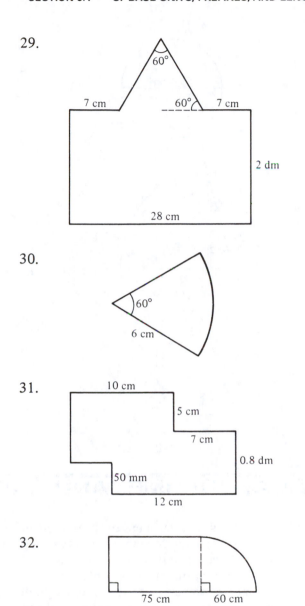

60°

7 cm   60°   7 cm

2 dm

28 cm

30.

60°

6 cm

31.

10 cm

5 cm

7 cm

0.8 dm

50 mm

12 cm

32.

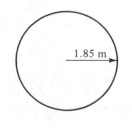

75 cm   60 cm

Calculator Exercises

Use a calculator to compute the circumference or perimeter in Exercises 33–38. Use $\pi = 3.1416$.

33.

1.85 m

34.

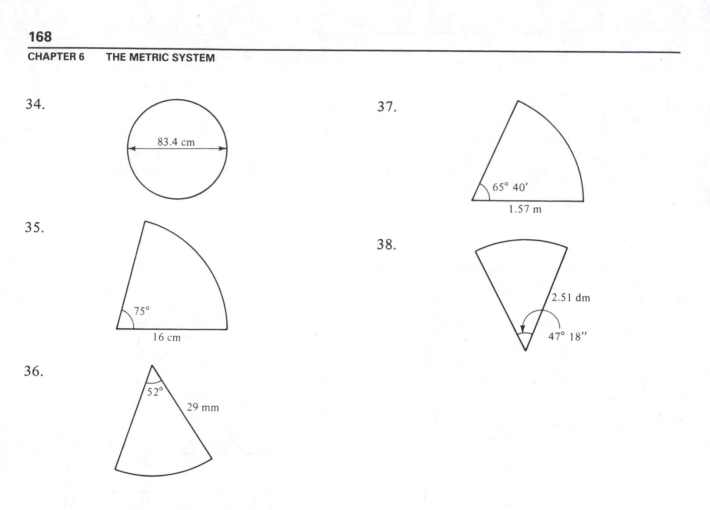

83.4 cm

35.

75°

16 cm

36.

52°

29 mm

37.

65° 40'

1.57 m

38.

2.51 dm

47° 18''

6.2
AREA, VOLUME, AND CAPACITY

The principles of area measure presented in Section 5.4 are the same for metric units as they are for English customary units. The standard metric units of area most commonly used are the **square meter, square decimeter,** and **square centimeter.** A square region one meter long and one meter wide has an area of one square meter (Figure 6.2). The SI symbol for a square meter is m^2.

A square region one decimeter long and one decimeter wide has an area of one square decimeter. The SI symbol for a square decimeter is dm^2. Since a meter is equal to 10 decimeters, a square meter is equal to 100 square decimeters (Figure 6.3).

A square region one centimeter long and one centimeter wide has an area of one square centi-

FIGURE 6.2

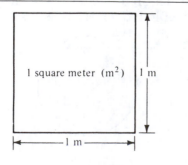

1 square meter (m^2) 1 m

1 m

FIGURE 6.3

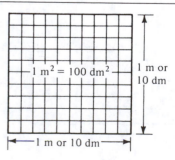

1 m^2 = 100 dm^2 1 m or 10 dm

1 m or 10 dm

meter (cm²). Since a decimeter is equal to 10 centimeters, 1 dm² = 100 cm². Since a meter is equal to 100 centimeters, 1m² = 10,000 cm².

The prefixes in Table 6.2, Section 6.1, are used to define other units of area measure such as square millimeter (mm²), square dekameter (dam²), square kilometer (km²), and so on. It is important to remember that the metric units of area measure are related to each other by **multiples of one hundred.**

A commonly used metric unit of land measure is the square hectometer, also called a **hectare (ha).** A hectare is about the size of two football fields side-by-side, or about 2.5 times as large as an acre.

$$1 \text{ ha} = 1 \text{ hm}^2 = 10,000 \text{ m}^2$$

EXAMPLE 1 A rectangular parking lot is 200 meters long and 150 meters wide. What is the area of the parking lot in hectares?

Solution   Area = (200 m) (150 m)
                = (2 hm) (1.5 hm)
                = 3 hm²
                = 3 hectares

EXAMPLE 2 A rectangular piece of sheet metal is 25 cm long and 18 cm wide. What is the area of the metal in square centimeters? What is the area in square meters?

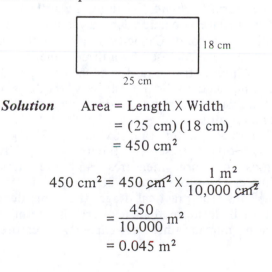

Solution   Area = Length × Width
                = (25 cm) (18 cm)
                = 450 cm²

$$450 \text{ cm}^2 = 450 \text{ cm}^2 \times \frac{1 \text{ m}^2}{10,000 \text{ cm}^2}$$

$$= \frac{450}{10,000} \text{ m}^2$$

$$= 0.045 \text{ m}^2$$

EXAMPLE 3 Find the lateral surface area of the concrete bridge support illustrated in the figure.

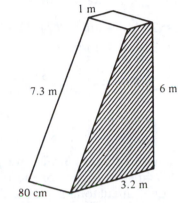

Solution

$$\text{Area} = (80 \text{ cm}) (7.3 \text{ m}) + (80 \text{ cm}) (6 \text{ m})$$

$$+ 2\left[\left(\frac{1}{2}\right)(6 \text{ m}) (3.2 \text{ m} + 1 \text{ m})\right]$$

$$= (0.8 \text{ m}) (7.3 \text{ m}) + (0.8 \text{ m}) (6 \text{ m})$$

$$+ 2\left[\left(\frac{1}{2}\right)(6 \text{ m}) (4.2 \text{ m})\right]$$

$$= 5.84 \text{ m}^2 + 4.8 \text{ m}^2 + 25.2 \text{ m}^2$$

$$= 35.84 \text{ m}^2$$

EXAMPLE 4 Find the total surface area of the oil drum shown in the figure in square decimeters. What is the area in square meters?

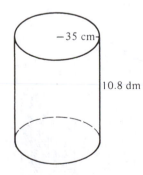

Solution The radius of the drum top is 35 cm, so the diameter is 70 cm, or 7 dm.

Lateral area = Circumference × Height
                = [(3.14)(7 dm)] (10.8 dm)
                = 237.384 dm²

$$\text{Area of base} = \pi \times (\text{radius})^2$$
$$= (3.14)(3.5 \text{ dm})^2$$
$$= 38.465 \text{ dm}^2$$

$$\text{Total area} = 237.384 \text{ dm}^2 + (2)(38.465 \text{ dm}^2)$$
$$= 314.314 \text{ dm}^2$$

$$314.314 \text{ dm}^2 \times \frac{1 \text{ m}^2}{100 \text{ dm}^2} = \frac{314.314}{100} \text{ m}^2$$
$$= 3.14314 \text{ m}^2$$

The principles of volume measure presented in Section 5.7 are the same for metric units as they are for English customary units. One of the standard metric units of volume is the **cubic meter**. A cube one meter long, one meter wide, and one meter high has a volume of one cubic meter. The SI symbol for cubic meter is m³. Other standard metric units of volume include the **cubic decimeter** (dm³), **cubic centimeter** (cm³), and so on.

Since a meter is equal to 10 decimeters, it would take 1000 cubic decimeters (10 dm × 10 dm × 10 dm) to fill a cubic meter (Figure 6.4). Likewise, it would take 1000 cubic centimeters to fill a cubic decimeter. All of the metric units of volume are related to each other by **multiples of one thousand**.

FIGURE 6.4

1 cubic meter (m³)

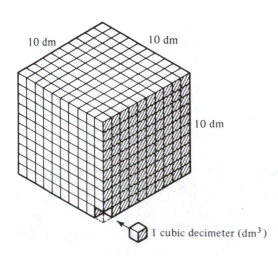

10 dm   10 dm

10 dm

1 cubic decimeter (dm³)

EXAMPLE 5   What is the volume of a rectangular box that is 35 cm long, 18 cm wide, and 12 cm high?

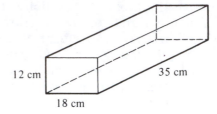

12 cm   35 cm

18 cm

Solution

$$\text{Volume} = \text{Length} \times \text{Width} \times \text{Height}$$
$$= (35 \text{ cm})(18 \text{ cm})(12 \text{ cm})$$
$$= 7560 \text{ cm}^3$$

EXAMPLE 6   What is the displacement in cubic centimeters of one piston with a diameter of 35.4 mm and a 4.02 cm stroke? Give the answer to the nearest thousandth of a cubic centimeter.

Solution

$$\text{Volume of a cylinder} = \pi r^2 h$$
$$= (3.14)(1.77 \text{ cm})^2(4.02 \text{ cm})$$
$$= 39.546 \text{ cm}^3$$

(*Note:* The radius of the cylinder is 17.7 mm or 1.77 cm.)

The term *volume* is used regardless of whether a space is occupied by a vacuum, a gas, a liquid, or a solid. **Capacity** is a term that is used for a space enclosed by a three-dimensional object. When metric units for capacity are used, there is no need to distinguish between dry and liquid measure, as must be done with English customary units.

The cubic meter (approximately 250 gallons) is too large to serve as the unit of capacity for most common measures, and the cubic centimeter (approximately 20 drops) is too small. Thus, for practical usage, the cubic decimeter (a little more than a quart) is the most common metric unit of capacity measure.

Another name for a cubic decimeter is the **liter** (also spelled **litre**). The symbol for liter is ℓ. Remember that dm³ and ℓ are symbols for the same unit, so

$$1 \text{ dm}^3 = 1 \text{ }\ell.$$

A cubic decimeter contains 1000 cubic centimeters. Therefore, a cubic centimeter is one-thousandth of a liter, or a cubic centimeter is equal to one milliliter. That is,

$$1 \text{ cm}^3 = 1 \text{ m}\ell$$

and

$$1 \text{ dm}^3 = 1000 \text{ m}\ell = 1 \text{ }\ell.$$

A cubic meter contains 1000 cubic decimeters. Therefore, a cubic meter is equal to one kiloliter. That is,

$$1 \text{ m}^3 = 1000 \text{ }\ell = 1 \text{ k}\ell.$$

The most commonly used units of capacity are **liter, kiloliter,** and **milliliter.** However, all of the prefixes listed in Table 6.2 can be used to define other units of capacity. For example, a deciliter is one-tenth of a liter, a hectoliter is one hundred liters, and so on.

EXAMPLE 7   A rectangular gasoline tank measures 3 dm by 5 dm by 2 dm. How many liters of gasoline will the tank hold?

Solution

$$\begin{aligned}
\text{Capacity} &= \text{Length} \times \text{Width} \times \text{Height} \\
&= (3 \text{ dm})(5 \text{ dm})(2 \text{ dm}) \\
&= 30 \text{ dm}^3
\end{aligned}$$

Since 1 dm³ = 1 ℓ, the tank will hold 30 liters.

EXAMPLE 8   A spherical water tank has a radius of 50 cm. How many liters of water will the tank contain when it is full?

Solution

$$\text{Capacity} = \frac{4}{3}\pi r^3$$

$$= \left(\frac{4}{3}\right)(3.14)(50 \text{ cm})^3$$

$$= 523{,}333.33 \text{ cm}^3$$

$$523{,}333.33 \text{ cm}^3 = \frac{1 \text{ }\ell}{1000 \text{ cm}^3} \doteq 523.33 \text{ }\ell$$

SUMMARY OF COMMON METRIC UNITS OF AREA, VOLUME, AND CAPACITY

1 m² = 100 dm² = 10,000 cm²

1 dm² = 100 cm²

1 hectare = 1 hm² = 10,000 m²

1 m³ = 1000 dm³ = 1,000,000 cm³

1 dm³ = 1000 cm³

1 dm³ = 1 ℓ = 1000 mℓ

1 cm³ = 1 mℓ

1 m³ = 1000 ℓ = 1 kℓ

Review Exercises

1. _____ m = 1.68 hm

2. 0.35 m = _____ mm

3. 13 cm = _____ m

4. _____ km = 47.5 m

5. Change $\frac{5}{8}$ to a percent.

6. Change $\frac{3}{5}$ to a percent.

7. $(1.2)^3 =$ _____

8. $(0.4)^3 =$ _____

9. $(0.9)(4.2)(3.6) =$ _____

10. $(2.8)(0.6)(14) =$ _____

11. $\sqrt{81} =$ _____

12. $\sqrt{100} =$ _____

EXERCISES
6.2

Find the area of the figures in Exercises 1–6. Give your answers in square centimeters.

Find the volume of the figures in Exercises 7–10. Give your answers in cubic centimeters.

1.
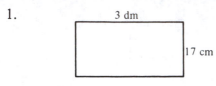

3 dm

17 cm

2.
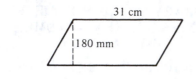

31 cm

180 mm

3.
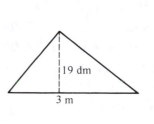

19 dm

3 m

4.
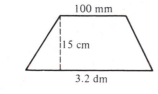

100 mm

15 cm

3.2 dm

5.
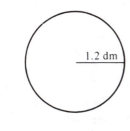

1.2 dm

6.
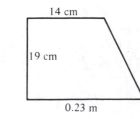

14 cm

19 cm

0.23 m

7.
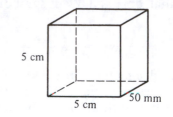

5 cm

5 cm

50 mm

8.
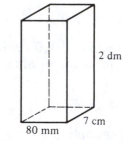

2 dm

7 cm

80 mm

9.
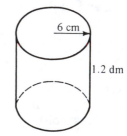

6 cm

1.2 dm

10.
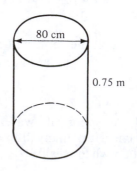

80 cm

0.75 m

11. What is the area of a square 15 cm on a side?

12. What is the area of a rectangle 12 cm wide and 19 cm long?

13. Sixteen triangular gussets like the one shown are cut from an iron plate. What is the area of one gusset? What is the total area of all sixteen gussets in square meters?

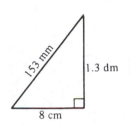

14. Twenty metal triangles to be used as corner braces are cut as shown. What is the area of one triangle? What is the total area of all twenty triangles in square meters?

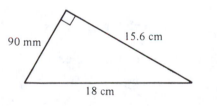

15. Two circles are cut from a face plate as shown. What is the area of the plate with the circles removed?

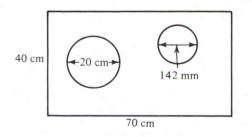

16. A circle and a triangle are cut from a face plate as shown. What is the area of the plate with the circle and triangle removed?

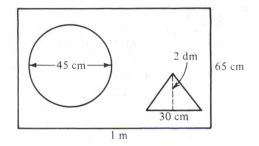

17. What is the area of the playground shown in the figure in hectares?

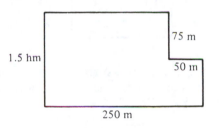

18. A new plant is to be built on a parcel of land as shown in the figure. What is the area of the land in hectares?

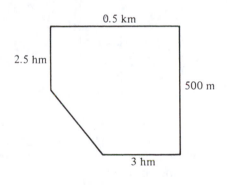

19. What is the total surface area of the box in square centimeters? What is the capacity of the box in liters?

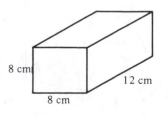

20. What is the total surface area of the container in square centimeters? What is the capacity of the container in milliliters?

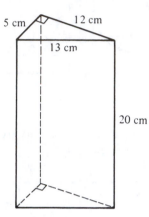

21. What is the lateral surface area of the drum shown here? What is the capacity of the drum in kiloliters?

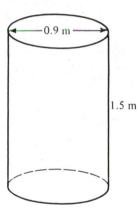

22. What is the lateral surface area of the container shown here? What is the capacity of the container in liters?

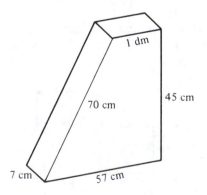

23. What is the volume of the brass bushing shown in the figure?

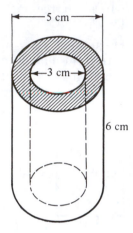

24. A 5-meter concrete sewer pipe has an outer radius of 18 cm and an inner radius of 125 mm. What is the volume of the concrete in the pipe?

Calculator Exercises

25. Find the area of the circle.

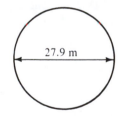

26. Find the area of the sector.

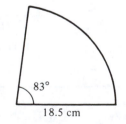

27. A sphere has a diameter of 8 dm. What is the surface area of the sphere? What is the capacity of the sphere in liters?

28. A sphere has a radius of 6 cm. What is the surface area of the sphere? What is the capacity of the sphere in liters?

29. What is the lateral area of the water tank? What is the capacity of the tank in liters?

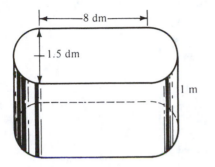

6.3
MASS AND TEMPERATURE

The **mass** of an object refers to a measure of the amount of material of which the object is composed. Since the amount of material in an object remains constant, the mass of an object, regardless of where it is measured, remains constant.

In physics and technology the term **weight** refers to the force that gravity exerts on the object. This force varies from place to place, so the weight of an object varies according to where it is measured. For our use the terms weight and mass will be synonymous. If force of gravity is implied, it will be so stated, and the term weight will not be used in this way.

The base SI unit for mass is the **kilogram (kg)**. A kilogram has the mass of one liter of pure water at 4 degrees Celsius (39.2 degrees Fahrenheit), the temperature at which water has the greatest density. A kilogram mass is approximately equal to 2.2 pounds. One cm^3 (milliliter) of cold water (4 °C) has a mass of one gram.

The prefixes in Table 6.2 are used to define other units of mass. The most commonly used units of mass are the **kilogram** (1000 grams), **gram**, and **milligram** (0.001 gram). Another common unit is the **metric ton (t)**, which is equal to 1000 kilograms.

$$1 \text{ t} = 1000 \text{ kg.}$$

The spelling *tonne* is also used for metric ton. A cubic meter of cold water (4 °C) has a mass of one metric ton.

EXAMPLE 1   A container holds 564 cm^3 of water. What is the weight of the water in kilograms?

Solution   One cm^3 of water weighs one gram.

$$564 \text{ g} = 0.564 \text{ kg}$$

Therefore, 564 cm^3 of water weighs 0.564 kg.

EXAMPLE 2 A spherical water tank has a radius of 6 m and a capacity of 904.78 kℓ. What is the weight of the water in metric tons when the tank is full?

Solution

$$1 \text{ kℓ} = 1 \text{ m}^3$$

One m^3 of water weighs one metric ton. The water weighs 904.78 t.

EXAMPLE 3 A rectangular cement block has dimensions 40 cm by 18 cm by 1 dm. Cement has a mass of 3 grams per cubic centimeter. What is the mass of the block?

Solution

$$\text{Volume} = (40 \text{ cm})(18 \text{ cm})(10 \text{ cm})$$
$$= 7200 \text{ cm}^3$$

$$\text{Mass of block} = (3)(7200)$$
$$= 21,600 \text{ g}$$
$$= 21.6 \text{ kg}$$

**SUMMARY OF COMMON
METRIC UNITS OF MASS**

1 kg = 1000 g = 1,000,000 mg

1 g = 1000 mg

1 liter of water weighs 1 kg

1 cm^3 of water weighs 1 g

1 metric ton = 1000 kg

In the metric system the Celsius scale (formerly called "centigrade") is used for temperature measurement. On the Celsius scale the freezing point for water is zero degrees (0 °C), and the boiling point for water is one hundred degrees (100 °C). Figure 6.5 shows several other familiar reference points to help you become more familiar with the Celsius scale.

FIGURE 6.5

Celsius temperature scale

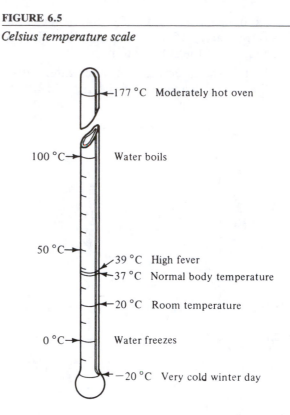

177 °C Moderately hot oven

100 °C → Water boils

50 °C →

39 °C High fever

37 °C Normal body temperature

20 °C Room temperature

0 °C → Water freezes

−20 °C Very cold winter day

Review Exercises

Perform the conversions in Exercises 1–6.

1. 1 dm^2 = _____ cm^2

2. _____ m^2 = hm^2

3. 3 cm^3 = _____ mℓ

4. 1 dm^3 = _____ cm^3

5. 1.4 m^2 = _____ dm^2

6. 2.8 ℓ = _____ dm^3

Find the product or quotient in Exercises 7–12.

7. $\left(\dfrac{3}{25}\right)\left(\dfrac{22}{3}\right)\left(\dfrac{30}{8}\right)$ 8. $\left(\dfrac{5}{12}\right)\left(\dfrac{4}{15}\right)\left(\dfrac{18}{8}\right)$

9. $0.801 \div 3.21$ 10. $7.54 \div 0.031$

11. $10.5 \div 1\dfrac{7}{8}$ 12. $18.4 \div 2\dfrac{5}{8}$

EXERCISES

6.3

Use the following constants for Exercises 1–14.

Substance	Mass per Cubic Centimeter	
Water	1	gram
Steel	7.834	grams
Cast iron	7.197	grams
Bronze	8.844	grams
Brass	8.354	grams
Cement	3	grams
Sand	2.3	grams

1. A rectangular container 20 cm by 13 cm by 17 cm is filled with water. What is the weight of the water in grams?

2. A rectangular container 30 cm by 9 cm by 17 cm is filled with water. What is the weight of the water in kilograms?

3. If the container shown in the figure is filled with water, how many kilograms of water does it hold?

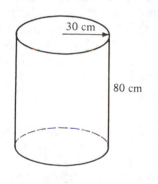

4. A platform holding a water tank like the one shown in the figure has an allowable weight limit of 3 metric tons. If the tank is filled with water how many kg under or over the limit is the weight of the water?

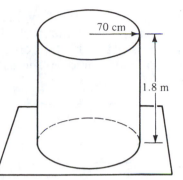

5. What is the weight of a piece of steel rod with dimensions as shown in the figure?

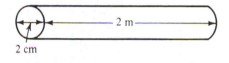

6. Three cubic meters of sand are loaded on a truck. How many metric tons of sand are in this load?

7. What is the weight in kilograms of a brass sphere with a radius of 6.8 cm?

8. A rectangular cast iron ingot has dimensions of 1 m by 30 cm by 10 cm. What is the mass of the ingot in kilograms?

9. The brass bar shown in the figure is cut into three equal-sized pieces. What is the weight of each piece?

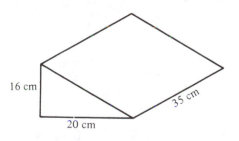

10. Steel spacers have dimensions as shown. What is the weight of 5000 spacers?

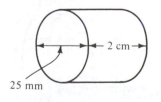

11. A chain hoist has a safety factor of one metric ton. If the cement block shown in the figure is lifted by the hoist, how many kilograms under or over the safety factor is this lift?

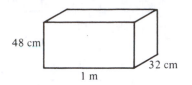

12. A cast iron angle iron is cut into six equal pieces. What is the weight of each piece?

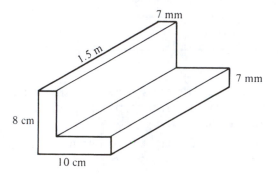

13. An order was received for two thousand bronze washers with the dimensions shown in the figure. What is the weight of this order?

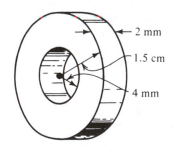

14. What is the weight of 750 brass pins like the one shown in the figure?

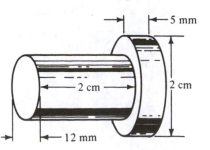

15. In a heater the temperature rises 5 °C for every 800 g of fuel that is added. If the temperature is 18 °C when the fire is lit and 2 kg of fuel are added every 10 minutes, how long will it take for the temperature of the heater to reach 105 °C?

16. A coolant added to a hot liquid will decrease the temperature 3 °C for each 50 milligrams of coolant. If the temperature of a liquid is 180 °C how many grams of coolant must be added to bring the temperature down to 50 °C?

17. A thermostat is fully open at 95 °C. When fully open the thermostat allows 0.75 ℓ of liquid per minute to flow through it. For each 1.5 °C reduction in temperature there is a decrease of 30 mℓ per minute in the flow. At what temperature is the thermostat fully closed?

18. A thermostat is fully closed when the temperature is 60 °C. The thermostat allows a flow of 4.5 mℓ per second for each 2 °C rise in temperature. When fully open the thermostat allows a flow of 5.4 ℓ per minute. At what temperature is the thermostat fully open?

19. At a temperature of 20 °C cement will dry at the rate of 2 mm thickness per hour. A rise in temperature of 2 °C increases the drying rate by 5 mm thickness per hour. At a temperature of 30 °C what is the drying rate per hour? If a cement driveway is 15 cm thick how long will it take to dry completely if the average temperature is 25 °C?

20. By adding a hardening agent to cement the drying rate is increased to 5 mm thickness per hour at 20 °C. For each 3 °C rise in temperature the drying rate is increased an additional 8 mm thickness per hour. If the drying agent is added how long will it take a cement driveway, 15 cm thick, to dry completely if the average temperature is 26 °C?

21. A weld is made at a temperature of 1580 °C. If the weld cools to room temperature in one hour, what is the cooling rate per minute for the weld?

22. The temperature of a metal compound is 1280 °C when it is poured to form ingots. The cooling rate is 0.3 °C per second. How long will it take an ingot to reach room temperature?

23. The boiling point of water is 100 °C. For each 640 milligrams of salt added to one liter of water the boiling point is decreased 0.1 °C. How much salt must be added to 5 liters of water to lower the boiling point to 95 °C?

24. A steel bar 8 cm thick is heated to 860 °C and treated with a hardening powder which penetrates the metal to a depth of 0.8 cm. For each 20 °C the temperature is raised the hardening powder will penetrate 0.3 mm deeper. What is the temperature needed to enable the powder to penetrate to a depth of 2.3 cm?

25. The temperature in an office was 28 °C. An air conditioning unit reduced the temperature 18% in two hours. What was the office temperature at the end of two hours?

26. On a summer day the temperature in an uninsulated attic is 55 °C. For each 17 mm of insulation the temperature is decreased by 1 °C. How many centimeters of insulation are needed to decrease the temperature in the attic to 28 °C?

28. What is the weight of a solid bronze cylinder as shown in the figure? Give the answer to the nearest tenth of a milligram.

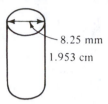

29. A cement block has dimensions 0.247 m by 0.35 m by 1.109 m. What is the mass of the block to the nearest hundredth of a kilogram? to the nearest tenth of a metric ton?

30. A rectangular metal box with dimensions 1.3 m by 0.525 m by 0.34 m is completely filled with sand. What is the mass of the sand to the nearest tenth of a kilogram? to the nearest hundredth of a metric ton?

Calculator Exercises

27. What is the weight of a rectangular cast iron spacer with dimensions as shown in the figure? Give the answer to the nearest tenth of a milligram.

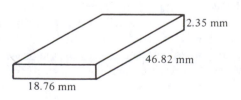

6.4
DERIVED UNITS

SI derived units are formed by simple multiplication and division of the SI base units without the introduction of any numerical factors. Table 6.4 includes some of the SI derived units. Examples that follow show how these units can be used to help solve certain types of problems.

TABLE 6.4 DERIVED SI UNITS

Quantity	Unit	Symbol	Expression in terms of other units
force	newton	N	$kg \cdot m/s^2$
energy	joule	J	$N \cdot m$
power	watt	W	J/s
voltage	volt	V	W/A
electric resistance	ohm	Ω	V/A
pressure	pascal	Pa	N/m^2

The **force** required to move an object depends on the mass of the object and how much you accelerate the object. A force of one **newton (N)** is required to accelerate a mass of one kilogram one meter per second squared. This is difficult to visualize. Another way to think of force is to think of what force the earth's gravity is exerting on an object. At sea level the force of gravity on a 100-gram object is one newton. The force of gravity on a 500-gram mass would be five newtons, and so on.

EXAMPLE 1   The force of gravity on an object at sea level is 2000 newtons. What is the mass of the object?

Solution   The force of gravity on a 100-gram mass is one newton.

$$2000 \times 100 \text{ g} = 200,000 \text{ g or } 200 \text{ kg}$$

The **energy** released when we exert a force through a distance is measured in **joules (J)**. A force of one newton through a distance of one meter is defined as one joule of energy or work.

> **Energy in joules = Force in newtons**
> **X Distance in meters**

This is symbolized as

> **J = N · m**

EXAMPLE 2   How much energy is required to exert a force of 10 N through a distance of 5 m?

Solution

Force in newtons X Distance in meters
= Energy in joules

$$10 \text{ N} \times 5 \text{ m} = 50 \text{ N} \cdot \text{m or } 50 \text{ joules}$$

EXAMPLE 3   How much work is done in lifting a 5-kg mass a distance of 2 m?

Solution   At sea level the force of gravity on 100 g is one newton. A mass of 5 kg is 5000 g or 50×100 g, so the force on 5 kg at sea level is 50 newtons. A force of one newton through a distance of one meter is one joule of work or energy. The energy required to move a force of 50 newtons through a distance of 2 meters is $50 \text{ N} \times 2 \text{ m} = 100 \text{ N} \cdot \text{m or } 100$ joules. Therefore, lifting a 5-kg mass through a distance of 2 m requires 100 joules of work.

Power, voltage, and **resistance** are familiar terms in electrical work. Power is the rate at which energy is used, and is measured in **watts (W)**, or joules per second. Voltage is measured in **volts (V)**, which is the amount of **power**

divided by the amperage, or **W/A**. Recall that amperage is a base unit in the metric system and is a measure of electric current. Electric resistance is measured in **ohms (Ω)**, which is the **voltage in the system divided by the amperage, or V/A.** A summary of the expressions which describe how to compute power, voltage, and resistance is shown here:

$$\text{Power} = \frac{\text{Joules}}{\text{Seconds}} \quad \left(W = \frac{J}{s}\right)$$

$$\text{Voltage} = \frac{\text{Power}}{\text{Amperage}} \quad \left(V = \frac{W}{A}\right)$$

$$\text{Resistance} = \frac{\text{Voltage}}{\text{Amperage}} \quad \left(\Omega = \frac{V}{A}\right)$$

EXAMPLE 4 How much power is required to expend 10 joules of energy in 5 seconds?

Solution

$$\frac{\text{Joules}}{\text{Seconds}} = \text{Power in watts}$$

$$\frac{10 \ \text{Joules}}{5 \ \text{Seconds}} = 2 \ \text{Watts}$$

EXAMPLE 5 A motor that requires 30,000 joules of energy per minute would be rated as requiring how many watts of power?

Solution $\dfrac{30,000 \ J}{60 \ s} = 500 \ W$

EXAMPLE 6 How many volts are necessary in a circuit of 1600 watts with a current of 25 amperes?

Solution $\dfrac{1600 \ \text{watts}}{25 \ \text{amperes}} = 64 \ \text{volts}$

EXAMPLE 7 A motor is rated at 3 kilowatts. If the current in a line is 30 A what voltage is required by the motor?

Solution

$$\frac{3000 \ W}{30 \ A} = 100 \ V$$

EXAMPLE 8 What is the resistance in a circuit that carries 220 volts and is drawing 20 amperes of current?

Solution

$$\frac{220 \ \text{volts}}{20 \ \text{amperes}} = 11 \ \text{ohms}$$

EXAMPLE 9 An electrical appliance connected to a 110-volt line draws 3.5 amperes of current. What is the internal resistance of the appliance in ohms?

Solution

$$\frac{110 \ V}{3.5 \ A} \doteq 31.429 \ \Omega$$

EXAMPLE 10 If a circuit drawing 5 amperes of current expends energy at the rate of 18,000 joules per minute what voltage is necessary in the circuit?

Solution

$$\text{Power in watts} = \frac{18,000 \ J}{60 \ s} = 300 \ W$$

$$\text{Voltage} = \frac{300 \ W}{5 \ A} = 60 \ V$$

Pressure is a quantity related to force. The SI unit for pressure is the **pascal (Pa)** which is a force of one newton per square meter. (This is like the force of gravity exerted on 100 grams of sand spread over an area of one square meter.) A pascal is a very small unit of pressure. In fact, the pressure in an ordinary inflated automobile tire is about 200,000 pascals. Since the pascal is so small, pressure is usually reported in **kilopascals (kPa)** or **megapascals (MPa)**. The pressure in an automobile tire is approximately 200 kPa, or 0.2 MPa. Another unit of pressure, which is particularly useful in meteorology, is the **bar (b)**, which is equal to 100,000 Pa.

EXAMPLE 11   Express 3 bars of pressure in kPa and MPa.

Solution

$$1 \text{ bar} = 100,000 \text{ Pa} = 100 \text{ kPa}$$
$$3 \text{ bars} = 300 \text{ kPa} = 0.3 \text{ MPa}$$

EXAMPLE 12   A force of 8 N is exerted on a rectangular plate 10 cm by 100 cm. What is the pressure on the plate in Pascals?

Solution

$$\text{Area of rectangle} = (10 \text{ cm})(100 \text{ cm})$$
$$= 1000 \text{ cm}^2$$

$$1 \text{ m}^2 = 10,000 \text{ cm}^2$$

$$\text{Pressure} = \frac{8 \text{ N}}{1000 \text{ cm}^2} = \frac{80 \text{ N}}{10,000 \text{ cm}^2}$$
$$= \frac{80 \text{ N}}{\text{m}^2} = 80 \text{ Pa}$$

Review Exercises

1. A dm³ of water has a mass of _____ kg.
2. Fifty cm³ of water has a mass of _____ g.
3. The area of the circle is _____ cm².

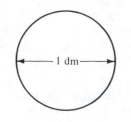

4. The area of the triangle is _____ dm².

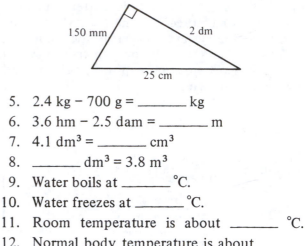

5. $2.4 \text{ kg} - 700 \text{ g} =$ _____ kg
6. $3.6 \text{ hm} - 2.5 \text{ dam} =$ _____ m
7. $4.1 \text{ dm}^3 =$ _____ cm³
8. _____ dm³ $= 3.8 \text{ m}^3$
9. Water boils at _____ °C.
10. Water freezes at _____ °C.
11. Room temperature is about _____ °C.
12. Normal body temperature is about _____ °C.

EXERCISES
6.4

1. What is the force of gravity in newtons on a mass of 800 grams?
2. What is the force of gravity in newtons on a mass of 1 kilogram?
3. What is the force in newtons needed to hold a mass of 250 grams in the air at sea level?
4. What is the force of gravity in newtons on a mass of 25 grams?
5. An object exerts a downward force of 1500 N at sea level. What is the mass of the object in kilograms?
6. The force of gravity on an object measured on a spring scale at sea level is 800 N. What is the mass of the object?
7. An automobile horn requires 0.4 amperes from a 12-volt battery. What is the resistance of the horn circuit?
8. The lighting system of an automobile draws a current of 4.7 A at a battery voltage of 10.3 V. What is the resistance of the lighting system?

9. A light uses 20 kilojoules of energy every minute. What is the power rating of the light?

10. How much power is used by a motor that expends 50 kJ of energy every 30 seconds?

11. How many volts are needed for a circuit of 2 kW with a current of 7.8 A?

12. What voltage is needed to cause a current of 3.2 amperes in a 500-watt appliance?

13. A force of 10,000 newtons is exerted over an area of 2.5 m². What is the pressure in pascals exerted by this force?

14. What is the pressure in kilopascals if a force of 30,000 N is exerted over an area of 0.5 m²?

15. A starting motor requires a current of 300 amperes with a 6-volt battery. What is the resistance of the starting motor?

16. What is the resistance of a blower motor in a fan if a voltage of 11.4 volts requires 1.3 amperes?

17. A circuit uses 9 kJ per minute with a current of 25 A. What is the voltage in the circuit?

18. The power used for running an air conditioning unit is 54 kJ per minute with a current of 45 amperes. What is the voltage required to run the unit?

Calculator Exercises

19. How many watts of power are required to expend 315.55 joules of energy in 2.75 seconds? Give the answer to the nearest thousandth of a watt.

20. An electric appliance uses 39.375 kilojoules of energy every 1.3 minutes. What is the power rating of the appliance to the nearest hundredth of a watt?

21. How many volts are needed for a circuit of 1.28 kW with a current of 12.23 A?

22. What is the resistance in a circuit that carries 118.48 volts and is drawing 15.034 amperes of current?

23. What force is necessary to exert a pressure of 478.29 kilopascals on a rectangular plate 8.32 cm by 15.07 cm?

24. A force of 739.5 N is exerted on a circular plate with a radius of 8.23 cm. What is the pressure on the plate in pascals?

EXTRA PROBLEMS
CHAPTER 6

1. Three pieces of stock measuring 2.7 decimeters, 11 centimeters and 58 millimeters in length are cut from a metal rod 0.8 meters long. If there are 2 millimeters of waste for each cut, how many centimeters long is the remaining piece of stock?

2. What is the circumference of a circle with a radius of 8 centimeters? Give your answer in millimeters.

3. Complete each metric conversion.

   a. 5.4 m = _____ cm

   b. 0.03 km = _____ dm

   c. _____ mm = 18 m

   d. _____ dam = 435 cm

   e. 250 mm = _____ cm

   f. 87.3 m = _____ km = _____ dm

   g. _____ cm = 4.5 km = _____ m

   h. _____ mm = _____ dam = 0.9 dm

4. A bar 1.5 meters long is cut into eight equal pieces. Each saw cut wastes 1.5 mm of the bar. How long is each of the eight pieces?

5. The diameter of a wheel is 36 cm. How far does the wheel roll if it turns through an angle of 110 degrees? Round off your answer to the nearest mm.

   In exercises 6 and 7 find the perimeter of each figure. Give each answer in centimeters and in meters.

6.

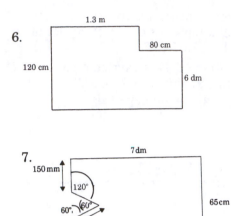

7.

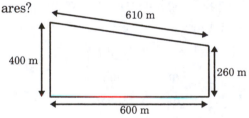

8. A rectangular storage tank measures 25 cm by 5 dm by 327 mm. If filled to the top, how many liters of water will the tank hold?

9. A building lot is shaped like a trapezoid. What is the area of the building lot in hectares?

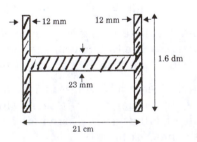

10. Find the cross sectional area of the I beam shown in the figure. Express your answer in square centimeters.

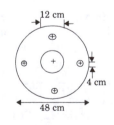

11. Find a) the surface area (in cm²) and b) the volume (in liters) of a spherical storage tank with a radius of 120 cm.

12. What is the area in square meters of a semi-circle with a radius of 170 centimeters?

13. Find the lateral area (in m²) and the capacity (in k ℓ) of a cylindrical storage tank with a radius of 3 m and a height of 19.4 m.

14. Find the total surface area of a closed rectangular box 65 cm long, 3.4 dm wide and 18 cm high.

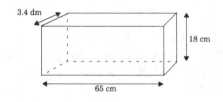

15. A solid cast-iron sphere has a diameter of 84 cm. What is the mass of the sphere in metric tons? (Mass of cast iron is 7.197 grams per cubic centimeter.)

16. The resistance of a wire 6.82 mm in diameter is reduced by 48% by replacing it with a wire that is twice the cross sectional area. What should be the diameter of the new wire?

17. A circular support base is shown in the figure. The base is cut from a steel plate that weighs 37.6 kilograms per square meter of surface area. Five circular holes are cut in the plate. What is the weight of a circular base plate?

18. Find the mass in kilograms of a hollow steel ball used as a buoy if the inside diameter is 2.28 m, the wall thickness is 15.0 mm, and the density of this type of steel is 8.50 grams per cubic centimeter.

19. A thermostat is fully open at 100°C. When fully open the thermostat allows 0.80 liters of liquid per minute to flow through it. For each 2°C decrease in temperature the flow of liquid through the thermostat is decreased by 25 milliliters per minute. At what temperature will the thermostat be fully closed?

20. Cement, treated with a hardening agent, will dry at a rate of 4 mm per hour at a temperature of 20°C. A rise in temperature of 1.5°C will increase the drying rate by 3 mm per hour. At an average temperature of 26°C, how long will it take a treated cement slab, that is 18 cm thick, to dry completely?

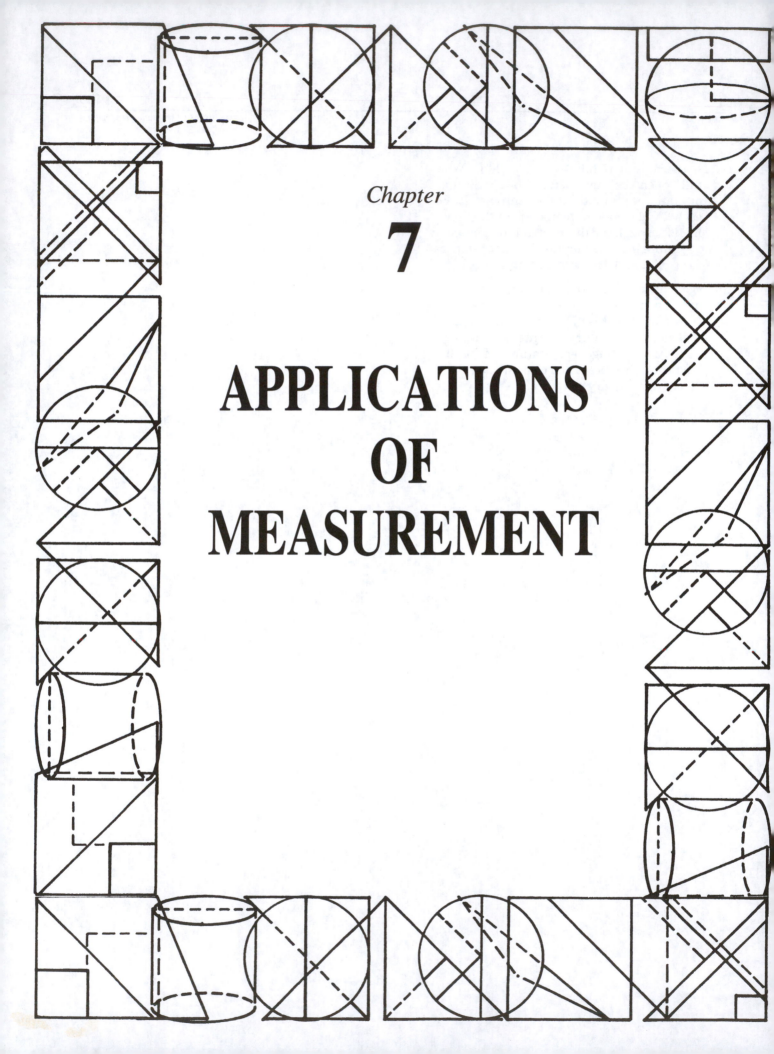

Chapter

7

APPLICATIONS
OF
MEASUREMENT

7.1
ROUNDING NUMBERS, SIGNIFICANT DIGITS, AND SCIENTIFIC NOTATION

Exact numbers are those numbers obtained by counting or by definition. All other numbers are **approximate**. Much of the work in industry and technology involves approximate numbers. In particular, all numbers obtained through measurements are approximate. Here are several examples of exact and approximate numbers.

Exact Numbers

350 people in a room	(counted)
18 books on a shelf	(counted)
An interest rate of 8.5%	(defined)
12 inches in a foot	(defined)

Approximate Numbers

A pencil $8\frac{3}{4}$ inches long	(measured)
The mass of a bolt is 37 grams	(measured)
A liter of gasoline	(measured)
An area of 2 square feet	(measured)

When working with approximate numbers in decimal form it may be necessary to round off the numbers. The rules for rounding off numbers are reviewed here:

Locate the digit in the place to which the expression is to be rounded, then:

Rule 1: If the digit to the right is less than five, leave the digit in the place to be rounded unchanged and change all other digits to the right of this place to zeros.

Rule 2: If the digit to the right is greater than five, or five followed by some non-zero digit, add one to the digit in the place to be rounded and change all other digits to the right of this place to zeros.

Rule 3: If the digit to the right is five followed by no other digits, or followed by zeros only, then if the digit in the place to be rounded is:

(a) *even,* leave it unchanged;

(b) *odd,* add one to the digit.

EXAMPLE 1 Round off each number to the nearest hundred.

34,620	4357
12,450	837,750

Solution

$34{,}620 \doteq 34{,}600$	(Rule 1)
$4357 \doteq 4400$	(Rule 2)
$12{,}450 \doteq 12{,}400$	(Rule 3a)
$837{,}750 \doteq 837{,}800$	(Rule 3b)

EXAMPLE 2 Round off each number to the nearest tenth.

25.35000	0.0538
8.0394	435.6500

Solution

$25.35000 \doteq 25.4$	(Rule 3b)
$0.0538 \doteq 0.1$	(Rule 2)
$8.0394 \doteq 8.0$	(Rule 1)
$435.6500 \doteq 435.6$	(Rule 3a)

All digits in an exact number are *significant* digits.

In an approximate number a digit is significant if:

> 1. it is a nonzero digit;
> 2. it is a zero between two significant digits; or
> 3. it is one of the terminal zeros on the right side of the decimal point.

The following chart lists several numbers and tells which digits are significant.

Number	Number of Significant Digits
450	two (4 and 5)
405	three (4, 0, and 5)
45.0	three (4, 5, and 0)
0.045	two (4 and 5)
70,350	four (7, 0, 3, and 5)
0.000679	three (6, 7, and 9)
5.00062	six (5, 0, 0, 0, 6, and 2)
35.200	five (3, 5, 2, 0, and 0)

A terminal zero on the left of the decimal point may be significant if information is available that identifies the digit as significant. For example, if the length of a pipe is 350 cm to the nearest centimeter, the number 350 has three significant digits. If the length of a pipe is 400 cm to the nearest ten centimeters, the number 400 has two significant digits. A way to show that terminal zeros on the left of the decimal point are significant is to place a short line under the last significant zero. Some examples are shown in the following chart.

Number	Number of Significant Digits
35,000	four (3, 5, 0, and 0)
8000	two (8 and 0)
400,000	six (4, 0, 0, 0, 0, and 0)
20,900	four (2, 0, 9, and 0)

Numbers representing very large or very small quantities can be written in a convenient exponential form called **scientific notation**.

> A number written in scientific notation is written as a product of a number between 1 and 10 and a power of 10.

You will remember that the number 10^3 means $10 \times 10 \times 10$ where 10 is the base and 3 is the exponent (see Section 6.1). The base 10 combined with any positive, negative or zero exponent is called a *power of ten*.

$10,000 = 10 \times 10 \times 10 \times 10 = 10^4$ (ten to the fourth power)

$1,000 = 10 \times 10 \times 10 = 10^3$ (ten to the third power)

$100 = 10 \times 10 = 10^2$ (ten to the second power)

$10 = 10^1$ (ten to the first power of ten)

When the numerator of a fraction is one and the denominator is a power of ten the fraction can be represented as a *negative power of ten*.

$$\frac{1}{10} = \frac{1}{10^1} = 10^{-1} \text{ (ten to the negative first power)}$$

$$\frac{1}{100} = \frac{1}{10 \times 10} = \frac{1}{10^2} = 10^{-2} \text{ (ten to the negative second power)}$$

$$\frac{1}{1000} = \frac{1}{10 \times 10 \times 10} = \frac{1}{10^3} = 10^{-3} \text{ (ten to the negative third power)}$$

Ten to the zero power is a special case and we define this number to be equal to one.

$$10^0 = 1$$

The following chart lists several numbers and shows how they are written in scientific notation.

Number		Scientific Notation
5346	$= 5.346 \times 1000$	$= 5.346 \times 10^3$
2,780,000	$= 2.78 \times 1,000,000$	$= 2.78 \times 10^6$
0.0036	$= 3.6 \times \frac{1}{1000}$	$= 3.6 \times 10^{-3}$
0.0000935	$= 9.35 \times \frac{1}{100,000}$	$= 9.35 \times 10^{-5}$
95	$= 9.5 \times 10$	$= 9.5 \times 10^1$
3.9	$= 3.9 \times 1$	$= 3.9 \times 10^0$

Note that the exponent of the power of 10 is exactly the same as the number of places the decimal point was moved in changing the number from the original form to scientific notation. If the decimal point is moved to the left the exponent is positive; and if the decimal point is moved to the right the exponent is negative. The number of significant digits in a number can be indicated by writing the number in scientific notation, as we see from the following chart.

Number	Number of Significant Digits	Scientific Notation
4000	1	4×10^3
4000	3	4.00×10^3
0.05040	4	5.040×10^{-2}
0.00009300	4	9.300×10^{-5}
8,000,000	7	8.000000×10^6
8,000,000	2	8.0×10^6

Note that all of the significant digits in the original number are included in the scientific notation for the number.

EXAMPLE 3 Write each number in scientific notation.

$$870,000 \qquad 0.46$$
$$35,004,000 \qquad 0.00002593$$
$$83 \qquad 5.43$$

Solution

$$870,000 = 8.7 \times 10^5$$
$$35,004,000 = 3.5004 \times 10^7$$
$$83 = 8.3 \times 10^1$$
$$0.46 = 4.6 \times 10^{-1}$$
$$0.00002593 = 2.593 \times 10^{-5}$$
$$5.43 = 5.43 \times 10^0$$

EXAMPLE 4 Round off each number to three significant digits and then write the number in scientific notation.

$$253,482 \qquad 0.000841$$
$$0.03847 \qquad 400.38$$
$$57.019 \qquad 8.04500$$

Solution

$$253,482 \doteq 253,000 = 2.53 \times 10^5$$
$$0.03847 \doteq 0.0385 = 3.85 \times 10^{-2}$$
$$57.019 \doteq 57.0 = 5.70 \times 10^1$$
$$0.000841 = 8.41 \times 10^{-4}$$
$$400.38 \doteq 400 = 4.00 \times 10^2$$
$$8.04500 \doteq 8.05 = 8.05 \times 10^0$$

Review Exercises

1. 3.8 m = _____ cm
2. 1.7 kg = _____ g
3. A motor that requires 3000 joules of energy per minute would be rated as requiring how many watts of power?
4. How many volts are necessary in a circuit of 800 watts with a current of 20 amperes?
5. 3 dm^3 = _____ ℓ
6. 450 mℓ = _____ cm^3
7. Normal body temperature is _____ °C.
8. Water boils at _____ °C.
9. What is the perimeter of the rectangle?

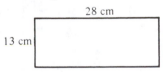

10. What is the perimeter of the parallelogram?

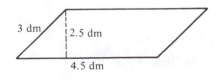

11. What is the area of the circle?

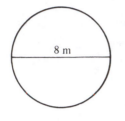

12. What is the area of the circle?

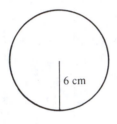

EXERCISES
7.1

In Exercises 1–10 indicate whether each number is exact or approximate.

1. The speed of a car is 50 mph.
2. The weight of a package is 1 pound 3 ounces.
3. There were 23 students in class.
4. The mass of steel is 7.834 g per cm³.
5. There are 10 meters in 1 dekameter.
6. One minute is equal to sixty seconds.
7. The state sales tax is 4%.
8. An alloy contains 25% copper.
9. A job takes 45 minutes to complete.
10. There are 250 parts in each packing case.

Round off each number in Exercises 11–22 to the indicated place.

11. 35, 264 (nearest thousand)
12. 2845 (nearest ten)

13. 503,094 (nearest hundred)
14. 104,968 (nearest ten thousand)
15. 5.329 (nearest hundredth)
16. 60.653 (nearest tenth)
17. 0.00275 (nearest ten-thousandth)
18. 0.05069 (nearest thousandth)
19. 80.726 (nearest unit)
20. 25.038 (nearest tenth)
21. 0.0439984 (nearest hundred-thousandth)
22. 604.597 (nearest hundredth)

Determine the number of significant digits in each of the approximate numbers in Exercises 23–36.

23. 25,000   24. 40,600
25. 3,084,0̲00   26. 67,40̲0,000
27. 93.005   28. 8.3050
29. 0.0217   30. 0.00047
31. 0.0009   32. 73.50̲0
33. 300,200   34. 73,500
35. 0.40̲0   36. 0.0206

37–50. Write each of the numbers in Exercises 23–36 in scientific notation.

51. Number 33 gauge wire has a diameter of 0.00708 inch. How many significant digits are used to report this measure? Write the measure of the diameter in scientific notation.

52. Number 20 gauge wire has a diameter of 0.0320 inch. How many significant digits are used to report this measure? Write the measure of the diameter in scientific notation.

53. Twelve and seven-eighths ounces is approximately equal to 365.0 grams. How many significant digits are there in the metric measure? Write the metric measure of this weight in scientific notation.

54. Ten and three-fifths ounces is approximately equal to 300.5 grams. How many significant digits are there in the metric measure?

Write the metric measure of this weight in scientific notation.

55. A gallon is approximately equal to 0.003785 kℓ. How many significant digits are there in the metric measure? Write the metric measure in scientific notation.

56. Seven-eighths of a cup is approximately equal to 207.015 mℓ. How many significant digits are there in the metric measure? Write the metric measure in scientific notation.

57. Eighty-four feet is approximately equal to 25.60 meters. How many significant digits are there in the metric measure? Write the metric measure in scientific notation.

58. A volume of 38.23 liters is equal to a volume of approximately 1.35 cubic feet. How many significant digits are there in the customary measure? Write the customary measure in scientific notation.

Calculator Exercises

Do the computations in Exercises 59-64 with your caculator and round off each answer to four significant digits.

59. 494×287 60. 13.9×15.5

61. $76 \div 15$ 62. $0.056 \div 206$

63. $7.9 \times 8.4 \times 6.32 \times 2.5$

64. $0.49 \times 83.4 \times 0.079 \times 0.006$

7.2
PRECISION AND GREATEST POSSIBLE ERROR

The following statements about **accuracy** are important to note.

> The *accuracy* of an approximate number refers to the number of significant digits in the number. The greater the number of significant digits the greater the accuracy of the number.

> The *precision* of an approximate number refers to the place value of the last significant digit in the number.

The precision of 750 feet is the tens place; that is, the measure is rounded to the nearest ten feet. The precision of 0.035 cm is the thousandths place; that is, the measure is rounded to the nearest thousandth of a centimeter.

EXAMPLE 1 Which is the more accurate measure: 27.5 inches or 350 inches?

Solution Since 27.5 has three significant digits and 350 has two significant digits, 27.5 is the more accurate number.

EXAMPLE 2 Which is the most accurate number: 16 m, 7300 m, 0.0138 m, or 1.460 m?

Solution The number 16 has two significant digits; 7300 has two significant digits; 0.0138 has three significant digits; and 1.460 has four significant digits. Therefore, 1.460 m is the most accurate measure.

EXAMPLE 3 Which is the more precise measure: 9.35 yd or 0.025 yd?

Solution Since 9.35 yd is expressed to the nearest hundredth of a yard, and 0.025 yd is expressed to the nearest thousandth of a yard, 0.025 yd is the more precise measure.

EXAMPLE 4 Which is the most *precise* measure: 3.510 kg, 29.2 kg, 0.75 kg, or 4345 kg? Which is the most *accurate* measure?

Solution The most precise measure is 3.510 kg. The precision is a thousandth of a kilogram. The most accurate measures are 3.510 kg and

4345 kg. Both measures have four significant digits.

The last significant digit of an approximate number is usually determined by estimation, or by rounding off to that place. Because of this, we know that approximate numbers may be in error. For example, an approximate measure of 15 feet, which has been rounded off to the nearest foot, could be any measure between 14.5 feet and 15.5 feet. Thus, the measure of 15 feet could be in error by $\frac{1}{2}$ foot at most. An approximate measure of 6.3 cm, which has been rounded off to the nearest tenth of a centimeter, could be any measure between 6.25 cm and 6.35 cm. Thus this measure of 6.3 cm could be in error by $\frac{1}{2}$ of a tenth of a centimeter (0.05 cm) at most. In all cases, the error of a measure is, at most, one-half of the unit of the place value to which it was rounded (the place value of the last significant digit).

The error of an approximate number is related to the precision of the number, as we see from the following important statement.

> The *greatest possible error* of a measure is equal to one-half of the precision of the measure.

The greatest possible error of the measure 15 feet is $\frac{1}{2}$ foot. The greatest possible error of the measure 6.3 cm is 0.05 cm. When the specification for a measurement is given, the **tolerance** of the measure is equal to plus or minus the greatest possible error, unless stated otherwise.

EXAMPLE 5   What is the precision and greatest possible error in a measure of 270 yards?

Solution   The precision is the nearest 10 yards. The greatest possible error is 5 yards. That is, the actual length could be any measure between 265 yards and 275 yards.

EXAMPLE 6   What is the precision and greatest possible error in a measure of 8.37 meters?

Solution   The precision of the measure is the nearest hundredth of a meter (0.01 m). The greatest possible error of the measure is 0.005 meter. (The actual measure is between 8.365 meters and 8.375 meters.)

With metric measures it is usually easier to determine the precision and greatest possible error if the measure is first converted to a smaller unit so that the measure is a whole number. In this case we can convert 8.37 m to 837 cm. The precision is the nearest centimeter, and the greatest possible error is 0.5 centimeter. Notice that these answers are equal to the answers expressed in meters.

EXAMPLE 7   The specified measure for the diameter of a cylindrical bearing is 0.37 inch. What is the tolerance for this measure?

Solution   Since no additional specifications are given, the tolerance of the measure is considered to be plus or minus the greatest possible error for the specified measure. The greatest possible error for a measure of 0.37 inch is 0.005 inch. Thus the tolerance is plus or minus 0.005 inch (±0.005 inch), or the allowable measures for the diameter of the bearing are between

$$0.37 - 0.005 \text{ in.} = 0.365 \text{ in.}$$

and

$$0.37 + 0.005 \text{ in.} = 0.375 \text{ in.}$$

Review Exercises

1. Round off 287.4352 to the nearest hundredth.
2. Round off 17,486 to the nearest thousand.
3. How many significant digits are there in the number 400.82?
4. How many significant digits are there in the number 5.0071?
5. Write 2,500,000 in scientific notation.
6. Write 0.000634 in scientific notation.
7. 13.5 cm = _____ dam
8. 84 hm = _____ mm
9. 0.49 _____ 49 dm

10. 83 ℓ = 0.083 _____
11. Fifteen is _____ % of 187.5.
12. Twenty-four is _____ % of 80.

EXERCISES
7.2

For the sets of approximate numbers in Exercises 1–14 identify the most accurate number and the most precise number.

1. 430, 27.9
2. 841, 7.9
3. 0.043, 4300
4. 6.800, 0.0068
5. 24.05, 0.734, 9480
6. 1.005, 39.86, 17.020
7. 500, 0.05, 50.0
8. 2040, 4200, 2.004
9. 15.07, 8003, 10.004
10. 9370, 73.09, 34.804
11. 6.3×10^2, 1.94×10^3, 4.1×10^{-1}
12. 8.9×10^{-3}, 7.3×10^5, 1.00×10^2
13. 18,000, 2.4×10^3, 0.009
14. 1.4×10^{-3}, 160, 2.08

For the approximate numbers in Exercises 15–26 determine the precision of the measure and the greatest possible error.

15. 17 inches
16. 24 mm
17. 8500 g
18. 93,000 m
19. 5.75 mℓ
20. 10.3 ounces
21. 0.006 cm
22. 4.008 feet
23. 8.30 mm
24. 48.0 cm²
25. 3.7×10^3 gallons
26. 6.25×10^4 ℓ

Determine the tolerance for the specified measures in Exercises 27–32.

27. 19 inches
28. 16 cm
29. 18.3 mm
30. 4.18 inches
31. 0.024 m
32. 0.0053 cm

33. A surveyor reported the length of a building lot to be approximately 400 feet to the nearest foot. What are the greatest and least possible lengths of the lot?

34. The specifications for a machine part call for a hole to be drilled and tapped with a diameter of 14.28 mm. What is the tolerance for this specification?

35. A carpenter measures the thickness of a piece of wood to be $\frac{5}{8}$ inch. What is the greatest possible error of this measurement?

36. A machinist measures the length of a bolt as $2\frac{11}{16}$ inches. What is the greatest possible error of this measurement?

37. A piece of metal measured as 0.143 cm thick and another piece measured as 0.576 cm thick are bolted together. What are the greatest and least possible combined thicknesses of the two metal pieces?

38. A student weighed two brass bars and recorded the measures as 15.73 g and 64.09 g. What are the greatest and least possible combined weights of the two bars?

39. The capacities of three different containers are measured and recorded as 340.5 mℓ, 18.75 mℓ, and 240.10 mℓ. What is the most accurate measure? the most precise?

40. The measures of the lengths of three metal rods are given as 40.3 cm, 8.625 cm, and 75.002 cm. Which is the most accurate measure? the most precise?

41. The area of a metal plate is measured as 545.38 cm². What is the precision of this measurement?

42. The floor area in a storage shed is measured as 50.4 square feet. What is the greatest possible error of this measure?

Calculator Exercises

Use your calculator to change each fraction to a decimal in Exercises 43–48. Then round off each decimal to three significant digits.

43. $\frac{7}{45}$
44. $\frac{3}{17}$
45. $\frac{2}{101}$
46. $\frac{8}{135}$
47. $\frac{17}{13}$
48. $\frac{43}{33}$

7.3
OPERATIONS WITH APPROXIMATE NUMBERS

When doing arithmetic operations with approximate numbers, the result cannot be of greater precision than the *least* precise of the approximate numbers used in the computation.

> When approximate numbers are added or subtracted, the precision of the result is the same as the precision of the least precise number used in the computation.

EXAMPLE 1   Three sections of pipe are welded together. The measures of the three lengths are 38.5 cm, 70.28 cm, and 45.05 cm. What is the length of the welded pipe?

Solution

$$
\begin{array}{r}
38.5 \text{ cm} \\
70.28 \text{ cm} \\
+\ \ 45.05 \text{ cm} \\
\hline
153.83 \text{ cm}
\end{array}
$$

The least precise of the three lengths is 38.5 cm. Thus, the answer must be rounded to the nearest tenth of a centimeter. The length of the welded pipe is approximately 153.8 cm.

EXAMPLE 2   The area of a metal plate is 284.5 square inches. A rectangular section with an area of 94.775 square inches is cut from the plate. What is the area of the plate that is left?

Solution   We find the difference, then round off the difference to tenths—the same precision as the least precise measure.

$$
\begin{array}{r}
284.500 \text{ in.}^2 \\
-\ \ 94.775 \text{ in.}^2 \\
\hline
189.725 \text{ in.}^2 \doteq 189.7 \text{ in.}^2
\end{array}
$$

The area of the metal plate that is left is approximately 189.7 in.^{2}.

EXAMPLE 3   What is the total thickness of three metal plates that are bolted together where

one plate is 0.432 cm thick, one is 1.364 mm thick, and one is 3.6 mm thick?

Solution

$$
\begin{array}{r}
4.32 \text{ mm} \\
1.364 \text{ mm} \\
+\ 3.6 \text{ mm} \\
\hline
9.284 \text{ mm}
\end{array}
$$

The total thickness of the three metal plates is approximately 9.3 mm, or 0.93 cm.

> When approximate numbers are multiplied or divided, the number of significant digits in the result is the same as the number of significant digits in the least accurate number (least number of significant digits).

EXAMPLE 4   What is the area of a rectangle with a length of 3.862 m and a width of 4.7 m?

Solution

$$\text{Area} = (3.862 \text{ m})(4.7 \text{ m}) = 18.1514 \text{ m}^2$$

Since the least accurate measure has only two significant digits, the area of the rectangle is rounded to two significant digits. Thus, the area is approximately 18 m^2. Note that the least possible value for the area is:

$$(3.8615 \text{ m})(4.65 \text{ m}) = 17.955975 \text{ m}^2,$$

and the greatest possible value for the area is:

$$(3.8625 \text{ m})(4.75 \text{ m}) = 18.346875 \text{ m}^2.$$

These values are the same when rounded to two significant digits, but would not be the same if rounded to more than two significant digits.

EXAMPLE 5   What is the volume of a cylinder with a radius of 6.25 cm and a height of 15.48 cm?

Solution

$$\text{Volume} = (3.14)(6.25 \text{ cm})^2(15.48 \text{ cm})$$
$$= 1898.7187 \text{ cm}^3$$

The least accurate measure has three significant digits, so the volume must be rounded to three significant digits. The volume of the cylinder is approximately 1900 cm³.

EXAMPLE 6 If 524.7 watts of power are used in 3.25 minutes, what is the rate of power use per minute?

Solution

$$524.7 \div 3.25 \doteq 161.44615$$

The least accurate number has three significant digits, so the quotient must be rounded to three significant digits. The rate of power use per minute is approximately 161 watts per minute.

EXAMPLE 7 If 6.236 mℓ of a liquid has a mass of 5.83 g, what is the mass of one milliliter of the liquid?

Solution

$$5.83 \div 6.236 \doteq 0.9348941$$

The mass of one milliliter of liquid is approximately 0.935 g. (The result of the division computation must be rounded to three significant digits.)

Review Exercises

1. What is the precision of a measure of 25,030 inches?

2. What is the precision of a measure of 47.30 kg?

3. What is the greatest possible error of a measure of 498 meters?

4. What is the greatest possible error of a measure of 2300 yards?

5. Round off 4935.086 to three significant digits.

6. Round off 0.04639 to two significant digits.

7. The measures of the diameters of three metal rods are 0.35 cm, 0.027 cm, and 0.406 cm. Which is the most accurate measure?

8. The weights of three boxes are 1830 g, 9.05 g, and 107.3 g. Which is the most accurate measure?

9. $8(3^2) =$ _____   10. $5(2^3) =$ _____

11. 8.5% of 424 = _____

12. 6.8% of 315 = _____

EXERCISES
7.3

Do the indicated operations in Exercises 1–20. All of the given numbers are approximate numbers.

1. $\begin{array}{r} 157.3 \\ 28.47 \\ + 306.086 \end{array}$ 2. $\begin{array}{r} 0.593 \\ 8.74 \\ + 164.6 \end{array}$

3. $86.9 + 49.26$ 4. $342.15 + 67.087$

5. $0.00364 + 8.437 + 0.9385 + 1.30$

6. $24.905 + 439.853 + 0.0009 + 16.411$

7. $\begin{array}{r} 529.43 \\ - 74.863 \end{array}$ 8. $\begin{array}{r} 1.0824 \\ - 0.165 \end{array}$

9. $16.2 - 9.351$ 10. $20.85 - 0.0472$

11. $75.472 + 8.39 - 19.357 + 5.009$

12. $472.238 - 13.8575 + 4.006 - 0.00053$

13. $(54.3)(6.2)$ 14. $(0.756)(1504)$

15. $(1700)(490)$ 16. $(85.4)(0.0063)$

17. $105 \div 3.5$ 18. $8.96 \div 0.16$

19. $45.00 \div 250.0$ 20. $65.0 \div 2.60$

The measures in Exercises 21–30 are approximate numbers.

21. What is the total weight of three pieces of steel if the weights of the three pieces are 2.486 g, 7.09 g, and 695 g?

22. What is the thickness of a four-ply piece of plywood if the thicknesses of the four individual pieces are 0.084 in., 0.13 in., 0.0942 in., and 0.108 in.?

23. What is the thickness of a metal pipe with an outer diameter of 2.338 cm and an inner diameter of 1.97 cm?

24. A metal rod with a radius of 1.525 cm is turned down on a lathe by making a cut of 4.836 mm. What is the radius of the finished rod?

25. If the circumference of a flywheel is 287.43 cm, what is the approximate diameter of the flywheel?

26. The weight of 200.5 cm³ of wrought iron is 1.57 kg. What is the weight of one cm³ of iron?

27. The weight of bronze is 0.3195 pound per cubic inch. What is the weight of 17.5 cubic inches of bronze?

28. What is the volume of a rectangular box with a length of 28.35 inches, a width of 16.7 inches, and a height of 18.05 inches?

29. A machine requires 3.87×10^3 watts per hour. What is the power required per second? (*Hint*: 1 hr = 3.600 sec or 1 hr = 3.600 $\times$ 10^3 sec.)

30. The distance to the moon is approximately 3.844×10^5 m. If a rocket travels 8.74×10^3 m per hour, how many hours would it take to reach the moon?

Calculator Exercises

Use your calculator to do Exercises 31–36. All of the given numbers are approximate numbers.

31. $78.43 \times 16.42 \times 1035$

32. $0.05426 \times 8.364 \times 98.3$

33. $(25.483 \times 7.4) \div 16.49$

34. $(86.01 \div 0.4675) \times 3407$

35. $(5.006 \times 7.539) \div (0.8426 \times 78.37)$

36. $(984,000 + 6749) \div 8.02$

7.4
RELATIVE ERROR OF MEASUREMENT

The **relative error** of a measurement is the decimal fraction obtained by dividing the greatest possible error by the actual measurement.

$$\text{Relative error} = \frac{\text{Greatest possible error}}{\text{Actual measurement}}$$

The relative error of a measurement is usually expressed as a percent. This represents the greatest possible error as a percent of the actual measurement.

EXAMPLE 1 The length of a shear pin is measured as 3.2 cm. What is the greatest possible error of the measurement? What is the relative error of the measurement?

Solution

Greatest possible error = 0.05 cm

$$\text{Relative error} = \frac{0.05 \text{ cm}}{3.2 \text{ cm}} \doteq 0.016$$

(*Note:* The decimal quotient is rounded off to the number of significant digits in the actual measurement.) The relative error expressed as a percent is 1.6%.

EXAMPLE 2 The mass of an iron casting is measured as 16 kg to the nearest kilogram. What is the relative error of the measurement?

Solution

Greatest possible error = 0.5 kg

$$\text{Relative error} = \frac{0.5 \text{ kg}}{16 \text{ kg}} \doteq 0.031 \text{ or } 3.1\%$$

EXAMPLE 3 The amount of hydraulic fluid in a brake line is measured as 850 mℓ rounded to the nearest 10 mℓ. What is the relative error of this measurement?

Solution

Greatest possible error = 5 mℓ

$$\text{Relative error} = \frac{5 \text{ mℓ}}{850 \text{ mℓ}} \doteq 0.0059 \text{ or } 0.59\%$$

EXAMPLE 4 If the measure in Example 3 is 850 mℓ to the nearest milliliter, what is the relative error?

Solution

Greatest possible error = 0.5 mℓ

$$\text{Relative error} = \frac{0.5 \text{ mℓ}}{850 \text{ mℓ}} \doteq 0.000588 \text{ or } 0.0588\%$$

EXAMPLE 5 The distance from Chicago to Indianapolis is approximately 160 miles to the nearest mile. The average height of an American male is 70 inches to the nearest inch. Which of these measures has the smaller relative error?

Solution The relative error of the distance from Chicago to Indianapolis is:

$$\text{Relative error} = \frac{0.5 \text{ mi}}{160 \text{ mi}} \doteq 0.00312 \text{ or } 0.312\%.$$

The relative error for the average height of an American male is:

$$\text{Relative error} = \frac{0.5 \text{ in.}}{70 \text{ in.}} \doteq 0.0071 \text{ or } 0.71\%.$$

The measure of the distance from Chicago to Indianapolis has the smaller relative error.

Review Exercises

1. What is the sum of the approximate measures 45.06 m, 28.5 m, and 350 m?

2. What is the sum of the approximate measures 0.009 in., 4.1 in., and 83.58 in.?

3. The length of a gauge is measured as 7.48 cm. What is the greatest possible error of this measure?

4. The measure of the area of a plate is 497.3 in.². What is the greatest possible error of this measure?

5. How many significant digits are there in the number 503,700?

6. How many significant digits are there in the number 0.00401?

7. The weights of three iron bars are 1875 g, 1530.6 g, and 980.03 g. Which is the most precise measure?

8. The lengths of three lead pipes are 17.4 cm, 9.05 cm, and 108 cm. Which is the most precise measure?

9. 15% of 408 is _____

10. 32% of 85 is _____

11. Write $\dfrac{3}{8}$ as a percent.

12. Write $\dfrac{27}{40}$ as a percent.

EXERCISES
7.4

Determine the relative error for the approximate numbers in Exercises 1–20. Give each answer as a decimal fraction and as a percent.

1. A length of 25 feet

2. A length of 5 inches

3. A mass of 35 grams

4. A mass of 27 kilograms

5. A capacity of 0.8 liters

6. A capacity of 4.2 quarts

7. A distance of 340 meters to the nearest 10 meters

8. A distance of 150 mm to the nearest 10 mm

9. An area of 753 cm²

10. A weight of 185 pounds

11. A length of 0.03 cm
12. A volume of 0.12 dm³
13. A height of 1.34 m
14. An area of 4.51 m²
15. A volume of 2400 cm³ to the nearest 100 cm³
16. A distance of 63,000 feet to the nearest 1000 feet
17. A mass of 0.024 kg
18. A capacity of 0.052 ℓ
19. A length of 3.20 km
20. A mass of 6.40 g

Calculator Exercises

Use your calculator to do Exercises 21–26. Give each answer as a decimal rounded off to five significant digits.

21. $\dfrac{8}{11} \times \dfrac{7}{13}$ 22. $\dfrac{6}{17} \times \dfrac{3}{14}$

23. $\dfrac{5}{21} \div \dfrac{9}{23}$ 24. $\dfrac{11}{15} \div \dfrac{5}{8}$

25. $\left(\dfrac{6}{7} \times \dfrac{4}{9}\right) \div \dfrac{7}{18}$ 26. $\left(\dfrac{7}{12} \times \dfrac{11}{8}\right) \div \dfrac{13}{17}$

7.5
THE PYTHAGOREAN THEOREM

The lengths of the three sides of a right triangle are related as follows:

> **The square of the length of the hypotenuse is equal to the sum of the squares of the lengths of the legs.**

This relationship is commonly known as the **Pythagorean Theorem** and can be symbolized as shown in Figure 7.1.

FIGURE 7.1

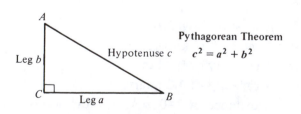

Pythagorean Theorem $c^2 = a^2 + b^2$

EXAMPLE 1   What is the length of the hypotenuse in right triangle *ABC*?

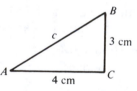

Solution

$$c^2 = a^2 + b^2$$
$$c^2 = 3^2 + 4^2$$
$$c^2 = 9 + 16$$
$$c^2 = 25$$

(*Note:* $a = 3$, the length of side *BC*; $b = 4$, the length of side *AC*; and c is the length of the hypotenuse.)

When a number squared equals 25, the number must be the square root of 25. Therefore,

$$c = \sqrt{25}$$
$$c = 5.$$

The length of the hypotenuse is 5 cm.

EXAMPLE 2   What is the length of the hypotenuse in right triangle *RST*? Round off the answer to three significant digits.

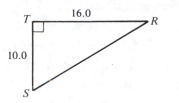

Solution

$$c^2 = a^2 + b^2$$
$$(RS)^2 = 10^2 + 16^2$$
$$(RS)^2 = 100 + 256$$
$$(RS)^2 = 356$$
$$RS = \sqrt{356}$$
$$RS \doteq 18.9$$

Use your hand calculator to get an estimate of $\sqrt{356}$. If you prefer, check the Appendices for an explanation of how you can interpolate an approximate value for the square root of a number.

EXAMPLE 3   What is the length of the diagonal brace in a metal rectangular frame with length 24.0 in. and width 20.0 in.? Round off the answer to three significant digits.

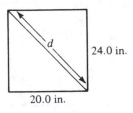

Solution

$$d^2 = 24.0^2 + 20.0^2$$
$$d^2 = 976$$
$$d = \sqrt{976}$$
$$d \doteq 31.2$$

The length of the diagonal is 31.2 inches.

When we know the lengths of the hypotenuse and one leg of a right triangle we can use the Pythagorean Theorem to compute the length of the other leg of the triangle.

EXAMPLE 4   Find the length of leg *AC* in right triangle *ABC*.

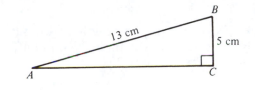

Solution

$$c^2 = a^2 + b^2$$
$$13^2 = 5^2 + b^2$$
$$169 = 25 + b^2$$
$$169 - 25 = 25 - 25 + b^2$$
$$144 = b^2$$
$$\sqrt{144} = b$$
$$12 = b$$

The length of leg *AC* is 12 cm.

EXAMPLE 5   What is the length of leg *PQ* in right triangle *PQR*? Round off the answer to two significant digits.

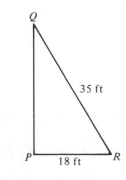

Solution

$$c^2 = a^2 + b^2$$
$$35^2 = (QP)^2 + 18^2$$
$$1225 = (QP)^2 + 324$$
$$1225 - 324 = (QP)^2 + 324 - 324$$
$$901 = (QP)^2$$
$$\sqrt{901} = QP$$
$$30 \doteq QP$$

EXAMPLE 6 What is the length of the run of the stairway shown in the figure? Round off the answer to three significant digits.

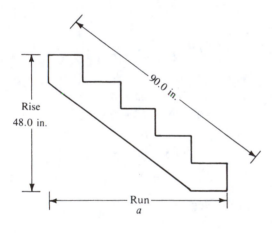

Solution

$$c^2 = a^2 + b^2$$
$$90^2 = a^2 + 48^2$$
$$8100 = a^2 + 2304$$
$$5796 = a^2$$
$$\sqrt{5796} = a$$
$$76.1 \doteq a$$

Review Exercises

1. What is the relative error of a measurement of 13.2 feet?

2. What is the relative error of a measurement of 4.25 km?

3. What is the difference of the approximate numbers 407.6 and 24.928?

4. What is the product of the approximate numbers 6.25 and 19.1?

5. Write 0.00852 in scientific notation.

6. Write 47,000,000 in scientific notation.

7. Convert 3 feet $4\frac{5}{8}$ inches to inches, to the nearest hundredth of an inch.

8. Convert 2 feet $7\frac{3}{16}$ inches to inches, to the nearest hundredth of an inch.

9. Convert 32.69 inches to feet and inches, to the nearest $\frac{1}{16}$ inch.

10. Convert 19.37 inches to feet and inches, to the nearest $\frac{1}{8}$ inch.

11. A distributor adds 35.8% to her cost to determine the selling price of each item. What will she sell an item for if the item costs her $7.80?

12. A merchant advertised a $3\frac{1}{2}$% discount if the customer paid cash. What was the cash price of an article that was marked $58.75?

EXERCISES
7.5

In Exercises 1–18 use the Pythagorean Theorem to find the missing side of right triangle *ABC*. Assume all the numbers are exact numbers. Round off each answer to three significant digits.

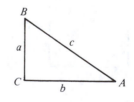

1. $a = 8$ ft, $b = 6$ ft
2. $a = 10$ cm, $b = 24$ cm
3. $a = 8$ in., $b = 15$ in.
4. $a = 12$ yd, $b = 9$ yd
5. $a = 7$ cm, $b = 3$ cm

6. $a = 9$ m, $b = 14$ m
7. $a = 6$ ft, $b = 7$ ft
8. $a = 11$ dm, $b = 16$ dm
9. $c = 25$ in., $a = 15$ in.
10. $c = 13$ m, $a = 5$ m
11. $c = 34$ cm, $b = 16$ cm
12. $c = 30$ yd, $b = 18$ yd
13. $a = 5$ in., $c = 10$ in.
14. $c = 11$ ft, $b = 8$ ft
15. $c = 30$ cm, $b = 12$ cm
16. $b = 19$ m, $c = 23$ m
17. $c = 16$ mm, $a = 11$ mm
18. $a = 9$ in., $c = 33$ in.

In Exercises 19-28, express your answers to two significant digits.

19. What is the length of a diagonal brace for a rectangular door frame 7 feet high and 3 feet wide?

20. What is the length of a diagonal brace between two rafters as shown in the figure?

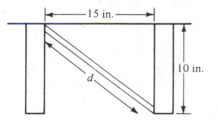

21. What is the rise of a slant roof if the rafter is 18 feet, the overhang is 2 feet, and the run is 14 feet?

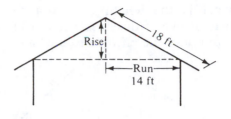

22. What is the center-to-center distance between holes A and C if the distance between the centers of A and B is 40 cm, and the distance between the centers of B and C is 18 cm?

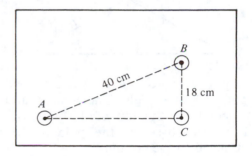

23. A gusset has the shape of a right triangle, and the length of each leg is equal to 16 cm. What is the length of the hypotenuse of the triangle?

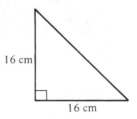

24. A piece of plywood in the shape of a trapezoid has the dimensions shown in the figure. What is the length of the plywood?

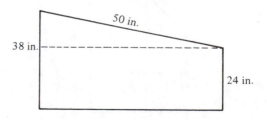

25. What is the length of a triangular steel plate if two of the sides measure 11 dm and 7 dm, and the altitude to the base is 5 dm?

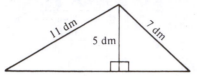

26. A concrete slab is formed in the shape of an isosceles triangle. The length of each of the equal sides is 12 feet, and the altitude to the base is 7 feet. What is the length of the base of the triangle?

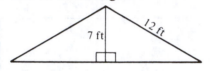

27. A guy wire 40 feet long is attached to a pole 30 feet high. The wire is attached 4 feet from the top of the pole. How far from the base of the pole is the wire attached to the ground?

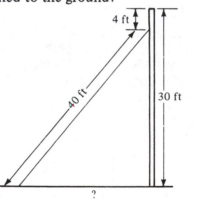

28. A face plate has dimensions as shown in the figure. What is the length of the slanted side of the plate?

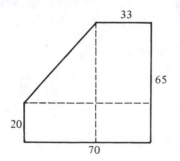

Calculator Exercises

Use the Pythagorean Theorem and your calculator to compute the length of the third side in each right triangle. Round off each answer to three significant digits.

29.

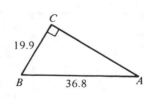

30.

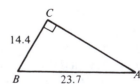

31.

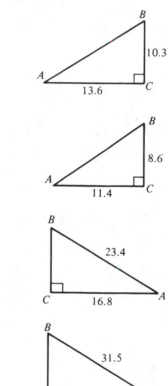

32.

33.

34.

EXTRA PROBLEMS
CHAPTER 7

1. a. What kinds of numbers are exact numbers?
    b. What kinds of numbers are approximate numbers?

2. Round off each number to the indicated place.
    a. 853.247 (nearest hundredth)
    b. 3548.357 (nearest ten)
    c. 90.07353 (nearest thousandth)
    d. 27,625.0849 (nearest unit)
    e. 5,384,628.94 (nearest thousand)
    f. 3,165.73846 (nearest ten thousandth)
    g. 327.059 (nearest hundred)
    h. 45.352 (nearest tenth)

3. Determine the number of significant digits in each approximate number.

    a. 0.023   e. 8.90017
    b. 18.500   f. 0.00620
    c. 103,500   g. 5000.020
    d. 369.004   h. 3,420,000

4. Write each number in scientific notation.

    a. 846,000   e. 5.03$\underline{0}$0
    b. 0.009301   f. 385 x 10^{-5}
    c. 295,40$\underline{0}$,000   g. 0.1029
    d. 493,002   h. 0.000007800

5. For each set of approximate numbers identify the most accurate number and the most precise number.

    a. 30,500; 0.0042; 83.510 x 10^3; 1.007

    b. 2.95 x 10^{-2}; 10,004.9; 786,000; 0.0004 x 10^3

    c. 250,$\underline{0}$00; 0.00801; 5.0 x 10^{-4}; 0.063000

    d. 9000.0; 8.5 x 10^{-3}; 70.04; 358,$\underline{0}$00

    e. 0.00749; 3,596,000; 15.0072; 9.8430; 8.35 x 10^6

    f. 8.19 x 10^{-3}; 10,040; 5.2810 x 10^5; 25.6 x 10^{-5}; 79.62 x 10^3

6. A machinist measures the length of an iron rod as 5.37 inches.
    a. What is the precision of this measurement?
    b. What is the greatest possible error of this measurement?

7. An inspector recorded measurements of the lengths of five metal bars as 0.78 cm, 0.192 cm, 5.02 cm, 11.6 cm and 103.48 cm.
    a. Of these five measurements which one was the most precise?
    b. What was the greatest possible error of the most precise measurement?

8. Specifications for a machine part call for a hole to be drilled and tapped with a diameter of 9.84 mm. What is the tolerance of error for this specification?

9. Do the indicated operations. All numbers are to be considered approximate numbers. Round off your answers according to the rules for operations with approximate numbers.

    a. 29.4" + 3.87" + 209.1"
    b. 832.90 cm − 25.046 cm
    c. (83.5 mm) (27.06 mm)
    d. 540.3 mℓ − (7.60 cm^2) (18.04 cm)
    e. (5.250 ft.) (6.4 ft.)
    f. 0.690 grams ÷ 52.7
    g. 9.56 x 10^3 miles ÷ 0.73 x 10^2
    h. (2.67 cm) (758.42 cm^2) ÷ 39.72 cm^2

10. What is the total weight of three iron castings if the weights of the individual pieces are 5.22 kg, 24.07 kg, and 130 kg.

11. What is the volume of a rectangular box with a length of 22.5 inches, a width of 13.09 inches, and a height of 9.025 inches?

12. The volume of a cylindrical storage tank is 2.080 kℓ. What is the height of the tank if the area of the base is 8035 cm^2?

13. The weight of 305.3 cm^3 of bronze is 2.70 kg. What is the weight of one cm^3 of bronze?

14. Determine the relative error for each approximate number. Give each answer as a decimal fraction and as a percent.

a. A length of 35.4 mm
b. An area of 5800 square feet
c. A mass of 65.06 grams
d. A volume of 300 cubic feet to the nearest 10 cubic feet
e. A capacity of 487 mℓ
f. A distance of 1,600,000 miles to the nearest hundred thousand miles

15. What is the length of a rafter for a roof that has a span of 30.0 feet and a rise of 7.50 feet? The overhang is 2.0 feet.

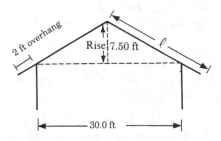

16. A ladder 8.3 m long is placed 4.5 m from the base of a wall. How far up the wall does the top of the ladder reach?

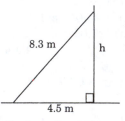

17. What is the length of the metal brace shown in the diagram?

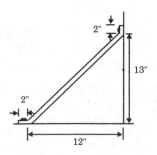

18. A derrick rig is 23.7 m high. A guy wire is attached 2.15 m from the top of the rig to a point 8.75 m from the center of the base of

the rig. How long is the guy wire between the two points of attachment?

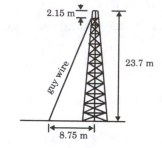

19. What is the length of the run of the stairway shown in the figure? Round off your answer to three significant digits.

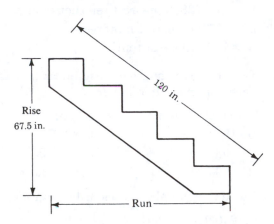

20. A pipe fitter has two parallel pipelines with a set (center axis to center axis) of 28.25". The two pipelines are to be connected using 45° fittings. What length of pipe (ℓ) is needed to connect the two pipelines?

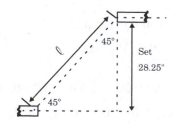

21. What is the area of a right triangle if the length of the hypotenuse is 35.0 cm and the length of the base is 28.0 cm?

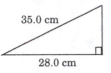

22. What is the area of the trapezoid shown in the figure?

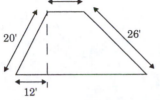

23. While remodeling a house a carpenter checks the squareness of a wall and the floor. A point A, 6.0 feet from the floor, is marked on the wall. A point B, 4.5 feet from the corner of the wall and the floor, is marked on the floor. The carpenter finds the measurement from point A to point B to be 7 feet 8 inches. Are the wall and the floor square? Show your computations to verify your answer.

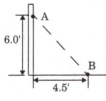

24. Can a triangular plate, with dimensions as shown in the figure, be cut from a rectangular sheet of metal that is 95 cm long and 46 cm wide? Show your computations to verify your answer.

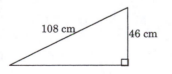

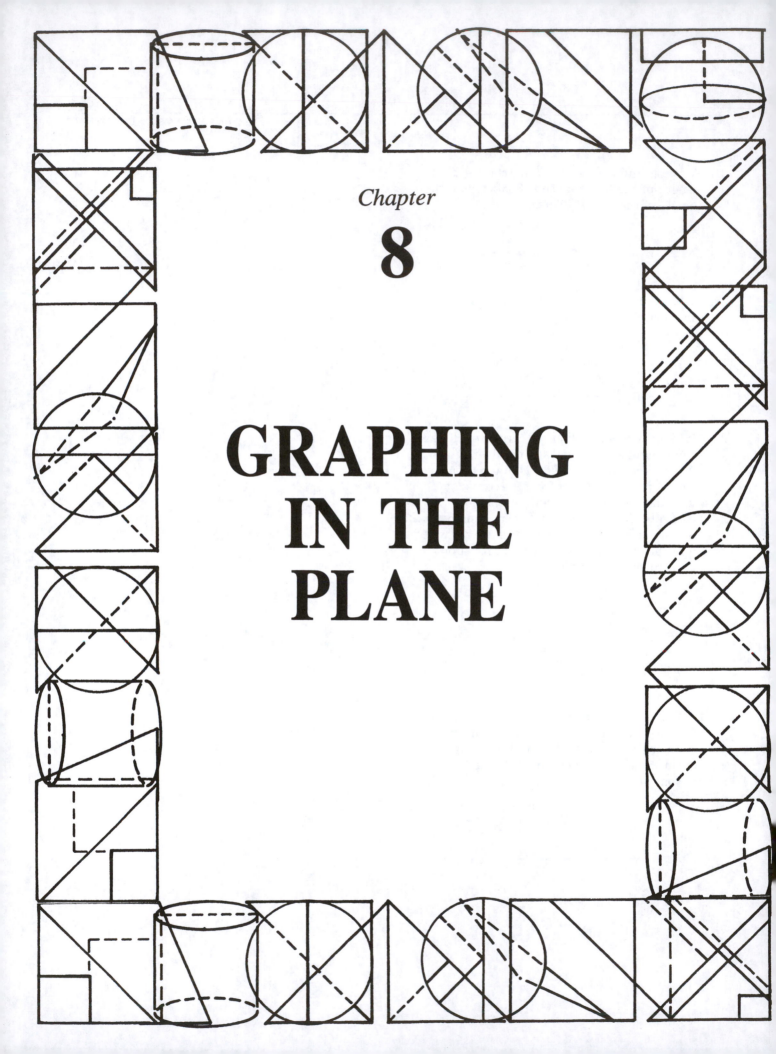

Chapter

8

GRAPHING IN THE PLANE

8.1
PICTURE AND CIRCLE GRAPHS

One of the most important ways of representing data and other factual information is by **graphing**. Graphs are often used in newspapers, magazines, technical reports, and handbooks to depict facts and relationships which are hard to describe in words or show in a table. A graph usually provides an easy way of seeing trends in data or making comparisons between sets of numbers. Because graphs help to analyze complicated sets of data they are used often in science and technology.

There are many different types of graphs, but most can be classified as either **picture graphs, circle graphs, bar graphs**, or **line graphs**. In this section we discuss picture and circle graphs. Attention is given to the methods of constructing graphs and how to read and interpret graphs. In the next section we discuss bar and line graphs.

A picture graph is usually easy to read but hard to draw. Look at Examples 1 and 2.

EXAMPLE 1 Answer the following questions from the information in the picture graph.

(a) What was the profit in April?

(b) What was the increase in profit from April to May?

(c) Which month showed the greatest profit? What was the profit for that month?

(d) What is the difference between the greatest profit and the least profit during the five-month period?

Profit for a Five-Month Period

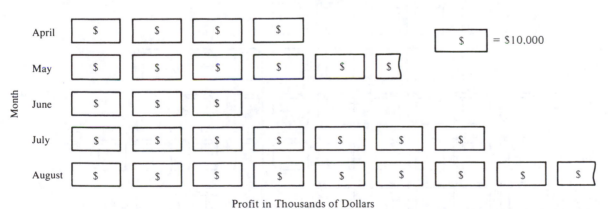

Profit in Thousands of Dollars

Solutions

(a) There are four "pictures" in the row for April. Each picture represents a profit of $10,000, so the total profit for April was $40,000.

(b) The profit for May was $55,000. (The "half-picture" represents $5,000.) Thus, the increase from April to May is:

$$\$55,000 - \$40,000 = \$15,000$$

(c) August approximately $87,500

(d) $87,500 − $30,000 = $57,500

EXAMPLE 2 Construct a picture graph of the data in the table.

Production of Packing Cases for One Week

Day	Number of Packing Cases
Monday	3000
Tuesday	4500
Wednesday	3750
Thursday	5500
Friday	2250

Solution   First we select an appropriate picture for the graph and decide what value each picture will represent. In this example, a box is an appropriate picture for representing packing cases. We will let each box represent the production of 500 packing cases. A smaller number would mean more boxes on any line of the graph.

Next, we determine what values we will place along the vertical axis and what values we will place along the horizontal axis. In this example we place the days of the week along the vertical axis and packing case production along the horizontal axis.

Finally, we draw the appropriate number of boxes and portions of boxes along each horizontal row to represent the specified production of packing cases for each day. For example, for Wednesday we draw 7 complete boxes and one-half of a box to represent the production of 3750 packing boxes for that day:

$$7\frac{1}{2} \times 500 = 3750.$$

Also, the graph is given a title, each axis is labeled, and the key for the value of each representative figure in the graph is given, e.g.,

= 500 packing cases.

Production of Packing Cases for One Week

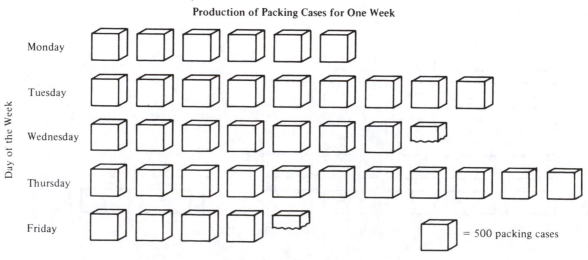

Production of Packing Cases

A circle graph is particularly useful in representing part-to-whole relationships. The interior of a circle represents the "whole," and the various parts of the whole are represented by pie-like pieces of the interior. To construct a circle graph we convert the data values to a percent of the whole, and then use the percent values to determine the number of degrees (% of 360°) each angle of the pie-like slices will contain.

EXAMPLE 3   Answer the following questions from the data given in the circle graph.

(a)   What percent of each tax dollar is used for education?

(b)   If your property tax is $895.50, how much of this amount will be used for debt retirement?

(c)   What percent of each tax dollar is used for general government and miscellaneous costs?

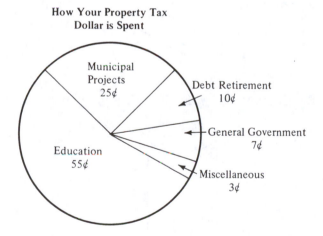

How Your Property Tax
Dollar is Spent

Solution

(a)   Fifty-five cents of each dollar, or 55%, is used for education.

(b)   Ten percent of each dollar is used for debt retirement.

$$10\% \text{ of } \$895.50 = \$89.55$$

(c)   Ten percent of each dollar is used for general government and miscellaneous costs.

EXAMPLE 4   Construct a circle graph to show the production costs for a small engine.

Production Costs

Item	Expenditure
Raw Materials	$200
Labor	$425
Administration & Engineering	$100
Sales & Service	$ 75
Total	$800

Solution   First compute what percent of the total cost each individual expenditure represents. Then compute the part of 360° each item will include in its portion of the graph.

Item	Percent of total cost	Part of 360°
Raw materials	$\frac{200}{800} = 25\%$	25% of $360° = 90°$
Labor	$\frac{425}{800} = 53\frac{1}{8}\%$	$53\frac{1}{8}\%$ of $360° = 191°$
Administration & Engineering	$\frac{100}{800} = 12\frac{1}{2}\%$	$12\frac{1}{2}\%$ of $360° = 45°$
Sales & Service	$\frac{75}{800} = 9\frac{3}{8}\%$	$9\frac{3}{8}\%$ of $360° = 34°$

Draw a circle and divide it into four pie-like pieces using the degrees in the right-hand column to determine the proper size for each part. The following figure shows this step of the process.

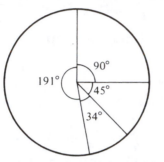

Give the graph a title, and label each part of the graph with the item name and percent of the total for each item. The completed graph is shown here:

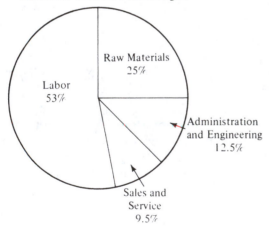

Production Costs for a Small Engine

Raw Materials 25%

Labor 53%

Administration and Engineering 12.5%

Sales and Service 9.5%

Review Exercises

1. 5.2m=_____cm

2. 3.92kg=_____g

3. Ninety-two is what percent of 700? (*Hint:* Write $\frac{92}{700}$ as a percent.)

4. Two hundred fifty-four is what percent of 950? (*Hint:* Write $\frac{254}{950}$ as a percent.)

5. Multiply: $(3.2)(4.6)(0.85) =$ _____ .

6. Multiply: $(16.1)(0.7)(3.04) =$ _____ .

7. Compute the area of the circle.

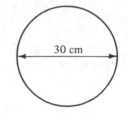

30 cm

8. Compute the circumference of the circle.

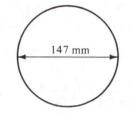

147 mm

9. Compute the volume of the container.

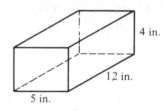

4 in.

12 in.

5 in.

10. Compute the volume of the cylinder.

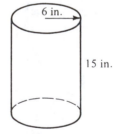

6 in.

15 in.

11. Compute the surface area of the open tool box shown in the figure.

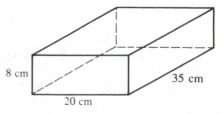

8 cm

20 cm

35 cm

12. Compute the surface area of the triangular container. Include the area of the top and bottom of the container.

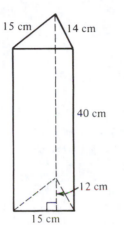

15 cm   14 cm

40 cm

12 cm

15 cm

EXERCISES
8.1

Use the figure shown here for Exercises 1–6.

1. How many skilled workers were employed in 1991?

2. How many skilled workers were employed in 1995?

3. How many more skilled workers were employed in 1997 than in 1989?

4. How many more skilled workers were employed in 1995 than in 1993?

5. In 1994 there were 75 skilled workers employed. How many figures would be needed to show this on the graph?

6. In 1998 there were 290 skilled workers employed. How many figures would be needed to show this on the graph?

Number of Skilled Workers Employed

= 20 workers

Number of Skilled Workers

Use the following figure for Exercises 7–10.

Dividend Per Share of Stock

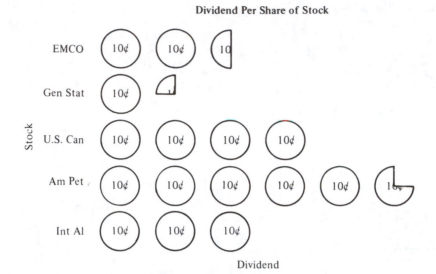

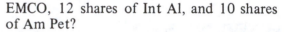

Stock

Dividend

7. What was the dividend on one share of Gen Stat stock?

8. What was the dividend on 100 shares of U.S. Can stock?

9. What was the total dividend on 5 shares of EMCO, 12 shares of Int Al, and 10 shares of Am Pet?

10. What was the total dividend on 20 shares of Gen Stat, 3 shares of Am Pet, and 16 shares of U.S. Can?

Refer to the following figure for Exercises 11–14.

U.S. Petroleum Production

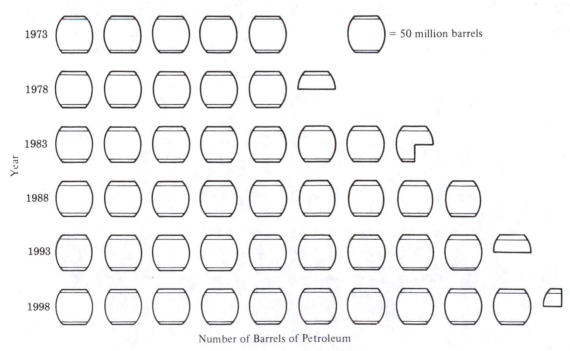

= 50 million barrels

Year

Number of Barrels of Petroleum

11. How many barrels of petroleum were produced in 1973?

12. How many barrels of petroleum were produced in 1983?

13. How many more barrels of petroleum were produced in 1988 than were produced in 1978?

14. How many more barrels of petroleum were produced in 1998 than were produced in 1993?

Refer to the following figure for Exercises 15–18.

Distribution of the Wholesale Price of $350 for a Major Appliance

15. What percent of the price of the appliance is to cover the cost of materials?

16. What percent of the price of the appliance is profit?

17. What percent of the price of the appliance is to cover the cost of labor?

18. What percent of the price of the appliance is to cover the overhead costs?

Use the following figure for Exercises 19–22.

Distribution of a Weekly Wage of $320

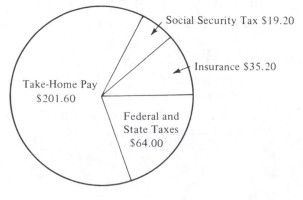

19. What percent of the week's wages is take-home pay?

20. What percent of the week's wages is paid toward insurance?

21. What percent of the week's wages is paid as social security tax?

22. What percent of the week's wages is for federal and state taxes?

Construct a picture graph for each set of data in Exercises 23–26.

23. **Production of Cement in the U.S.**

1987	45,000,000 tons
1989	52,500,000 tons
1991	67,500,000 tons
1993	60,000,000 tons
1995	80,000,000 tons
1997	94,000,000 tons

24. **Profits for the First Six Months of the Year**

January	$4500
February	$8000
March	$6250
April	$5000
May	$9500
June	$7750

25. **Number of People Employed in a Manufacturing Company**

Administration	50
Skilled Workers	150
Unskilled Workers	325
Maintenance and Service	75
Sales	100

26. **State Gasoline Taxes in Selected States**

Texas	5¢
Oregon	7¢
Maine	9¢
Idaho	8.5¢
Oklahoma	6.5¢
Montana	7.75¢
Connecticut	10¢

27. Complete the following table and then construct a circle graph for the data.

Distribution of Costs for a New Home

Item		Cost	Percent of total cost	Part of 360° (Number of degrees)
Land		$ 7,000	_____	_____
House		$40,000	_____	_____
Landscaping		$ 3,000	_____	_____
Furnishings		$10,000	_____	_____
	Total	$60,000	_____	_____

28. Complete the following table and then construct a circle graph for the data.

Average Work Day

Item	Time	Percent of total time	Part of 360°
Work	8 hours	_____	_____
Sleep	7 hours	_____	_____
Meals	2 hours	_____	_____
Recreation	3 hours	_____	_____
Miscellaneous	4 hours	_____	_____

Construct a circle graph for each set of data in Exercises 29–32.

29. **Where the Federal Government Obtains Each Budget Dollar**

Individual Income Taxes	39¢
Social Security Taxes	29¢
Corporation Income Taxes	13¢
Borrowing	11¢
Excise Taxes	4¢
Others	4¢

30. **How the Federal Government Spends Each Budget Dollar**

Benefit Payments to Individuals	38¢
National Defense	26¢
Grants to States and Localities	16¢
Interest on Debts	7¢
Federal Operations	13¢

31. **Distribution of Receipts of a Machine Company**

Wages	$135,000
Raw Materials	$ 81,000
Interest and Taxes	$ 24,000
Maintenance	$ 45,000
Miscellaneous	$ 15,000

32. **Production of Liquid Petroleum Products (number of barrels)**

Motor Fuel	2,000,000,000
Fuel Oil	1,500,000,000
Jet Fuel	800,000,000
Liquid Gases	300,000,000
Others	400,000,000

Calculator Exercises

33. What is 758.3% of 749.4 to the nearest tenth?

34. What is 105.03% of 49.76 to the nearest hundredth?

35. What is 0.0035% of 178.3 to five decimal places?

36. What is $4\frac{7}{8}\%$ of $3\frac{4}{5}$ to four decimal places?

37. What percent is 25.8 of 117.32 to the nearest thousandth of a percent?

38. What percent is 5.783 of 846.004 to the nearest thousandth of a percent?

8.2
BAR AND LINE GRAPHS

A **bar graph** uses bars (shaded rectangular regions) to show the relative magnitude of a set of data.

EXAMPLE 1 Answer the following questions from the information illustrated by the bar graph.

(a) What information is shown along the horizontal axis?

(b) What information is shown along the vertical axis?

(c) In what month was power use the greatest?

(d) Approximately how much power was used during May?

(e) What is the total amount of power used during the first three months of the year?

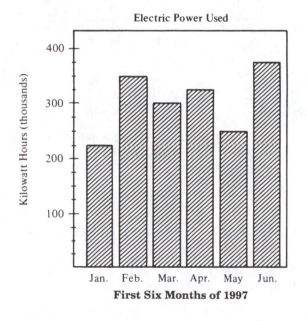

Electric Power Used

First Six Months of 1997

Solution

(a) The scale along the horizontal axis lists the months of the year.

(b) The scale along the vertical axis shows the amount of power used in thousands of kilowatt hours.

(c) The greatest amount of power was used during the month of June—approximately 375,000 kilowatt hours.

(d) Approximately 250,000 kilowatt hours of power were used during May.

(e) During the first three months of 1992 approximately (225 + 350 + 300), or 875,000 kilowatt hours of power were used.

EXAMPLE 2 Construct a bar graph to show the auto sales from January through May.

Month	Amount of Sales
Jan	$6,500,000
Feb	7,200,000
March	7,000,000
April	7,800,000
May	8,500,000

Solution

(a) Determine what information will be shown along the horizontal axis and what information will be shown along the vertical axis. In this example we place the *years* along the horizontal axis and the *amount of sales,* in millions of dollars, along the vertical axis. With this choice, the bars of the graph will be vertical. (It is also possible to interchange the data on the two axes so that the bars of the graph run horizontally.)

(b) Determine the scale along each axis so that the full range of data can be shown on the graph, yet keep the graph to an appropriate size. The units along a given scale must all be the same size. The units on one scale do not have to be the same size as the units on the other scale. Note that it is not necessary to start at zero on a scale if only high values of the scale are to be used.

(c) Draw bars to represent the information given in the table. Usually the bars are shaded and a space is left between bars.

(d) Give the graph a title and label each axis.

The completed graph for Example 2 is shown here. (*Note:* The jagged lines at the bottom of the graph are used to show that a portion of the graph from 0.0 to 5.0 has not been drawn.)

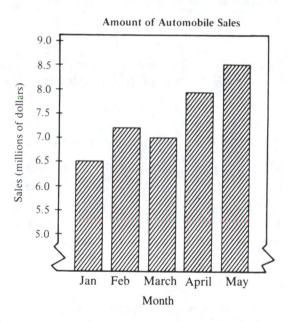

A line graph can be used to show the trend of a set of data that is continuous. Two or more sets of data can be compared by plotting both sets of data on the same line graph.

EXAMPLE 3 Answer the following questions from the information shown by the line graph.

(a) At what speed does the automobile get the highest mileage per gallon (mpg)?

(b) What is the approximate mpg at 30 miles per hour (mph)?

(c) What is the approximate mpg at 55 miles per hour?

(d) At what speed(s) does the automobile get approximately 20 mpg?

(e) What is the decrease in the number of miles per gallon between 60 mph and 70 mph?

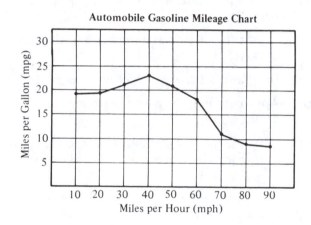

Solution

(a) The automobile gets the highest mileage per gallon (approximately 23 mpg) at 40 miles per hour.

(b) At 30 mph the mileage is approximately 21 miles per gallon.

(c) At 55 mph the mileage is approximately 19 miles per gallon.

(d) The automobile gets 20 mpg at approximately 23 mph and 52 mph.

(e) The decrease in the number of miles per gallon between 60 mph and 70 mph is approximately (18 mph − 11 mpg) or 7 mpg.

EXAMPLE 4 This example shows two sets of data on the same line graph. The solid line represents the supply of motor fuel. The dashed line represents the demand for motor fuel. Answer the following questions from the information shown in the graph.

(a) What was the *supply* of motor fuel in 1978?

(b) What was the *demand* for motor fuel in 1993?

(c) During what year(s) was the supply of motor fuel greater than the demand?

(d) In 1988 what was the difference between the supply and demand for motor fuel?

(e) During what year(s) did the supply of motor fuel equal the demand for motor fuel?

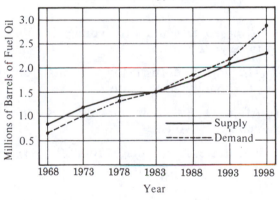

U.S. Motor Fuel Supply and Demand

Solution

(a) The supply of motor fuel in 1978 was approximately 1,400,000 barrels.

(b) The demand for motor fuel in 1993 was approximately 2,200,000 barrels.

(c) The supply of motor fuel was greater than the demand from 1968 to 1983.

(d) In 1988 the demand for motor fuel was approximately 1,000,000 more than the supply.

(e) The supply and demand for motor fuel were equal in 1983.

EXAMPLE 5 Using the following data, construct a line graph to show the outdoor temperature during an eight-hour period on February 20, 1997.

Time	Temperature
8:00 A.M.	−7 °C
9:00 A.M.	−5 °C
10:00 A.M.	−1 °C
11:00 A.M.	0 °C
12:00 noon	4 °C
1:00 P.M.	5 °C
2:00 P.M.	2 °C
3:00 P.M.	0 °C

Solution

(a) Determine what information will be shown along the horizontal axis and what information will be shown along the vertical axis. In this example we place *time of day* along the horizontal axis and *temperature* along the vertical axis.

(b) Determine the scale along each axis so the full range of data can be shown. The units on a given scale must be equal to each other. However, a unit on the horizontal scale does not have to be equal to a unit on the vertical scale. The scales do not have to start at zero.

(c) Locate the pairs of numbers on the horizontal and vertical scales and place a dot on the graph to represent each pair of values.

(d) Connect the dots with line segments.

(e) Give the graph a title and label each axis.

The completed graph for Example 5 is shown here.

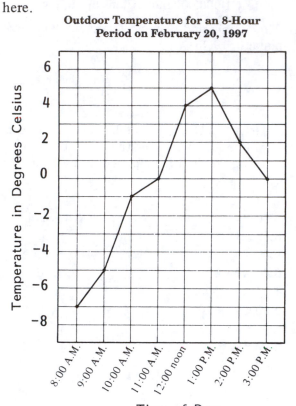

Outdoor Temperature for an 8-Hour Period on February 20, 1997

A line graph may be drawn as a series of connected line segments (broken line), as shown in the previous two examples, or it may be drawn as a smooth curve. Example 6 shows a curved-line graph.

EXAMPLE 6 Use the following graph to answer questions (a) – (d).

(a)   What is the approximate voltage after 10 test hours?

(b)   After how many hours does the voltage drop below 12 volts?

(c)   What is the approximate voltage after 30 test hours?

(d)   What is the decrease in voltage from 20 test hours to 25 test hours?

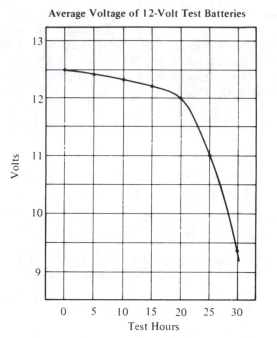

Average Voltage of 12-Volt Test Batteries

Solution

(a)   After 10 test hours the voltage is approximately 12.3 volts.

(b)   The voltage drops below 12 volts after approximately 20 test hours.

(c)   After 30 test hours the voltage is approximately 9.4 volts.

(d)   The decrease in voltage from 20 to 25 test hours is approximately 1 volt.

Review Exercises

1.   What is the area of the triangle?

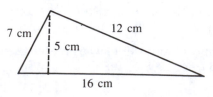

2.   What is the area of the trapezoid?

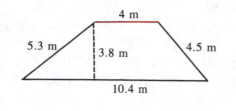

3. $\dfrac{3}{5} \div \dfrac{8}{7} =$ _____

4. $\dfrac{4}{3} \div \dfrac{12}{16} =$ _____

5.   17.94 X 0.082 = _____

6.   0.035 X 7.04 = _____

7.   Eight is _____ % of 160.

8.   Nine is _____ % of 45.

9.   0.054 ÷ 4.5 = _____

10.   8.4 ÷ 0.012 = _____

11.   Convert 8°43′ to degrees, to the nearest thousandth of a degree.

12.   Convert 13°17′ to degrees, to the nearest thousandth of a degree.

EXERCISES
8.2

Use the following graph for Exercises 1–6.

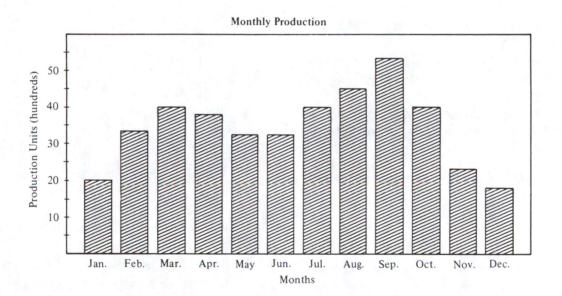

1.   Which month had the greatest production? Approximately how many units were produced that month?

2.   Which month had the least production? Approximately how many units were produced that month?

3.   During which month(s) was production about one-third of the production during September?

4.   During which month(s) was production about twice the production during November?

5.   What was the approximate average production for the first three months of the year?

6.   What was the approximate average production for the months of July, August, and September?

Use the following graph for Exercises 7–10.

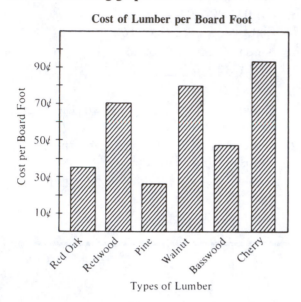

Cost of Lumber per Board Foot

7. What is the approximate cost of 1000 board feet of walnut?

8. What is the approximate cost of 100 board feet of red oak?

9. What is the difference in the price of one board foot of cherry and one board foot of redwood?

10. What is the difference in the price of one board foot of pine and one board foot of basswood?

Use the following graph to answer Exercises 11–18.

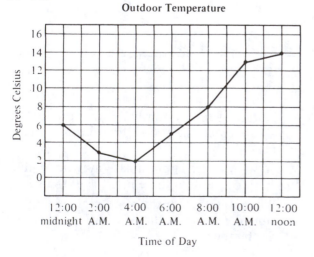

Outdoor Temperature

11. What was the lowest temperature during the time period?

12. What was the highest temperature during the time period?

13. What was the temperature at 6:00 A.M.?

14. What was the temperature at 10:00 A.M.?

15. What was the temperature at 1:00 A.M.?

16. What was the temperature at 5:00 A.M.?

17. At what time of day was the temperature 10 °C?

18. At what time of day was the temperature 8 °C?

Use the following figure for Exercises 19–24.

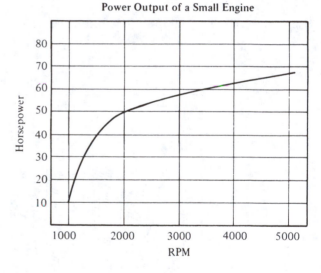

Power Output of a Small Engine

19. What is the horsepower at 2000 rotations per minute (rpm)?

20. What is the horsepower at 3000 rpm?

21. What is the horsepower at 1500 rpm?

22. What is the horsepower at 4500 rpm?

23. At what rpm will the engine produce 45 horsepower?

24. At what rpm will the engine produce 60 horsepower?

Use the following graph for Exercises 25–32. Answer each of the questions either True or False.

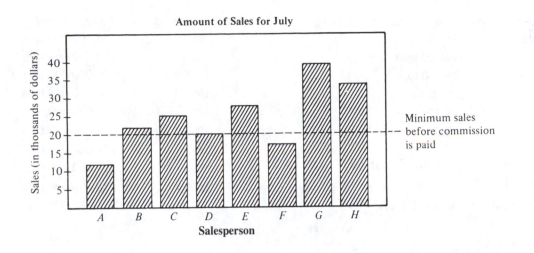

Amount of Sales for July

25. Three salespersons did not receive sales commissions for the month of July.

26. Salesperson H had commission sales greater than the commission sales of salespersons C and E combined.

27. The average sales of salespersons A and B was less than $15,000.

28. The average sales of salespersons C, D, and E was approximately $24,000.

29. Salesperson G had sales greater than twice the sales of salesperson B.

30. Six salespersons received commissions on part of their sales for the month.

31. Salesperson A had sales less than one-half of the sales of salesperson E.

32. The average sales of salespersons F, G, and H was greater than $27,000.

Construct a bar graph for each set of data in Exercises 33–38.

33. **Length of Underwater Vehicular Tunnels in the U.S.**

Callahan Tunnel	5000 feet
Baytown Tunnel	4000 feet
Posey Tube	3500 feet
Bankhead Tunnel	3000 feet
Harvey Tunnel	1000 feet

34. **Number of Stories in Notable New York Buildings**

Empire State Bldg.	102
Chrysler Bldg.	77
RCA Bldg.	70
Woolworth Bldg.	60
Union Carbide Bldg.	52

35. **U.S. Pig Iron Production (millions of tons)**

1963	45
1968	50
1973	60
1978	75
1983	65
1988	85
1993	90
1998	98

36. **Production of Hydroelectric Energy in the U.S. (billions of kilowatt hours)**

1968	75
1973	100
1978	125
1983	140
1988	200
1993	250
1998	315

37. **Density of Gases**

Air	1.3
Ammonia	0.8
Butane	2.5
Neon	0.9
Fluorine	1.7
Acetylene	1.2

38.   <u>**Mineral Production in the U.S. in 1995**</u>

Asbestos	125,000 tons
Barite	850,000 tons
Calcium	625,000 tons
Fluorspar	250,000 tons
Perlite	450,000 tons
Sulfur	900,000 tons

Construct a broken line graph for each set of data in Exercises 39–44.

39.   **Cost Per Unit on Multiple Orders**

Number of Units	Cost per Unit
1	80¢
5	78¢
10	75¢
20	73¢
25	70¢

40.   **Efficiency of an Engine at Different Operating Temperatures**

Temperature	Percent Efficiency
150 °F	40%
175 °F	48%
200 °F	62%
225 °F	66%
250 °F	59%
275 °F	50%

41.   **Amount of Natural Gas Used in a Single Family Home**

January	350 THERMS
February	260 THERMS
March	225 THERMS
April	150 THERMS
May	90 THERMS
June	25 THERMS
July	35 THERMS
August	50 THERMS
September	80 THERMS
October	125 THERMS
November	300 THERMS
December	325 THERMS

42.   **Number of Tons of Asphalt Used**

1979	74,000,000 tons
1981	80,000,000 tons
1983	86,000,000 tons
1985	90,000,000 tons
1987	84,000,000 tons
1989	87,000,000 tons
1991	94,000,000 tons
1993	105,000,000 tons
1995	108,000,000 tons
1997	116,000,000 tons

43.   **Number of BTU Needed to Raise the Temperature of a Metal**

Number of BTU	Temperature
500	20 °C
750	40 °C
1000	60 °C
1250	80 °C
1750	100 °C
2250	120 °C

44.   **Torque of a High-Compression Engine**

RPM	Foot Pounds of Torque
500	100
1000	112
1500	118
2000	116
2500	107
3000	105
3500	95
4000	80

In Exercises 45 and 46 graph both sets of data on the same line graph.

45.
Production of Kitchen Ranges
(millions of units)

Year	Electric	Gas
1989	1.8	2.4
1990	2.0	2.6
1991	2.0	1.7
1992	2.4	1.8
1993	3.0	2.5
1994	2.6	3.0
1995	2.8	3.5
1996	3.0	3.8
1997	3.2	3.9
1998	3.6	4.1

46.
Supply and Demand of
New Single Family Housing Units
(millions of units)

Year	Supply	Demand
1989	2.8	3.0
1990	2.8	3.2
1991	3.4	4.0
1992	3.6	4.1
1993	3.9	3.9
1994	3.2	3.5
1995	2.8	2.2
1996	2.3	1.8
1997	2.4	2.0
1998	2.7	2.6

Calculator Exercises

47. What is $0.02\frac{3}{4}\%$ of 75.893 to five decimal places?

48. What is the cash price if the listed price of $4975.50 is discounted by $15\frac{7}{8}\%$?

49. Find the total cost of 39 meals at $4.85 each plus gratuity of 15% and sales tax of 4%?

50. First an add-on of 22.5% was applied to $789.75, then a discount of 18.3% was applied to the balance. What is the final amount to the nearest cent?

51. What is 0.035% of 79.34% of 17.093 to six decimal places?

52. What is one-fifth of one-fourth of three-eighths of 9898.76 to four decimal places?

8.3
GRAPHING POINTS IN THE PLANE

Another type of graphing system that is useful in science and technology is called a **rectangular coordinate system**. We use this system to graph data that is mathematically related. In this section we develop the basic system and discuss how to graph on a coordinate plane.

Two number lines which are drawn perpendicular to each other, and which intersect at their respective zero points, form the reference lines for a rectangular coordinate system. Points, corresponding to ordered pairs of numbers, can then be located in the coordinate plane of the two number lines. A rectangular coordinate system (also called a "coordinate plane") is represented on a grid as shown in Figure 8.1.

FIGURE 8.1

Rectangular Coordinate System

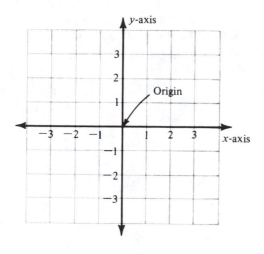

We call the point of intersection of the two number lines the **origin** and let it represent zero on both lines. The horizontal number line is called the **x-axis**, and the vertical number line is called the **y-axis**.

A point in a coordinate plane is named by an ordered pair of numbers. For example, in Figure 8.2, the ordered pair (3,5) refers to the point three units to the right of the origin and five units up (point A). The ordered pair (6,2) refers to the point six units to the right of the origin and two units up (point B). Point C is represented by the ordered pair (−3,2) and is three units to the left of the origin and two units up. Point D is represented by the ordered pair (−5,−3) and is five units to the left of the origin and three units down. The ordered pair (2,−5) refers to the point two units to the right of the origin and five units down (point E).

FIGURE 8.2

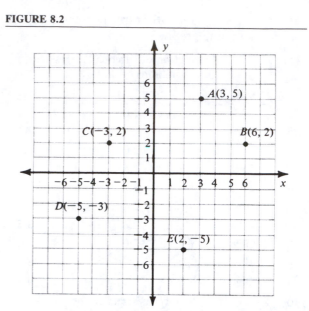

In any ordered pair of numbers (x, y) that corresponds to a point in a coordinate plane, the first number, x, is called the **x-coordinate**. The x-coordinate indicates direction along the x-axis (to the right of the origin if positive and to the left of the origin if negative), and distance from the origin. The second number, y, is called the **y-coordinate** and indicates direction along the y-axis (up from the origin if positive and down from the origin if negative), and distance from the origin. The two numbers in the ordered pair are called the **coordinates** of the corresponding point in the plane. (See Figure 8.3.)

FIGURE 8.3

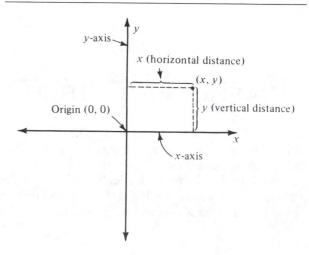

EXAMPLE 1   Identify the coordinates of each point in the graph.

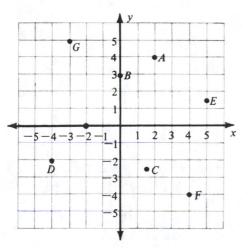

Solution

A	(2,4)	E	(5,1.5)
B	(0,3)	F	(4, −4)
C	(1.5,−2.5)	G	(−3,5)
D	(−4,−2)	H	(−2,0)

EXAMPLE 2   Graph the following points in the coordinate plane: $A(-4,1)$, $B(1.5,3)$, $C(0,-4)$, $D(-3,-5)$, $E(3,-2.5)$, $F(0,0)$, $G(-1,5)$, $H(5.5,0)$, $I(3.5,3.5)$, and $J(-5,-3)$.

Solution

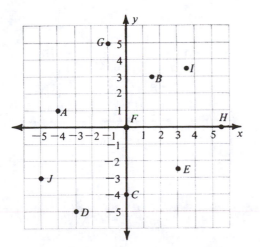

Review Exercises

1. A metal box has a volume of 1.5 cubic feet. How many liters will this box hold?

2. The weight of an iron bar is 3.2 kilograms. What is the cost of the bar if the price of iron is 12¢ an ounce?

3. A speed of 50 miles per hour is the same as how many kilometers per hour?

4. The area of a metal plate is 14 square inches. What is the area of the plate in square centimeters?

5. $12\frac{1}{4} \div 3\frac{1}{6} =$ _____

6. $8\frac{3}{5} \div 3\frac{1}{2} =$ _____

7. $2\frac{1}{3} \times \frac{3}{5} =$ _____

8. $\frac{7}{10} \times 3\frac{5}{8} =$ _____

9. Convert 35.385° to degrees and minutes to the nearest minute.

10. Convert 17.829° to degrees and minutes to the nearest minute.

11. $28\frac{1}{2} \div 3\frac{4}{5} =$ _____

12. $108\frac{1}{8} \div 8\frac{3}{4} =$ _____

EXERCISES
8.3

List the coordinates of each point on the graphs
in Exercises 1–8.

1.

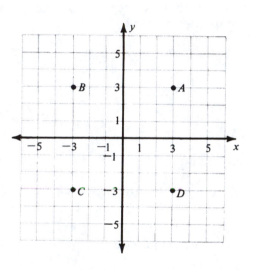

3.

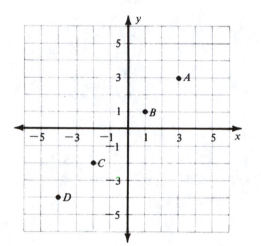

2.

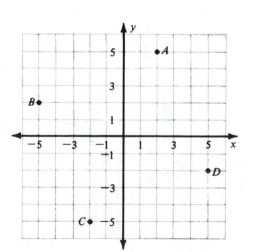

4.

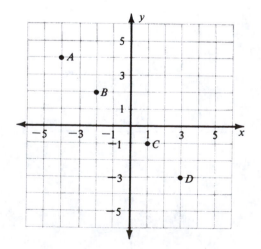

5.

7.

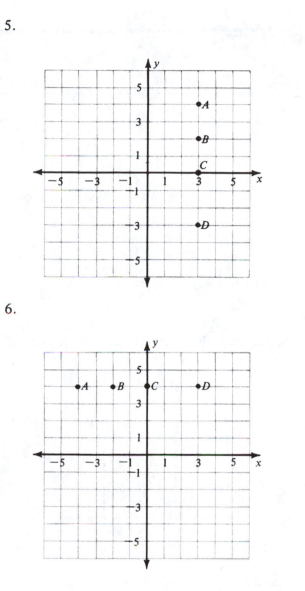

6.

8.

Calculator Exercises

9. Graph the following points on a coordinate plane: $A(5,6)$, $B(-2,-4)$, $C(3,-1.5)$, $D(-\frac{1}{2},4)$, $E(7,-3.5)$, $F(0,-3)$, $G(4,0)$, and $H(-2\frac{1}{2},-2\frac{1}{2})$.

10. Graph the following points on a coordinate plane: $A(-3,4)$, $B(\frac{5}{2},-2)$, $C(-1,0)$, $D(\frac{1}{2},-4\frac{1}{2})$, $E(6,4)$, $F(-\frac{3}{2},-\frac{5}{2})$, $G(0,6)$, and $H(5,0)$.

Perform the indicated operations and write the answer as decimals, rounded off to four decimal places.

11. $27\frac{15}{32} - 12\frac{43}{64}$

12. $15\frac{8}{15} + 7\frac{13}{21}$

13. $19\frac{3}{8} \times 76\frac{2}{3}$

14. $14\frac{5}{9} \times 23\frac{1}{6}$

15. $219\frac{3}{8} \div 17\frac{2}{3}$

16. $105\frac{3}{16} \div 11\frac{1}{7}$

8.4
DISTANCE BETWEEN TWO POINTS

To find the distance from the origin of a co-ordinate plane to a point on one of the axes, we simply count the number of units along the axis from the origin to the point. In Figure 8.4 the distance from the origin to $P_1(3, 0)$ is three units. The distance from the origin to $P_2(0, 2)$ is two units. The distance from the origin to $P_3(-4, 0)$ is four units. The distance from the origin to $P_4(-3, 0)$ is three units. Notice that, for a point on the x-axis, the distance from the origin to the point is equal to the *positive* value of the x-coordinate. If a point is on the y-axis, the distance from the origin to the point is equal to the *positive* value of the y-coordinate.

FIGURE 8.4

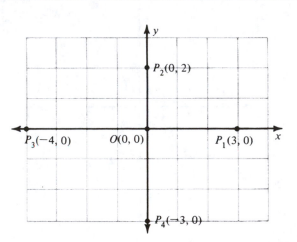

To find the distance from the origin to a point $P(x, y)$ not on either axis, construct a right triangle with OP as hypotenuse, one leg on the x-axis, and the other leg a perpendicular line segment from the point P to the x-axis. The point Q on the x-axis has coordinates (x, O). The length of OQ will be x units (positive value of the x-coordinate of $P(x, y)$), and the length of PQ will be y units (positive value of the y-coordinate of $P(x, y)$). The distance from O to P can now be found by using the **Pythagorean Theorem**. (See Figure 8.5.)

FIGURE 8.5

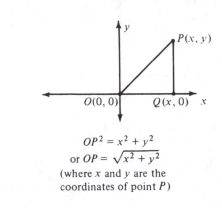

$$OP^2 = x^2 + y^2$$
$$\text{or } OP = \sqrt{x^2 + y^2}$$
(where x and y are the coordinates of point P)

EXAMPLE 1   Find the distance from the origin to the point $P(6, 8)$.

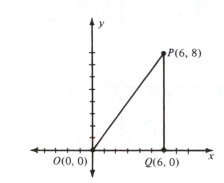

Solution

$$OP^2 = 6^2 + 8^2$$
$$OP^2 = 36 + 64$$
$$OP^2 = 100$$
$$OP = \sqrt{100}$$
$$OP = 10$$

EXAMPLE 2   Find the distance from the origin to the point $M(-7, 4)$. Round off the answer to two significant digits.

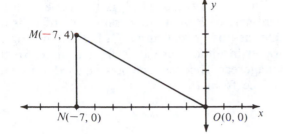

Solution

$$OM^2 = 7^2 + 4^2$$
$$OM^2 = 49 + 16$$
$$OM^2 = 65$$
$$OM = \sqrt{65}$$
$$OM \doteq 8.1$$

EXAMPLE 3 Find the distance from the origin to the point $Z(10, -7)$. Round off the answer to the nearest tenth of a unit.

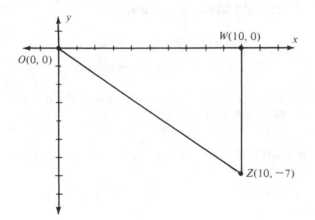

Solution

$$OZ^2 = 10^2 + 7^2$$
$$OZ^2 = 100 + 49$$
$$OZ^2 = 149$$
$$OZ = \sqrt{149}$$
$$OZ \doteq 12.2$$

EXAMPLE 4 Find the distance from the origin to the point $K(-8, -8)$. Round off the answer to three significant digits.

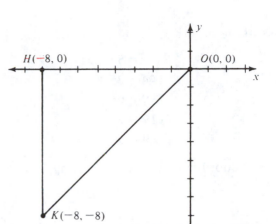

Solution

$$OK^2 = 8^2 + 8^2$$
$$OK^2 = 64 + 64$$
$$OK^2 = 128$$
$$OK = \sqrt{128}$$
$$OK \doteq 11.3$$

EXAMPLE 5 The distance from the origin to a point is 12 units. If the x-coordinate of the point is 5, what is the y-coordinate? Round off the answer to three significant digits.

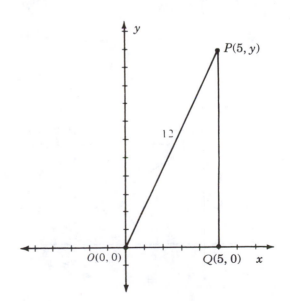

Solution

$$12^2 = 5^2 + y^2$$
$$144 = 25 + y^2$$
$$119 = y^2$$
$$\sqrt{119} = y$$
$$10.9 \doteq y$$

The *y*-coordinate of point *P* is approximately 10.9.

The distance between any two points in the coordinate plane can be found in a similar manner. If we have two distinct points, P_1 and P_2, in the coordinate plane we can label the coordinates of P_1 as (x_1, y_1) and the coordinates of P_2 as (x_2, y_2). If we draw a line through P_1 parallel to the *y*-axis (vertical line) and a line through P_2 parallel to the *x*-axis (horizontal line), these two lines will meet at a third point, P_3, as shown in the diagrams in Figure 8.6.

FIGURE 8.6

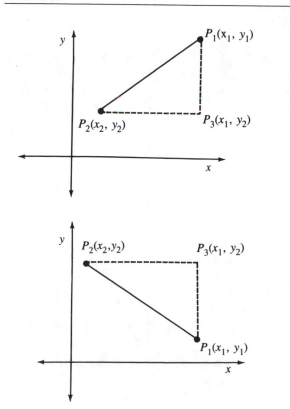

The *x* coordinate of P_3 is the same as the *x* coordinate of P_1 and the *y* coordinate of P_3 is the same as the *y* coordinate of P_2. The figure formed in either case is a right triangle so the Pythagorean Theorem can now be used to find the distance from P_1 to P_2.

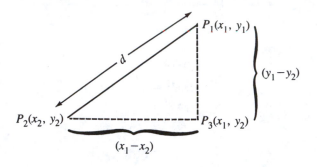

The distance from P_1 to P_3 is represented by $(y_1 - y_2)$ and the distance from P_2 to P_3 by $(x_1 - x_2)$. Since *d* is the length of the hypotenuse of the right triangle we have

$$d^2 = (x_1 - x_2)^2 + (y_1 - y_2)^2, \text{ or}$$
$$d = \sqrt{(x_1 - x_2)^2 + (y_1 - y_2)^2}$$

where d is the positive square root of the number under the radical sign.

EXAMPLE 6   Find the distance between P_1 (4, 8) and P_2 (1, 3).

Solution

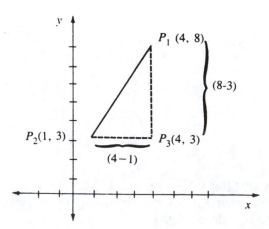

Using $(4, 8) = (x_1, y_1)$ and $(1, 3) = (x_2, y_2)$ we can find the distance from P_1 to P_2 as follows:

$$d = \sqrt{(x_1 - x_2)^2 + (y_1 - y_2)^2}$$

$$d = \sqrt{(4 - 1)^2 + (8 - 3)^2}$$

$$d = \sqrt{3^2 + 5^2}$$

$$d = \sqrt{9 + 25}$$

$$d = \sqrt{34}$$

$$d \doteq 5.83$$

EXAMPLE 7   Find the distance between $(-1, 4)$ and $(3, -2)$

Solution

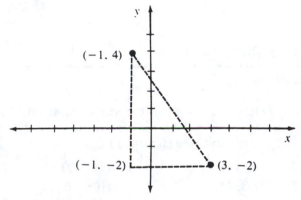

Using $(x_1, y_1) = (3, -2)$ and $(x_2, y_2) = (-1, 4)$ we have

$$d = \sqrt{[3 - (-1)]^2 + (-2 - 4)^2}$$

$$d = \sqrt{4^2 + (-6)^2}$$

$$d = \sqrt{16 + 36}$$

$$d = \sqrt{52}$$

$$d \doteq 7.21$$

Note that the same result would be found if we used $(x_1, y_1) = (-1, 4)$ and $(x_2, y_2) = (3, -2)$.

$$d = \sqrt{(-1 - 3)^2 + [4 - (-2)]^2}$$

$$d = \sqrt{(-4)^2 + 6^2}$$

$$d = \sqrt{16 + 36}$$

$$d = \sqrt{52}$$

$$d \doteq 7.21$$

In the previous discussion it was assumed that the points (x_1, y_1) and (x_2, y_2) did not both lie on a line parallel to the x-axis or parallel to the y-axis. However, the distance formula will still apply even if both points are on the same vertical or horizontal line.

If both points are on the same vertical line the coordinates can be represented as (x_1, y_1) and (x_1, y_2) since the x coordinate is the same for all points on a vertical line. This is illustrated in Figure 8.7a. Thus,

$$d = \sqrt{(x_1 - x_1)^2 + (y_1 - y_2)^2}$$

$$d = \sqrt{0^2 + (y_1 - y_2)^2}$$

$$d = \sqrt{(y_1 - y_2)^2}$$

Therefore d is equal to $y_1 - y_2$ or $y_2 - y_1$, whichever difference is the *positive square root*.

If both points are on the same horizontal line then the coordinates can be represented by (x_1, y_1) and (x_2, y_1). This is illustrated in Figure 8.7b. Thus,

$$d = \sqrt{(x_1 - x_2)^2 + (y_1 - y_1)^2}$$

$$d = \sqrt{(x_1 - x_2)^2 + 0^2}$$

$$d = \quad (x_1 - x_2) \text{ or } (x_2 - x_1), \text{ whichever is the positive value.}$$

FIGURE 8.7

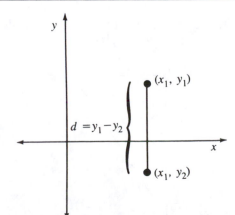

(a) Distance between two points on the same vertical line

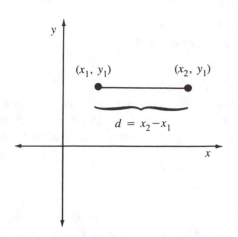

(b) Distance between two points on the same horizontal line

3. The measure of a length of conduit is 850 centimeters to the nearest centimeter. What is the relative error of this measurement?

4. The measure of the length of a bolt is 34.0 millimeters. What is the relative error of this measurement?

5. What is the product of the approximate numbers 35.2 and 0.038?

6. What is the sum of the approximate numbers 206.4, 5.08, 27.394, and 5700?

7. $850 \text{ cm}^2 = $ _____ m^2

8. $6 \text{ dm}^2 = $ _____ cm^2

9. $\dfrac{2}{3} + \dfrac{4}{5} = $ _____

10. $\dfrac{3}{4} + \dfrac{5}{6} = $ _____

11. $\dfrac{3}{5} \times 2\dfrac{1}{3} = $ _____

12. $\dfrac{5}{8} \times 4\dfrac{3}{5} = $ _____

Review Exercises

1. What is the length of leg AC to the nearest tenth of a foot?

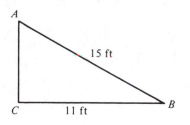

2. What is the length of leg HK, rounded off to two significant digits?

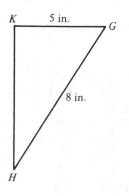

EXERCISES
8.4

In Exercises 1–14 find the distance from the origin to each of the points given. Round off each answer to three significant digits.

1. $P(5, 12)$ 2. $A(9, 12)$
3. $C(-8, 15)$ 4. $Q(-8, 6)$
5. $M(4, -8)$ 6. $S(6, -6)$
7. $Z(-7, -8)$ 8. $H(-10, -6)$
9. $R(-8, 11)$ 10. $W(16, -9)$
11. $K(13, 8)$ 12. $N(-15, 20)$
13. $B(-12, -12)$ 14. $V(7, 2)$

Complete the table below for Exercises 15–24. Give each answer to the nearest tenth of a unit.

	Distance from origin to point	x-coordinate	y-coordinate
15.	5 units	3	_____
16.	17 units	15	_____
17.	34 units	_____	16
18.	30 units	_____	18

19.	14 units	7	_____
20.	9 units	6	_____
21.	12 units	_____	−7
22.	8 units	−3	_____
23.	18 units	13	_____
24.	22 units	_____	−14

31. (-3, 4), (8, -7)   32. (5, -6), (-11, 2)
33. (8, 5), (-2, 5)   34. (-3, 7), (-3, 18)
35. (-2, -5), (-5, 3)   36. (5, -3), (-5, -7)

Calculator Exercises

Find the distance from the origin to each of the points with the sets of coordinates shown in Exercises 37-42. Give the answers to four decimal places.

In exercises 25-36 find the distance between the two points. Round off each answer to the nearest tenth of a unit.

25. (2, 2), (5, 6)   26. (7, 2), (2, 14)
27. (-3, 3), (7, 11)   28. (-2, 1), (1, 13)
29. (-4, -3), (1, 9)   30. (5, 2), (-3, -4)

37. (5.6, 8.3)   38. (7.5, 4.9)
39. (14.2, 9.8)   40. (11.3, 18.4)
41. (−7.23, 6.45)   42. (18.33, −8.25)

8.5
SLOPE OF A LINE

A line drawn in the plane on a rectangular coordinate system is a graph of an infinite set of points. Every point on the line can be named by a pair of coordinates (x, y) as illustrated in Figure 8.8. The points on any given line satisfy a set of conditions that describe that particular line.

One condition that helps to describe a line is how steep or flat the line is. The steepness or tilt of a line is called its *slope*. The slope of a line can be defined as a ratio of rise to run:

$$\text{Slope} = \frac{\text{rise}}{\text{run}}$$

Figure 8.9 illustrates both a positive and negative slope.

FIGURE 8.8

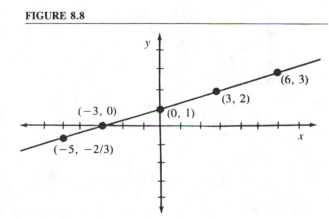

FIGURE 8.9

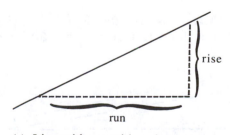

(a) Line with a positive slope

(b) Line with a negative slope

The rise of a line is considered to be positive if the line goes up as you move from left to right. The rise is considered to be negative if the line goes down as you move from the left to the right.

The following examples illustrate how slope is used in some practical situations.

EXAMPLE 1   What is the slope (pitch) of a house roof if the gable has a rise of 4 feet and a span of 30 feet? The highest point of the roof is at the midpoint of its span.

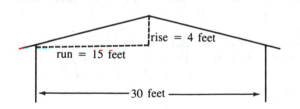

Solution

$$\text{Slope} = \frac{\text{rise}}{\text{run}} = \frac{4}{15}$$

EXAMPLE 2   What is the slope of a roadbed on a hill that declines 15 feet for each 350 feet of the roadway?

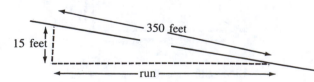

Solution   We use the Pythagorean Theorem to compute the run of the decline.

$$\text{run} = \sqrt{350^2 - 15^2}$$
$$= \sqrt{122500 - 225}$$
$$\doteq 349.68$$

$$\text{Slope} = \frac{\text{rise}}{\text{run}} = \frac{-15}{349.68} \doteq \frac{-1}{23.31}$$

Figure 8.10 illustrates how we can find the slope of a line given its graph.

FIGURE 8.10

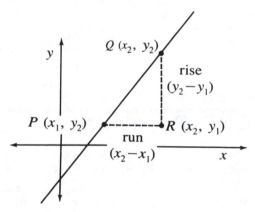

(a) Positive Slope

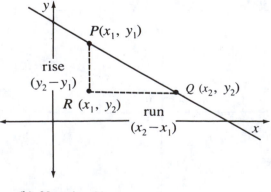

(b) Negative Slope

Selecting two arbitrary points on the graph, $P(x_1, y_1)$ and $Q(x_2, y_2)$, we find a third point R which is the intersection of a vertical line through P and a horizontal line through Q. The vertical distance between points, called the rise, is the change in y coordinates of P and Q, $(y_2 - y_1)$. The horizontal distance between points, called the run, is the change in x coordinates of P and Q, $(x_2 - x_1)$. We often use the letter m to represent the slope of a line. Thus,

$$\text{Slope of line } PQ = m = \frac{\text{rise}}{\text{run}} = \frac{y_2 - y_1}{x_2 - x_1}$$

EXAMPLE 3   Find the slope of line PQ.

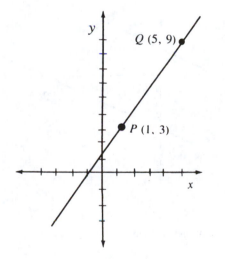

Solution

Let $(x_1, y_1) = (1, 3)$ and $(x_2, y_2) = (5, 9)$

$$m = \frac{y_2 - y_1}{x_2 - x_1} = \frac{9 - 3}{5 - 1} = \frac{6}{4} = \frac{3}{2}$$

The slope of line PQ is $\frac{3}{2}$.

EXAMPLE 4   Find the slope of a line passing through the points $(-3, 5)$ and $(5, -2)$.

Solution   We can see from a sketch of the graph of the line that the slope will be negative.

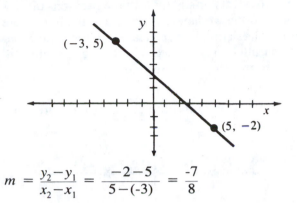

$$m = \frac{y_2 - y_1}{x_2 - x_1} = \frac{-2 - 5}{5 - (-3)} = \frac{-7}{8}$$

The slope of any given line is a constant, therefore any two points on the line can be used to calculate the value of the slope.

EXAMPLE 5   Find the slope of line L.

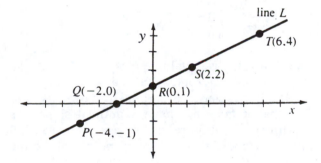

Solution

Case 1   Let $(x_1, y_1) = P(-4, -1)$ and $(x_2, y_2) = S(2, 2)$

$$m = \frac{y_2 - y_1}{x_2 - x_1} = \frac{2 - (-1)}{2 - (-4)} = \frac{3}{6} = \frac{1}{2}$$

Case 2   Let $(x_1, y_1) = Q(-2, 0)$ and $(x_2, y_2) = T(6, 4)$

$$m = \frac{4 - 0}{6 - (-2)} = \frac{4}{8} = \frac{1}{2}$$

Case 3   Let $(x_1, y_1) = R(0, 1)$ and $(x_2, y_2) = P(-4, -1)$

$$m = \frac{-1 - 1}{-4 - 0} = \frac{-2}{-4} = \frac{1}{2}$$

In each case we obtain the same ratio for the slope of line L. We would obtain the same slope using any two points on the line. In the previous discussion of slope we have assumed that the lines were neither horizontal or vertical. If a line is horizontal (parallel to the x-axis), then $\frac{y_2-y_1}{x_2-x_1}$ will still give us the slope of the line.

EXAMPLE 6   Find the slope of the horizontal line H.

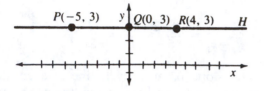

Solution   Every point on the line H will have the same y coordinate, 3, since H is parallel to the x-axis and three units above the axis. Let P (-5, 3) $= (x_1, y_1)$ and R (4, 3) $= (x_2, y_2)$.

$$m = \frac{y_2-y_1}{x_2-x_1} = \frac{3-3}{4-(-5)} = \frac{0}{9} = 0$$

The slope of horizontal line H is zero. In fact, the slope of any horizontal line will be zero.

If a line is vertical (parallel to the y-axis), we find that it does not have a defined slope. As illustrated in Figure 8.11, on any vertical line the x coordinates of every point will be the same. By our definition of slope we would have $m = \frac{y_2-y_1}{x_2-x_1} = \frac{y_2-y_1}{0}$ for any vertical line.

FIGURE 8.11

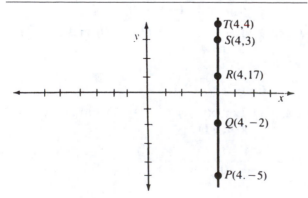

Since division by zero is undefined in our mathematics system we say that a vertical line has no slope, or its slope is undefined.

Figure 8.12 shows two parallel nonvertical lines in the same plane. If two nonvertical lines are parallel, then they have the same slope. Conversely, if two nonvertical lines have the same slope, then they are parallel lines. It is also true that two lines with no defined slope are vertical lines and vertical lines are parallel.

FIGURE 8.12

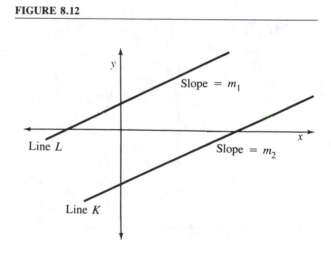

Line L is parallel to Line K
$m_1 = m_2$

Figure 8.13 shows two perpendicular lines in the same plane. Note that neither line is a vertical line. If two nonvertical lines are perpendicular, then the product of their slopes is a negative one. Conversely, if the product of the slopes of two lines is equal to -1, then the lines are perpendicular. It is also true that a horizontal line, with a slope of 0, is perpendicular to any intersecting vertical line, which has no defined slope.

FIGURE 8.13

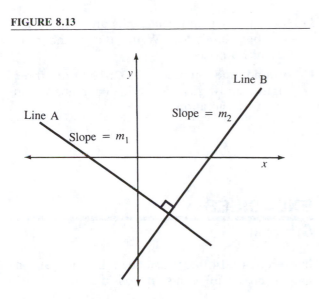

Line *A* is perpendicular to Line *B*
$(m_1)\,(m_2) = -1$

Note: If $(m_1)\,(m_2) = -1$ then we can also say that

$$m_1 = \frac{-1}{m_2} \text{ or } m_2 = \frac{-1}{m_1}$$

EXAMPLE 7 Line *PQ* contains the points (2, −1) and (−4, 8). Line *RS* contains the points (4, −12) and (−2, −3). Are lines *PQ* and *RS* parallel, perpendicular, or neither?

Solution The slope of line *PQ* is

$$m_1 = \frac{8-(-1)}{-4-2} = \frac{9}{-6} = -\frac{3}{2}$$

The slope of line *RS* is

$$m_2 = \frac{-3-(-12)}{-2-4} = \frac{9}{-6} = -\frac{3}{2}$$

Since $m_1 = m_2$ the lines are parallel.

EXAMPLE 8 Line *AB* contains the points (5, 1) and (−10, −5). Line *DE* contains the points (2, −3) and (−4, 12). Are lines *AB* and *DE* parallel, perpendicular, or neither?

Solution The slope of line *AB* is

$$m_1 = \frac{1-(-5)}{5-(-10)} = \frac{6}{15} = \frac{2}{5}$$

The slope of line *DE* is

$$m_2 = \frac{12-(-3)}{-4-2} = \frac{15}{-6} = -\frac{5}{2}$$

Since $(m_1)(m_2) = (\frac{2}{5})(-\frac{5}{2}) = -1$, the lines *AB* and *DE* are perpendicular.

Review Exercises

Use the following figure for questions 1 and 2.

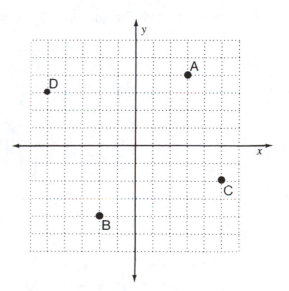

1. What are the coordinates of points *A* and *B*?
2. What are the coordinates of points *C* and *D*?
3. What is the distance from the origin of a rectangular coordinate system to the point *P* (9, 12)?
4. What is the distance from the origin of a rectangular coordinate system to the point *Q* (−4, 7)?

5. What is the length of leg *CB* to the nearest tenth of an inch?

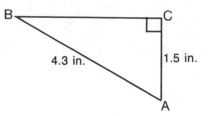

6. What is the length of the hypotenuse *XZ* to the nearest tenth of a foot?

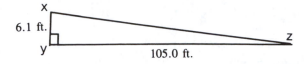

Use the following graph for questions 7 and 8.

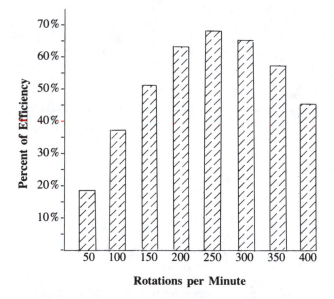

Rotations per Minute

7. What is the highest efficiency of the engine and at what speed does this occur?

8. What is the difference of the efficiency rates at 100 rpm and 350 rpm?

9. 18 is _____ % of 300.

10. 14 is _____ % of 2000.

11. How many significant digits are there in the number 0.00520? Write the number in scientific notation.

12. Round the number 0.030162 to three significant digits and write your answer in scientific notation.

EXERCISES

8.5

In exercises 1-14 find the slope of the line that contains each set of points, if possible.

1. *P* (2,5); *Q* (6,11)	2. *P* (3,7); *Q* (6,16)
3. *P* (-3,1); *Q* (6,2)	4. *P* (8,14); *Q* (-3,9)
5. *P* (4,-1); *Q* (8,-7)	6. *P* (5,3); *Q* (-2,4)
7. *P* (-7,-4); *Q* (-7,11)	8. *P* (5,13); *Q*(-3,13)
9. *P* (-8,15); *Q* (-3,-9)	10. *P* (-4,-9); *Q* (16,-3)
11. *P* (-3,-5); *Q* (-4,-7)	12. *P* (0,5); *Q* (0,-5)
13. *P* (6,-5); *Q* (-6,-5)	14. *P* (-4,3); *Q* (4,-3)

In exercises 15−20 locate two points on each line and find the slope of the line.

15.

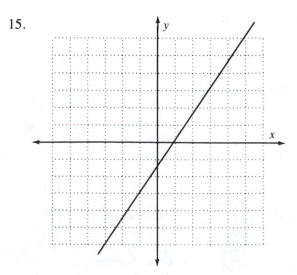

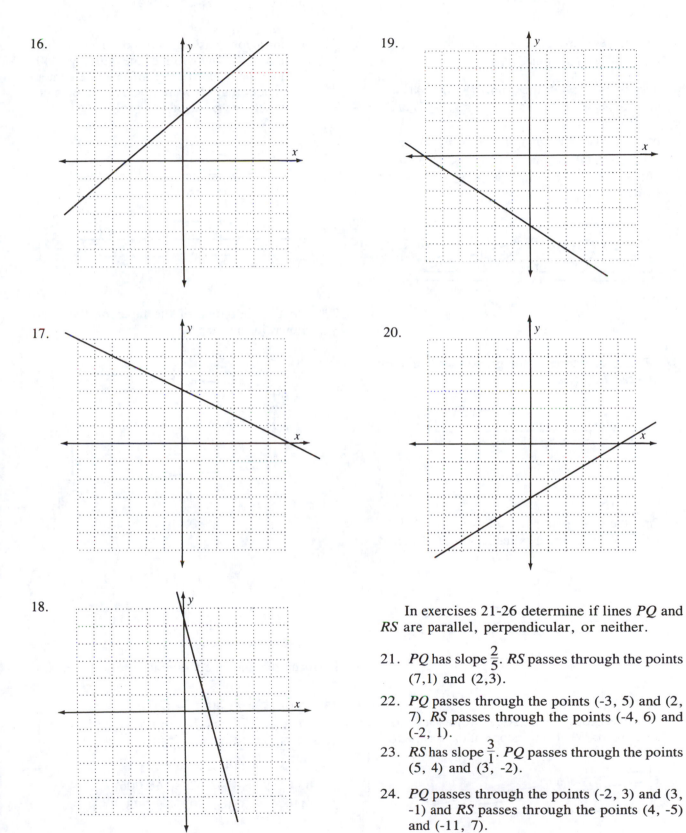

16.

17.

18.

19.

20.

In exercises 21-26 determine if lines *PQ* and *RS* are parallel, perpendicular, or neither.

21. *PQ* has slope $\frac{2}{5}$. *RS* passes through the points (7,1) and (2,3).

22. *PQ* passes through the points (-3, 5) and (2, 7). *RS* passes through the points (-4, 6) and (-2, 1).

23. *RS* has slope $\frac{3}{1}$. *PQ* passes through the points (5, 4) and (3, -2).

24. *PQ* passes through the points (-2, 3) and (3, -1) and *RS* passes through the points (4, -5) and (-11, 7).

25. PQ passes through the points (-5, 3) and (-1, -3). RS has a slope of $\frac{2}{3}$.

26. RS passes through the points (8, -2) and (4, 6). PQ passes through the points (4, 7) and (6, 6).

27. A roof truss, like that shown in the figure, has a pitch (slope) of $\frac{1}{4}$. What is the height of the truss at the midpoint of the horizontal span?

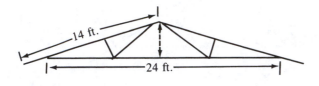

28. What is the slope of the grade shown in the diagram?

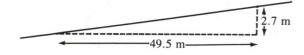

29. Find the span of a roof with a pitch of $\frac{1}{3}$, a rise of 6 feet and a rafter length of 21 feet.

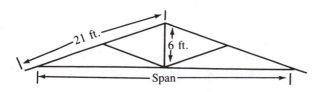

30. A roadbed goes down 220 feet for each 2400 feet of road. What is the slope of the roadbed?

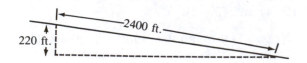

31. A contractor is asked to build a roof with a pitch of 0.36. The width of the building is 40 feet and the peak of the roof is at the center of the building. How high above the outside walls will the peak of the roof be?

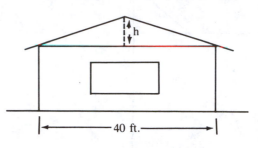

32. What is the slope of the taper on the plug shown in the diagram?

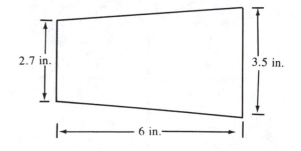

Calculator Exercises

Perform the indicated operations in Exercises 33-38.

33. (6.0875 x 23.402) ÷ 0.046

34. (234.69 − 89.073) x (40.004 − 26.738)

35. $\sqrt{493} + \sqrt{2906} + \sqrt{879}$

36. $(1.05)^3 + (14.65)^2 - (3.72)^3$

37. 83.0569 ÷ (143.82 ÷ 65.006)

38. (419 ÷ 8.35) + (0.54 x 278.309)

8.6
INDIRECT REASONING

A fascinating set of problems is made up of those sometimes called "reasoning" problems. The basic approach when solving such problems is to use indirect reasoning.

Indirect reasoning, in this case, can be one of two types:

a)   You consider all the possibilities for each happening and then use reasoning to eliminate all but one; or

b)   when there are exactly two possibilities and you don't know which is true, *assume* that one is true and use logical reasoning to see if this creates a contradiction. If so, your assumption was false.

The first type can be illustrated in the solution of the following problem.

EXAMPLE 1   Brian, Kati, and Jim were comparing food likes and dislikes. Their conversation went something like this.

Jim: I do not like pizza.
Brian: I do not like M & M's.
Kati: I do not like popcorn.
Jim: I agree with Kati.

Each likes only one of the foods named. Which food does each person like?

Solution   The various possibilities can be shown in a chart like the one shown here. Re-reading the problem information makes it possible to mark out various possibilities. When doing a simple problem like this one, this leads to only one of each type of possibility remaining.

	Pizza	M & M's	Popcorn
Jim	X		
Brian		X	
Kati			X

Each person indicating his or her dislike can be symbolized by the X's in the diagram. The last statement by Jim is that he agrees with Kati, and placing an X appropriately results in the grid looking like this:

	Pizza	M & M's	Popcorn
Jim	X		X
Brian		X	
Kati			X

This leaves Jim liking M & M's; and placing an X in Kati's row under M & M's leaves her liking pizza; and, of course, that leaves Brian liking popcorn. The final grid looks like this:

	Pizza	M & M's	Popcorn
Jim	X		X
Brian	X	X	
Kati		X	X

EXAMPLE 2   A more sophisticated problem, one that demonstrates the strategies related to indirect reasoning, is explained below.

Nine men, Brown, White, Adams, Miller, Green, Hunter, Knight, Jones, and Smith, make up a baseball team. Determine from the following facts the position played by each.

1.   Smith and Brown each won $10 playing poker with the pitcher.
2.   Hunter is taller than Knight and shorter than White, but each of the three weighs more than the first baseman.
3.   The third baseman lives across the corridor from Jones in the same apartment house.
4.   Miller and the outfielders play bridge in their spare time.

5. White, Miller, Brown, and the right fielder and center fielder are bachelors, and the rest are married.

6. Of Adams and Knight, one plays in the outfield.

7. The right fielder is shorter than the center fielder.

8. The third baseman is the pitcher's wife's brother.

9. Green is taller than the infielders, and also taller than the battery (pitcher & catcher), except for Jones, Smith, and Adams.

10. The second baseman beat Jones, Brown, Hunter, and the catcher at cards.

11. The third baseman, the shortstop, and Hunter each made money speculating on U.S. Steel.

12. The second baseman is engaged to Miller's sister.

13. Adams lives in the same house as his sister, but he dislikes the catcher.

14. Adams, Brown, and the shortstop each lost money speculating on the stock market.

15. The catcher has three daughters living at home with him and his wife; the third baseman has three sons living at home with him and his wife; but Green has no children and is being sued for divorce.

Solution   Begin by creating a diagram like the following:

	C	P	1B	2B	SS	3B	LF	CF	RF
Brown									
White									
Adams									
Miller									
Green									
Hunter									
Knight									
Jones									
Smith									

Then, go through each numbered clue carefully and eliminate squares in the grid as a result of what the clue says. Inside each eliminated square, mark the number of the clue you used. If you use a com-

bination of two clues to eliminate a square in the grid, mark both their numbers in the grid. This record keeping system helps you review why you eliminated particular squares.

Once you have gone through each clue once, your grid should look like the one shown below.

	C	P	1B	2B	SS	3B	LF	CF	RF
Brown	10	1		10	14			5	5
White		2						5	5
Adams	13				14		9	9	9
Miller	.						4	4	4
Green	9	9	9	9	9	9			
Hunter	10		2	10	11	11			
Knight		2							
Jones	10			10		3	9	9	9
Smith		1					9	9	9

Clues 6 and 9 can be *combined* to eliminate Knight from being C, P, 2B, SS, and 3B. These will be entered on the next grid.

First, however, it is necessary to make lists of bachelors and married men. Clues 5, 8, 12, and 16 yield the following:

Bachelors	**Marrieds**
White	Catcher
Miller	3B
Brown	Green
RF	P
CF	
———	———
2B	

The 2B under bachelors is an added position because nine men have already been indicated above the lines. This means that the second baseman is one of White, Miller, or Brown.

The following conclusions can be drawn from the table:

(1a) White not C, 3B, P

(2a) Miller not C, 3B, P

(3a) Brown not C, 3B, P

(4a) Green not 2B, RF, CF

Adding this information as well as the married/ bachelor information to the grid yields:

	C	P	1B	2B	SS	3B	LF	CF	RF
Brown	10	1		10	14	3a	x	5	5
White	1a	1a	2			1a	x	5	5
Adams	13	x	x	x	14		9	9	9
Miller	1a	1a	x			1a	4	4	4
Green	9	9	9	9	9	9		4a	4a
Hunter	10		2	10	11	11	x		
Knight	9/6	9/6	2	9/6	9/6	9/6	x		
Jones	10		x	10		3	9	9	9
Smith		1	x	x	x	x	9	9	9

The x's were placed in the grid because Green was left being the left fielder, so no one else could be. It should be noted that the grid now shows that Brown has to be the first baseman. Entering appropriate x's leaves Smith as the catcher; and then more appropriate x's leaves Adams as the third baseman.

Next comes a little bit of the second type of indirect reasoning. The grid shows that 2B is either Miller or White. Assume that it is Miller. Clue 12 contradicts that assumption because a person would not be engaged to his own sister. Therefore, White has to be the second baseman.

The final breakthrough to complete the grid and solve the problem is to reason as follows:

> The grid should now show that Hunter and Knight have to be the two outfielders, RF and CF. Clue 2 yields enough information to conclude without doubt that Hunter has to be the center fielder.

The problem is now solved, with the following solution:

> Brown is first baseman. White is second baseman. Adams is third baseman. Miller is shortstop. Green is left fielder. Hunter is center fielder. Knight is right fielder. Jones is the pitcher. Smith is the catcher.

Complicated "reasoning problems" often require several grids for each problem, and/or several separate tables. Working several problems helps you become rather proficient at setting up indirect reasoning schemes that will lead to the desired solutions.

In addition to the three problems at the end of this chapter, there are several "reasoning problems" included in Appendix C that will challenge even the most sophisticated problem solver.

EXERCISES
8.6

1. A cub reporter on assignment in New York City to interview six professional men wrote: "I found the men walking up Fifth Avenue six abreast. Mr. Thompson was on the outside next to the doctor. The lawyer was between Mr. Jones and the engineer. Brewster was not next to the man on the inside. Mr. Harvey is engaged to the engineer's sister. The architect was walking between Mr. Harvey and the doctor. Mr. Fish had nobody on his right. The author had never met Mr. Thompson before. Babeson was not the teacher. Identify the profession of each of the six men.

2. (This problem appeared in a summer, 1989 issue of Delta Airlines' magazine) Five representatives of the International Palindrome Society arrived at a conference on Sunday, Monday, and Tuesday. They came from five different countries, Argentina, Brazil, Germany, Italy, and Peru. Determine the first name (Bob, Eve, Nan, Otto, or Mairiam), the last name (Civic, Gereg, Hannah, Mallum, or Stuts), arrival day, and the home country of each if:

   1. Stuts (who came from South America) arrived the day before Hannah, who is from Europe.
   2. Bob and Gereg arrived on the same day, a day after Civic arrived.
   3. The day Nan arrived, no one else did.
   4. No more than two came on any one day. On the days when two arrived together, they came from different continents.
   5. Mairiam arrived on the same day as the person who came from Germany.

6. Eve and Hannah (whose first name isn't Bob) arrived the day before the one from Brazil.

7. The ones who came from Peru and Italy arrived on the same day.

3. In a row of five houses live Mr. and Mrs. Green, Mr. and Mrs. Brown, Mr. and Mrs. Jones, Mr. and Mrs. Smith, and Mr. and Mrs. Cook. There are five tradesmen who call at these houses, namely a grocer, a coalman, a butcher, a baker, and a milkman, whose names are Green, Brown, Smith, Jones, and Cook, but not respectively.

1. The butcher's married sister lives at Number 1.

2. Mr. Jones lives in the second house from the coalman's namesake.

3. The milkman's namesake has no relatives.

4. The butcher's namesake lives at Number 2.

5. Mr. Jones goes to work with the butcher's brother-in-law.

6. Mr. Brown helps the coalman's namesake in the garden.

7. Mr. Smith lives in the second house from the milkman's namesake.

8. Mrs. Green and Mrs. Jones are the only women involved who are sisters.

9. The baker's namesake has only one brother-in-law, who lives at Number 3.

10. Mr. Cook lives next door to the coalman's namesake. What is the name of each tradesman?

Solutions Do not look at these answers until you have tried to solve all the problems. If any one of your answers is incorrect, **forget the answer given here** and try again to solve the problem.

1. Thompson, teacher; Brewster, doctor; Jones, architect; Harvey, lawyer; Babeson, engineer; Fish, author.

2. Tuesday, Bob Mallum from Germany; Monday, Eve Civic from Peru; Sunday, Ann Stuts from Argentina; Monday, Otto Hannah from Italy; Tuesday, Mairiam Gereg from Brazil.

3. Brown, grocer; Smith, butcher; Green, coalman; Cook, milkman; Jones, baker.

EXTRA PROBLEMS
CHAPTER 8

1.

LABOR FORCE IN AN AREA LABOR MARKET

NUMBER EMPLOYED

= 750 MALE WORKERS   = 750 FEMALE WORKERS

a. How many workers are in the labor market?

b. What percent of the workers are female?

c. How many male workers are under 36 years of age?

d. What percent of employed males are over 55 years of age?

e. What percent of employed females are between the ages of 35 and 56?

f. What percent of employed men are between the ages of 35 and 56?

2. In a certain town the average property taxes for a given year were $2800.00. A circle graph, like the graph below, was published in the local paper to illustrate how those tax dollars were distributed.

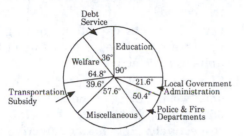

a. What percent of the property taxes was used for education and welfare?

b. If your tax bill for the year was $3350.00, what amount was used to help pay for debt service?

c. If your tax bill for the year was $1525.00, how much of this amount was used to help pay for local government administration costs?

d. What percent of the property taxes was used to help pay for police and fire department expenses and local transportation subsidies?

3. Construct a picture graph to illustrate the commissions earned by a salesperson for the first six months of the year.

Month	Commission
January	$1100
February	$1350
March	$1200
April	$ 800
May	$1500
June	$1050

4. Draw a circle graph to illustrate the sources and proportional shares of the electric energy produced in the United States for a year.

Energy Source	Percent of Total Energy Produced
Gas	32%
Oil	25%
Coal	20%
Hydropower	15%
Nuclear	8%

5. Draw a circle graph to illustrate research and development expenditures of 35.3 billion dollars by selected R & D agencies for a year.

R & D Agency	Expenditures
Federal Government	4.58 billion
State Governments	1.05 billion
Universities & Colleges	5.30 billion
Medical Institutions	3.90 billion
Private Industry	20.47 billion

6.

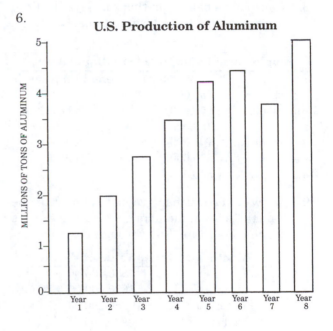

a. How many tons of aluminum were produced in the fifth year?

b. How many more tons of aluminum were produced in the seventh year than in the second year?

c. What is the percent of increase in production from the third year to the eighth year?

d. What is the total number of tons of aluminum produced during the eight years?

e. What is the percent of decrease in production from the sixth year to the seventh year?

f. What is the percent of increase in production from the first year to the eighth year?

7. Construct a horizontal bar graph for the data in the following table.

Unit Production Totals (Thousands of Units)	
Factory A	72
Factory B	78
Factory C	83
Factory D	75
Factory E	80
Factory F	89
Factory G	79

8. Construct a bar graph that displays all of the information in the following table.

Employment in Major Occupational Areas		
Area	Male Workers	Female Workers
Clerical	350,000	1,100,000
Professional	850,000	500,000
Service	450,000	650,000
Administration	700,000	200,000
Sales	350,000	400,000
Technology	500,000	150,000

9. Draw a vertical bar graph to illustrate the value of a manufacturing company's exports for the months of July through December.

Month	Dollar Value of Exports
July	$850,000.00
August	$600,000.00
September	$1,250,000.00
October	$900,000.00
November	$1,300,000.00
December	$850,000.00

10.

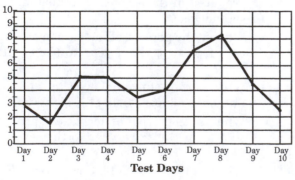

Percent of Defective Units for a Ten Day Trial Period

a. What is the greatest percent of defective units produced on a given day during the test period?

b. What is the smallest percent of defective units produced on a given day during the test period?

c. On the seventh day of the testing 3580 units were produced. How many of these units were defective?

d. The number of units produced on days one, two and three of the test period were 3590 units, 2000 units and 4750 units respectively. How many defective units were produced over these three days?

e. On day eight, 3000 units were produced. On day ten, 4350 units were produced. What is the percent of decrease in the number of defective units produced on day ten compared with day eight?

11. Construct a line graph for the data below showing current and resistance in a line with a constant power output of 150 watts.

Current in Amperes	Resistance in Ohms
5.4	5
3.8	10
3.0	15
2.6	20
2.4	25
2.2	30

12. Construct a line graph for the temperature readings in degrees Celsius over a ten hour period.

Time	Temperature °C
12:00 noon	20
1:00 P.M.	23
2:00 P.M.	25
3:00 P.M.	24
4:00 P.M.	22
5:00 P.M.	20
6:00 P.M.	18.5
7:00 P.M.	19
8:00 P.M.	17
9:00 P.M.	15

13. Construct a multiple (two) line graph on the same set of axes that will show the number of assembled components produced per hour by two different production systems.

Hours of Production	Number of Components Produced	
	System A	System B
40	400	600
45	650	630
50	670	720
55	710	780
60	780	840
65	800	900
70	830	950
75	870	950
80	900	1000

14. Identify the coordinates for each point on the graph.

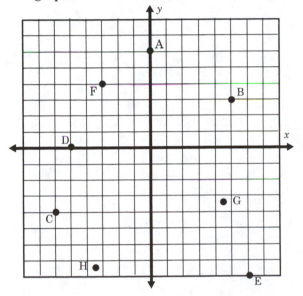

15. Locate the following points on a graph of the coordinate plane: A(5, 6); B(–4, 3.5); C(0, –7); D(–2.5, –4.5); E(6, 0); F(–6, 6.5); G(4.5, –5); H(0,0)

16. Find the distance from the origin to each point. Round off your answers to three significant digits.

    a. X(0, 7)   d. Q(–5, –18)
    b. Y(–13, 5)   e. M(2.5, 4.7)
    c. P(4, –9)   f. N(–3.9, –11.2)

17. Find the distance between the two points in each set. Round off your answers to three significant digits.

    a. (2, 3), (8, 11)   d. (–7, 12), (8, 4)
    b. (–4, 6), (2, –5)   e. (3.5,–2.4),(–1.4,–5.9)
    c. (7, –2), (–9, –15)   f. (6.3,–1.8),(6.3,–7.4)

18. What kind of triangle is formed by the three points A (–2, 2), B(2, 5) and C(1, 6)?

19. PQRS is a rectangle. The coordinates of three of the vertices of the rectangle are P(0, 5), Q(3, 7) and R(5, 4). Opposite sides of a rectangle are equal in length. What are the coordinates of point S so that the segment PQ is equal in length to segment SR?

20. Use the Pythagorean Theorem and the distance formula to show that the points A(2, 5), B(3, –4) and C(–2, 0) are the vertices of a right triangle.

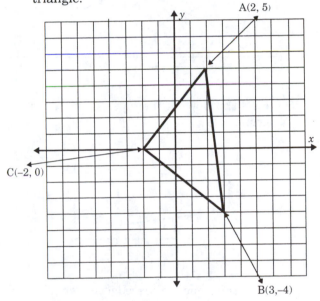

21. Determine if the lines XY and AB are parallel, perpendicular or neither.

    a. XY has a slope of $-\frac{4}{3}$. AB passes through the points (–1, 5) and (3, 2).

    b. XY passes through the points (–3, 4) and (–8, 2). AB passes through the points (–2, 5) and (3, 7).

    c. AB passes through the points (5, 11) and (0, 9). XY has a slope of $-\frac{5}{2}$.

    d. AB passes through the points (–1, –5) and (–5, 12). XY passes through the points (3, 7) and (4, –6).

22. Find the slope of the line that contains each set of points, if possible.

    a. X(9, –7), Y(–2, –12)   d. P(5, –3), Q(5, 8)
    b. A(–5, –6), B(13, –6)   e. R(–8, 0), S(8, –6)
    c. M(13, –3), N(6, –5)

23. The grade on a mountain road is a drop of 3.5 meters for each 50 meters of roadbed. What is the slope of this grade?

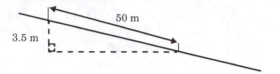

24. What is the pitch (slope) of the roof truss shown in the diagram?

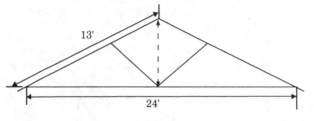

25. Find the slope of the taper shown in the diagram.

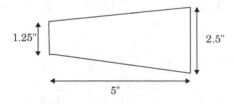

26. Find the slope of the wedge shown in the diagram.

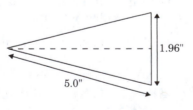

27. A line AB passes through the points A(-1, 1) and B(5, -3). A line XY is perpendicular to line AB and passes through the point X (-2, 6). What are the coordinates of a point Y if it is on the line AB? (XY crosses the line AB at point Y.)

INTRODUCTION
TO
ALGEBRA

9.1
ADDING POSITIVE AND NEGATIVE NUMBERS

All the numbers you will study in this course can be placed on a number line. These numbers are called **real numbers**. Any number that can be written as a decimal is a real number. Figure 9.1 shows some real numbers.

FIGURE 9.1

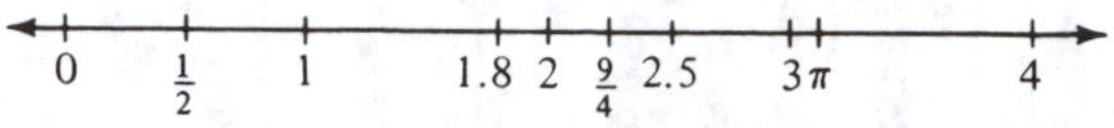

$$
\begin{array}{ccccccccc}
0 & \tfrac{1}{2} & 1 & 1.8\ 2\ \tfrac{9}{4}\ 2.5 & 3\ \pi & 4
\end{array}
$$

The numbers to the right of zero are called **positive numbers**. The numbers to the left of zero are called **negative numbers**. The number -1 is read "negative one." The number $-\tfrac{9}{4}$ is read "negative nine-fourths." Every positive

number (to the right of zero) has an opposite, or negative, to the left of zero. Each number and its opposite are the same distance away from zero on the number line, as shown in Figure 9.2. The following three rules are helpful to keep in mind:

1. The opposite of a positive number is a negative number.
    For example: The opposite of 2 is -2; and the opposite of 1.8 is -1.8.

2. The opposite of a negative number is a positive number.
    For example: The opposite of -1 is 1; and the opposite of $\tfrac{-5}{2}$ is $\tfrac{5}{2}$.

3. The opposite of 0 is 0.

FIGURE 9.2

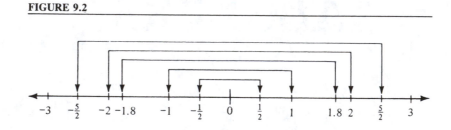

> **The sum of any number and its opposite is zero.**

For example,

$$9 + (-9) = 0$$
$$-14 + 14 = 0$$
$$\frac{8}{9} + \left(-\frac{8}{9}\right) = 0$$
$$-3.2 + 3.2 = 0.$$

> **The sum of any number and zero is that number.**

For example,

$$57 + 0 = 57$$
$$0 + (-19) = -19$$
$$-15.7 + 0 = -15.7.$$

The sums of other positive and negative numbers can be viewed as yards gained (+) and lost (−) on a football field.

EXAMPLE 1   Sixteen yards gained, followed by 7 yards gained, is a 23-yard gain.

$$16 + 7 = 23$$

EXAMPLES

2 Sixteen yards gained, followed by 7 yards lost, is a 9-yard gain.

$$16 + (-7) = 9$$

3 Sixteen yards lost, followed by 7 yards gained, is a 9-yard loss.

$$-16 + 7 = -9$$

4 Sixteen yards lost, followed by 7 yards lost, is a 23-yard loss.

$$-16 + (-7) = -23$$

Finding the sum of two positive numbers requires no new information.

EXAMPLES

5 $57 + 15 = 72$

6 $14.9 + 3.68 = 18.58$

7 $\dfrac{2}{3} + \dfrac{1}{6} = \dfrac{5}{6}$

> The sum of two negative numbers is the negative of the sum of their "opposites."

EXAMPLES

8 $(-8) + (-7) = -(8 + 7) = -15$ — Negative of the sum of the "opposites" of -8 and -7

9 $-21 + (-13) = -(21 + 13) = -34$

10 $(-2.7) + (-5.8) = -(2.7 + 5.8) = -8.5$

11 $-\dfrac{5}{8} + \left(-\dfrac{3}{16}\right) = -\left(\dfrac{5}{8} + \dfrac{3}{16}\right) = -\dfrac{13}{16}$

12 $\dfrac{-1}{4} + \left(\dfrac{-3}{8}\right) = -\left(\dfrac{1}{4} + \dfrac{3}{8}\right) = \dfrac{-5}{8}$

Note in Examples 11 and 12 that, when using a negative fraction, the "negative sign" can be either in the numerator of the fraction or in front of the fraction. (Usually the negative sign does not appear in the denominator of the fraction.) For example,

$$-\dfrac{5}{8} = \dfrac{-5}{8}, \; not \; \dfrac{5}{-8}.$$

> The sum of a positive number and a negative number can be found by renaming one number and then adding opposites.

EXAMPLES

13 $8 + (-3) = (5 + 3) + (-3)$ — 8 can be written as $5 + 3$.
$\qquad = 5 + [3 + (-3)]$
$\qquad = 5 + 0$ — The sum of a number and its opposite is zero.
$\qquad = 5$

14 $-12 + 5 = [(-7) + (-5)] + 5$ — The number -12 can be written as $(-7) + (-5)$.
$\qquad = -7 + [(-5) + 5]$
$\qquad = -7 + 0$
$\qquad = -7$

15 $4.16 + (-12.5)$
$\qquad = 4.16 + [(-4.16) + (-8.34)]$
$\qquad = [4.16 + (-4.16)] + (-8.34)$
$\qquad = 0 + (-8.34)$
$\qquad = -8.34$

In the first step of Example 15, notice that -8.34 can be found quickly by subtracting 4.16 from 12.5 and then using a "negative sign." This leads to a more efficient method of adding positive and negative numbers. Table 9.1 shows this method with examples.

TABLE 9.1

Step	$19 + (-37)$	$4.16 + (-12.5)$
1. Determine which number is further from zero on the number line. The sign (positive or negative) will be the sign of the answer.	The number -37 is furthest from zero on the number line. The answer will be negative.	The number -12.5 is furthest from zero on the number line. The answer will be negative.
2. Consider them both as positive.	37 19	12.5 4.16
3. Subtract the smaller from the larger.	$\begin{array}{r} 37 \\ -\ 19 \\ \hline 18 \end{array}$	$\begin{array}{r} 12.50 \\ -\ 4.16 \\ \hline 8.34 \end{array}$
4. Affix the sign from step one.	-18	-8.34

A more formal version of the intuitive method noted in Table 9.1 uses the concept of absolute value. The absolute value of a number can be viewed two ways:

1. The **absolute value** of a number is its distance away from zero on a number line. Both 5 and -5 are 5 units from zero on a number line. Therefore, the absolute value of 5 is 5 and the absolute value of -5 is 5. This can be written as

$$|5| = 5 \text{ and } |-5| = 5.$$

2. The **absolute value** of a number is the value of the number without regard to its sign. Again, the absolute value of 5 is $5(|5| = 5)$ and the absolute value of -5 is $5(|-5| = 5)$.

Note that every number and its opposite have the same absolute value.

Table 9.2 summarizes the rules for adding two numbers using the concept of absolute value.

TABLE 9.2

Signs alike + and + – and –	Add their absolute values.	Give result the common sign.
Signs unlike + and – – and +	Find the difference between their absolute values.	Give result the sign of the number with greater absolute value.

When adding more than two numbers, we can add in order from left to right.

EXAMPLE 16

$$
\begin{aligned}
-6 + 15 + (-12) &= (-6 + 15) + (-12) \\
&= (15 - 6) + (-12) \\
&= 9 + (-12) \\
&= -(12 - 9) \\
&= -3
\end{aligned}
$$

EXAMPLE 17

$$-4.8 + 9 + (-2.2) + (-5)$$
$$= (-4.8 + 9) + (-2.2) + (-5)$$
$$= (9 - 4.8) + (-2.2) + (-5)$$
$$= 4.2 + (-2.2) + (-5)$$
$$= [4.2 + (-2.2)] + (-5)$$
$$= (4.2 - 2.2) + (-5)$$
$$= 2 + (-5)$$
$$= -(5 - 2)$$
$$= -3$$

Changing the order of the numbers being added may make the computing easier.

EXAMPLE 18

$$8 + (-13) + (-8) = [8 + (-8)] + (-13)$$
$$= 0 + (-13)$$
$$= -13$$

EXAMPLE 19

$$-4.8 + 9 + (-2.2) + (-5)$$
$$= [-4.8 + (-2.2)] + [9 + (-5)]$$
$$= -(4.8 + 2.2) + (9 - 5)$$
$$= -7 + 4$$
$$= -(7 - 4)$$
$$= -3$$

Review Exercises

1. $14.9 - 3.24 = $ _____
2. $26.7 - 2.91 = $ _____
3. $\left(\dfrac{14}{5}\right)(5) = $ _____
4. $(9)\left(\dfrac{29}{9}\right) = $ _____
5. 0.087 km = _____ m
6. _____ km = 92 m
7. Change 0.34 to a percent.
8. Change 1.5 to a percent.

9. 42% of 60 = _____
10. 78% of 55 = _____
11. What is the perimeter of a square 8 cm on a side?
12. What is the perimeter of a rectangle 9 cm wide and 11 cm long?

EXERCISES

9.1

Place the numbers in Exercises 1–6 in their approximate positions on a number line.

1. -5
2. 2.4
3. -3.6
4. $\dfrac{-15}{4}$
5. $-\dfrac{1}{3}$
6. $\dfrac{27}{6}$

State the opposite of each number in Exercises 7–12.

7. 72
8. -15
9. 0
10. -5.9
11. $\dfrac{15}{16}$
12. $-4\dfrac{5}{6}$

Find the sum in Exercises 13–40.

13. $-13 + (-18)$
14. $-12 + (-15)$
15. $-13 + 5$
16. $12 + (-27)$
17. $15 + (-9)$
18. $29 + (-13)$
19. $175 + (-200)$
20. $175 + (-150)$
21. $-4.8 + (-5.9)$
22. $-8.1 + (-4.2)$
23. $5.6 + (-4)$
24. $-6 + 12.8$
25. $-8.3 + 4.24$
26. $9.6 + (-3.89)$
27. $-4.7 + 2$
28. $9 + (-15.1)$
29. $-8 + (-7) + 5$
30. $10 + (-3) + (-11)$
31. $-6 + 13 + (-17) + 8$
32. $-15 + (-8) + (-3)$
33. $-4 + 7 + (-9) + 12$
34. $18 + (-6) + (-7)$
35. $-17 + 32 + (-25)$

36. $-68 + (-17) + 4 + 23$
37. $-432 + 578 + (-125)$
38. $735 + (-5710) + 2$
39. $-78 + 43 + (-16)$
40. $63 + (-87) + -52$

What number should be added to each of the following to get 0?

41. -8 42. 9
43. 16   44. -27

Name a positive or negative number for the measurements in Exercises 45–50.

45. The lowest point in North America is located in Death Valley, California. It is 282 feet below sea level.
46. The referee called a 5-yard penalty on the home team.
47. National Telephone and Telegraph stock went up $1\frac{1}{8}$ points yesterday.
48. The temperature is 8 degrees below zero (Fahrenheit).
49. Bigby's profit last year was $107,000.
50. It is 15 seconds until launch time.

Use addition of positive and negative numbers to answer Exercises 51–55.

51. If the temperature at 6:00 PM is 8^0, and if it drops 17^0 in the next 6 hours, what is the temperature at midnight?
52. At what elevation would some hikers be if they started from a point with elevation 76 feet below sea level and climbed 203 feet?
53. What is the net profit (loss) of the Jones Company for two months if it recorded a profit of $135.17 in January and a loss of $226.37 in February?
54. What is the net profit (loss) of a $313.38 profit and a $298.96 loss?
55. What is the sum of these currents: 4.8 amperes, -3 amperes, -1.8 amperes, and 1.6 amperes?

Calculator Exercises

Use a calculator to determine the answers to Exercises 56–58.

56. $1468 + -386 + 1032 + -3684$
57. $-49.8 + -2.65 + 14.9 + -18.06$
58. The stock of Consolidated Corporation went up 2 points on Monday; down $3\frac{1}{2}$ points on Tuesday; down $\frac{1}{8}$ point on Wednesday; down $2\frac{1}{4}$ points on Thursday; and up $3\frac{3}{8}$ points on Friday. What was the change in price of the stock for the week?

9.2
SUBTRACTING POSITIVE AND NEGATIVE NUMBERS

In adding positive numbers, you may have noticed that:

$$5 + (-3) = 5 - 3$$
$$18 + (-6) = 18 - 6$$
$$27 + (-5) = 27 - 5.$$

You could also say that:

$$27 - 5 = 27 + (-5)$$
$$18 - 6 = 18 + (-6)$$
$$5 - 3 = 5 + (-3).$$

From these examples we see that *subtracting* a number is the same as *adding its opposite. This is true for all numbers.*

Subtracting a number means adding its opposite

or

$$a - b = a + (-b).$$

The opposite of a number needs more consideration before we begin subtracting:

1. The opposite of zero is zero.
2. The opposite of a positive number is a negative number.
    For example: The opposite of 7 is −7. This is usually written $-(7) = -7$.
3. The opposite of a negative number is a positive number.
    For example: The opposite of −13 is 13. This is usually written $-(-13) = 13$.

Other examples are shown here:

$$-(24) = -24$$
$$-(-39) = 39$$
$$-(-57) = 57.$$

It is helpful to remember that, in general:

$-(-a) = a$ for any number a.

Here are several examples of subtracting positive and negative numbers using the definition $a - b = a + (-b)$.

EXAMPLES

1
$$-9 - 6 = -9 + [-(6)] \longleftarrow$$
$$= -9 + (-6)$$
$$= -(9 + 6)$$
$$= -15$$

Subtracting 6 means adding the opposite of 6.

2
$$-9 - 23 = -9 + [-(23)]$$
$$= -9 + (-23)$$
$$= -(9 + 23)$$
$$= -32$$

3
$$14 - 29 = 14 + [-(29)]$$
$$= 14 + (-29)$$
$$= -(29 - 14)$$
$$= -15$$

4
$$-9 - (-4) = -9 + [-(-4)]$$
$$= -9 + 4$$
$$= -(9 - 4)$$
$$= -5$$

5
$$-9 - (-34) = -9 + [-(-34)]$$
$$= -9 + 34$$
$$= (34 - 9)$$
$$= 25$$

Subtracting decimals and fractions is done in the same way. The subtraction is performed by *adding* the opposite.

EXAMPLE 6

$$-3.4 - 6.09 = -3.4 + [-(6.09)]$$
$$= -3.4 + (-6.09)$$
$$= -(3.4 + 6.09)$$
$$= -9.49$$

When subtracting more than one number remember that operations are computed from left to right.

EXAMPLE 7

$$8 - 4 - 13 - (-5) = 8 + (-4) + (-13) + [-(-5)]$$
$$= 8 + (-4) + (-13) + 5$$
$$= 4 + (-13) + 5$$
$$= -9 + 5$$
$$= -4$$

It is true that numbers can be *added* in any order:

$$8 + 3 = 3 + 8.$$

But numbers cannot be *subtracted* in any order:

$$8 - 3 \neq 3 - 8.$$

So, when adding and subtracting in the same problem, remember to do the operations in order from *left to right*.

EXAMPLE 8

$$-3 + 9 - 14 - (-7) + 4$$
$$= -3 + 9 + (-14) + [-(-7)] + 4$$
$$= -3 + 9 + (-14) + 7 + 4$$
$$= 6 + (-14) + 7 + 4$$
$$= -8 + 7 + 4$$
$$= -1 + 4$$
$$= 3$$

Review Exercises

1. $25.9 + 14.68 =$ _____
2. $100 + 4.6 =$ _____
3. 8 in. = _____ ft
4. 9 in. = _____ ft
5. 65 mm = _____ m
6. _____ km = 2400 m
7. $(2.5)^2 =$ _____
8. $(1.6)^2 =$ _____
9. $12 + (-18) + 5 =$ _____
10. $-3 + (-8) + 11 =$ _____
11. Change $\frac{19}{8}$ to a decimal.
12. Change $\frac{34.6}{1000}$ to a decimal.

EXERCISES
9.2

Find the value of the numbers in Exercises 1–8.

1. $-(6)$
2. $-(-82)$
3. $-(0)$
4. $-(8)$
5. $-(-39)$
6. $-\left(-\frac{3}{5}\right)$
7. $-(-6.94)$
8. $-(-r)$

Compute the sums and/or differences in Exercises 9–46.

9. $7 - 4$
10. $10 - (-15)$
11. $-12 - 6$
12. $9 - (-6)$
13. $-8 - (-6)$
14. $15 - (-7)$
15. $9 - (-9)$
16. $8 - 8$
17. $12 - 17$
18. $-15 - 22$
19. $0 - (-7)$
20. $-27 - 43$
21. $5.8 - 6.7$
22. $4.5 - 6.3$
23. $2.04 - 5.3$
24. $-16.5 - 5.84$
25. $3 - 5.6$
26. $-7.6 - 4$
27. $3 - 4 - 9$
28. $-2 - 6 - 1$
29. $3 - 9 - 4$
30. $7 - 13 - 14$
31. $10 - 2 - 12$
32. $-2 - 5 - 6$
33. $-14 - 1 - (-8)$
34. $-21 + 12 - (-9)$
35. $-9 - (-6) + 4$
36. $4 - (-7) + 8$
37. $-19 + 16 + 5$
38. $7 + 16 - 20$
39. $14 - 17 + 3 - (-8)$
40. $5 - (-9) + 6 - 3$
41. $-18 + 8 - 6 - 5$
42. $-16 - 2 + 5 - 6$
43. $4 - (5 - 13) + 9$
44. $-6 + 7 - (15 - 8)$
45. $14 - 20 - (4 - 8)$
46. $-3 - (4 - 9) - 13$
47. The change in temperature at a particular time is found by calculating $-3 - 28$. Find the value.
48. The current of a certain circuit in amperes is found by calculating $4.8 - 6 - 2$. Find the value.
49. The terminal velocity in meters per second of a particular object is found by calculating $24 - 35$. Find the value.
50. To find the profit (loss) of a certain company for a month calculate $840.60 - 1089.50$.

Calculator Exercises

Use a calculator to determine the answers to Exercises 51–54.

51. $-27.9 - 15.6 + 24.98$
52. $98 - 46.5 + 31.7$
53. $-674 + 4.07 - 115 + 14.2$
54. $42.6 - 98.48 - 63 - 4.7$

9.3
MULTIPLYING AND DIVIDING POSITIVE AND NEGATIVE NUMBERS

The products of positive and negative numbers can be viewed as future (+) and past (−) yardage gained (+) and lost (−) in a football game.

EXAMPLES

1 Assume a 4-yard gain per play. Three plays in the future the team will have 12 yards more than they do now.

$$(4)(3) = 12$$

2 Assume a 4-yard gain per play. Three plays in the past the team had 12 yards less than they do now.

$$(4)(-3) = -12$$

3 Assume a 4-yard loss per play. Three plays in the future the team will have 12 yards less than they do now.

$$(-4)(3) = -12$$

4 Assume a 4-yard loss per play. Three plays in the past the team had 12 yards more than they do now.

$$(-4)(-3) = 12$$

Finding the product of two positive numbers requires no new information.

EXAMPLES

5 $(14)(3) = 42$

6 $(1.7)(6.28) = 10.676$

7 $\left(\dfrac{3}{8}\right)(4) = \dfrac{3}{2}$

The product of a negative number and a positive number is a negative number.

EXAMPLES

8 $(4)(-9) = -36$

9 $(-10)(8) = -80$

10 $(-2.1)(5) = -10.5$

11 $\left(\dfrac{1}{2}\right)\left(-\dfrac{3}{4}\right) = -\dfrac{3}{8}$

The product of two negative numbers is a positive number.

EXAMPLES

12 $(-8)(-4) = 32$

13 $(-5)(-9) = 45$

14 $(-6)(-2.7) = 16.2$

15 $(-5)\left(-\dfrac{7}{10}\right) = \dfrac{7}{2}$

Table 9.3 summarizes the rules for multiplying two numbers.

TABLE 9.3

Sign of Factors		Sign of Product	
(+)(+)	=	(+)	Signs alike; product is positive.
(−)(−)	=	(+)	
(+)(−)	=	(−)	Signs different; product is negative.
(−)(+)	=	(−)	

Multiplying any positive number by zero is zero. For example, $(0)(7) = 0$ and $(6)(0) = 0$.

Multiplying any negative number by zero is also zero. For example, $(0)(-5) = 0$ and $(-8)(0) = 0$.

The next three examples illustrate the procedure for multiplying several numbers, including numbers with exponents. Recall that an exponent means that the number is to be multiplied by itself.

EXAMPLES

16 $(-3)(-8)(-5) = [(-3)(-8)](-5)$
$$= (24)(-5)$$
$$= -120$$

17 $(-4)(-2)^3(5)(-3) = (-4)(-2)(-2)(-2)(5)(-3)$
$$= 8(-2)(-2)(5)(-3)$$
$$= (-16)(-2)(5)(-3)$$
$$= 32(5)(-3)$$
$$= 160(-3)$$
$$= -480$$

18 $4(-3)^2(-5)(-1)^3$
$$= 4(-3)(-3)(-5)(-1)(-1)(-1)$$
$$= -12(-3)(-5)(-1)(-1)(-1)$$
$$= (-12)(15)(-1)(-1)(-1)$$
$$= (-12)(15)(-1)$$
$$= (-12)(-15)$$
$$= 180$$

Every number except zero has a **reciprocal**. (Reciprocals were discussed earlier, in Chapter 2.)

EXAMPLES

19 The reciprocal of 4 is $\frac{1}{4}$.

20 The reciprocal of -6 is $-\frac{1}{6}$.

21 The reciprocal of $\frac{1}{8}$ is 8.

22 The reciprocal of $-\frac{1}{5}$ is -5.

23 The reciprocal of $\frac{4}{3}$ is $\frac{3}{4}$.

24 The reciprocal of $-\frac{2}{3}$ is $-\frac{3}{2}$.

25 The reciprocal of 0 does not exist.

> **The product of any number (except zero) and its reciprocal is one.**

From Examples 19–24, we see that this is true:
$$(4)\left(\frac{1}{4}\right) = 1$$
$$(-6)\left(-\frac{1}{6}\right) = 1$$
$$\left(\frac{1}{8}\right)(8) = 1$$
$$\left(-\frac{1}{5}\right)(-5) = 1$$
$$\left(\frac{4}{3}\right)\left(\frac{3}{4}\right) = 1$$
$$\left(-\frac{2}{3}\right)\left(-\frac{3}{2}\right) = 1$$

(*Note:* This does not hold true for the number zero. No number can be multiplied by zero to equal one. The product of any number and zero is zero. This is the reason why zero does not have a reciprocal.)

Dividing by a number gives the same result as multiplying by the number's reciprocal. For example,
$$3 \div 4 = 3\left(\frac{1}{4}\right).$$

Because zero has no reciprocal, division by zero is not possible. Since the reciprocal of a positive number is positive, and the reciprocal of a negative number is negative, the sign ($+$ or $-$) of a quotient follows the same patterns as for multiplication.

Zero divided by any number (except zero) is zero. For example,
$$\frac{0}{-4} = 0\left(-\frac{1}{4}\right) = 0.$$

Numbers cannot be divided by zero because zero has no reciprocal.

> **The quotient of a positive number divided by a positive number is a positive number.**

EXAMPLE 26 The number 14 divided by 2 is 7. This can be written as

$$14 \div 2 = 7,$$
$$\frac{14}{2} = 7, \text{ or}$$
$$14\left(\frac{1}{2}\right) = 7.$$

The quotient of a negative number and a positive number is a negative number.

EXAMPLES

27 $(16) \div (-4) = -4$

28 $(-15) \div (5) = -3$

29 $\dfrac{27}{-3} = -9$

30 $\dfrac{-32}{4} = -8$

31 $\dfrac{-48}{0.2} = -240$

32 $-5 \div \dfrac{2}{3} = (-5)\left(\dfrac{3}{2}\right)$ — The reciprocal of $\frac{2}{3}$ is $\frac{3}{2}$.

$\quad\quad = \dfrac{-15}{2}$

33 $\dfrac{8}{9} \div (-2) = \dfrac{8}{9}\left(-\dfrac{1}{2}\right)$ — The reciprocal of -2 is $-\frac{1}{2}$.

$\quad\quad = -\dfrac{4}{9}$

The quotient of two negative numbers is a positive number.

EXAMPLES

34 $(-45) \div (-9) = 5$

35 $\dfrac{-56}{-8} = 7$

36 $\dfrac{-1.25}{-0.5} = 2.5$

37 $\left(-\dfrac{3}{4}\right) \div \left(-\dfrac{3}{5}\right) = \left(-\dfrac{3}{4}\right)\left(-\dfrac{5}{3}\right)$ — The reciprocal of $-\frac{3}{5}$ is $-\frac{5}{3}$.

$\quad\quad = \dfrac{5}{4}$

Note from the preceding examples of division that multiplying by the reciprocal is especially helpful when working with fractions.

It is true that numbers can be multiplied in any order:

$$6 \cdot 8 = 8 \cdot 6.$$

This is not true for division:

$$14 \div 2 \neq 2 \div 14.$$

When multiplication and division computations occur in the same problem do the computations in order from left to right.

EXAMPLE 38

$$(8) \div (4)(-2) = [8 \div 4](-2)$$
$$= 2(-2)$$
$$= -4$$

Review Exercises

Perform the indicated computations in Exercises 1–10.

1. $\left(\dfrac{2}{5}\right)(14.5)$ 2. $(27.6)\left(\dfrac{3}{4}\right)$

3. $9.8 + 7 + 2.13$ 4. $4.08 + 5 + 3.2$

5. $\left(\dfrac{3}{4}\right) \div \left(\dfrac{9}{7}\right)$ 6. $5 \div \left(\dfrac{5}{6}\right)$

7. $87 + (-100) + (-3)$

8. $-49 + 53 + (-60)$

9. $13 - 20$ 10. $53 - 89$

11. Change 0.7 feet to inches, to the nearest fourth of an inch.

12. Change 0.3 feet to inches, to the nearest eighth of an inch.

EXERCISES
9.3

Write the reciprocal of each number in Exercises 1–6.

1. 4

2. −3

3. $\dfrac{7}{8}$

4. $\dfrac{1}{2}$

5. $-\dfrac{1}{9}$

6. $-\dfrac{4}{5}$

Multiply as indicated in Exercises 7–20.

7. $(-13)(5)$

8. $(-6)(-17)$

9. $(14)(-9)$

10. $(-5)(-8)$

11. $\left(-\dfrac{2}{3}\right)(3)$

12. $(5)\left(-\dfrac{4}{5}\right)$

13. $\left(-\dfrac{2}{3}\right)\left(-\dfrac{6}{5}\right)$

14. $\dfrac{9}{4}(-2)$

15. $(-3)\left(\dfrac{14}{15}\right)$

16. $\left(-\dfrac{3}{16}\right)\left(-\dfrac{1}{3}\right)$

17. $(-2.7)(3)$

18. $2(-4.5)$

19. $(-6.8)(-2.5)$

20. $3.4(-6.54)$

Divide as indicated in Exercises 21–42.

21. $24 \div (-6)$

22. $(-49) \div 7$

23. $(-72) \div (-8)$

24. $(-42) \div (-3)$

25. $\dfrac{-8}{-8}$

26. $\dfrac{9}{-9}$

27. $\dfrac{-56}{4}$

28. $\dfrac{20}{-5}$

29. $\dfrac{4}{3} \div (-2)$

30. $-\dfrac{5}{6} \div 10$

31. $-4 \div \left(-\dfrac{2}{3}\right)$

32. $-8 \div \left(-\dfrac{3}{4}\right)$

33. $\dfrac{-4.8}{2}$

34. $\dfrac{6.9}{-3}$

35. $\dfrac{-200}{-2.5}$

36. $\dfrac{-51}{0.3}$

37. $\dfrac{0.48}{-2.4}$

38. $\dfrac{-5.7}{-0.3}$

39. $-\dfrac{4}{5} \div \dfrac{4}{3}$

40. $\dfrac{5}{16} \div \left(-\dfrac{1}{2}\right)$

41. $-\dfrac{4}{3} \div \dfrac{1}{8}$

42. $-\dfrac{2}{3} \div \left(-\dfrac{1}{4}\right)$

What number should each of the following be divided or multiplied by to get one?

Example: −6

Solution: Divide by −6 or multiply by $-\dfrac{1}{6}$.

43. 4

44. −8

45. $-\dfrac{1}{2}$

46. $\dfrac{2}{3}$

47. $-\dfrac{3}{8}$

48. $-\dfrac{1}{16}$

In Exercises 49–66 perform the indicated computations.

49. $(-5)(0)(-6)$

50. $(-8)(-4)(3)(0)$

51. $(-3)(5)(-6)$

52. $(-4)(-6)(-1)$

53. $(-2)\left(-\dfrac{3}{2}\right)(-3)(-8)$

54. $\left(-\dfrac{4}{5}\right)(5)(-1)(-4)$

55. $(-4)^3$

56. $(-6)^3$

57. $(-3)^2\left(\dfrac{1}{3}\right)$

58. $-\dfrac{1}{2}(-2)^3$

59. $16(-2) \div 4$

60. $(-15)(6) \div (-3)$

61. $(-5)(-4) \div (-2)$

62. $(14)(3) \div (-2)$

63. $\dfrac{(-5)(6)}{-2}$

64. $\dfrac{(-8)(-7)}{4}$

65. $\dfrac{(-2)^2(6)}{-4}$

66. $\dfrac{3(-2)^3}{-3}$

67. The magnification of a certain lens is found by calculating

$$-\left(\dfrac{-20}{5}\right).$$

Find the value.

68. The pressure (in lb/in.^{2}) required to compress 200 in.^{3} of water by 4 in.^{3} is found by calculating

$$\frac{(-330000)(-4)}{200}$$

Find the value.

Calculator Exercises

Use a calculator to determine each of the following:

69. $\dfrac{(-16.4)(-2.65)}{1.6}$

70. $\dfrac{256(-15.9)}{-3.25}$

71. $\dfrac{1}{-5.8}(46.4)$

72. $(-4.24)\left(-\dfrac{1}{13.25}\right)$

9.4
ORDER OF OPERATIONS

Much of the arithmetic and algebra that is used in the technical area involves computations with several different operations. Consider this complex calculation:

$$6(-9) - \frac{12(-4)}{(-3)8} - 18 + 4.$$

It would be difficult if various occupations, countries, or even textbooks followed different rules for computations like the example above. Fortunately, wherever mathematics is used, the same rules for determining the order of operations are followed. We will practice using the correct order of operations with positive and negative numbers in this section.

> Computations within parentheses are always done first.

EXAMPLES

1 $6 - (7 - 11) = 6 - (-4)$
$$= 6 + 4$$
$$= 10$$

2 $-6(8 - 13) = -6(-5)$
$$= 30$$

The fraction bar, indicating division, is often a special case of "parentheses." That is, when there are addition or subtraction computations to be performed in the numerator or denominator of a fraction, these computations should be performed first. After the addition or subtraction has been done, then the numerator is divided by the denominator.

EXAMPLES

3 $\dfrac{45}{-6 - 4} = \dfrac{45}{-10}$
$$= -\frac{9}{2} \text{ or } -4.5$$

4 $\dfrac{14 - 30}{2} = \dfrac{-16}{2}$
$$= -8$$

> After all computations with parentheses are done, you go back to the left. Starting from the left, perform all multiplications and divisions in order as they occur.

Leave any addition and subtraction for later. In this way, doing a long series of calculations is like reading a book. One works from left to right.

EXAMPLES

5 $-12 \div (-3)(-5) = 4(-5)$
$$= -20$$

6 $4(-8) + -3(-9) = -32 + 27$
$$= -5$$

7 $\dfrac{-36}{-3-6} - (-3)(6-8) = \dfrac{-36}{-9} - (-3)(-2)$
$$= 4 - 6$$
$$= -2$$

Now, after all computations within parentheses (including those grouped by a fraction bar), followed by all multiplications and divisions from left to right, are done, you go back to the left again.

> Moving along again from the left, do all additions and subtractions in order as they appear.

EXAMPLES

8 $3 + 2(-6) = 3 + -12$
$$= -9$$

9 $-5(-3) - 4(2) - (12 - 9) = -5(-3) - 4(2) - 3$
$$= 15 - 8 - 3$$
$$= 7 - 3$$
$$= 4$$

Since a number raised to a power means multiplying the number by itself, exponents imply multiplication.

EXAMPLES

10 $\dfrac{4(-3)^2}{(-6)2^3} = \dfrac{4(-3)(-3)}{(-6)(2)(2)(2)}$
$$= \dfrac{36}{-48}$$
$$= -\dfrac{3}{4} \text{ or } -0.75$$

11 $2^2 - 6^2 = (2)(2) - (6)(6)$
$$= 4 - 36$$
$$= -32$$

> A summary of the order of operations is:
> 1. **Compute within grouping symbols, which are usually parentheses.**
> 2. **Compute powers.**
> 3. **Compute multiplication and division in order from left to right.**
> 4. **Compute addition and subtraction in order from left to right.**

Review Exercises

Perform the indicated operations in Exercises 1–8.

1. $8.7 + (-5.9)$ 2. $-6.3 + (-2.46)$
3. $-14 - (-9)$ 4. $-33 - (-14)$
5. $(-15)(-8)$ 6. $(-5)(-13)$
7. $\left(-\dfrac{3}{5}\right)(15)$ 8. $(-6)\left(\dfrac{3}{8}\right)$

9. What is the area of a square 8 cm on a side?
10. What is the area of a rectangle 9 cm wide and 82 cm long?

11. Find the total surface area.

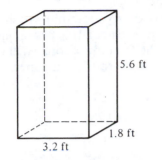

5.6 ft

1.8 ft

3.2 ft

12. Find the total surface area.

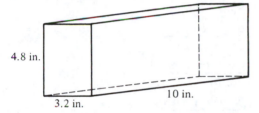

4.8 in.

3.2 in.   10 in.

EXERCISES
9.4

In Exercises 1–46, compute using the correct order of operations.

1. $4 + 6(-5)$
2. $5 + 7(-8)$
3. $-5 - (8 - 9)$
4. $-9 - (-8 + 4)$
5. $14 - (-4 + 5)$
6. $2 - (7 - 12)$
7. $-5(-3) + 2(-6)$
8. $4(-8) - 13(2)$
9. $-3(6) - (-2)(4)$
10. $-9(-3) + 4(-8)$
11. $4^2 + 3^2$
12. $5^2 + (12)^2$
13. $5^2 - 3^2$
14. $(13)^2 - 5^2$
15. $2^2 - 9^2$
16. $5^2 - 7^2$
17. $4(-5) + 2(6)^2$
18. $-3(-7) - 4(2)^3$
19. $5(6 - 10)$
20. $-8(17 - 5)$
21. $4(-5) - 2$
22. $-6(8) - (-4)$
23. $16 - (-8)(-4)$
24. $-18 - 9(-3)$
25. $\dfrac{15 - 10}{-5}$
26. $\dfrac{8 - 12}{2}$
27. $\dfrac{-12}{14 - 10}$
28. $\dfrac{45}{-3 - 6}$

29. $\dfrac{5 + 9}{1 - 8}$
30. $\dfrac{-5 - 7}{14 - 9}$
31. $\dfrac{-18}{2(3)}$
32. $\dfrac{56}{-2(7)}$
33. $\dfrac{-42(-4)}{-6(5)}$
34. $\dfrac{(-6)(8)}{4 + 6}$
35. $\dfrac{7(-9)}{-6 - 3}$
36. $\dfrac{(15)(-4)}{(-3)(8)}$
37. $\dfrac{4(-6)^2}{12}$
38. $\dfrac{-5(3^2)}{2(-1)^2}$
39. $\dfrac{12}{(-3)^2}(15)$
40. $\dfrac{72}{(-2)^2}(-5)$
41. $6(-8) - (4 + 2) + 3^2$
42. $-(6 - 9) + 6(2) - 3 + 5$
43. $-1 + 8 + 4(-6) - (3 - 9)$
44. $-6 - 2(4) - 2^3$
45. $\dfrac{4(-3)}{6} - 4(5 - 6) + 8(-2)$
46. $-(8 - 4) - \dfrac{32}{2(-8)} - 4^2$

47. The bottom diameter of a certain gear can be found by calculating
$$12 - 2(1.25 + 0.5).$$
Find the value.

48. The height of a certain object after being propelled upward is found by calculating
$$(96)(2) - 16(2)^2.$$
Find the value.

49. The sum of $1 + 2 + 3 + 4 + 5 + 6 + 7 + 8 + 9 + 10$ may be found by calculating
$$\frac{1}{2}(10)(10 + 1).$$
Find the value.

50. The correction for sag in a certain tape is found by calculating
$$-\frac{3^2(160)}{24(10)^2}.$$
Find the value.

51. A man has $800 in the bank. If he withdraws $40 each week for 8 weeks, what will his balance be at the end of 8 weeks?

52. One pipe fills a tank at a rate of 200 gallons per hour. Another fills the tank at a rate of 175 gallons per hour. How many gallons will be in the tank if both pipes are open for 3 hours?

53. A pipe has been filling a tank for 4 hours at 350 gallons per hour. As the tank continues filling, a drain is opened which empties at a rate of 200 gallons per hour. How many gallons will be in the tank after 2 more hours?

54. The perimeter of a rectangle is 48 feet. What is its length if the width is 6 feet?

55. A car travels 4 hours at 55 mph, and then $\frac{1}{2}$ hour at 30 mph. What is the total distance traveled?

56. The frequency of sound in a certain situation is found by calculating

$$\frac{30(3.35)}{3.35 - 2.4}.$$

Find the value to the nearest hundredth.

Calculator Exercises

Use a calculator to determine each of the following. If your calculator has a memory, try to do all the computations without writing down intermediate results on paper.

57. $(48.6 + 3.98)123$

58. $48.6 + (3.98)(123)$

59. $679.2 - 46.8 - 68.73$

60. $679.2 - (46.8 - 68.73)$

61. $(24.5)(62.34) - (29.3)^2$

62. $\dfrac{4.2 - 6.7}{3.2}$

63. $\dfrac{48.7 - 28.368}{(-34)(9.2)}$

64. $\dfrac{(28.5)(-6.09)}{190.38 - 19.38}$

65. $\dfrac{(75.9)(-12.46)}{(-8.9)(6.6)}$

66. $\dfrac{(8.7)(3.6)^2}{6.51 - 30}$

67. $-(6.5)^2 - \dfrac{52.8}{(13.2)(6.4)} - (87.6 - 32.1)$

9.5
ALGEBRAIC EXPRESSIONS

In Chapter 5, letters of the alphabet were used to represent unknown numbers. Letters of the alphabet or other symbols used when a number is not known, or when a number has yet to be chosen, are called **variables**. A collection of numbers, variables, and symbols of operation is called an **algebraic expression**. Table 9.4 shows some of the algebraic expressions you used in Chapter 5.

Variables are sometimes also called **unknowns**, or even just **letters**. Writing variables for unknown numbers in an algebraic expression is little different than writing the numbers themselves. One big difference, however, is in combining numbers and variables, or variables and

TABLE 9.4

	The meaning of the variables	Algebraic expression
Perimeter of a rectangle	a = length b = width	$2a + 2b$
Area of a circle	r = radius	πr^2
Perimeter of a square	s = length of a side	$4s$
Surface area of a rectangular prism	ℓ = length w = width h = height	$2\ell w + 2wh + 2\ell h$
Volume of a sphere	r = radius	$\frac{4}{3}\pi r^3$

variables, without using a symbol of operation:

1. 23 means twenty-three
2. $4s$ means 4 times s
3. ℓwh means ℓ times w times h

A number and variable, or two variables, written next to each other means that they are to be multiplied. Different variables in the same expression do not necessarily name different numbers. For example, the algebraic expression $a + b$ could mean any of the following and many more:

$$2 + 7, \quad 9 + 9, \quad 8.7 + 3.15, \quad 6.4 + 6.4,$$

$$\frac{4}{5} + \frac{7}{8}, \quad 123 + 2.67.$$

Note that, since a and b are *any* numbers, they might both be the same number.

In different expressions, the same letter can be used to represent different quantities. For example, the expression rt means "r times t," where often r is a rate and t is a time. The expression $2\pi r$ means "2 times π times r," where r is the radius of a circle.

It is important in the technical areas to change phrases written with words into algebraic expressions. In the following examples the words used for different operations are the keys to writing the expression.

EXAMPLE 1 Write the product of two numbers as an algebraic expression.

Solution We choose some variable, say a, for one number. There is nothing known about either number, so we choose a different letter, say b, for the other number. A product of two numbers is one number multiplied by another. The algebraic expression, then, is ab. This can also be written as $(a)(b)$, $a(b)$, $(a)b$, or $a \cdot b$.

EXAMPLE 2 One number is twice another number. Write each number as an algebraic expression.

Solution We choose any letter for one of the numbers, say y. Twice means "times two" or

"multiplied by two." So the other number is $2y$. The two numbers are y and $2y$.

Example 1 showed that multiplication can be written in different ways. Division is another operation that can be written in various ways.

EXAMPLE 3 Write two times a number divided by three as an algebraic expression.

Solution Choose any letter as the number, say m. Then, two times that number is $2m$. Two times that number divided by three is written as $\frac{2m}{3}$ or $2m \div 3$. Note here that two-thirds a number is two-thirds *times* a number (say m again), or $\frac{2}{3}m$. The three expressions

$$\frac{2m}{3}, \quad 2m \div 3, \text{ and } \frac{2}{3}m$$

all represent the same number regardless of the value of m.

EXAMPLE 4 Write six times a number, plus another number, all divided by six, as an algebraic expression.

Solution This can be written as

$$\frac{6x + y}{6} \text{ or } x + \frac{y}{6}.$$

Review Exercises

Perform the indicated operations in Exercises 1–10.

1. $4.5 - 9$
2. $-8.7 - 8$
3. $\left(-\frac{3}{5}\right)(15)$
4. $(-6)\left(\frac{3}{8}\right)$
5. $\frac{-84}{-7}$
6. $\frac{54}{-9}$
7. Change $53°21'$ to degrees.
8. Change $132°36'$ to degrees.
9. $89.5 \text{ m} = \underline{\hspace{2cm}} \text{ cm}$
10. $\underline{\hspace{2cm}} \text{ m} = 38 \text{ cm}$

11. Find the perimeter.

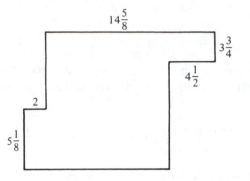

12. Find the perimeter.

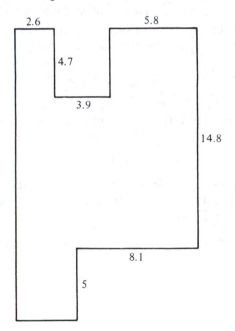

EXERCISES
9.5

Write each of the expressions in Exercises 1–26 as an algebraic expression.

1. the sum of x and y
2. m multiplied by n
3. c divided by b 4.   4 more than x
5. 8 less than y 6. p times q
7. 6 added to x
8. 19 subtracted from y
9. the quotient of p divided by q
10. the product of g and h
11. x minus y 12. m less 8
13. y squared   14. x cubed
15. m to the fourth power
16. n to the fifth power
17. x increased by 6
18. y decreased by 18
19. 8 more than the product of 3 and x
20. 14 less than the product of 8 and z
21. the sum of m and n divided by 4
22. the difference of m minus n all divided by 6
23. 13 divided by the product of p and q
24. 46 divided by the sum of s and t
25. the product of r and s divided by the sum of 4 and x
26. the sum of a and b divided by the product of c and d

Use the given variable to express each of the following (Exercises 27–44).

27. the number which is twice some number p
28. the number which is three times some number x
29. the number which is half some number y
30. the number which is two-thirds some number s
31. the number of bottles in m cases if each case contains 24 bottles
32. the number of machine parts in x boxes if they are packed 18 to a box
33. the cost of m feet of molding if the molding costs 79¢ per foot
34. the cost of s square feet of sheet metal if it costs \$3 per square foot
35. the number of inches in b feet
36. the number of feet in d yards
37. the number of yards in x feet
38. the number of feet in y inches
39. the number of cubic feet in m cubic yards
40. the number of cubic inches in n cubic feet
41. the number of cubic yards in s cubic feet

42. the number of cubic feet in t cubic inches

43. the perimeter of a rectangle whose length is r centimeters and whose width is s centimeters

44. the area of a square with length of a side q inches

Choose variables and write algebraic expressions for each of the following (Exercises 45–54).

45. One number is two more than another number. What are the numbers?

46. One number is six less than another number. What are the numbers?

47. One number is three times another number. What are the numbers?

48. One number is three-eighths another number. What are the numbers?

49. One number is eight less than twice another number. What are the numbers?

50. One number is five more than four times another number. What are the numbers?

51. One of two equal sides of an isosceles triangle is twice the base. What is the perimeter?

52. The height of a triangle is five meters more than the base. What is the area?

53. The length of a rectangular frame is three times the width. What is the perimeter? What is the area?

54. The length of a rectangular pool is seven feet longer than the width. What is the perimeter? What is the area?

9.6
EVALUATING EXPRESSIONS

The **value** of an algebraic expression is the number obtained after the variable is replaced by a number. The process of finding the value of an algebraic expression can be called **evaluating** the expression.

EXAMPLE 1 Evaluate $a + b + c$ if $a = 5$, $b = -11$, and $c = 8$.

Solution First replace each letter (variable) with its value.

$$a + b + c$$

becomes

$$5 + (-11) + 8.$$

Then add the three numbers. The result is 2, so the value of the expression is 2.

Replacing an unknown (variable) with a number is also called substituting values for the variables.

EXAMPLE 2 Evaluate $8 + 3x$ if $x = -9$.

Solution
$$8 + 3x = 8 + 3(-9) \longleftarrow \text{Substitute } -9 \text{ for } x.$$
$$= 8 + (-27)$$
$$= -19$$

The value of the expression is -19.

Note that, in Example 2 and in the next two examples, the operations must be performed in the correct order after substituting for the variables.

EXAMPLE 3   Evaluate $x - (y - z)$ if $x = -8$, $y = 4$, and $z = -12$.

Solution

$$x - (y - z) = -8 - [4 - (-12)] \longleftarrow \text{Substi-}$$
$$= -8 - (4 + 12) \quad \text{tute given}$$
$$= -8 - 16 \quad \text{values for}$$
$$= -8 - 16 \quad \text{the variables.}$$
$$= -24$$

The value of the expression is -24.

EXAMPLE 4   Evaluate $\dfrac{3.5f}{s - 3.5}$ if $f = 600$ and $s = 11$.

Solution

$$\frac{3.5f}{s - 3.5} = \frac{3.5(600)}{11 - 3.5} \longleftarrow \text{Substitute}$$
$$\text{given values}$$
$$= \frac{2100}{7.5} \quad \text{for the}$$
$$\text{variables.}$$
$$= 280$$

The value of the expression is 280.

If an algebraic expression containing exponents is first rewritten in terms of multiplication, confusion can sometimes be avoided. For example, the expression ab^2 can be rewritten as abb. Note that only the b is squared. If the product of a and b is to be squared, the expression must be written as $(ab)^2$.

EXAMPLE 5   Evaluate xy^3 if $x = 5$ and $y = -2$.

Solution   The expression is rewritten in terms of multiplication before substituting the values for the variables.

$$xy^3 = xyyy \qquad\qquad xy^3 = 5(-2)^3$$
$$= 5(-2)(-2)(-2) \quad \text{or} \qquad = 5(-2)(-2)(-2)$$
$$= -40 \qquad\qquad\qquad = -40$$

The value of the expression is -40.

EXAMPLE 6   Evaluate $2x^2 - xy + y$ if $x = -4$ and $y = 7$.

Solution

$$2x^2 - xy + y = 2xx - xy + y$$
$$\text{or } 2(-4)^2 - (-4)(7) + 7$$
$$= 2(-4)(-4) - (-4)(7) + 7$$
$$= 32 - (-28) + 7$$
$$= 32 + 28 + 7$$
$$= 67$$

The value of the expression is 67.

Review Exercises

Perform the operations indicated in Exercises 1–8.

1. $\left(-\dfrac{4}{3}\right)\left(-\dfrac{9}{8}\right)$ 2. $\left(\dfrac{-3}{4}\right)\left(\dfrac{4}{5}\right)$

3. $\dfrac{-3}{5} \div 6$ 4. $-4 \div \left(-\dfrac{1}{2}\right)$

5. $5 - 6(-3)$ 6. $4 + 8(-2)$

7. $-3 - (8 - 11)$ 8. $4 - (6 - 12)$

9. Change 0.8 feet to inches, to the nearest sixteenth of an inch.

10. Change 0.2 feet to inches, to the nearest half inch.

11. How many cubic feet of space are there in a storage space 8 feet long, 6 feet wide, and $7\frac{1}{2}$ feet high?

12. How many cubic feet of space are there in a rectangular room 14 feet 6 inches by 10 feet by 12 feet 9 inches?

EXERCISES

9.6

Find the value of the expression $16 - 4y$ if:

1. $y = 2$ 2. $y = 6$

3. $y = -5$ 4. $y = -8$

Find the value of the expression $a - (b + c)$ if:

5. $a = -5, b = 6, c = 3$
6. $a = 15, b = -2, c = 4$
7. $a = 2, b = 9, c = -4$
8. $a = -8, b = -10, c = 6$

Evaluate the expression $\frac{m + n}{4d}$ if:

9. $m = 8, n = -9, d = -2$
10. $m = -6, n = -2, d = 2$
11. $m = -12, n = 8, d = -1$
12. $m = 14, n = -5, d = -3$

Evaluate the expression $2xy^2$ if:

13. $x = 3, y = 4$ 14. $x = 5, y = 6$
15. $x = 8, y = -2$ 16. $x = 6, y = -1$

Evaluate the expression $x^2 + \frac{x}{y} - xy$ if:

17. $x = 4, y = 2$ 18. $x = 6, y = 3$
19. $x = -3, y = -1$ 20. $x = 5, y = -1$

Find the value of $a(b - c)$ if:

21. $a = 6, b = 4.2, c = 2.9$
22. $a = 8, b = 9.84, c = 7.46$
23. $a = 4, b = \frac{1}{2}, c = \frac{1}{4}$
24. $a = 8, b = \frac{3}{4}, c = \frac{5}{8}$

Find the value of $\frac{xy}{z}$ if:

25. $x = 2, y = 4.68, z = 0.5$
26. $x = 1.4, y = 36, z = 4.2$

27. $x = \frac{3}{4}, y = 2, z = 3$

28. $x = 5, y = \frac{5}{8}, z = \frac{3}{4}$

Evaluate each expression if $x = 4$, $y = 0.4$, and $z = 2.5$.

29. y^3 30. $2z^2$
31. $x + yz$ 32. $(x + y)z$

33. $\frac{x}{yz}$ 34. $\frac{xy}{z}$

35. $\frac{x - y}{z}$ 36. $\frac{z - y}{x}$

37. xyz 38. $x + y + z$

Evaluate each expression if $a = 3$, $b = \frac{1}{4}$, and $c = \frac{3}{8}$.

39. $4b^2$ 40. b^3
41. $(b + a)c$ 42. $b + ac$

43. $\frac{ab}{c}$ 44. $\frac{c}{ab}$

45. $\frac{a}{c - b}$ 46. $\frac{a}{c + b}$

47. Under certain conditions distance (in miles) is found to be rt. Evaluate if $r = 55$ and $t = 2.75$.

48. Work (in foot pounds) is fs. Evaluate if $f = 20.5$ and $s = 16$.

49. In a particular situation heat (in joules) is found to be RI^2. Evaluate this if $R = 12$ and $I = 3$.

50. Under certain conditions, distance (in feet) is found to be kt^2. Evaluate this if $k = 16$ and $t = 5$.

51. Power (in foot pounds per second) is $\frac{fs}{t}$. Evaluate if $f = 20$, $s = 5$, and $t = 16$.

52. The speed of a driven pulley (in rpm) is $\frac{DS}{d}$. Evaluate if $S = 400$, $D = 8$, and $d = 5$.

53. The number of threads per inch of a screw is $\frac{1}{p}$. Evaluate if $p = 0.05$.

54. The force required to trim forging is $\frac{PTS}{2000}$. Evaluate if $P = 30$, $T = \frac{1}{8}$, and $S = 60,000$.

55. The section modulus for rectangular beams is $\frac{bd^2}{6}$. Evaluate if $b = 8$ and $d = 6$.

56. Horsepower under certain conditions is found to be $0.04d^2n$. Evaluate if $n = 8$ and $d = 4$.

57. Under certain conditions, profit (in dollars) is $n(p - c)$. Evaluate if $n = 10$, $p = 16.95$, and $c = 12.14$.

58. Fahrenheit temperature (in degrees) is $\frac{9}{5}C + 32$. Evaluate when $C = 15$.

59. Celsius temperature (in degrees) is $\frac{5}{9}(F - 32)$. Evaluate when $F = 86$.

60. In a particular situation, velocity (in meters per second) is found to be $v + at$. Evaluate if $v = 40$, $a = 5$, and $t = 6$.

61. Evalute the expression in Exercise 60 if $v = 22$, $a = -4$, and $t = 5$.

62. The combined resistance of two electrical resistors in a parallel circuit is $\frac{ab}{a+b}$. Evaluate if $a = 4.5$ and $b = 3.5$.

63. In particular situations, the voltage in a circuit can be found using $IA + \frac{P}{I} + IB$. Evaluate if $I = 6$, $A = 4$, $B = 5$, and $P = 9$.

Calculator Exercises

Use a calculator to determine each of the following.

64. Evaluate the expression in Exercise 58 when $C = 23$. Express the answer to the nearest tenth.

65. Evaluate the expression in Exercise 59 when $F = 60$. Express the answer to the nearest hundredth.

66. Evaluate the expression in Exercise 62 when $a = 5.6$ and $b = 4.7$. Express the answer to the nearest tenth.

67. Evaluate the expression in Exercise 62 when $a = 7.32$ and $b = 8.69$. Express the answer to the nearest hundredth.

68. Evaluate the expression in Exercise 63 when $I = 5$, $A = 10.7$, $P = 12$, and $B = 4.8$. Express the answer to the nearest tenth.

9.7
MULTIPLYING AND DIVIDING EXPRESSIONS

In algebra we work with numbers without always knowing what the numbers are. Numbers can be added, subtracted, multiplied, and divided, and so can algebraic expressions. The number of feet in 5 yards is:

$$5 \text{ yd} \left(3\frac{\text{ft}}{\text{yd}}\right) \text{ or } 5(3) \text{ ft.}$$

We would, however, say 15 feet, rather than $5(3)$ feet. The number of feet in $4y$ yards is:

$$4y \text{ yd} \left(3\frac{\text{ft}}{\text{yd}}\right) \text{ or } 4y(3) \text{ ft.}$$

Even though this is an algebraic expression and there is a variable, we would still do the multiplication

$$(3)(4)y \text{ ft,}$$

and say $12y$ feet. We work with y as a number without knowing its value.

Numbers can be multiplied in any order:

$$8 \cdot 3 = 3 \cdot 8$$

and numbers can be multiplied in any grouping:

$$8(3 \cdot 2) = (8 \cdot 3)2.$$

Because of these two facts, algebraic expressions containing multiplication can be simplified.

EXAMPLE 1   Multiply (or simplify): $5(2x)$.

Solution   Since numbers can be multiplied in any grouping,

$$5(2x) = (5 \cdot 2)x$$
$$= 10x.$$

EXAMPLE 2   Multiply (or simplify): $(3y)\left(\frac{2}{3}\right)$.

Solution Since numbers can be multiplied in any grouping and in any order,

$$(3y)\left(\frac{2}{3}\right) = (3)\left(\frac{2}{3}\right)y$$
$$= 2y.$$

EXAMPLE 3 Multiply (or simplify): $4\left(\frac{x+5}{4}\right)$.

Solution The computation $x + 5$ cannot be made without knowing the value of x. To show that the x and the 5 are to be added first (when the value of x is known) parentheses are kept around the quantity. We write $(x + 5)$. So

$$4\left(\frac{x+5}{4}\right) = \frac{4}{1}\left(\frac{x+5}{4}\right)$$
$$= \frac{4(x+5)}{4}$$
$$= (x + 5).$$

In this case, there is no chance for misinterpretation at the end, so the answer can be written $x + 5$.

Table 9.5 compares multiplication of numbers with multiplication of expressions. In each illustration, the numbers have been computed and the expressions have been simplified.

Dividing expressions follows directly from previous work with fractions.

EXAMPLE 4 Divide (or simplify): $\frac{-7x}{-7}$.

Solution

$$\frac{-7}{-7} = 1.$$

Therefore,

$$\frac{-7x}{-7} = 1x \text{ or } x.$$

EXAMPLE 5 Divide (or simplify): $\frac{30x}{15}$.

Solution

$$\frac{30}{15} = 2.$$

Therefore,

$$\frac{30x}{15} = 2x.$$

Table 9.6 compares division of numbers with division of expressions.

TABLE 9.5

Numbers	Computed	Expressions	Simplified
$7\left(\frac{9}{7}\right)$	9	$7\left(\frac{x}{7}\right)$	x
$\left(\frac{5\cdot 3}{4}\right)\left(\frac{4}{5}\right)$	3	$\left(\frac{5x}{4}\right)\left(\frac{4}{5}\right)$	x
$24\left(\frac{5}{16}\right)$	$\frac{15}{2}$	$24\left(\frac{x}{16}\right)$	$\frac{3x}{2}$
$5\left(\frac{13-6}{5}\right)$	7	$5\left(\frac{x-6}{5}\right)$	$x - 6$
$\frac{4\cdot 3}{9}9$	12	$\frac{4x}{m}m$	$4x$
$(2\cdot 5)\frac{4}{3}$	$\frac{40}{3}$	$(2m)\frac{x}{3}$	$\frac{2xm}{3}$

TABLE 9.6

Numbers	Computed	Expressions	Simplified
$\frac{18(2)}{18}$	2	$\frac{18x}{18}$	x
$\frac{-14(5)}{21}$	$\frac{-10}{3}$	$\frac{-14x}{21}$	$\frac{-2x}{3}$
$\frac{2(3^2)}{2}$	9	$\frac{2x^2}{2}$	x^2

Consider the following:

$$5(3 + 4) = 5(7) \qquad 5(3 + 4) = 5(3) + 5(4)$$
$$= 35 \qquad\qquad\qquad = 15 + 20$$
$$= 35$$

Careful study of the above should convince you that, *to multiply a number by a sum, the number can be multiplied by each part of the sum and then these products can be added.* In general, this is stated:

> $a(b + c) = ab + ac$ for any numbers a, b, and c.

This important relationship is called the **distributive property of multiplication over addition**. The next example shows how the distributive property is used in multiplying expressions.

EXAMPLE 6 Multiply (or simplify): $2(3x + 4)$.

Solution Although we cannot compute within parentheses, we can use the distributive property to help simplify the expression.

WATCH OUT! $2(3x + 4)$ is *not* $2(3x) + 4$.

$$2(3x + 4) = 2(3x) + 2(4) \longleftarrow a(b + c) = ab + ac$$
$$= 2(3x) + 8$$
$$= 6x + 8$$

By convention, $6x + 8$ is simpler than the original because it does not contain parentheses.

Review Exercises

1. $(-3.2)(1.5) =$ _____
2. $(4.8)(-2.5) =$ _____
3. $(-62) \div (-0.2) =$ _____
4. $2.7 \div (-5) =$ _____
5. 79 mm = _____ m
6. _____ mm = 6.8 cm
7. Change 48% to a decimal.
8. Change 76% to a decimal.
9. Change 2.75% to a decimal.
10. Change 7.25% to a decimal.
11. What is the circumference of a circle with a radius of 0.3 feet?
12. What is the circumference of a circle with a diameter of 17 inches?

EXERCISES
9.7

Simplify the multiplications and divisions in Exercises 1–40.

1. $4(5x)$
2. $3(2x)$
3. $(8x)3$
4. $(7x)2$
5. $-6(-3x)$
6. $-2(8x)$
7. $4.2(5x)$
8. $6.8(10x)$
9. $(8.7x)3$
10. $(3.4x)6$
11. $\dfrac{6x}{6}$
12. $\dfrac{13x}{13}$
13. $\dfrac{-5x}{-5}$
14. $\dfrac{-9x}{-9}$
15. $\dfrac{8x}{4}$
16. $\dfrac{10x}{2}$
17. $\dfrac{mx}{m}$
18. $\dfrac{nx}{n}$
19. $\dfrac{-6x}{3}$
20. $\dfrac{-15x}{3}$
21. $\dfrac{8x}{6}$
22. $\dfrac{6x}{4}$
23. $\dfrac{15x}{24}$
24. $\dfrac{27x}{30}$
25. $3\left(\dfrac{x}{3}\right)$
26. $\left(\dfrac{x}{5}\right)5$
27. $\left(\dfrac{x}{8}\right)4$
28. $3\left(\dfrac{x}{6}\right)$
29. $12\left(\dfrac{x}{18}\right)$
30. $\left(\dfrac{x}{16}\right)24$
31. $\left(\dfrac{2}{3}x\right)\dfrac{3}{2}$
32. $\dfrac{4}{5}\left(\dfrac{5}{4}x\right)$
33. $\dfrac{-8}{3}\left(\dfrac{-3x}{8}\right)$
34. $\left(\dfrac{-16x}{3}\right)\left(\dfrac{-3}{16}\right)$
35. $\left(\dfrac{2x}{6}\right)\dfrac{3}{2}$
36. $\dfrac{4}{5}\left(\dfrac{5x}{6}\right)$
37. $\left(\dfrac{x + 7}{3}\right)3$
38. $9\left(\dfrac{x - 6}{9}\right)$
39. $d\left(\dfrac{m - 4}{d}\right)$
40. $\left(\dfrac{n + 5}{t}\right)t$

Multiply the expressions in Exercises 41–54.

41. $2(x + 9)$

42. $3(x - 8)$

43. $(x - 6)4$

44. $(x + 3)7$

45. $4(3x - 2)$

46. $5(2x + 3)$

47. $(2x + 5)(4.5)$

48. $(4x - 7)(6.2)$

49. $-2(x + 4)$

50. $-3(x + 8)$

51. $d(2x + 6)$

52. $m(3x - 4)$

53. $(4m + 9)\frac{2}{3}$

54. $(5n - 8)\frac{3}{4}$

Simplify each of the following expressions from different technical areas.

55. $\dfrac{IR}{I}$ (electricity)

56. $\dfrac{RI^2}{I^2}$ (heat)

57. $\ell\left(\dfrac{v}{\ell}\right)$ (frequency)

58. $\left(\dfrac{p}{f}\right)f$ (architecture)

59. $\dfrac{8}{15}\left(\dfrac{15}{8}h\right)$ (volume)

60. $\left(S\dfrac{I}{C}\right)\dfrac{C}{I}$ (mechanical design)

61. $\dfrac{250S}{550(4)}$ (power)

62. $\left(\dfrac{550}{62.4f}\right)H$ (power)

63. $\dfrac{t}{s}\left(\dfrac{fs}{t}\right)$ (power)

64. $\dfrac{33000}{F}\left(\dfrac{FV}{33000}\right)$ (mechanical design)

Write an expression for each of the following, and simplify.

65. Find the perimeter of a square if one side is $3y$ feet.

66. Find the number of cubic feet in $4m$ cubic yards.

67. Find the number of square feet in $120s$ square inches.

68. If the number of items produced by a certain machine in one hour is $2x$, find the number of items produced in 16 hours.

69. There are $135x$ pieces of equipment in one shipment packed in 18 boxes. How many are in each box?

9.8
ADDING AND SUBTRACTING EXPRESSIONS

Parts of an algebraic expression separated by + and − signs are called **terms**. Terms with the same variable and exponent are called **similar terms**, or **like terms**. Table 9.7 shows some algebraic expressions that have like terms and some that have unlike terms.

Similar or like terms can be added and subtracted by adding and subtracting the numbers that are multiplied by the variable. This is often called *combining like terms*.

TABLE 9.7

Like terms	Unlike terms
$3x + 7x$	$3x + 4$
$-4y - 8y$	$5 - 2x$
$5x^2 + 2x^2$	$3x + 2y$
$m + 3m$	$2r + s$

EXAMPLES

1 $3x + 8x = (3 + 8)x$
$\qquad = 11x$

2 $-4y - 8y = (-4 - 8)y$
$\qquad = -12y$

The numbers that are multiplied by the variables are called **coefficients**. The coefficient of $3x$ is three; the coefficient of $-8y$ is negative eight. A variable standing alone can always be considered as being multiplied by one. That is,

$x = 1x$ **for any number** x.

Thus, the coefficient of x is one.

EXAMPLES

3 $6x + x = 6x + 1x$
$\qquad = (6 + 1)x$
$\qquad = 7x$

4 $5x^2 - x^2 = 5x^2 - 1x^2$
$\qquad = (5 - 1)x^2$
$\qquad = 4x^2$

Fractions can always be added and subtracted even when the terms cannot be combined as in the preceding examples.

EXAMPLE 5

$$\frac{2}{3}x + \frac{a}{4} = \frac{2x}{3} + \frac{a}{4}$$
$$= \frac{4}{4}\left(\frac{2x}{3}\right) + \frac{3}{3}\left(\frac{a}{4}\right)$$
$$= \frac{8x}{12} + \frac{3a}{12}$$
$$= \frac{8x + 3a}{12}$$

Expressions from technical areas often need to be rewritten. This can be done using the distributive property.

EXAMPLE 6

$$r + ar = 1r + ar$$
$$= (1 + a)r$$

(*Note:* It is necessary to keep the parentheses since the quantity $1 + a$ is being multiplied by r.)

Three or more like terms are added or subtracted in the same way as two terms are added or subtracted.

EXAMPLE 7

$$x - 8x + 6x = 1x - 8x + 6x$$
$$= (1 - 8 + 6)x$$
$$= -1x$$
$$= -x$$

(*Note:* $-x = -1x$ for any number x.)

The next example shows an algebraic expression in which some terms can be added or subtracted and some cannot. Only like terms can be added or subtracted.

EXAMPLE 8

$$5x + 2y - 8x - y = 5x - 8x + 2y - y$$
$$= 5x - 8x + 2y - 1y$$
$$= (5 - 8)x + (2 - 1)y$$
$$= -3x + 1y$$
$$= -3x + y$$

In Example 8 we combine x term with x term and y term with y term. The x and y terms cannot be combined because they are not *similar* terms.

EXAMPLE 9 Write and simplify an expression for the perimeter of this figure.

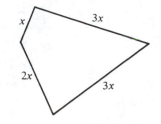

Solution The perimeter is the distance around the figure, or the sum of the lengths of the sides. Thus the perimeter is:

$$x + 2x + 3x + 3x.$$

This expression can be simplified:

$$x + 2x + 3x + 3x = 1x + 2x + 3x + 3x$$
$$= (1 + 2 + 3 + 3)x$$
$$= 9x.$$

The perimeter is $9x$.

In general, expressions have been simplified when :
1. there are no parentheses, and
2. similar terms are combined.

Review Exercises

1. $-3(-4) + 5(-2) =$ _____
2. $5(-6) - 4(2) =$ _____
3. $3^2 - 8^2 =$ _____   4. $2^3 + 3^2 =$ _____
5. $\dfrac{20 - 25}{10} =$ _____   6. $\dfrac{7}{4 - 18} =$ _____
7. Change 0.3 to a percent.
8. Change 0.254 to a percent.
9. Change 0.036 to a percent.
10. Change 2.1 to a percent.
11. Find the area in square feet.

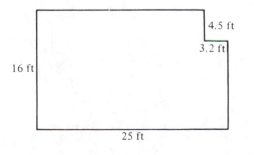

12. Find the area in square feet.

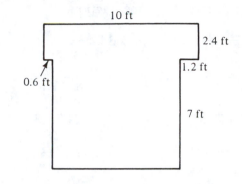

EXERCISES
9.8

In Exercises 1–12 identify each pair of terms as being like or unlike.

1. $3x, 5x$ 2. $4y, -2x$
3. $y, -8y$ 4. $-7m, -9m$
5. $6m, 8n$ 6. $12y, -y$
7. $4x^2, 5x$ 8. $6x, 6$
9. $5x, 6$ 10. $4y^2, -8y^2$
11. aq, rq 12. dt, t

Combine similar terms to simplify the expressions in Exercises 13–32.

13. $5a + a$ 14. $8c - 2c$
15. $4t - 8t$ 16. $x + 4x$
17. $x + x$ 18. $y - y$
19. $-3y + 2y$ 20. $3s - 6s$
21. $4x^2 - 2x^2$ 22. $6y^2 + 9y^2$
23. $x^2 - x^2$ 24. $x^2 + x^2$
25. $4.8x - 2.9x$ 26. $5.7y + 2.04y$
27. $\dfrac{1}{2}t + \dfrac{3}{4}t$ 28. $\dfrac{3}{8}s - \dfrac{3}{16}s$
29. $4q + \dfrac{3}{2}q$ 30. $6r - \dfrac{1}{4}r$
31. $x - \dfrac{3x}{2}$ 32. $\dfrac{3m}{4} + m$

In Exercises 33–42 add or subtract to write each sum or difference as one fraction.

33. $\dfrac{2x}{3} - \dfrac{y}{4}$

34. $\dfrac{5m}{6} + \dfrac{n}{3}$

35. $\dfrac{r}{2} + \dfrac{s}{8}$

36. $\dfrac{a}{8} - \dfrac{b}{16}$

37. $\dfrac{1}{3}q + \dfrac{1}{2}p$

38. $\dfrac{1}{3}c - \dfrac{1}{4}d$

39. $\dfrac{3z}{8} + w$

40. $x + \dfrac{3y}{4}$

41. $a - \dfrac{5c}{8}$

42. $\dfrac{5h}{4} - k$

Use the distributive property to rewrite each expression in Exercises 43–52.

43. $ry + sy$

44. $mx - nx$

45. $3mn - xn$

46. $pt + 2qt$

47. $at + t$

48. $vt - t$

49. $b - rb$

50. $d + ad$

51. $2mx - x$

52. $y - ary$

Combine similar terms in Exercises 53–70.

53. $4m + 2m + 3m$

54. $y + 2y + 9y$

55. $t - t + t$

56. $7b - 8b - 2b$

57. $-3s - 8s + 5s$

58. $3x + 3x + 3x$

59. $3y + 4y + 1$

60. $3y + 2x + 4y + 2x$

61. $8a + 9b + 2b + 4a$

62. $9m - 2m + 6$

63. $3s + 5 + 3s$

64. $5t + 9 + 3t$

65. $5b - (4b + b)$

66. $12p - (8p + 2p)$

67. $2(3x) + 2x$

68. $2y + 2(4y)$

69. $8y + 8(2y)$

70. $16(3x) + 16x$

Find and simplify an expression for each of the following:

71. The fencing needed to enclose a rectangular lot that is twice as long as it is wide.

72. The cost of molding, at $1 per foot, to go around a rectangular room that is three times as long as it is wide.

73. The perimeter of an equilateral triangle if each side has length y feet.

74. The total distance traveled by a car which travels 55 mph for $2t$ hours and then travels 25 mph for t hours.

Find the perimeter of the figures in Exercises 75 and 76.

75.

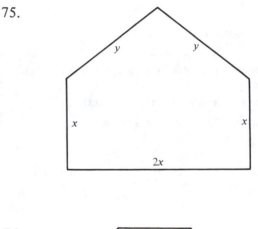

76.

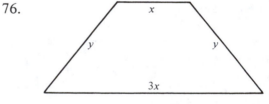

77. A square shaft is milled from a round steel bar. Write and simplify an expression for the diameter of the bar.

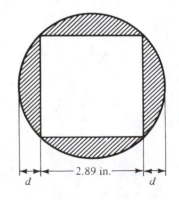

78. Seven small pipes fit snugly into a larger pipe. Ignoring the thickness of the pipes, find and simplify an expression for the circumference of the large pipe. The radius of each small pipe is r inches.

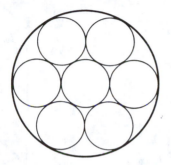

Calculator Exercises

79. Evaluate $4.3q + 7.9q$ if $q = 8.73$.

80. Evaluate $14.7m - 6.34m$ if $m = 15.62$.

81. Evaluate $\frac{3}{7}s + \frac{4}{9}s$ if $s = 5.329$. Give the result to five decimal places.

82. Evaluate $\frac{5}{8}p + \frac{8}{11}p$ if $p = 19.76$. Give the result to five decimal places.

83. Evaluate $0.04d^2n$ if $d = 3.956$ and $n = 12.026$.

84. Evaluate $IA + \frac{P}{I} + IB$ if $I = 14.8$, $A = 7.35$, $P = 108.28$, and $B = 2.009$.

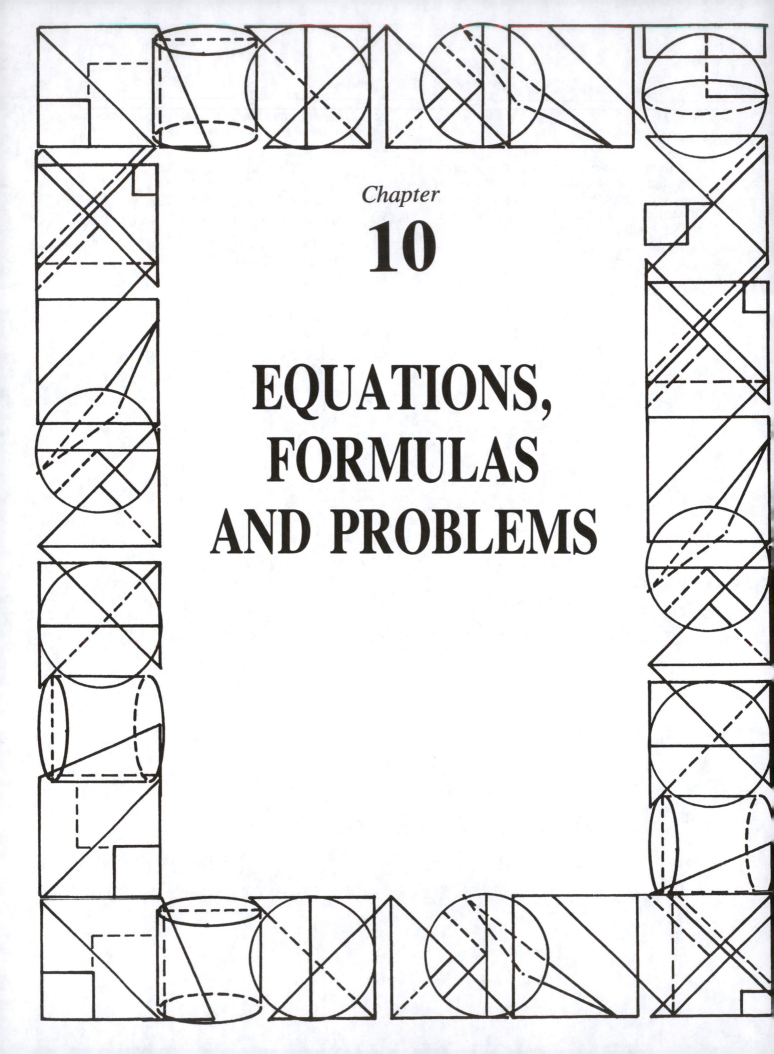

Chapter

10

EQUATIONS, FORMULAS AND PROBLEMS

10.1
USING FORMULAS (I)

Formulas express general facts or rules in terms of symbols. In Chapter 5 we learned several rules written with symbols. For example,

$$P = 2\ell + 2w \text{ (perimeter of a rectangle)},$$

$$A = \ell w \text{ (area of a rectangle), and}$$

$$C = \pi d \text{ (circumference of a circle)}$$

are all formulas. Table 10.1 lists several other formulas, along with the meaning of each symbol and a "translation" of the formula into words.

Each formula in Table 10.1 is an expression of one variable in terms of other variables. For example, $V = \ell wh$ is an expression of V in terms of ℓ, w, and h. The formula $A = \pi r^2$ is an expression for the area of a circle, A, in terms of the radius, r.

To find the value of one variable:

1. **substitute values for the other variables into the formula; and**

2. **perform the indicated operations.**

TABLE 10.1

$d = r \cdot t$	d = distance r = rate t = time	The distance traveled equals the rate of travel times the time.
$V = I \cdot R$	V = voltage I = current R = resistance	The voltage in an electrical circuit equals the current in the circuit times the resistance.
$I = p \cdot r \cdot t$	I = interest p = principal r = rate t = time	Simple interest equals the amount of principal times the rate of interest times the length of time it is invested.
$V = \ell \cdot w \cdot h$	V = volume ℓ = length w = width h = height	The volume of a rectangular solid equals its length times its width times its height.

EXAMPLE 1 Use the formula $d = rt$ to find the distance, d, when $r = 55$ miles per hour and $t = 3$ hours.

Solution

$$d = rt$$

$$d = \left(\frac{55 \text{ mi}}{\text{hr}}\right)(3 \text{ hr}) \longleftarrow \text{Substitute given values for the variable.}$$

$$d = (55)(3) \text{ mi}$$

$$d = 165 \text{ mi}$$

EXAMPLE 2 Use the formula $A = \ell w$ to find the area of a rectangle when $\ell = 6$ yards and $w = 15$ feet.

Solution Since

$$6 \text{ yd} = \left(\frac{3 \text{ ft}}{\text{yd}}\right)(6 \text{ yd})$$

$$= (3)(6) \text{ ft}$$

$$= 18 \text{ ft},$$

then,

$$A = \ell w$$
$$A = (18 \text{ ft})(15 \text{ ft}) \longleftarrow \text{Substitute given values}$$
$$A = (18)(15) \text{ ft}^2 \qquad \text{for the variable.}$$
$$A = 270 \text{ ft}^2$$

EXAMPLE 3 Use the formula $V = IR$ to find V in volts when $I = 6.4$ amperes and $R = 8.7$ ohms.

Solution

$$V = IR$$
$$V = (6.4 \text{ amperes})(8.7 \text{ ohms}) \longleftarrow \text{Substitute}$$
$$V = (6.4)(8.7) \text{ volts} \qquad \text{given values for the}$$
$$V = 55.68 \text{ volts} \qquad \text{variable.}$$

EXAMPLE 4 Find the interest on $3,500 at 7.6% yearly interest for 3½ years.

Solution The formula for simple interest is $I = prt$. To use 7.6% in calculation, we change it to a decimal:

$$7.6\% = \frac{7.6}{100} = 0.076.$$

Then,

$$I = prt$$
$$\text{Substitute.}$$
$$I = \left(3500 \text{ dollars}\right)\left(0.076 \ \frac{1}{yr}\right)(3.5 \ yr) \longleftarrow$$
$$I = (3500)(0.076)(3.5) \text{ dollars}$$
$$I = 931 \text{ dollars.}$$

EXAMPLE 5 Use the formula $V = \frac{1}{3}Bh$ to find the volume of a pyramid when $B = 18$ square centimeters and $h = 2.1$ meters.

Solution Since 2.1 meters = 210 centimeters,

$$V = \frac{1}{3}Bh$$
$$V = \frac{1}{3}(18 \text{ cm}^2)(210 \text{ cm}) \longleftarrow \text{Substitute.}$$
$$V = \frac{1}{3}(18)(210) \text{ cm}^3$$
$$V = 1260 \text{ cm}^3.$$

EXAMPLE 6 Use the formula $P = 4s$ to find the perimeter of a square with a side measuring 2.3 feet. Express the answer in feet and inches to the nearest $\frac{1}{8}$ inch.

Solution

$$P = 4s$$
$$P = 4(2.3 \text{ ft})$$
$$P = 9.2 \text{ ft} \qquad\qquad (9 \text{ ft})$$

Change 0.2 feet to inches:

$$(0.2 \text{ ft})\left(\frac{12 \text{ in.}}{ft}\right) = 2.4 \text{ in.} \qquad (2 \text{ in.})$$

Change 0.4 inches to the nearest $\frac{1}{8}$ inch:

$$(0.4 \text{ in.})\left(\frac{8}{8}\right) = \frac{3.2}{8} \text{ in.} \quad (\tfrac{3}{8} \text{ in. rounded})$$

The perimeter is 9 feet $2\frac{3}{8}$ inches.

EXAMPLE 7 A triangle has a height of 3 feet 5 inches and a base measuring 4 feet 3 inches. Find the area of the triangle. Express the answer in square feet and square inches.

Solution

$$4 \text{ ft } 3 \text{ in.} = 4\frac{3}{12} \text{ or } 4\frac{1}{4} \text{ ft}$$

$$3 \text{ ft } 5 \text{ in.} = 3\frac{5}{12} \text{ ft}$$

Then,

$$A = \frac{1}{2}bh$$

$$A = \frac{1}{2}\left(4\frac{1}{4}\ \text{ft}\right)\left(3\frac{5}{12}\ \text{ft}\right)$$

$$A = \frac{1}{2}\left(\frac{17}{4}\right)\left(\frac{41}{12}\right)\ \text{ft}^2$$

$$A = \frac{697}{96}\ \text{ft}^2$$

$$A = 7\frac{25}{96}\ \text{ft}^2 \qquad\qquad (7\ \text{ft}^2)$$

Change $\frac{25}{96}$ square feet to square inches:

$$\left(\frac{25}{96}\ \text{ft}^2\right)\left(\frac{144\ \text{in.}^2}{1\ \text{ft}^2}\right) = \frac{(25)(144)}{96}\ \text{in.}^2$$

$$= 37.5\ \text{in.}^2 \quad (38\ \text{in.}^2,\ \text{rounded})$$

The area of the triangle is 7 square feet 38 square inches.

EXAMPLE 8 Engine power in horsepower can be found using the formula:

$$H = (0.4)(d^2)(n),$$

where H is the horsepower, d is the diameter of the pistons in inches, and n is the number of cylinders. Find H when $d = 4$ inches and $n = 6$ cylinders.

Solution

$$\begin{aligned}
H &= (0.4)(d^2)(n)\\
&= (0.4)(4^2)(6)\\
&= (0.4)(16)(6)\\
&= 38.4\ \text{horsepower}
\end{aligned}$$

Review Exercises

Simplify the expressions in Exercises 1–10.

1. $1500 + 375$

2. $\frac{1}{2}(14 + 8)$

3. $1500 + (0.15)(8)$

4. $2(14.7) + 2(15)$

5. $32 + \frac{9}{5}(20)$

6. $\dfrac{3(650)(2.4)}{5000}$

7. $\dfrac{8x}{8}$

8. $\dfrac{-10x}{-10}$

9. $\left(\dfrac{x}{6}\right)(6)$

10. $(-8)\left(\dfrac{x}{-8}\right)$

11. When $x = -2$ and $y = 4$, then $xy^2 = $ _____ .

12. When $x = 4$ and $y = -3$, then $xy^2 = $ _____ .

EXERCISES

10.1

Use the formula $d = rt$ to find the distance traveled when:

1. $r = 30$ miles per hour, $t = 3$ hours.
2. $r = 55$ kilometers per hour, $t = 4$ hours.
3. $r = 65$ kilometers per hour, $t = 2\frac{1}{2}$ hours.
4. $r = 40$ miles per hour, $t = 3\frac{1}{2}$ hours.
5. $r = 60$ miles per hour, $t = 1$ hour 20 minutes.
6. $r = 40$ miles per hour, $t = 4$ hours 40 minutes.
7. $r = 15$ kilometers per hour, $t = 45$ minutes.
8. $r = 25$ kilometers per hour, $t = 15$ minutes.

Use the formula $A = \frac{1}{2}bh$ to find the area of a triangle when:

9. $b = 4$ inches, $h = 5$ inches.
10. $b = 6$ inches, $h = 9$ inches.
11. $b = 3$ feet, $h = 16$ inches.
12. $b = 15$ feet, $h = 2$ yards.
13. $b = 3.2$ meters, $h = 62$ centimeters.
14. $b = 30$ centimeters, $h = 0.4$ meters.

Use the formula $P = 4s$ to find the perimeter of each square in Exercises 15–20. Express your answers in feet and inches to the nearest $\frac{1}{8}$ inch.

15. $s = 2$ feet 3 inches.
16. $s = 2$ feet 6 inches.
17. $s = 4.1$ feet.
18. $s = 2.8$ feet.
19. $s = 1$ foot 7 inches.
20. $s = 2$ feet 11 inches.

Use the formulas in this section to answer Exercises 21–34.

21. What is the voltage in a circuit if the resistance is 8 ohms and the current is 6.5 amperes?

22. What is the voltage in a circuit if the resistance is 432 ohms and the current is 0.025 amperes?

23. What is the interest on $5000 at $7\frac{3}{4}\%$ yearly interest for 3 years?

24. What is the interest on $3600 at 6.8% yearly interest for 4 years?

25. Find the distance traveled in 3 hours 45 minutes at 60 miles per hour.

26. Find the distance traveled in 1 hour 20 minutes at 600 miles per hour.

27. What is the area in square meters of a rectangle if its width is 75 cm and its length is 4.8 m?

28. What is the volume in cubic meters of a rectangular solid if its length is 24 m, its width is 3.2 m, and its height is 65 cm?

29. What is the volume in cubic yards of a rectangular solid if its length is 8 feet, its width is 4 feet, and its height is 12 feet?

30. Find the area in square feet of a rectangle if its length is 14 yards and its width is 4 feet 6 inches.

31. Find the area in square feet and square inches of a rectangle if its length is 2 feet 4 inches and its width is 1 foot 7 inches.

32. What is the area in square feet and square inches of a triangle if its height is 3 feet 5 inches and the length of its base is 2 feet 9 inches?

33. Find the area in square yards and square feet of a triangle if its height is 3 yards 1 foot and the length of its base is 2 yards 2 feet.

34. What is the area of a rectangle in square yards and square feet if its length is 5 yards 2 feet and its width is 2 yards 1 foot?

In Exercises 35–42 use the formula and the given information to find the unknown. Use 3.14 for π.

35. $H = RI^2t$ (heat); $R = 14$ ohms, $I = 3$ amperes, $t = 1$ second   (H = heat in joules)

36. $d = kt^2$ (distance); $k = 16$ feet per second^2, $t = 4$ seconds

37. $N = \frac{0.6495}{D}$ (mechanics); $D = 0.036$ inches (N = number of threads per inch)

38. $P = \frac{fs}{t}$ (power); $f = 15$ lb, $s = 4$ ft, $t = 10$ sec

39. $V = \frac{1}{3}\pi r^2 h$ (volume); $r = 2$ cm, $h = 6$ cm

40. $V = \pi r^2 h$ (volume); $r = 4$ inches, $h = 2$ inches

41. $V = e^3$ (volume); $e = 4$ cm

42. $H = 0.04d^2n$ (power); $n = 6$, $d = 3$ inches

43. The formula giving the distance an object falls, d, after t seconds with gravity at g feet per second squared is

$$d = \frac{1}{2}gt^2.$$

Find d when g is 32 feet per second squared and t is 4 seconds.

44. A formula giving the heat H measured in joules in an electric circuit of current I with resistance R is

$$H = RI^2t.$$

Find H when R is 14 ohms, I is 3.4 amperes, and t is 1 second.

45. How many pounds of pressure, p, are exerted on the wings of an airplane traveling 100 miles per hour? (Use the formula $p = 0.0005s^3$, where s is the speed.)

46. The total displacement of an engine, d, is given by the formula

$$d = \pi r^2 sn$$

where

r = radius of the bore

s = stroke

n = number of cylinders.

Find the displacement of an engine with a 3-inch radius of bore, 3-inch stroke, and 6 cylinders. Use 3.14 for π.

47. The speed of a driven pulley, s, is given by the formula

$$s = \frac{DS}{d}$$

where

D = diameter of driving pulley

d = diameter of driven pulley

S = speed of driving pulley.

Find the speed of a driven pulley when the speed of the driving pulley is 900 rpm, the diameter of the driving pulley is 14 inches, and the diameter of the driven pulley is 7 inches.

48. How many miles of travel can a 24-inch automobile tire withstand? (Use the formula $m = 6545d$ where d is the diameter of the tire in inches.)

Calculator Exercises

Use a calculator to find the value of V in Exercises 49–54. Use 3.1416 for π. Round each answer to the nearest hundredth.

49. $V = \frac{4}{3}\pi r^3, r = 5.7$

50. $V = \frac{1}{3}\pi r^2 h, r = 2.6, h = 4.1$

51. $V = \pi r^2 h, r = 5\frac{1}{2}, h = 5\frac{3}{4}$

52. $V = \frac{1}{3}\pi r^2 h, r = 7\frac{1}{4}, h = 6\frac{1}{2}$

53. $V = \frac{4}{3}\pi r^3, r = 6\frac{3}{8}$

54. $V = \pi r^2 h, r = 5\frac{5}{8}, h = 6\frac{1}{2}$

10.2
SOLVING EQUATIONS (I)

An **equation** is a mathematical sentence, usually with at least one variable. For example, the formula $d = rt$ is an equation. To find the rate of speed of a car which traveled 135 miles over $2\frac{1}{2}$ hours, we use this formula:

$$d = rt.$$

Substituting for d and t:

$$135 = r\left(2\frac{1}{2}\right).$$

In this section we discuss how to solve this type of equation.

Solving an equation means finding a value for the unknown (variable) that makes the equation true. To do this we rewrite the equation to get the unknown (variable) alone on one side of the equal sign. Equations like:

$$3x = 4$$
$$7 = 5x$$
$$2.3x = -3.5$$
$$\frac{2}{3} = -2x$$

are equations of the form $ax = b$, where x is the variable and a and b are constants.

> **When you multiply or divide both sides of an equation by the same number, the results are still equal.**

To solve any equation of the form $ax = b$ we multiply or divide *both* sides of the equation by some number. We choose that number so that, after the multiplication or division is performed, the variable will be alone on one side of the equation.

The following chart shows three ways to solve the equation $3x = 6$.

$3x = 6$	$3x = 6$	$3x = 6$
$\dfrac{3x}{3} = \dfrac{6}{3}$	$3x \div 3 = 6 \div 3$	$\dfrac{1}{3}(3x) = \dfrac{1}{3}(6)$
$x = 2$	$x = 2$	$x = 2$
Divide both sides by 3.	Divide both sides by 3.	Multiply both sides by $\frac{1}{3}$.

If both sides are divided by 3, x is alone on one side of the equation, because a non-zero number divided by itself is equal to one. If both sides are multiplied by $\frac{1}{3}$, x is alone on one side of the equation, because a non-zero number multiplied by its reciprocal is equal to one.

We *check* the solution (value for x) by substituting it back into the original equation. If the sentence is true, the solution is correct. If the sentence is false, the solution is incorrect. To check the equation above:

$$3x \overset{?}{=} 6$$
$$3(2) \overset{?}{=} 6$$

$6 = 6$ is true, so the solution checks.

EXAMPLE 1   Solve $\frac{2}{3}x = 10$ for x.

Solution   We multiply both sides by the reciprocal of $\frac{2}{3}$ since a number multiplied by its

reciprocal is one.

$$\frac{2}{3}x = 10$$
$$\frac{3}{2}\left(\frac{2}{3}x\right) = \frac{3}{2}(10)$$
$$x = 15$$

The value for x that makes the equation true is 15. We check by substituting 15 for x in the original equation.

$$\frac{2}{3}x = 10$$
$$\frac{2}{3}(15) \overset{?}{=} 10$$
$$10 = 10 \text{ is true.}$$

EXAMPLE 2   Solve $8 = 0.4x$ for x.

Solution   We divide both sides by 0.4 since a number divided by itself equals one.

$$8 = 0.4x$$
$$\frac{8}{0.4} = \frac{0.4x}{0.4}$$
$$20 = x$$

Check:

$$8 \overset{?}{=} 0.4x$$
$$8 \overset{?}{=} 0.4(20)$$
$$8 = 8 \text{ is true, so the solution is 20.}$$

EXAMPLE 3   Solve $14 = \frac{x}{7}$ for x.

Solution

$$14 = \frac{x}{7}$$
$$7(14) = 7\left(\frac{x}{7}\right) \leftarrow \text{Multiply both sides by 7.}$$
$$7(14) = \frac{7x}{7}$$
$$98 = x$$

The solution of the equation is 98.

Check:

$$14 = \frac{98}{7} \text{ is true.}$$

EXAMPLE 4 Solve $\frac{4x}{5} = \frac{2}{3}$ for x.

Solution

$$\frac{4x}{5} = \frac{2}{3}$$

$$\frac{5}{4}\left(\frac{4x}{5}\right) = \frac{5}{4}\left(\frac{2}{3}\right) \longleftarrow \text{Multiply both sides by } \frac{5}{4}.$$

$$x = \frac{5}{6}$$

Check:

$$\frac{4\left(\frac{5}{6}\right)}{5} \stackrel{?}{=} \frac{2}{3}$$

$$\frac{\frac{10}{3}}{5} \stackrel{?}{=} \frac{2}{3}$$

$$\frac{10}{3}\left(\frac{1}{5}\right) \stackrel{?}{=} \frac{2}{3}$$

$$\frac{2}{3} = \frac{2}{3} \text{ is true.}$$

EXAMPLE 5 Solve $-5x = 18$ for x.

Solution

$$-5x = 18$$

$$\frac{-5x}{-5} = \frac{18}{-5} \longleftarrow \text{Divide both sides by } -5.$$

$$x = -3.6 \text{ or } x = -3\frac{3}{5}$$

EXAMPLE 6 Solve $6x = 0$ for x.

Solution

$$6x = 0$$

$$\frac{6x}{6} = \frac{0}{6} \longleftarrow \text{Divide both sides by 6.}$$

$$x = 0$$

In Examples 1–6 we multiplied or divided both sides of the equation by the *same* number. The number chosen was one that would result in the variable being alone on one side. We divided by the coefficient of x or multiplied by its reciprocal, whichever was easier.

Sometimes variable terms must be combined before solving the equation. Recall from Chapter 9 that *like terms* can be added or subtracted.

EXAMPLE 7

$$14 = 3x + 4x$$

$$14 = 7x \longleftarrow \text{Combining like terms}$$

$$\frac{14}{7} = \frac{7x}{7}$$

$$2 = x$$

EXAMPLE 8

$$8x + x = 6$$

$$8x + 1x = 6$$

$$9x = 6 \longleftarrow \text{Combining like terms}$$

$$\frac{9x}{9} = \frac{6}{9}$$

$$x = \frac{2}{3}$$

Review Exercises

Simplify the expressions in Exercises 1–8.

1. $3(5x)$
2. $6(-8x)$
3. $4x(-5)$
4. $7x(6)$
5. $6x + x - 9x$
6. $y - 3y + 4y$
7. $5m + 3 - 6m$
8. $4r - 8r + 5$
9. If $a = -3$ and $b = 4$, then $a^2 - b = $ _____ .
10. If $a = 2$, $b = 3$, and $c = -5$, then $a(b - c) = $ _____ .
11. Change $40\frac{1}{2}\%$ to a decimal.
12. Change $3\frac{5}{8}\%$ to a decimal.

EXERCISES
10.2

Solve the equations in Exercises 1–50. Check the solutions.

1. $4x = 20$
2. $2x = 16$
3. $6x = 12$
4. $8x = 24$
5. $42 = 7x$
6. $48 = 4x$
7. $2x = 15$
8. $3x = 20$
9. $3x = 0$
10. $5x = 0$
11. $\dfrac{12}{5} = 6x$
12. $\dfrac{14}{9} = 2x$
13. $8x = 50.4$
14. $5x = 14.5$
15. $-2x = 2$
16. $-3x = 3$
17. $-18 = 3x$
18. $-35 = 7x$
19. $-18 = -2x$
20. $-54 = -6x$
21. $\dfrac{3}{4}x = 27$
22. $\dfrac{5}{2}x = 25$
23. $\dfrac{2x}{7} = -3$
24. $\dfrac{9x}{2} = -4$
25. $\dfrac{x}{8} = 14$
26. $\dfrac{x}{3} = 5$
27. $\dfrac{3}{4} = \dfrac{-5x}{2}$
28. $\dfrac{4}{3} = \dfrac{-2x}{5}$

29. $\dfrac{-1}{4}x = -2$
30. $\dfrac{-1}{15}x = -4$
31. $0.8x = 5$
32. $0.4x = 5$
33. $15 = 0.6x$
34. $19 = 0.2x$
35. $2x + 3x = 30$
36. $5x + 3x = 40$
37. $12 = 6x - 3x$
38. $36 = 8x - 2x$
39. $2x - 4x = 8$
40. $8x - 12x = 4$
41. $6 = x + 2x$
42. $18 = 5x + x$
43. $3x - x = 2$
44. $8x - x = 21$
45. $\dfrac{x}{2} + \dfrac{x}{4} = 1$
46. $\dfrac{x}{6} + \dfrac{x}{3} = 1$
47. $4 = \dfrac{2}{3}x + x$
48. $2 = \dfrac{4}{5}x + x$
49. $5x - x + 2x = 48$
50. $4x - 3x + x = 10$

Calculator Exercises

Use a calculator to solve the equations in Exercises 51–58.

51. $168x = 504$
52. $213x = 852$
53. $2.7x = 4.05$
54. $3.9x = 9.36$
55. $2 = -0.08x$
56. $2.1 = -0.06x$
57. $8.5 = 1.25x$
58. $28.5 = 3.75x$

10.3
SOLVING FORMULAS USING NUMBERS (I)

We can now quickly answer the question asked in the last section: *What was the rate of speed of a car which traveled 135 miles over $2\frac{1}{2}$ hours?* Substituting into the formula:

$$d = rt$$
$$135 = r\left(2\frac{1}{2}\right)$$
$$135 = r\left(\frac{5}{2}\right).$$

We multiply both sides by $\frac{2}{5}$:

$$(135)\left(\frac{2}{5}\right) = r\left(\frac{5}{2}\right)\left(\frac{2}{5}\right)$$
$$54 = r.$$

Since $d = rt$, we have:

$$\text{miles} = (r) \text{ hours}$$

or

$$\frac{\text{miles}}{\text{hour}} = \frac{r \text{ hours}}{\text{hours}}$$

so

$$r = \frac{\text{miles}}{\text{hour}}.$$

The answer is 54 miles per hour.

Whenever we know a formula and the values for all but one variable or letter, we can find a value for the unknown by completing the following steps:

> **We find the unknown quantity in a formula by:**
>
> 1. **substituting the known values into the formula,**
> 2. **simplifying if necessary, and**
> 3. **solving the equation for the unknown (variable).**

EXAMPLE 1 Find the width of a rectangle if the length is 24 meters and the area is 324 square meters.

Solution We use the formula for the area of a rectangle.

$$A = \ell w$$
$$324 = \ell(24)$$
$$\frac{324}{24} = \frac{\ell(24)}{24}$$
$$13.5 = \ell$$

For units we have:

$$A = \ell w$$
$$\text{square meters} = \ell(\text{meters})$$
$$(\text{meters})(\text{meters}) = \ell(\text{meters})$$
$$\frac{(\text{meters})(\text{meters})}{\text{meters}} = \frac{\ell(\text{meters})}{\text{meters}}$$
$$\text{meters} = \ell.$$

Therefore, the length is 13.5 meters.

It is also possible to find the numerical answer and the units at the same time. For this example we would have:

$$324 \text{ square meters} = \ell(24 \text{ meters})$$
$$\frac{324 \text{ square meters}}{24 \text{ meters}} = \frac{\ell(24 \text{ meters})}{24 \text{ meters}}$$
$$13.5 \text{ meters} = \ell.$$

EXAMPLE 2 Use $h = \dfrac{V}{b^2}$ to find V if $h = 12$ feet and $b = 3$ feet.

Solution

$$h = \frac{V}{b^2}$$
$$12 \text{ ft} = \frac{V}{(3 \text{ ft})^2}$$
$$12 \text{ ft} = \frac{V}{9 \text{ ft}^2}$$
$$(12 \text{ ft})(9 \text{ ft}^2) = \frac{V}{9 \text{ ft}^2}(9 \text{ ft}^2)$$
$$108 \text{ ft}^3 = V$$

Since we had (feet)(square feet), the answer is 108 *cubic feet*.

EXAMPLE 3 The formula $DS = ds$ relates the speeds and diameters of two pulleys where

D = diameter of a driving pulley

S = angular velocity of a driving pulley

d = diameter of the driven pulley

s = angular velocity of the driven pulley.

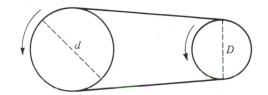

Find D when $S = 300$ rpm, $s = 450$ rpm, and $d = 24$ cm.

Solution

$$DS = ds$$
$$D(300 \text{ rpm}) = (24 \text{ cm})(450 \text{ rpm})$$
$$D(300 \text{ rpm}) = 10800(\text{cm})(\text{rpm})$$
$$\frac{D(300 \text{ rpm})}{300 \text{ rpm}} = \frac{10800(\text{cm})(\text{rpm})}{300 \text{ rpm}}$$
$$D = 36 \text{ cm}$$

Thus, $D = 36$ centimeters.

The next two examples show how to *simplify* an equation after substituting the given values and *before solving* the equation.

EXAMPLE 4   Use $H = \frac{fs}{550t}$, where

H = horsepower

f = force in pounds

s = distance in feet

t = time in seconds,

to find s when $f = 275$ pounds, $t = 3$ seconds, and $H = 10$ horsepower.

Solution

$$H = \frac{fs}{550t}$$

$$10 = \frac{275s}{550(3)}$$

$$10 = \frac{s}{6} \longleftarrow \text{Reducing the fraction}$$

$$10(6) = \frac{s}{6}(6)$$

$$60 = s$$

The distance, s, is 60 feet.

EXAMPLE 5   Use $V = \frac{1}{3}\pi r^2 h$ to find h to the nearest tenth if $r = 3$ and $V = 132$. Use $\frac{22}{7}$ for π.

Solution

$$V = \frac{1}{3}\pi r^2 h$$

$$132 = \left(\frac{1}{3}\right)\left(\frac{22}{7}\right)(3^2)h$$

$$132 = \frac{66}{7}h \longleftarrow \text{Simplifying the right side of the equation}$$

$$\frac{7}{66}(132) = \frac{7}{66}\left(\frac{66}{7}h\right)$$

$$14 = h$$

This last example shows an equation with terms that must be combined before the equation can be solved. Recall that like terms can be combined by adding the coefficients.

EXAMPLE 6   Find p in the formula $A = p + prt$ if $r = 8.5\%$, $t = 4$ years, and $A = \$1608$.

Solution   Note that, before substituting, 8.5% must be changed to a decimal.

$$A = p + prt$$

$$1608 = p + p(0.085)(4) \quad \text{(Since 8.5\% = 0.085)}$$

$$1608 = p(1) + p(0.34) \quad \text{(Since } p = 1p = p(1))$$

$$1608 = p(1 + 0.34)$$

$$1608 = p(1.34)$$

$$\frac{1608}{1.34} = \frac{p(1.34)}{1.34}$$

$$1200 = p$$

Therefore, the principal, p, is \$1200.

Review Exercises

1. $3.04 + (-6.7) = $ _____

2. $-11.5 + 3.46 = $ _____

3. $-3 - 5 - (-13) = $ _____

4. $6 - (-11) - 8 = $ _____

5. 20% of 55 = _____

6. 36% of 400 = _____

7. Six is _____ % of 30.

8. Three is _____ % of 150.

9. When $x = 0.6$, then $3x^2 = $ _____ .

10. When $x = -4.5$ and $y = 3$, then $\frac{2x}{y} = $ _____ .

11. Rewrite $x + bx$ using the distributive property.

12. Rewrite $ary - y$ using the distributive property.

EXERCISES
10.3

Given the formula $d = rt$, where d = distance, r = rate, and t = time, and the known quantities, solve for the unknown in Exercises 1–6.

1. $d = 180$ miles, $r = 60$ miles per hour: Solve for t.

2. $d = 200$ miles, $t = 4$ hours: Solve for r.

3. $d = 595$ kilometers, $r = 85$ kilometers per hour: Solve for t.

4. $d = 370$ kilometers, $t = 5$ hours: Solve for r.

5. $d = 270$ miles, $t = 4\frac{1}{2}$ hours: Solve for r.

6. $d = 240$ kilometers, $r = 45$ kilometers per hour: Solve for t.

Given the formula $V = \ell wh$, where V = volume, ℓ = length, w = width, and h = height, and the known quantities, solve for the unknown in Exercises 7–12.

7. $V = 36$ cubic inches, $\ell = 6$ inches, $w = 3$ inches: Solve for h.

8. $V = 96$ cubic feet, $\ell = 8$ feet, $w = 3$ feet: Solve for h.

9. $V = 72$ cubic meters, $w = 4$ meters, $h = 3$ meters: Solve for ℓ.

10. $V = 48$ cubic centimeters, $w = 2$ centimeters, $h = 8$ centimeters: Solve for ℓ.

11. $V = 105$ cubic centimeters, $\ell = 7$ centimeters, $h = 5$ centimeters: Solve for w.

12. $V = 88$ cubic meters, $\ell = 4$ meters, $h = 11$ meters: Solve for w.

Given the formula $I = prt$, where I = interest, p = principal, r = rate, and t = time, and the known quantities, solve for the unknown in Exercises 13–18.

13. $I = \$240$, $r = 6\%$, $p = \$1000$: Solve for t.

14. $I = \$375$, $t = 2$ years, $p = \$2500$: Solve for r.

15. $I = \$1430$, $t = 10$ years, $r = 5.5\%$: Solve for p.

16. $I = \$168$, $t = 3\frac{1}{2}$ years, $r = 8\%$: Solve for p.

17. $I = \$264$, $t = 5\frac{1}{2}$ years, $p = \$800$: Solve for r.

18. $I = \$2160$, $r = 9\%$, $p = \$3000$: Solve for t.

19. If $V = IR$, solve for I when $V = 54$ volts and $R = 6$ ohms.

20. If $V = IR$, solve for R when $V = 75$ volts and $I = 5$ amperes.

21. If $A = \ell w$, solve for ℓ when $A = 42$ square feet and $w = 12$ feet.

22. If $A = \ell w$, solve for w when $A = 84$ square meters and $\ell = 16$ meters.

23. If $h = \dfrac{V}{b^2}$, solve for V when $h = 4$ feet and $b = 2$ feet.

24. If $h = \dfrac{V}{b^2}$, solve for V when $h = 5$ feet and $b = 6$ feet.

25. If $DS = ds$, solve for S when $D = 9$ inches, $d = 6$ inches, and $s = 210$ rpm.

26. If $DS = ds$, solve for d when $D = 8$ inches, $S = 150$ rpm, and $s = 100$ rpm.

27. If $H = \dfrac{fs}{550t}$, solve for f in pounds when $H = 12$ horsepower, $s = 40$ feet, and $t = 8$ seconds.

28. If $H = \dfrac{fs}{550t}$, solve for s in feet when $H = 20$ horsepower, $f = 15$ pounds, and $t = 3$ seconds.

29. If $A = \frac{1}{2}bh$, solve for b when $A = 38$ square feet and $h = 20$ feet.

30. If $A = \frac{1}{2}bh$, solve for h when $A = 46$ square meters and $b = 4$ meters.

31. If $C = 2\pi r$, solve for r when $C = 88$ inches. (Use $\frac{22}{7}$ for π.)

32. If $C = 2\pi r$, solve for r when $C = 44$ inches. (Use $\frac{22}{7}$ for π.)

33. If $P = \dfrac{fs}{t}$, solve for f when $P = 60$ ft lb/sec, $s = 15$ feet, and $t = 3$ seconds.

34. If $P = \dfrac{fs}{t}$, solve for s when $P = 80$ ft lb/sec, $f = 25$ pounds, and $t = 5$ seconds.

35. If $pN = 1$, solve for p when $N = 16$.

36. If $H = Ri^2$, solve for R when $H = 108$ joules and $i = 3$ amperes.

37. If $T = rt$, solve for r to the nearest hundredth when $T = 35$ and $t = 9$.

38. If $A = p + prt$, solve for p when $A = \$2200$, $t = 2$ years, and $r = 5\%$.

39. If $A = p + prt$, solve for p when $A = \$2360$, $t = 3$ years, and $r = 6\%$.

40. If $V = \pi r^2 h$, solve for h when $r = 7$ feet and $V = 308$ cubic feet. (Use $\frac{22}{7}$ for π.)

41. If $V = \frac{1}{3}\pi r^2 h$, solve for h when $r = 3$ feet and $V = 33$ cubic feet. (Use $\frac{22}{7}$ for π.)

42. If $V = \frac{1}{3}\pi r^2 h$, solve for h to the nearest hundredth when $V = 25$ cubic feet and $r = 3$ feet. (Use 3.14 for π.)

Calculator Exercises

Use a calculator to solve for the unknown in Exercises 43–48. Use 3.1416 for π. Give answers to the nearest thousandth.

43. If $V = \pi r^2 h$, solve for h when $V = 30$ cubic inches and $r = 2$ inches.

44. If $T^2 = 4\pi^2 \dfrac{L}{g}$, solve for L when $g = 32$ and $T = 5$.

45. If $H = 0.4d^2 n$, solve for n when $d = 3.5$ and $H = 39.2$.

46. If $m = 6545d$, solve for d when $m = 117{,}810$.

47. If $Q = \frac{1}{2}RI^2$, solve for R when $Q = 50$ and $I = 2.3$.

48. If $P = \frac{1}{2}sh^2$, solve for s when $h = 5.7$ and $P = 65$.

10.4
SOLVING FORMULAS USING LETTERS (I)

In all of the formulas we have seen so far, there was only one unknown (variable). We knew the values for all but one variable, and we solved the formula for that one unknown. Many times you will need to solve a formula for a certain letter without knowing the values of the other letters. In this section we look at how to do this.

The procedures for solving a formula for a particular letter (variable) are the same as the procedures for solving a formula or equation for an unknown number, except that we do not substitute numbers for the other variables in the formula.

To solve an equation you may:

1. **multiply both sides by the same number (or letter), or**

2. **divide both sides by the same number (or letter) except zero.**

EXAMPLE 1   Solve $d = rt$ for t.

Solution

$$d = rt$$
$$\frac{d}{r} = \frac{rt}{r} \quad \longleftarrow \substack{\text{Divide both sides}\\ \text{by } r.}$$
$$\frac{d}{r} = t$$

Notice that, in Example 1 and in the following examples, we treat letters just like numbers. It just happens that we don't know exactly what numbers the letters represent.

EXAMPLE 2   Solve $DS = ds$ for D.

Solution

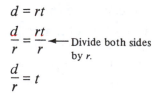

$$DS = ds$$
$$\frac{DS}{S} = \frac{ds}{S} \quad \longleftarrow \text{Divide both sides by } S.$$
$$D = \frac{ds}{S}$$

We can divide by more than one letter or number, as the next example shows.

EXAMPLE 3   Solve $V = \pi r^2 h$ for h.

Solution

$$V = \pi r^2 h$$

$$\frac{V}{\pi r^2} = \frac{\pi r^2 h}{\pi r^2} \quad \longleftarrow \text{ Divide both sides by } \pi r^2.$$

$$\frac{V}{\pi r^2} = h$$

In the next example the formula $A = \frac{1}{2}bh$ is solved for b in two steps. The first step is to multiply both sides by 2. The second step is to divide both sides by h.

EXAMPLE 4   Solve $A = \frac{1}{2}bh$ for b.

Solution

$$A = \frac{1}{2}bh$$

$$2A = 2\left(\frac{1}{2}bh\right) \quad \longleftarrow \text{ Multiply both sides by 2.}$$

$$\frac{2A}{h} = \frac{bh}{h} \quad \longleftarrow \text{ Divide both sides by } h.$$

$$\frac{2A}{h} = b$$

It is possible to solve a formula like the one in Example 4 in one step, as the next example shows.

EXAMPLE 5   Solve $P = \frac{fs}{t}$ for s.

Solution

$$P = \frac{fs}{t}$$

$$\left(\frac{t}{f}\right)P = \left(\frac{t}{f}\right)\frac{fs}{t} \quad \longleftarrow \text{ Multiply both sides by } \frac{t}{f}.$$

$$\frac{tP}{f} = \left(\frac{tf}{ft}\right)s$$

$$\frac{tP}{f} = s$$

The formula $P = \frac{fs}{t}$ can also be solved in two steps, as was the formula $A = \frac{1}{2}bh$ in Example 4.

$$P = \frac{fs}{t}$$

$$tP = t\left(\frac{fs}{t}\right) \quad \longleftarrow \text{ Multiply both sides by } t.$$

$$tP = fs$$

$$\frac{tP}{f} = \frac{fs}{f} \quad \longleftarrow \text{ Divide both sides by } f.$$

$$\frac{tP}{f} = s$$

It does not matter how you choose to solve the formula for s. The first way is shorter, but the second way may be easier for you.

EXAMPLE 6   Solve $T^2 = 4\pi^2\dfrac{L}{g}$ for L.

Solution   The formula $T^2 = 4\pi^2\dfrac{L}{g}$ can be written as:

$$T^2 = \frac{4\pi^2 L}{g}.$$

$$gT^2 = g\left(\frac{4\pi^2 L}{g}\right) \quad \longleftarrow \text{ Multiply both sides by } g.$$

$$gT^2 = 4\pi^2 L$$

$$\frac{gT^2}{4\pi^2} = \frac{4\pi^2 L}{4\pi^2} \quad \longleftarrow \text{ Divide both sides by } 4\pi^2.$$

$$\frac{gT^2}{4\pi^2} = L.$$

This formula can also be solved for L as follows:

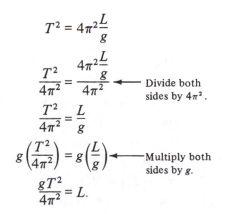

$$T^2 = 4\pi^2\frac{L}{g}$$

$$\frac{T^2}{4\pi^2} = \frac{4\pi^2\frac{L}{g}}{4\pi^2} \quad \longleftarrow \text{ Divide both sides by } 4\pi^2.$$

$$\frac{T^2}{4\pi^2} = \frac{L}{g}$$

$$g\left(\frac{T^2}{4\pi^2}\right) = g\left(\frac{L}{g}\right) \quad \longleftarrow \text{ Multiply both sides by } g.$$

$$\frac{gT^2}{4\pi^2} = L.$$

Review Exercises

1. $9 \text{ ft}^2 = \underline{\hspace{1.5cm}} \text{ yd}^2$
2. $2 \text{ yd}^2 = \underline{\hspace{1.5cm}} \text{ ft}^2$
3. $4.8 + (-20) + 0.06 = \underline{\hspace{1.5cm}}$
4. $-2.7 + 15 + (-4.3) = \underline{\hspace{1.5cm}}$
5. $(-31.8) \div (-12) = \underline{\hspace{1.5cm}}$
6. $-46 \div 2.5 = \underline{\hspace{1.5cm}}$
7. $200\% \text{ of } 7 = \underline{\hspace{1.5cm}}$
8. $120\% \text{ of } 5 = \underline{\hspace{1.5cm}}$
9. $0.4\% \text{ of } 30 = \underline{\hspace{1.5cm}}$
10. $0.75\% \text{ of } 40 = \underline{\hspace{1.5cm}}$
11. 54 is $\underline{\hspace{1.5cm}} \%$ of 27
12. 21 is $\underline{\hspace{1.5cm}} \%$ of 14

EXERCISES

10.4

Solve the equations in Exercises 1–12 for the indicated letter.

1. $ab = c$ for b
2. $d = ef$ for f
3. $g = \dfrac{h}{i}$ for h
4. $\dfrac{j}{k} = m$ for j
5. $\dfrac{f}{g}x = b$ for x
6. $c = \dfrac{m}{n}x$ for x
7. $p = \dfrac{3y}{q}$ for y
8. $\dfrac{5y}{r} = s$ for y
9. $3x = \dfrac{a}{b}$ for x
10. $\dfrac{p}{4} = rx$ for x
11. $\dfrac{ym}{b} = \dfrac{c}{d}$ for y
12. $\dfrac{e}{f} = \dfrac{yd}{b}$ for y

Solve the formulas in Exercises 13–40 for the indicated letter.

13. $A = \ell w$ for w (area)
14. $d = rt$ for r (distance)
15. $W = fs$ for s (work)
16. $V = IR$ for I (electricity)
17. $I = prt$ for r (interest)
18. $V = \ell wh$ for w (volume)
19. $E = mc^2$ for m (energy)
20. $P = I^2R$ for R (electricity)
21. $H = RI^2$ for R (heat)
22. $d = kt^2$ for k (distance)
23. $f = \dfrac{v}{\ell}$ for v (frequency)
24. $P = \dfrac{W}{t}$ for W (power)
25. $DS = ds$ for D (pulleys)
26. $PV = RT$ for V (chemistry)
27. $A = \tfrac{1}{2}bh$ for h (area)
28. $V = \tfrac{1}{3}Bh$ for h (volume)
29. $V = \tfrac{4}{3}\pi r^3$ for π (volume)
30. $V = \tfrac{1}{3}\pi r^2 h$ for h (volume)
31. $d = \tfrac{1}{2}gt^2$ for g (distance)
32. $H = \tfrac{1}{2}RI^2$ for R (heat)
33. $E = \dfrac{mv^2}{2}$ for m (energy)
34. $F = \dfrac{wa}{g}$ for w (physics: force)
35. $d = \pi r^2 sn$ for n (engines)
36. $H = 0.04d^2 n$ for n (engines)
37. $H = \dfrac{62.4fd}{550}$ for d (power)
38. $H = \dfrac{ND^2}{2.5}$ for N (power)
39. $A = \dfrac{M}{fjd}$ for M (mechanical design)
40. $v^2 = \dfrac{2GM}{r}$ for G (physics)

Calculator Exercises

In Exercises 41–46 use 3.1416 for π. Give the answers to four decimal places.

41. If $DS = ds$, solve for S when $D = 18.72$, $d = 7.83$, and $s = 148$.
42. If $A = \tfrac{1}{2}bh$, solve for h when $A = 308.254$ and $b = 15.37$.

43. If $V = \pi r^2 h$, solve for h when $r = 9.65$ and $V = 1847.93$.

44. If $d = \pi r^2 sn$, solve for n when $r = 3.5$, $s = 105$, and $d = 8000$.

45. If $A = p + prt$, solve for p when $r = 0.098$, $t = 2.25$, and $A = 1933.27$.

46. If $H = \dfrac{fs}{550t}$, solve for f when $s = 405.4$, $t = 3.8$, and $H = 360$.

10.5
PERCENT APPLICATIONS

Finding a given **percent** of a number means multiplying the number by the decimal value of the given percent (Chapter 3). The number is often called the **base**. The percent is often called the **rate**. The percent of a number is referred to as the **percentage**. The *percentage* is a percent (*rate*) of the *base*. This relationship can be written as the formula:

Percentage = (Rate)(Base)
or   **P = RB**
where R is expressed as a decimal.

Note in the formula that "is" becomes "equals," and "of" becomes "times."

In a formula containing three variables it is possible to find any one of the three variables when the other two are given. This is illustrated in the following examples.

EXAMPLE 1   What number is 35% of 240?

Solution

$$R = 35\%$$
(The rate is the percent.)

$$B = 240$$
(The base follows word "of.")

Substituting into the formula:

$$P = RB$$
$$P = (35\%)(240)$$
$$P = (0.35)(240) \leftarrow \text{Change the percent to a decimal.}$$

The equation is then solved by doing the computation:

$$P = 84.$$

The number is 84. (*Note:* When computing, all percents must be changed to decimals, as we learned in Chapter 3.)

EXAMPLE 2   What percent of 250 is 2?

Solution Since the percent is the rate, R is the unknown.

$$B = 250 \quad \text{(The base follows "of.")}$$
$$P = 2$$

Substituting into the formula:

$$P = RB$$
$$2 = R(250)$$
$$\frac{2}{250} = \frac{R(250)}{250}$$
$$0.008 = R.$$

Changing 0.008 to a percent, the answer becomes 0.8%.

EXAMPLE 3   One hundred thirty percent of what number is 97.5?

Solution Since the base follows "of," B is the unknown.

$$R = 130\%$$
$$P = 97.5$$

Substituting into the formula:

$$P = RB$$
$$97.5 = (130\%)B$$
$$97.5 = (1.3)B \longleftarrow \text{Change the percent}$$
$$\frac{97.5}{1.3} = \frac{1.3B}{1.3} \qquad \text{to a decimal.}$$
$$75 = B.$$

The number is 75.

Other examples using the formula $P = RB$ follow. In each case it will be important to identify which two of P, B, and R are given and which one is the unknown.

EXAMPLE 4 An engine lathe is billed at $1,960 less 2% discount for payment within 30 days. How much is saved by paying within the 30 days?

Solution The unknown is the amount saved. The amount saved is 2% of $1,960, so the percentage P is unknown. (Percentage is the percent of a number.) Then

$$B = \$1960$$
$$R = 2\%.$$

The formula $P = RB$
becomes $P = (2\%)(1960)$
$$P = (0.02)(1960)$$
$$P = 39.20.$$

Therefore, $39.20 is saved by paying within 30 days.

EXAMPLE 5 In the manufacture of hexagonal steel nuts and bolts, 84 were found to be defective one Tuesday. It is known that 1.5% of those produced on any given day are defective. How many nuts and bolts were made on Tuesday?

Solution The problem states "1.5% of those produced . . ." Therefore, the unknown is the base, since the base follows the word "of."

$$R = 1.5\%$$
$$P = 84$$

The formula $P = RB$
becomes $84 = (1.5\%)B$
$$84 = 0.015B$$
$$\frac{84}{0.015} = \frac{0.015B}{0.015}$$
$$5600 = B.$$

Therefore, 5600 nuts and bolts were made on Tuesday.

The efficiency of a machine may be expressed as the quotient "output divided by input." The quotient is usually written as a percent and is thus called the percent of efficiency for the machine. The equation $P = RB$ is sometimes written as the formula

$$o = ei$$

for this purpose, where

$$o = \text{output}$$
$$e = \text{efficiency}$$
$$i = \text{input}.$$

EXAMPLE 6 An electric motor has an output of 96 watts for an input of 108 watts. Find the percent efficiency of the motor to the nearest tenth of a percent.

Solution We use the percent formula

$$o = ei.$$

Substituting, $96 = e(108).$
Solving, $\dfrac{96}{108} = \dfrac{e(108)}{108}$
$$0.889 = e \quad \text{(rounded)}.$$

Changing 0.889 to a percent, the percent efficiency is 88.9%.

"Adding on" a percent of a number to the number involves finding the given percent of the number and adding it to the number (Chapter 3). This can be represented by

$$B + RB = A$$

where B = the base (original number), R = the rate (percent of "add on"), and A = the total (original number plus the amount of "add on").

EXAMPLE 7 Fourteen thousand square feet of floorspace is to be increased by 30%. What is the new amount of floorspace?

Solution This is an "add on" problem with the base and rate given. The formula we use is $B + RB = A$.

$$B + RB = A$$
$$14{,}000 + (30\%)(14{,}000) = A$$
$$14{,}000 + (0.3)(14{,}000) = A$$
$$14{,}000 + 4200 = A$$
$$18{,}200 = A$$

The new floorspace is 18,200 square feet.

The formula $B + RB = A$ can be modified for discount and decrease problems. In such problems the given percent of the number is *subtracted* from the number (Chapter 3). Thus, the formula is,

$$B - RB = A.$$

EXAMPLE 8 An order of drawing instruments cost $72 after a trade discount of 20%. What was the list price?

Solution We know the rate, R, and the amount after discount, A. We are looking for the list price, or original price, B. Using the formula $B - RB = A$ and substituting,

$$B - (20\%)B = 72$$
$$B - (0.2)B = 72$$
$$1B - 0.2B = 72$$
$$0.8B = 72$$
$$\frac{0.8B}{0.8} = \frac{72}{0.8}$$
$$B = 90.$$

The list price was $90.

Review Exercises

1. $54{,}300 \text{ cm}^2 = \underline{\hspace{1cm}} \text{ m}^2$
2. $4.1 \text{ m}^2 = \underline{\hspace{1cm}} \text{ cm}^2$
3. $(0.2)(-5.6)(-2.5) = \underline{\hspace{1cm}}$
4. $(-3.1)(42)(-0.06) = \underline{\hspace{1cm}}$
5. $(-4.2)^2 = \underline{\hspace{1cm}}$
6. $(-0.23)^2 = \underline{\hspace{1cm}}$
7. $(-6.49)(1000) = \underline{\hspace{1cm}}$
8. $(100)(-42.8) = \underline{\hspace{1cm}}$
9. $(-429) \div (10) = \underline{\hspace{1cm}}$
10. $(-53.4) \div (1000) = \underline{\hspace{1cm}}$
11. $\sqrt{121} = \underline{\hspace{1cm}}$
12. $\sqrt{144} = \underline{\hspace{1cm}}$

EXERCISES
10.5

In Exercises 1–18, identify the two given numbers as the base, rate, or percentage. Then answer the question.

1. What number is 12% of 80?
2. What number is 82% of 75?
3. Two hundred fifty-eight is 86% of what number?
4. Eleven is 22% of what number?
5. Twenty-six is what percent of 40?
6. Fifty-three is what percent of 265?
7. What number is 125% of 15?
8. What number is 250% of 38?

9. Six is 150% of what number?

10. Twenty-two is 200% of what number?

11. What percent of 500 is 1325?

12. What percent of 17 is 51?

13. What number is 0.2% of 75?

14. What number is $\frac{3}{4}$% of 400?

15. Three is $\frac{1}{4}$% of what number?

16. Eighteen is 0.9% of what number?

17. Fifteen is what percent of 3000?

18. Seven is what percent of 7000?

Use the increase (add-on) and decrease formulas for Exercises 19–24.

19. Seventy is 40% more than what number?

20. One hundred twenty is 40% less than what number?

21. One hundred sixty is 60% less than what number?

22. Fifty-four is 60% more than what number?

23. Twenty-eight is 250% more than what number?

24. Five hundred three is 0.6% more than what number?

25. A metal alloy contains 8% tin. How many pounds of tin are needed to make 675 pounds of the alloy?

26. A copper sheet containing 300 square centimeters is used for stamping in a punch press. After the sheet passes through the press, 8% of the sheet remains as scrap. How many square centimeters are unused?

27. John puts $11 of his $220 weekly paycheck into a long-term savings program. What percent of his income is being saved?

28. Of 4600 machine screws produced in one day, 279 were rejected for not conforming to specifications. What percent of the total output was rejected?

29. A contractor hired five new men, increasing his work crew by 25%. How many people were on the crew before the hiring?

30. It is known that 2% of the resistors made by a company did not pass inspection. If 35 did not pass inspection, what was the total number made?

31. The efficiency of a generator is 85%. If the input is 78,000 watts, what is its output?

32. Find the output of a motor with 87% efficiency if the input is 25 watts.

33. Find the efficiency of a generator if the input is 30,000 watts and the output is 24,000 watts.

34. The input to an electrical motor is 150 watts. The actual output is 126 watts. What is the percent efficiency of the motor?

35. An electric motor operates at 17.6 watts, which is 88% of the input. What is the input in watts?

36. Find the input of a generator with 75% efficiency whose output is 7500 watts.

37. A new car cost $4500. The first year it depreciated about 23%. What was its value after one year?

38. A set of drawing instruments lists for $85. What is the purchase price with a 20% trade discount?

39. A company is billed $1284 for wrought iron pipe. This was 7% over the original estimate. What was the original estimate?

40. During a power shortage the speed of a motor decreased to 3800 rpm. This was 5% less than its speed at full power. What was the full power speed of the motor?

41. A company producing 540 refrigeration units per month increases production by 15%. How many units per month are now produced?

42. A new generator must increase power by 20%. The old generator operates at 120 watts. What must the new generator operate at?

43. A certain alloy shrinks 2% during cooling. If the cast length of a project is 3.4 centimeters, what is its cooled length?

44. The shrinkage rate of a metal is 3%. How long must a metal rod be made so that it is 12 centimeters long after cooling (to the nearest hundredth)?

45. Employee A earns $690 a month, which is 15% more than employee B earns. What is employee B's monthly salary?

46. If an agent's 6% commission for selling amounted to $34.80, what was the total amount of sales?

47. Bill wishes to buy a new house. The bank is willing to lend him 85% of the purchase price. Bill has the $5175 necessary for the down payment. What is the price of the house?

48. Find the output in watts of a motor with 87% efficiency. Its input is 237.5 watts.

Calculator Exercises

Use a calculator to answer Exercises 49–54.

49. How many pounds of nickel are in 3655 pounds of an alloy containing 57.8% nickel?

50. Find the percent efficiency of a motor with 5560-watt output and 7440-watt input. Express your answer to the nearest hundredth of a percent.

51. A man earning $13,575 per year has $3962 deducted for taxes and insurance. What percent, to the nearest tenth of a percent, of his earnings were deducted?

52. A board measures 92.75 centimeters after shrinking $12\frac{3}{4}\%$ while drying. What was the original length of the board to the nearest hundredth of a centimeter?

53. What is the new hourly wage of a worker presently receiving $7.85 per hour after a 5.5% increase?

54. A new machine costing $23,440 will be purchased. A trade discount of 1.75% will be given. What will be the actual cost of the machine?

10.6
WRITTEN PROBLEMS (I)

As we have learned, many problems in the technical area can be solved by using given formulas. Other problems must be solved without being *given* any formulas. In problems where no formula is given, the data must first be translated into an equation or formula. In this section we will look at several such problems and learn to solve them when no formula is given.

EXAMPLE 1 Seven times a certain number is 42. What is the number?

Solution Let x be the number. The equation is then:

$$7x = 42.$$

Next we solve the equation:

$$7x = 42$$

$$\frac{7x}{7} = \frac{42}{7}$$

$$x = 6.$$

Therefore, the certain number is 6. To check, $7(6) = 42$ is true.

EXAMPLE 2 Find the length of a side of a square if the perimeter is 66 feet.

Solution The formula for the perimeter of a square is:

$$P = 4s.$$

Since the perimeter is given as 66 feet, we substitute 66 for P:

$$66 = 4s.$$

Then we solve the equation for s:

$$66 = 4s$$
$$\frac{66}{4} = \frac{4s}{4}$$
$$16.5 = s.$$

Thus, the side of the square is 16.5 feet, or 16 feet 6 inches.

EXAMPLE 3 A board 60 inches long is cut so that the length of the longer piece is three times the length of the shorter piece. How long is the shorter piece of board?

Solution Let x be the length in inches of the shorter piece of board. Then, $3x$ is the length in inches of the longer piece of board. The sum of the two lengths, $(3x + x)$, is the length of the entire board, 60 inches, so the equation is:

$$3x + x = 60.$$

Then
$$3x + 1x = 60$$
$$4x = 60$$
$$\frac{4x}{4} = \frac{60}{4}$$
$$x = 15.$$

Therefore, the shorter piece of board is 15 inches long.

Sometimes we already know a formula that can help us solve a problem, as in Example 2. Sometimes we need to make up an equation or formula, as in Example 3. In the following two examples note that it is important to express with variables what you are looking for.

EXAMPLE 4 The length of a rectangle is twice its width. Its perimeter is 74 feet. Find the dimensions of the rectangle.

Solution Let x be the width in feet of the rectangle. Then $2x$ is the length in feet. The equation is:

$$P = 2l + 2w$$
$$74 = 2(2x) + 2x$$
$$74 = 4x + 2x$$
$$74 = 6x$$
$$\frac{74}{6} = \frac{6x}{6}$$
$$12\frac{1}{3} = x.$$

The width of the rectangle, x, is $12\frac{1}{3}$ feet, or 12 feet 4 inches. The length is $2x$, or $2(12\frac{1}{3})$, which is $24\frac{2}{3}$ feet, or 24 feet 8 inches.

Use the following steps in solving written problems:

Step 1:	Read the problem.
Step 2:	Express the unknowns with variables.
Step 3:	Make a chart or diagram if helpful.
Step 4:	Write an equation or find the right formula.
Step 5:	Solve the equation or formula.
Step 6:	Answer the question.

The following example illustrates this six-step process. *Step 1* is to *read the problem.*

EXAMPLE 5 Two airplanes leave terminals located 1900 miles apart and travel toward one another. The average speeds of the two airplanes are 600 miles per hour and 350 miles per hour. At what time will the two planes pass each other?

Solution

Step 2: Let t be the amount of time it takes for the planes to meet. Then, using the formula $d = rt$, we can set up the following chart.

Step 3:

	Time	Rate	Distance
Plane *A*	*t*	600	600*t*
Plane *B*	*t*	350	350*t*

A sketch also helps in writing the equation.

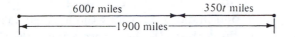

Step 4: Since the sum of the distances is 1900 miles, the equation is:

$$600t + 350t = 1900.$$

Step 5:

$$950t = 1900$$

$$\frac{950t}{950} = \frac{1900}{950}$$

$$t = 2$$

Step 6: Therefore, it takes the planes 2 hours to meet or pass each other.

Review Exercises

1. If $x = -2$, $y = -3$, and $z = -4$, then $xyz =$ _____ .

2. If $x = -2$, $y = -3$, and $z = -4$, then $x - y - z$ = _____ .

3. Rewrite $ax + 3x$ using the distributive property.

4. Rewrite $-2y + py$ using the distributive property.

5. Change 0.2 square feet to square inches.

6. Change 0.8 square feet to square inches.

7. What is the area of a triangle if its base is 14 centimeters and its altitude is 15 centimeters?

8. What is the area of a triangle if its base is 19 inches and its altitude is 2 feet?

9. What is the area of a circle if its radius is 4.3 inches?

10. What is the area of a circle if its diameter is 2.8 feet?

11. Find d using $d = kt^2$ if $k = 16$ ft/sec² and $t = 3$ sec.

12. Find d using $d = kt^2$ if $k = 16$ ft/sec² and $t = 2$ sec.

EXERCISES
10.6

Write the necessary equation and find the number in each of Exercises 1–12.

1. Twice a number equals 124.

2. Three times a number equals 168.

3. A number times four is thirty-six.

4. Five times a number is seventy-five.

5. A number divided by four is eleven.

6. A number divided by six is thirteen.

7. One-fourth of a number equals 16.

8. One-half of a number equals 13.

9. Two-fifths of a number is 32.

10. Four-thirds of a number is 16.

11. Four times a number divided by 5 is 20.

12. Twice a number divided by 3 is 18.

13. After an increase in salary, Joe is earning twice as much as Jerome. If Joe's salary is $17,000, find Jerome's salary.

14. A crew was laying a pipeline between two cities. The engineer's report showed that during the first week they laid 148 miles of pipe, or two-thirds of the distance between the cities. How far apart are the cities?

15. During one year a company showed a sales increase of three-eighths over the previous year. The amount of increase was $45,000. What were the sales during the previous year?

16. A 15-foot board is cut into two pieces, one of which is twice as long as the other. What are the lengths of the two pieces?

17. One engine has twice as much horsepower as a second engine. Together they have 24 horsepower. What is the power of each?

18. One electronic data card sorter operates three times as fast as another. Together the two machines sort 2040 cards per minute. What is the sorting rate of each machine?

19. Oil tank B's capacity is twice that of tank A. Tank C has a capacity three times that of tank A. The three tanks together have a capacity of 3600 gallons. What is the capacity of each tank?

20. Together, two machines cost $8400 per month to rent. If one costs three times as much as the other, what is the monthly cost of each?

21. The electric current in one transistor is twice that in another transistor. If the sum of the currents is 0.018 amperes, what is the current in each?

22. Two furnaces require a total of 5600 watts for their operation. The first furnace needs three-fifths the number of watts that the second furnace needs. How many watts does each furnace require?

23. Find the length of a side of a square with perimeter 15 feet. Express your answer in feet and inches.

24. Find the length of a side of a square with perimeter 29 feet. Express your answer in feet and inches.

25. The length of a rectangle is twice its width. The perimeter is 168 meters. What are the dimensions of the rectangle?

26. The length of a rectangle is three times its width. The perimeter is 120 centimeters. What are the dimensions of the rectangle?

27. A traveler drove 9 hours to go from city A to city B. He then drove 6 hours from city B to city C. The total distance traveled was 795 miles. What was his average speed?

28. Two cars, 630 miles apart, start toward each other. One travels at 55 miles per hour, and the other at 50 miles per hour. When will they meet?

29. Two cars, originally 735 kilometers apart, start at the same time and travel toward each other. One travels at 60 kilometers per hour and the other at 45 kilometers per hour. When will they meet?

30. At 2:00 P.M., two airplanes leave terminals located 2500 miles apart and begin traveling toward each other. The average speed of one plane is $1\frac{1}{2}$ times the average speed of the other. If they pass each other at 6:00 P.M., what is the average speed of each plane?

10.7
USING FORMULAS (II)

Most of the formulas used so far in this chapter involve only multiplication and/or division. Often the formulas you will work with involve *several* of the operations of addition, subtraction, multiplication, and division. Table 10.2 contains some examples of such formulas, along with the meaning of each symbol and a "translation" of the formula into words.

TABLE 10.2

Formula	Symbols	Translation
$P = n(p - c)$	P = total profit n = number of items sold p = selling price c = cost	The total profit is the number of items sold times the difference between the selling price and the cost.
$C = \dfrac{r_1 r_2}{r_1 + r_2}$	C = combined resistance r_1 = resistance of one resistor r_2 = resistance of other resistor	The combined resistance of two electrical resistors in a parallel circuit is the quotient of their product and their sum.
$E = \dfrac{I - P}{I}$	E = engine efficiency I = heat input P = heat output	Engine efficiency is the difference of heat input and heat output divided by heat input.
$F = \dfrac{9}{5}C + 32$	F = Fahrenheit temperature C = Celsius temperature	The Fahrenheit temperature is $\frac{9}{5}$ times the Celsius temperature plus 32.
$v_2 = v_1 + at$	v_2 = terminal velocity v_1 = initial velocity a = acceleration t = time	The terminal velocity is the initial velocity added to the product of the acceleration and the time.

A capital, or upper-case, letter and a small, or lower-case, letter can have different meanings in a formula. This can be seen in the formula $P = n(p - c)$, where P equals total profit and p equals selling price.

Letters with different **subscripts** also have different meanings in a formula. They are considered different variables or letters. The formula

$$C = \frac{r_1 r_2}{r_1 + r_2}$$

uses subscripts. A subscript is the small symbol at the lower right-hand corner of a variable. For instance, in the symbols r_1 and r_2, 1 and 2 are the subscripts. The symbol r_1 is read "r sub one," and the symbol r_2 is read "r sub two."

We now look at finding the value of a variable in a formula. It will be important to keep in mind the order of operations.

EXAMPLE 1   Use the formula $v_2 = v_1 + at$ to find v_2 if $v_1 = 30$ feet/second, $a = 4$ feet/second^2, and $t = 6$ seconds.

Solution

$$v_2 = v_1 + at$$

$$v_2 = 30\,\frac{\text{ft}}{\text{sec}} + \left(\frac{4\ \text{ft}}{\text{sec}^2}\right)(6\ \text{sec}) \quad \text{Substitute given values for the variable.}$$

$$v_2 = 30\,\frac{\text{ft}}{\text{sec}} + (4)(6)\left(\frac{\text{ft}}{\text{sec}\ \text{sec}}\right)(\text{sec})$$

$$v_2 = 30\,\frac{\text{ft}}{\text{sec}} + 24\,\frac{\text{ft}}{\text{sec}}$$

$$v_2 = 54\,\frac{\text{ft}}{\text{sec}}$$

The terminal velocity, therefore, is 54 feet/second.

We find the value of a formula (value of a variable alone on one side of the equation) by:
1. substituting values for the other variables into the formula, and
2. performing the indicated operations.

EXAMPLE 2 Use the formula $E = \dfrac{I - P}{I}$ to find the engine efficiency to the nearest hundredth if the heat input is 18,400 calories and the heat output is 5,700 calories.

Solution

$$E = \frac{I - P}{I}$$

$$E = \frac{18{,}400 \text{ calories} - 5700 \text{ calories}}{18{,}400 \text{ calories}} \leftarrow \text{Substitute given values for the variables.}$$

$$E = \frac{12{,}700 \text{ calories}}{18{,}400 \text{ calories}}$$

$$E = 0.69 \qquad \text{(to nearest hundredth)}$$

The engine efficiency to the nearest hundredth is 0.69.

(*Note:* Recall here that the large fraction bar acts as parentheses, so the top of the fraction is computed before the division is done.)

EXAMPLE 3 Use $F = \frac{9}{5}C + 32$ to find the Fahrenheit temperature if the Celsius temperature is 25°.

Solution

$$F = \frac{9}{5}C + 32$$

$$F = \frac{9}{5}(25) + 32 \leftarrow \text{Substitute a given value for the variable.}$$

$$F = 45 + 32$$

$$F = 77$$

The Fahrenheit temperature is 77°.

(*Note:* Examples 1 and 3 both illustrate the fact that, in the order of operations, multiplication is done before addition.)

EXAMPLE 4 If $\ell = 4.9$ feet and $w = 5.3$ feet, use the formula $P = 2\ell + 2w$ to find the perimeter of a rectangle to the nearest $\frac{1}{16}$ inch.

Solution

$$P = 2\ell + 2w$$

$$P = 2(4.9 \text{ ft}) + 2(5.3 \text{ ft})$$

$$P = 9.8 \text{ ft} + 10.6 \text{ ft}$$

$$P = 20.4 \text{ ft} \qquad\qquad (20 \text{ ft})$$

Change 0.4 feet to inches:

$$(0.4 \text{ ft})\left(\frac{12 \text{ in.}}{1 \text{ ft}}\right) = 4.8 \text{ in.} \qquad (4 \text{ in.})$$

Change 0.8 inches to the nearest $\frac{1}{16}$ inch:

$$(0.8 \text{ in.})\left(\frac{16}{16}\right) = \frac{12.8}{16} \text{ in.}$$

$$\left(\tfrac{13}{16} \text{ in., rounded}\right)$$

The perimeter is 20 feet $4\frac{13}{16}$ inches.

EXAMPLE 5 If $r = 3$ feet and $h = 5$ feet, use the formula $S = 2\pi r^2 + 2\pi rh$ to find the surface area in square feet and square inches of a right circular cylinder.

Solution It will be easier to leave π as "π" at first. We will substitute 3.14 for π as one of the last steps in the computation.

$$S = 2\pi(3 \text{ ft})^2 + 2\pi(3 \text{ ft})(5 \text{ ft})$$

$$S = 2\pi(9 \text{ ft}^2) + 2\pi(15 \text{ ft}^2)$$

$$S = 18\pi \text{ft}^2 + 30\pi \text{ft}^2$$

$$S = 48\pi \text{ft}^2$$

We approximate here, using 3.14 for π:

$$S = 48(3.14) \text{ft}^2$$

$$S = 150.72 \text{ ft}^2 \qquad\qquad (150 \text{ ft}^2)$$

Change 0.72 square feet to square inches:

$$(0.72 \text{ ft}^2)\left(\frac{144 \text{ in.}^2}{1 \text{ ft}^2}\right) = 103.68 \text{ in.}^2$$

$$(104 \text{ in.}^2, \text{ rounded})$$

The surface area is 150 square feet 104 square inches.

Use the following summary of the order of operations when simplifying formulas:

> 1. **Compute within grouping symbols.**
> 2. **Compute powers.**
> 3. **Compute multiplications and divisions in order from left to right.**
> 4. **Compute additions and subtractions in order from left to right.**

Review Exercises

Find the value of x in Exercises 1–6.

1. $8x = 48$
2. $54 = 6x$
3. $40 = 7x + x$
4. $5x - 8x = 30$
5. $0.5x = 8$
6. $7 = 0.2x$

Simplify the expressions in Exercises 7 and 8.

7. $5x + 7 - 8x + 2$
8. $-3x - 3y - 4y + x$
9. Change $\frac{7}{8}$ to a percent.
10. Change $\frac{5}{16}$ to a percent.
11. Change $70\frac{1}{6}\%$ to a decimal.
12. Change $1\frac{5}{6}\%$ to a decimal.

EXERCISES
10.7

Use the formula $v_2 = v_1 + at$ to find the terminal velocity v_2 when:

1. $v_1 = 20$ feet/second, $a = 9$ feet/second^2, and $t = 4$ seconds.

2. $v_1 = 15$ feet/second, $a = 8$ feet/second^2, and $t = 6$ seconds.
3. $v_1 = 22$ feet/second, $a = -10$ feet/second^2, and $t = 6$ seconds.
4. $v_1 = 36$ meters/second, $a = -9$ meters/second^2, and $t = 4$ seconds.
5. $v_1 = 18$ meters/second, $a = -3$ meters/second^2, and $t = 5$ seconds.
6. $v_1 = 24$ meters/second, $a = 5$ meters/second^2, and $t = 7$ seconds.

Using the length and width measurements given in Exercises 7–12 and the formula $P = 2\ell + 2w$, find the perimeter of each rectangle. Express your answers in feet and inches to the nearest $\frac{1}{16}$ of an inch.

7. $\ell = 3$ feet, $w = 2$ feet 9 inches
8. $\ell = 6$ feet 3 inches, $w = 4$ feet
9. $\ell = 5.2$ feet, $w = 3.1$ feet
10. $\ell = 6.8$ feet, $w = 2.6$ feet
11. $\ell = 5$ feet 4 inches, $w = 2$ feet 6 inches
12. $\ell = 8$ feet 6 inches, $w = 4$ feet 8 inches

If $P = S - C$, where P is the profit, S is the selling price, and C is the cost, find P in each of Exercises 13–16.

13. $S = \$38, C = \$31$
14. $S = \$54, C = \$43$
15. $S = \$278, C = \$179.95$
16. $S = \$460, C = \$288.75$

Use the formulas in this section to answer Exercises 17–26.

17. Find the combined resistance of two electrical resistors in a parallel circuit if the resistance of one resistor is 6 ohms and the resistance of the other is 2 ohms.

18. Find the combined resistance of two electrical resistors in a parallel circuit if the resistance of one resistor is 6 ohms and the resistance of the other is 4 ohms.

19. Find the Fahrenheit temperature if the Celsius temperature is 10°.

20. Find the Fahrenheit temperature if the Celsius temperature is 35°.

21. Find the profit in selling 130 items if the selling price is $4.95 per item and the cost is $3.00 per item.

22. Find the profit in selling 200 items if the selling price is $19.95 per item and the cost is $15.00 per item.

23. Find the efficiency of an engine if the heat input is 23,600 calories and the heat output is 8000 calories.

24. Find the efficiency of an engine if the heat input is 18,000 calories and the heat output is 7500 calories.

25. Find the terminal velocity if the initial velocity is 14 meters/second, the acceleration is 4 meters/second^2, and the time is 10 seconds.

26. Find the terminal velocity if the initial velocity is 32 feet/second, the acceleration is -5 feet/second^2, and the time is 10 seconds.

The distance W between the parallel sides of a bolt head is given by the formula $W = 1.5D + 0.125$, where D is the diameter in inches of the bolt. Find W when:

27. $D = \frac{3}{4}$ inch.

28. $D = 1\frac{1}{4}$ inches.

29. $D = \frac{5}{8}$ inch.

30. $D = 1$ inch.

31. $D = 1\frac{1}{8}$ inches.

32. $D = \frac{7}{8}$ inch.

Find the area in each of Exercises 33–38 by using the given area formula and measurements. Express answers in square feet and square inches. Use 3.14 for π.

33. $A = \frac{1}{2}h(b_1 + b_2); h = 4$ feet, $b_1 = 3$ feet, $b_2 = 5$ feet

34. $S = 2\ell w + 2wh + 2\ell h; h = 3$ feet, $\ell = 4$ feet, $w = 2$ feet

35. $S = 2\ell w + 2wh + 2\ell h; h = 2\frac{1}{2}$ feet, $w = 3\frac{1}{2}$ feet, $\ell = 4$ feet

36. $A = \frac{1}{2}h(b_1 + b_2); h = 3$ feet, $b_1 = 8$ feet, $b_2 = 7$ feet

37. $S = 2\pi r^2 + 2\pi rh; r = 4$ feet, $h = 6$ feet

38. $S = \pi rh + \pi r^2; r = 5$ feet, $h = 7$ feet

In Exercises 39–46 use the given information to find the value of the left variable in the formula.

39. $I = I_1 + I_2 + I_3$ (electricity); $I_1 = 2$ amperes, $I_2 = 1.5$ amperes, $I_3 = 4.7$ amperes (I = current in amperes)

40. $S = L + 2B$ (area); $L = 14$ square meters, $B = 8$ square meters (S = area in square meters)

41. $F = k(R - L)$ (physics; force); $R = 0.4, L = 0.2, k = 45$

42. $h = \frac{a}{1 - a}$ (electronics); $a = 0.95$

43. $P = \frac{N + 2}{D}$ (mechanics); $N = 57, D = 4$

44. $E = \frac{T_2}{T_1 - T_2}$ (heat); $T_1 = 150, T_2 = 100$

45. $I = V\left(\frac{1}{R_1} + \frac{1}{R_2}\right)$ (electricity); $V = 6$ volts, $R_1 = 2$ ohms, $R_2 = 3$ ohms (I = current in amperes)

46. $V = IR_1 + \frac{P}{I} + IR_2$ (power); $I = 2$ amperes, $R_1 = 3.5$ amperes, $R_2 = 1$ ohm, $P = 6$ watts (V = voltage)

47. The formula giving the taper per inch, T, for a taper of length L, where the diameter of the large end is D and the diameter of the small end is d, is:

$$T = \frac{D - d}{L}.$$

Find T if D is $2\frac{1}{2}$ inches, d is $1\frac{1}{2}$ inches, and L is 8 inches.

48. The formula giving the height h of an object after t seconds when propelled upward with velocity v is:

$$h = vt - 16t^2.$$

Find h when v is 84 feet per second and t is 2 seconds.

49. The formula giving the sum S of the first n positive whole numbers is:

$$S = \tfrac{1}{2}n(n+1).$$

Find the sum of the first 8 positive whole numbers.

50. The average A of three numbers a, b, and c is given by the formula:

$$A = \frac{a+b+c}{3}.$$

Find the average of 18, 76, and 53.

51. The formula for the combined resistance C of a parallel circuit with three resistors is:

$$C = \frac{1}{\dfrac{1}{r_1}+\dfrac{1}{r_2}+\dfrac{1}{r_3}},$$

where r_1, r_2, and r_3 are the three resistances, Three resistors of 2 ohms, 3 ohms, and 6 ohms are connected in parallel. What is their combined resistance?

Calculator Exercises

Use a calculator to answer Exercises 52–56.

52. If $f = \dfrac{f_s u}{u - v_s}$ (sound), where $f_s = 28$, $u = 3.35$, and $v_s = 1.8$, find f to the nearest hundredth.

53. Using the formula $r = \dfrac{24F}{MN(N+1)}$ (interest), where $F = \$15$, $M = \$15$, and $N = 12$, find r. Express your answer as a percent to the nearest hundredth of a percent.

54. Find the combined resistance to the nearest thousandth of a parallel circuit of three resistors of 4 ohms, 5 ohms, and 8 ohms. (See Exercise 51.)

Use the formula $S = 2\pi r(r + h)$ to find the surface area of the cylinders in Exercises 55 and 56. Use 3.1416 for π. Express your answers in square yards, square feet, and square inches.

55. $r = 1.2$ yards, $h = 5.6$ yards
56. $r = 2$ yards 1 foot, $h = 4$ yards 2 feet

10.8
SOLVING EQUATIONS (II)

We have seen how to solve equations for a particular variable when that variable has been multiplied or divided by a number or letter. In this section solving for the variable will require adding or subtracting a number or letter.

> When you add or subtract the same number or letter from both sides of an equation, the results are still equal.

There are two ways to solve the equation $x + 5 = 9$. The first way is to *subtract* 5 from both sides:

$$x + 5 = 9$$
$$x + 5 - 5 = 9 - 5$$
$$x + 0 = 4$$
$$x = 4.$$

The second way is to add -5 to both sides:

$$x + 5 = 9$$
$$x + 5 + (-5) = 9 + (-5)$$
$$x + 0 = 4$$
$$x = 4.$$

Subtracting 5 from both sides leaves x alone on one side of the equation, because a number subtracted from itself is zero. Adding -5 to both sides leaves x alone on one side of the equation, because a number added to its opposite is zero.

EXAMPLE 1   Solve $x - 6 = 11$ for x.

Solution   We add 6 to both sides:

$$x - 6 = 11$$
$$x - 6 + 6 = 11 + 6$$
$$x = 17.$$

We check to be sure that 17 is the solution, or the number that makes the sentence true, by substituting it into the original equation:

$$x - 6 \overset{?}{=} 11$$
$$17 - 6 \overset{?}{=} 11$$
$$11 = 11 \text{ is true.}$$

EXAMPLE 2   Solve $4 = x + 2.8$ for x.

Solution

$$4 = x + 2.8$$
$$4 - 2.8 = x + 2.8 - 2.8 \quad \longleftarrow \text{ Subtract 2.8}$$
$$1.2 = x \qquad\qquad\qquad \text{from both sides.}$$

The solution is 1.2.

$$\textit{Check:} \quad 4 \overset{?}{=} x + 2.8$$
$$4 \overset{?}{=} 1.2 + 2.8$$
$$4 = 4 \text{ is true.}$$

Sometimes equations can be simplified before they are solved, as we see in the next example.

EXAMPLE 3   Solve $4 - 8 + x = 14$ for x.

Solution   Before solving for x, the equation can be simplified by doing the computation $4 - 8$.

$$4 - 8 + x = 14$$
$$-4 + x = 14$$
$$-4 + 4 + x = 14 + 4 \quad \longleftarrow \text{ Add 4 to}$$
$$x = 18 \qquad\qquad\qquad \text{both sides.}$$

To check, remember that, in the order of operations, additions and subtractions are done in order from left to right.

$$\textit{Check:} \quad 4 - 8 + x \overset{?}{=} 14$$
$$4 - 8 + 18 \overset{?}{=} 14$$
$$-4 + 18 \overset{?}{=} 14$$
$$14 = 14 \text{ is true.}$$

The order of operations is important in solving equations. In the next example we use the fact that the fraction bar acts like parentheses, and thus compute $7 + 18$ first.

EXAMPLE 4   Solve $6 = \dfrac{12x}{7 + 18}$ for x.

Solution   First we compute $7 + 18$.

$$6 = \frac{12x}{7 + 18}$$
$$6 = \frac{12x}{25}$$

After simplifying the bottom of the fraction (the denominator), the equation can be solved as follows:

$$6 = \frac{12x}{25}$$
$$25(6) = 25\left(\frac{12x}{25}\right) \quad \longleftarrow \text{Multiply both}$$
$$\qquad\qquad\qquad\qquad \text{sides by 25.}$$
$$150 = 12x$$
$$\frac{150}{12} = \frac{12x}{12} \quad \longleftarrow \text{ Divide both}$$
$$\qquad\qquad\qquad \text{sides by 12.}$$
$$12.5 = x.$$

The next example is much like the preceding one. First both sides are multiplied by 7. Then 3 is subtracted from both sides.

EXAMPLE 5 Solve $\dfrac{x+3}{7} = 2$ for x.

Solution

$$\frac{x+3}{7} = 2$$

$$7\left(\frac{x+3}{7}\right) = 7(2) \longleftarrow \text{Multiply both sides by 7.}$$

$$x + 3 = 14$$

$$x + 3 - 3 = 14 - 3 \longleftarrow \text{Subtract 3 from both sides.}$$

$$x = 11$$

Note in Example 5 that, in the original equation, x is added to 3 and then the *quantity* is *divided* by 7. Therefore, when solving, we must first *multiply* the *quantity* by 7. (Both sides are multiplied by 7.) Then 3 is subtracted from both sides.

EXAMPLE 6 Solve $4 = \dfrac{x-8}{2}$ for x.

Solution

$$2(4) = 2\left(\frac{x-8}{2}\right) \longleftarrow \text{Multiply both sides by 2.}$$

$$8 = x - 8$$

$$8 + 8 = x - 8 + 8 \longleftarrow \text{Add 8 to both sides.}$$

$$16 = x$$

Again, notice that, in Example 6, both sides are *first* multiplied by 2, since the quantity $x - 8$ was divided by 2. *Then* 8 is added to both sides.

Review Exercises

Simplify the expressions in Exercises 1–12.

1. $9 - 12 - 13 + 6$
2. $-5 + 2 - 15 - 6$
3. $3 + 5(-6)$
4. $4 + 3(-2)$
5. $3 + 4x + 5x$
6. $3 - 2x + 9x$
7. $8x - 5y + x - y$
8. $x - y - 4x + 6y$
9. $\frac{1}{2}(14.89)$
10. $\frac{1}{4}(65.5)$
11. $\frac{1}{3}(25.8)$
12. $\frac{1}{6}(5.82)$

EXERCISES
10.8

In Exercises 1–58 solve the equation for x.

1. $x + 15 = 46$
2. $x + 22 = 53$
3. $5 + x = 19$
4. $18 + x = 30$
5. $14 = x + 6$
6. $25 = x + 9$
7. $27 = 19 + x$
8. $34 = 18 + x$
9. $x - 11 = 4$
10. $x - 8 = 16$
11. $38 = x - 15$
12. $23 = x - 12$
13. $x + 26 = 26$
14. $x - 14 = 0$
15. $x - 46 = 46$
16. $15 = x - 15$
17. $x + 7 = 3$
18. $x + 15 = 8$
19. $-2 = x - 4$
20. $-6 = x - 9$
21. $6 + x = 1$
22. $16 + x = 4$
23. $x + 2.5 = 7.9$
24. $x + 6.9 = 8.7$
25. $x - 3.2 = 5$
26. $x - 5.3 = 7$
27. $4.3 = x - 4.3$
28. $13.7 = x - 13.7$
29. $x + 5 - 7 = 8$
30. $x + 10 - 13 = 4$
31. $x - 7 + 15 = 0$
32. $x - 3 + 12 = 9$
33. $4 - 19 + x = 7$
34. $5 - 11 + x = -2$
35. $\frac{5}{8} + x = 7$
36. $\frac{3}{5} + x = 8$
37. $14 + x + 6 = 4$
38. $3 + x + 17 = 9$
39. $15 + x - 8 = 16$
40. $20 + x - 17 = 5$
41. $18 - 7 + 2 + x = 4$
42. $23 - 15 + 4 + x = 6$
43. $\dfrac{x}{7-4} = 6$
44. $\dfrac{x}{8+3} = 2$
45. $\dfrac{2x}{5+1} = 3$
46. $\dfrac{3x}{8-6} = 6$
47. $\dfrac{x}{5-9} = 1$
48. $\dfrac{x}{7-11} = 1$
49. $\dfrac{3x}{4-3.2} = 6$
50. $\dfrac{2x}{5-2.7} = 4$
51. $\dfrac{x+2}{8} = 2$
52. $\dfrac{x+3}{6} = 4$
53. $\dfrac{x-6}{2} = 5$
54. $\dfrac{x-1}{4} = 9$
55. $\dfrac{x+4}{5.6} = 7$
56. $\dfrac{x+3}{4.8} = 9$

57. $\dfrac{x-5}{1.2} = 3$ 58. $\dfrac{x-2}{1.6} = 3$

Calculator Exercises

Use a calculator to solve the following equations in Exercises 59–66.

59. $x - 39.87 = -2.6$

60. $x + 60.4 = 3.85$
61. $0.876 + x = -3.1$
62. $-4.965 + x = -6.8$
63. $3.9 - 6.7 + x = 2$
64. $6 + x - 9.8 = -3.87$
65. $14.67 + x - 4.8 + 3.2 = 6.9$
66. $8 + x - 17 + 9.68 = 2.1$

10.9
SOLVING FORMULAS USING NUMBERS (II)

In Sections 10.7 and 10.8 you learned how to solve equations involving several operations: addition, subtraction, multiplication, and division. It is important to know how to solve such equations because many formulas similar to such equations occur in technical areas. You will often need to solve formulas for a certain variable or letter. In this section you will solve formulas using several operations. You will be given the values for all but one letter in each formula.

EXAMPLE 1   Use $p = s - c$ to find the selling price s of an item costing \$4.67 if the desired profit p is \$1.12.

Solution   Substitute the two known quantities into the formula:

$$p = s - c$$
$$\$1.12 = s - \$4.67.$$

Then add \$4.67 to both sides:

$$\$1.12 + \$4.67 = s - \$4.67 + \$4.67$$
$$\$5.79 = s.$$

The selling price is \$5.79.

EXAMPLE 2   Use $v_2 = v_1 + at$ to find the initial velocity v_1 when the terminal velocity v_2 is 24

meters/second, the acceleration a is 3 meters/second^2, and the time t is 5 seconds.

Solution

$$v_2 = v_1 + at$$

$$24\,\frac{m}{sec} = v_1 + \left(3\,\frac{m}{sec^2}\right)(5\ sec)$$

$$24\,\frac{m}{sec} = v_1 + \left(3\,\frac{m}{(sec)(sec)}\right)(5\ sec)$$

$$24\,\frac{m}{sec} = v_1 + (3)(5)\,\frac{m}{sec}$$

$$24\,\frac{m}{sec} = v_1 + 15\,\frac{m}{sec}$$

$$24\,\frac{m}{sec} - 15\,\frac{m}{sec} = v_1 + 15\,\frac{m}{sec} - 15\,\frac{m}{sec}$$

$$9\,\frac{m}{sec} = v_1$$

The initial velocity is 9 meters/second.

Note that we are using the same rules here that we used in Section 10.3. To find the unknown quantity in a formula:

1. **Substitute the known values into the formula.**
2. **Simplify if necessary.**
3. **Solve the equation for the unknown (variable).**

EXAMPLE 3 Find the length of the third side of a triangle if the lengths of two sides are 10.2 feet and 7.5 feet and the perimeter is 42 feet.

Solution We use the formula for the perimeter of a triangle:

$$P = a + b + c,$$

where P is the perimeter, and a, b, and c are the lengths of the sides.

$$P = a + b + c$$
$$42 = 10.2 + 7.5 + c$$
$$42 = 17.7 + c \qquad \text{(Simplifying)}$$
$$42 - 17.7 = 17.7 - 17.7 + c$$
$$24.3 = c$$

Since all values were given in feet, the length of the third side is 24.3 feet, or approximately 24 feet $3\frac{1}{2}$ inches.

EXAMPLE 4 The formula relating the frequency of sound waves, f_s, the frequency of sound waves detected, f, the velocity of the source of sound, v_s, and the velocity of sound, u, is:

$$f = \frac{f_s u}{u - v_s}.$$

Find f_s if

$$f = 140 \text{ per second}$$
$$u = 340 \text{ meters/second}$$
$$v_s = 85 \text{ meters/second}.$$

Solution

$$f = \frac{f_s u}{u - v_s}$$

$$\frac{140}{\text{sec}} = \frac{f_s \left(340 \frac{\text{m}}{\text{sec}}\right)}{340 \frac{\text{m}}{\text{sec}} - 85 \frac{\text{m}}{\text{sec}}}$$

$$\frac{140}{\text{sec}} = \frac{f_s \left(340 \frac{\text{m}}{\text{sec}}\right)}{255 \frac{\text{m}}{\text{sec}}}$$

$$\frac{140}{\text{sec}} = \frac{f_s \left(340 \frac{\cancel{\text{m}}}{\cancel{\text{sec}}}\right)}{255 \frac{\cancel{\text{m}}}{\cancel{\text{sec}}}} \qquad \text{(Cancelling units)}$$

$$\frac{140}{\text{sec}} = \frac{f_s (4)}{3} \qquad \text{(Simplifying)}$$

$$\frac{140}{\text{sec}} \left(\frac{3}{4}\right) = \frac{f_s (4)}{3} \left(\frac{3}{4}\right)$$

$$\frac{105}{\text{sec}} = f_s$$

Therefore, f_s is 105 per second.

The next example shows that much simplifying can be done before actually solving the equation. When numbers are being multiplied, the order in which the numbers are multiplied does not matter. For example,

$$(2)(3)(7) = (6)(7)$$
$$= 42$$

or

$$(2)(3)(7) = (2)(7)(3)$$
$$= (14)(3)$$
$$= 42.$$

EXAMPLE 5 Find the height of a trapezoid if its area is 162 square feet and the length of its two bases are 14 feet and 16 feet.

Solution Use the formula for the area of a trapezoid:

$$A = \frac{1}{2}h(b_1 + b_2).$$

$$162 = \frac{1}{2}h(14 + 16)$$

$$162 = \frac{1}{2}h(30) \qquad \text{(Simplifying)}$$

$$162 = \frac{1}{2}(30)h$$

$$162 = 15h$$

$$\frac{162}{15} = \frac{15h}{15}$$

$$10.8 = h$$

Since the units in $\frac{162}{15}$ were feet² and feet $\left(\frac{162 \text{ ft}^2}{15 \text{ ft}}\right)$, the unit for h is feet. The height is 10.8 feet, or about 10 feet 10 inches.

EXAMPLE 6 The armature current, I, flowing through a motor armature is:

$$I = \frac{E_a - E_g}{R},$$

where

E_a = applied voltage
E_g = counter-voltage
R = resistance.

Find the applied voltage, E_a, if $I = 3.4$ amperes, $E_g = 18$ volts, and $R = 30$ ohms.

Solution

$$I = \frac{E_a - E_g}{R}$$

$$3.4 = \frac{E_a - 18}{30}$$

$$30(3.4) = 30\left(\frac{E_a - 18}{30}\right) \leftarrow \text{Multiply both sides by 30.}$$

$$102 = E_a - 18$$

$$102 + 18 = E_a - 18 + 18 \leftarrow \text{Add 18 to both sides.}$$

$$120 = E_a$$

Therefore, the applied voltage is 120 volts.

In this last example it was important *first* to multiply both sides by 30, *then* to add 18 to both sides.

Review Exercises

Simplify the expressions in Exercises 1–8.

1. $(-9.1) - (-7)$
2. $7.2 - (-6)$
3. $(3.85)(4.6)$
4. $(5.96)(0.35)$

5. $\left(\frac{-4}{5}\right)\left(\frac{10}{9}\right)$
6. $\left(\frac{-27}{8}\right)\left(\frac{-4}{9}\right)$
7. $\left(\frac{4}{3}x\right)\left(\frac{3}{4}\right)$
8. $\left(\frac{-5}{6}x\right)\left(\frac{-6}{5}\right)$

Write the expressions in Exercises 9–12 as algebraic expressions.

9. 6 more than y
10. 8 less than x
11. m increased by 8
12. n decreased by 11

EXERCISES

10.9

Given the formula $p = s - c$ and the known quantities, solve for the unknown in Exercises 1–4. (p = profit, s = selling price, c = cost)

1. $p = \$6.95$, $c = \$4.50$: Solve for s.
2. $p = \$8.89$, $c = \$5.32$: Solve for s.
3. $p = \$389$, $c = \$257.50$: Solve for s.
4. $p = \$495$, $c = \$368.75$: Solve for s.

Given the formula $v_2 = v_1 + at$ and the known quantities, solve for the unknown in Exercises 5–8. (v_2 = terminal velocity, v_1 = initial velocity, a = acceleration, t = time)

5. $v_2 = 15$ meters/second, $a = 3$ meters/second², $t = 4$ seconds: Solve for v_1.
6. $v_2 = 24$ meters/second, $a = 5$ meters/second², $t = 3$ seconds: Solve for v_1.
7. $v_2 = 0$, $a = -10$ feet/second², $t = 6$ seconds: Solve for v_1.
8. $v_2 = 0$, $a = -8$ feet/second², $t = 5$ seconds: Solve for v_1.

Given the formula $P = a + b + c$ and the known quantities, solve for the unknown in Exercises 9–14. Express your answers in feet and inches to the nearest half-inch. (P = perimeter of triangle; a, b, and c = lengths of sides)

9. $P = 16$ feet 10 inches, $a = 5$ feet 7 inches, $b = 8$ feet 10 inches: Solve for c.
10. $P = 8$ feet 4 inches, $b = 3$ feet 9 inches, $c = 1$ foot 11 inches: Solve for a.

11. $P = 17.8$ feet, $a = 3.6$ feet, $b = 8$ feet: Solve for c.

12. $P = 12.1$ feet, $b = 4.8$ feet, $c = 5$ feet: Solve for a.

13. $P = 10$ feet, $a = 2.5$ feet, $c = 3.1$ feet: Solve for b.

14. $P = 11$ feet, $a = 5.4$ feet, $c = 3$ feet: Solve for b.

Given the formula $P = \dfrac{N + 2}{D}$ (gears) and the known quantities, solve for the unknown in Exercises 15–18. (P = diametral pitch, N = number of teeth, D = outside diameter)

15. $P = 5$, $D = 10$: Solve for N.

16. $P = 4.75$, $D = 8$: Solve for N.

17. $P = 15.5$, $D = 4$: Solve for N.

18. $P = 9.125$, $D = 16$: Solve for N.

Given the formula $A = \frac{1}{2}h(b_1 + b_2)$ and the known quantities, solve for the unknown in Exercises 19–24. Express your answers in feet and inches. (A = area, h = height, b_1 and b_2 = lengths of bases)

19. $A = 24$ square feet, $b_1 = 5$ feet, $b_2 = 7$ feet: Solve for h.

20. $A = 180$ square feet, $b_1 = 18$ feet, $b_2 = 12$ feet: Solve for h.

21. $A = 102.5$ square feet, $b_1 = 12$ feet, $b_2 = 8$ feet: Solve for h.

22. $A = 51$ square feet, $b_1 = 4$ feet, $b_2 = 8$ feet: Solve for h.

23. $A = 63$ square feet, $b_1 = 9.8$ feet, $b_2 = 7$ feet: Solve for h.

24. $A = 34.5$ square feet, $b_1 = 5$ feet, $b_2 = 7$ feet: Solve for h.

25. Given $P = n(p - c)$, solve for n if $P = \$40.50$, $p = \$6.95$, and $c = \$4.25$.

26. Given $P = n(p - c)$, solve for n if $P = \$32$, $p = \$4.65$, and $c = \$3.05$.

27. Given $P = \ell + 2w$, solve for ℓ if $P = 32$ meters and $w = 8$ meters.

28. Given $P = \ell + 2w$, solve for ℓ if $P = 16$ meters and $w = 5$ meters.

29. Given $I = I_1 + I_2 + I_3$, solve for I_3, if $I = 12$ amperes, $I_1 = 6$ amperes, and $I_2 = 2.4$ amperes.

30. Given $I = I_1 + I_2 + I_3$, solve for I_1 if $I = 10$ amperes, $I_2 = 4.6$ amperes, and $I_3 = 2$ amperes.

31. Given $F = k(R - L)$, solve for k if $F = 5.4$, $R = 0.5$, and $L = 0.3$.

32. Given $F = k(R - L)$, solve for k if $F = 96$, $R = 8$, and $L = 2$.

33. Given $A = \dfrac{a + b}{2}$, solve for b if $A = 79$ and $a = 68$.

34. Given $A = \dfrac{a + b}{2}$, solve for a if $A = 14.5$ and $b = 11$.

35. Given $I = \dfrac{V}{r_1 + r_2}$, solve for V if $I = 2.5$ amperes, $r_1 = 2$ ohms, and $r_2 = 4$ ohms. (V = voltage)

36. Given $I = \dfrac{V}{r_1 + r_2}$, solve for V if $I = 7$ amperes, $r_1 = 9$ ohms, and $r_2 = 6$ ohms. (V = voltage)

37. Given $I = V\left(\dfrac{1}{r_1} + \dfrac{1}{r_2}\right)$, solve for V if $I = 5$ amperes, $r_1 = 2$ ohms, and $r_2 = 3$ ohms. (V = voltage)

38. Given $I = V\left(\dfrac{1}{r_1} + \dfrac{1}{r_2}\right)$, solve for V if $I = 6$ amperes, $r_1 = 2$ ohms, and $r_2 = 4$ ohms. (V = voltage)

39. Given $T = \dfrac{D - d}{L}$, solve for D if $T = \frac{7}{64}$ (taper per inch), $d = 1\frac{5}{8}$ inches, and $L = 8$ inches. (D = diameter of large end of taper in inches)

40. Given $T = \dfrac{D - d}{L}$, solve for D if $T = 0.3$ (taper per inch), $d = 0.25$ inch, and $L = 2.5$ inches. (D = diameter of large end of taper in inches)

41. Given $VI = I^2R_1 + P + I^2R_2$, solve for P if $V = 12$ volts, $I = 2$ amperes, $R_1 = 1$ ohm, and $R_2 = 3.5$ ohms. (P = power in watts)

42. Given $VI = I^2R_1 + P + I^2R_2$, solve for P if $V = 24$ volts, $I = 3$ amperes, $R_1 = 2$ ohms, and $R_2 = 4$ ohms. (P = power in watts)

43. Given $S = L + 2B$, solve for L in square feet if $S = 2$ square yards 1 square foot and $B = 1.5$ square feet. (L = lateral area)

44. Given $I = \dfrac{E_a - E_g}{R}$, solve for E_a if $I = 2.8$ amperes, $R = 4.5$ ohms, and $E_g = 2.4$ volts. (E_a = applied voltage)

45. Given $f = \dfrac{f_s u}{u - v_s}$, solve for f_s if $u = 340$ meters/second, $v_s = 170$ meters/second, and $f = 450$ per second. (f_s = frequency per second)

Calculator Exercises

Use a calculator to solve for the unknown in each of Exercises 46–49.

46. Given $r = \dfrac{24F}{MN(N + 1)}$, solve for F if $r = 8.25\%$, $N = 36$ months, and $M = \$117.75$. ($F$ = finance charges in dollars)

47. Given $VI = I^2R_1 + P + I^2R_2$, solve for P (to the nearest tenth) if $I = 15.6$ amperes, $V = 220$ volts, $R_1 = 5.2$ ohms, and $R_2 = 8.7$ ohms. (P = power in watts)

48. Given $T = \dfrac{D - d}{L}$, solve for D if $T = 0.0864$ taper per inch, $d = 0.5028$ inch, and $L = 4$ inches.

49. Given $T = \dfrac{D - d}{L}$, solve for D if $T = 0.3$ taper per inch, $d = \frac{7}{8}$ inch, and $L = 3\frac{3}{4}$ inches.

10.10
SOLVING FORMULAS USING LETTERS (II)

Let us turn again to solving formulas for a certain letter without knowing the values of the other letters. This is a continuation of Section 10.4, where you solved formulas containing multiplication and division. Here we look at other formulas you will need to solve which contain addition, subtraction, multiplication, and division.

Notice again in this section that letters are treated as if they were numbers. The next example shows that more than one letter can be added to or subtracted from both sides of a formula.

EXAMPLE 1 Solve $P = S - C$ for S.

Solution

$$P = S - C$$
$$P + C = S - C + C \leftarrow \text{Add } C \text{ to both sides.}$$
$$P + C = S$$

EXAMPLE 2 Solve $P = a + b + c$ for b.

Solution

$$P = a + b + c$$
$$P - c = a + b + c - c \leftarrow \text{Subtract } c \text{ from both sides.}$$
$$P - c = a + b$$
$$P - c - a = a - a + b \leftarrow \text{Subtract } a \text{ from both sides.}$$
$$P - c - a = b$$

There are several other ways to solve this formula for b. They all have the same main steps: (1) c must be subtracted from both sides; and (2) a must be subtracted from both sides. Another possible way to solve for b is as follows:

$$P = a + b + c$$
$$P = b + a + c \quad \longleftarrow \text{Changing the order of the terms.}$$
$$P - c = b + a + c - c$$
$$P - c = b + a$$
$$P - c - a = b + a - a$$
$$P - c - a = b.$$

The following example is really no different from those in Section 10.4. Although part of the formula contains addition, the addition cannot be performed, so the quantity is treated as a single number using parentheses.

EXAMPLE 3   Solve $I = \dfrac{V}{r_1 + r_2}$ for V.

Solution

$$I = \frac{V}{r_1 + r_2}$$
$$(r_1 + r_2)I = (r_1 + r_2)\left(\frac{V}{r_1 + r_2}\right) \longleftarrow \begin{array}{l}\text{Multiply both}\\\text{sides by } r_1 + r_2.\end{array}$$

Note that the quantity $r_1 + r_2$ is treated as a single number. Since a number divided by itself is one,

$$(r_1 + r_2)I = \frac{(r_1 + r_2)}{(r_1 + r_2)}(V)$$
$$(r_1 + r_2)I = V.$$

It is important to have parentheses around $r_1 + r_2$ to show that the *entire quantity* is being multiplied by I.

EXAMPLE 4   Solve $f = \dfrac{f_s u}{u + v_s}$ for f_s.

Solution

$$f = \frac{f_s u}{u + v_s}$$
$$(u + v_s)f = (u + v_s)\left(\frac{f_s u}{u + v_s}\right) \longleftarrow \begin{array}{l}\text{Multiply both}\\\text{sides by } u + v_s.\end{array}$$
$$(u + v_s)f = f_s u$$
$$\frac{(u + v_s)f}{u} = \frac{f_s u}{u} \longleftarrow \begin{array}{l}\text{Divide both}\\\text{sides by } u.\end{array}$$
$$\frac{(u + v_s)f}{u} = f_s$$

Only multiplication and division are involved in the two preceding examples. The addition could not be computed. In Example 4, the expression which remained on the right after both sides were multiplied by $u + v_s$ was u *multiplied by* f_s. To get f_s alone on one side of the equation, we *divided* both sides by u.

In the following example, after both sides are multiplied by D, what remains on the right is 2 *added* to N. To get N alone on one side of the equation, we must *subtract* 2 from both sides.

EXAMPLE 5   Solve $P = \dfrac{N + 2}{D}$ for N.

Solution

$$P = \frac{N + 2}{D}$$
$$DP = D\left(\frac{N + 2}{D}\right) \longleftarrow \begin{array}{l}\text{Multiply both}\\\text{sides by } D.\end{array}$$
$$DP = N + 2$$
$$DP - 2 = N + 2 - 2 \longleftarrow \text{Subtract 2}$$
$$DP - 2 = N \qquad\quad \begin{array}{l}\text{from both}\\\text{sides.}\end{array}$$

Review Exercises

1. What number is 175% of 14?
2. What number is 220% of 37?
3. What number is 0.8% of 65?
4. What number is $\frac{1}{2}$% of 20?

5. "Add on" 37% of 40 to 40.
6. "Add on" 53% of 250 to 250.
7. 95 yd² = _____ ft²
8. _____ yd² = 108 ft²
9. _____ ft² = 288 in.²
10. 0.75 ft² = _____ in.²
11. Change 2 feet 8 inches to feet. Give your answer to the nearest hundredth of a foot.
12. Change 1 foot 7 inches to feet. Give your answer to the nearest hundredth of a foot.

EXERCISES
10.10

In Exercises 1–28 solve each equation for the indicated letter.

1. $a + c = b$ for c
2. $e = d + c$ for d
3. $e = x - c$ for x
4. $x - f = g$ for x
5. $h = x + j$ for x
6. $m = k + x$ for x
7. $5 = \dfrac{z}{a + b}$ for z
8. $\dfrac{k}{m + n} = 4$ for k
9. $\dfrac{x}{y - 4} = d$ for x
10. $a = \dfrac{y}{k - b}$ for y
11. $x + m + n = f$ for x
12. $a + x + b = c$ for x
13. $k = a + x - c$ for x
14. $j = 3 - f + x$ for x
15. $x - b + f = g$ for x
16. $x - k + s = m$ for x
17. $c = \dfrac{4y}{m + n}$ for y
18. $j = \dfrac{3y}{k + p}$ for y
19. $\dfrac{ab}{m - 4} = d$ for a
20. $\dfrac{rs}{a - 3} = k$ for r
21. $\dfrac{r + 3}{w} = k$ for r
22. $\dfrac{r + a}{b} = m$ for r
23. $d = \dfrac{x - q}{4}$ for x
24. $m = \dfrac{4 + y}{k}$ for y
25. $a = x + bc$ for x
26. $mn + x = k$ for x
27. $y - kb = p$ for y
28. $d = y - ab$ for y

In Exercises 29–47 solve each formula for the indicated letter.

29. $V = \dfrac{\pi DN}{12}$ (gears)   for D
30. $F = \dfrac{PN}{120}$ (electricity)   for N
31. $R = \dfrac{kL}{D^2}$ (electricity)   for L
32. $N = \dfrac{12C}{\pi D}$ (mechanics)   for C
33. $v_2 = v_1 + at$ (velocity)   for v_1
34. $P = \ell + 2w$ (perimeter)   for ℓ
35. $I = I_1 + I_2 + I_3$ (electricity)   for I_1
36. $R = R_1 + R_2 + R_3$ (electricity)   for R_2
37. $F = k(R - L)$ (physics: force)   for k
38. $P = n(p - c)$ (profit)   for n
39. $S = L + 2B$ (area)   for L
40. $T = \dfrac{D - d}{L}$ (mechanics)   for D
41. $A = \dfrac{a + b}{2}$ (average)   for a
42. $A = \dfrac{a + b + c}{3}$ (average)   for a
43. $S_0 = \dfrac{D - d}{2}$ (mechanics)   for D
44. $I = \dfrac{E_a - E_g}{R}$ (electricity)   for E_a
45. $r = \dfrac{24F}{MN(N + 1)}$ (interest)   for F
46. $VI = I^2 R_1 + P + I^2 R_2$ (electricity)   for P
47. $L = L_0(1 + at)$ (temperature expansion) for L_0

Calculator Exercises

48. Given $A = \dfrac{a + b}{2}$, solve for a if $b = 13.56$ and $A = 19.34$.
49. Given $F = k(R - L)$, solve for k if $R = 8.7$, $L = 3.04$, and $F = 47.261$.

50. Given $I = \dfrac{V}{r_1 + r_2}$, solve for V if $I = 8.35$, $r_1 = 14.02$, and $r_2 = 6.535$.

51. Given $I = V\left(\dfrac{1}{r_1} + \dfrac{1}{r_2}\right)$, solve for V if $r_1 = 2.5$, $r_2 = 12.5$, and $I = 18.95$.

52. Given $A = \frac{1}{2}h(b_1 + b_2)$, solve for h if $b_1 = 9.8$, $b_2 = 13.64$, and $A = 74.39$.

53. Given $P = \dfrac{N+2}{D}$, solve for D if $N = 86$ and $P = 6.432$.

10.11
WRITING FORMULAS

Much of this chapter has involved working with formulas. Most of the time the formula was given. However, often in your work you will need to write a formula from a given relationship. You will practice writing formulas in this section.

Words describing the operations of arithmetic are important in writing formulas. Some of these are contained in the following chart.

Product of a and b	ab
Sum of x and y	$x + y$
Difference between m and n	$m - n$
y squared (or the square of y)	y^2
Quotient of j divided by k	$\dfrac{j}{k}$

EXAMPLE 1 Write the formula for the sum of the lengths in feet (S) of a board m feet long and a board x yards long.

Solution There are 3 feet in a yard, so a board x yards long is:

$$(x \ \cancel{yd})\left(3 \ \frac{ft}{\cancel{yd}}\right) = 3x \ \text{ft}.$$

Since a sum is the result of adding, the formula is

$$S = m + 3x.$$

EXAMPLE 2 The heat energy created by an electric current passing through a resistance for a given period of time is the product of the resistance (in ohms), the square of the current in the circuit (in amperes), and the time (in seconds). Write this relationship as a formula.

Solution Different letters must be used for different quantities in a formula. It is common practice to use the first letter of the quantity as the variable for that quantity. (Using I for current departs from this practice but it is most often used. The letter I as a variable for current comes from the French word "Intensité" which is the word used by Ampère for describing current.) So we can let

$$H = \text{heat energy (joules)}$$
$$R = \text{resistance (ohms)}$$
$$I = \text{current (amperes)}$$
$$t = \text{time (seconds)}.$$

The square of a number is that number multiplied by itself, so the square of the current is $(I)(I)$, or I^2. A product is the result of multiplying, so the formula is:

$$H = RI^2t.$$

Key words are not always given. The operations then must be determined from the context of the problem, as the next example shows.

EXAMPLE 3 Find the cost C of putting molding around the ceiling of a rectangular room if the dimensions of the room are x feet by y feet, if the molding costs c cents per inch.

Solution Since there are 12 inches in a foot, molding which costs c cents per inch will cost:

$$\left(c\,\frac{\text{cents}}{\text{inch}}\right)\left(12\,\frac{\text{inches}}{\text{foot}}\right) = 12c \text{ cents/foot.}$$

The molding is to go around the edges of the ceiling, so we need the perimeter of the ceiling. The formula for perimeter is:

$$P = 2\ell + 2w$$

so we have

$$P = 2x + 2y.$$

The total cost C is found by multiplying the cost per foot by the number of feet.

$$\left(12c\,\frac{\text{cents}}{\text{foot}}\right)\left[(2x + 2y)\text{feet}\right]$$

$$= 12c(2x + 2y)\left(\frac{\text{cents}}{\text{foot}}\right)(\text{feet})$$

$$= 12c(2x + 2y) \text{ cents.}$$

Hence, the formula is $C = 12c(2x + 2y)$.

In distance problems, the distance d is the product of the rate r and the time t, or $d = rt$. This is one example of multiplying a rate by a time. Multiplying rate by time is often helpful in writing other formulas. This is shown in the next example.

EXAMPLE 4 One machine produces m articles per hour. A newer machine produces twice as many articles per hour. Find a formula for the total number of articles produced by both machines together in an eight-hour day.

Solution To find the number of articles produced by a machine we multiply its rate by its time. The rates of the two machines are m and $2m$. The number of articles produced by each in an eight-hour day, then, is $8m$ and $8(2m)$.

If T = the total number of articles, then the formula is:

$$T = 8m + 8(2m).$$

This can be simplified as follows:

$$T = 8m + 8(2m)$$
$$T = 8m + 16m$$
$$T = 24m.$$

Many formulas are used in determining perimeters, areas, and volumes of geometric figures. Your knowledge of the basic geometric formulas will be important in most technical areas.

EXAMPLE 5 Find a formula for the area A of the side of the house (excluding the window) shown here.

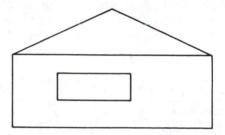

Solution No single area formula will give the required area. Rather, the area must be broken up into a triangle and two rectangles.

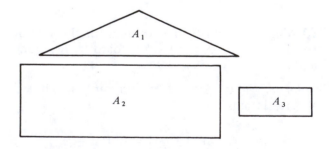

If we find the area of each of these three parts, then the area of the side of the house, A, will be:

$$A = A_1 + A_2 - A_3.$$

Let w be the width of the house and let h be the height. Then $A_2 = hw$.

New variables are needed for the dimensions of the window. Let x be the length of the window and let y be the height. Then $A_3 = xy$.

The roof portion is a triangle, and its area is $\frac{1}{2}$ times the base times the height. The base of this triangle is the width of the house, which is w. Let a be the height of the roof. (The height of the roof has no relation to the other dimensions, so it must be represented by a different variable.) Then $A_1 = \frac{1}{2}wa$. The following figure shows all the dimensions.

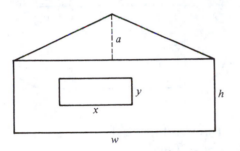

The area of the side of the house, then, is:

$$A = A_1 + A_2 - A_3$$

or

$$A = \frac{1}{2}wa + hw - xy.$$

If all dimensions are in feet, the area is in square feet.

Review Exercises

Solve the equations in Exercises 1–12 for x.

1. $ax = m$
2. $k = rx$
3. $18 = 3x + 6x$
4. $5x - 2x = 12$
5. $mx + nx = 5$
6. $q = 3x - ax$
7. $14 = \frac{px}{q}$
8. $\frac{cx}{d} = 6$
9. $\frac{x+b}{7} = 5$
10. $\frac{x-3}{8} = g$
11. $9 = \frac{3x}{r+s}$
12. $\frac{3x}{p+q} = 15$

EXERCISES
10.11

Write the statements in Exercises 1–22 as formulas.

1. Work in foot pounds is the product of the distance an object is moved and the force exerted in pounds.

2. The distance an object falls due to gravity is one-half times the acceleration due to gravity times the square of the time the object falls.

3. The energy of a mass is the product of the mass and the square of the speed of light.

4. The average of four numbers is the sum of the four numbers divided by four.

5. The current in a circuit with resistor R and a battery with voltage V and resistance r is the voltage divided by the sum of the resistances.

6. The area of a trapezoid is one-half times the height times the sum of the bases.

7. The profit is the selling price minus the cost.

8. The pressure in pounds per square inch is the force in pounds divided by the area in square inches.

9. The sum of the areas in square feet, A, of a square with side s feet and a square with side g yards.

10. The amount of fence, P, needed to enclose three sides of a rectangular lot x feet long and y feet wide.

11. The cost of tiling a rectangular room m feet by n feet with tile costing \$2 per square foot tile.

12. The area of a circle, A, with diameter m feet.

13. The number of machine parts, N, in a carload of j boxes, each box containing 18 machine parts.

14. The surface area, S, in feet of the six faces of a box with dimensions x feet by 1 yard by y feet.

15. The total distance d covered in miles by a

car traveling 40 mph for 3 hours and 55 mph for *t* hours.

16. The cost *c* of fencing to enclose a rectangular lot with dimensions *m* feet by *n* feet, if fencing costs *x* dollars per yard.

17. The number of gallons of paint, *G*, needed to cover a rectangular surface *x* feet by *y* feet. A gallon of paint will cover 300 square feet.

18. The time *t* needed to travel 380 miles at a speed of $x + 1$ miles per hour.

19. The number of feet of molding, *N*, needed for the edges of the floor of a rectangular room *a* feet by *b* feet. The room has two doors each $2\frac{3}{4}$ feet wide.

20. The cost *C* of metal edging for a circular table of diameter *d* feet, if edging costs *x* dollars per yard.

21. The number of items produced in a 40-hour week by machines *A* and *B* together. Machine *B* produces *j* items per hour. The rate of machine *A* is three times that of machine *B*.

22. One pipe drains *k* gallons per hour and another drains *b* gallons per hour. How much is drained by the two pipes together in 6 hours?

Find a formula for the perimeter of the figures in Exercises 23 and 24.

23.

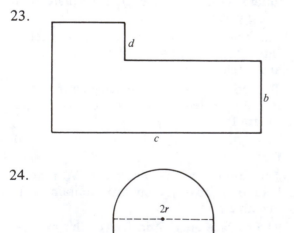

24.

Find a formula for the area of the figures in Exercises 25 and 26.

25.

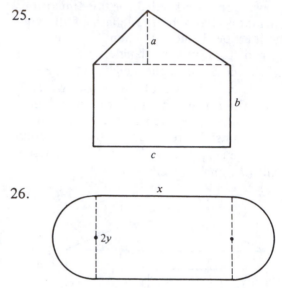

26.

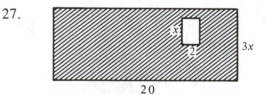

Find a formula for the shaded area in the figures in Exercises 27 and 28.

27.

28.

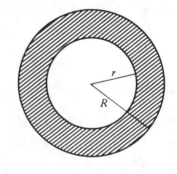

Calculator Exercises

29. If $T = \dfrac{D-d}{L}$, solve for *T* (taper per inch) when $D = 0.836$ inch, $d = 0.245$ inch, and $L = 7.294$ inches.

30. If $I = \dfrac{E_a - E_g}{R}$, solve for I (amperes) when $E_a = 105.75$ volts, $E_g = 8.83$ volts, and $R = 7.292$ ohms.

31. If $r = \dfrac{24F}{MN(N + 1)}$, solve for r (rate of interest) when $F = \$25.65$, $M = \$325.80$, and $N = 2.83$.

32. If $f = \dfrac{f_s u}{u - v_s}$, solve for f (frequency) when $f_s = 550.5$, $u = 275.34$ m/sec, and $v_s = 132.87$ m/sec.

10.12
WRITTEN PROBLEMS (II)

This section is a continuation of Section 10.6. We will practice solving some problems in which a formula or an equation must be written first.

EXAMPLE 1 Five more than a certain number is 22. What is the number?

Solution Let x be the number. The equation, then, is:

$$x + 5 = 22.$$

Next we solve the equation:

$$x + 5 = 22$$
$$x + 5 - 5 = 22 - 5$$
$$x = 17.$$

Therefore, the number x is 17. To check, $5 + 17 = 22$ is true.

EXAMPLE 2 Nine less than a certain number is 15. What is the number?

Solution Let x be the number. The equation, then, is:

$$x - 9 = 15.$$

We solve the equation:

$$x - 9 + 9 = 15 + 9$$
$$x = 24.$$

Thus, the number x is 24. To check, $24 - 9 = 15$ is true.

EXAMPLE 3 The long side of a rectangular lot is on a lake. Two hundred fifty meters of fencing were used to enclose the lot. (The lake side is not fenced.) If the lot is 23 meters deep, what is the lake frontage?

Solution Let x be the lake frontage in meters.

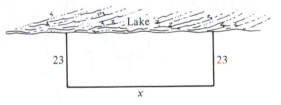

The fencing used was 250 meters. With that and the sketch, the equation is:

$$23 + x + 23 = 250$$
$$x + 46 = 250$$
$$x + 46 - 46 = 250 - 46$$
$$x = 204.$$

Therefore, the lake frontage is 204 meters.

EXAMPLE 4 A man has two kinds of stocks, which have a total value of \$1570. One of the stocks is worth $1 per share and the other is worth $5 per share. If he has 165 shares of the $5 stock, how many shares does he have of the $1 stock?

Solution   Let x be the number of shares of $1 stock. The value of the $1 stock is $1x$ dollars. The value of the $5 stock is $5(165)$ dollars. The total value is 1570 dollars, so the equation is:

$$1x + 5(165) = 1570$$
$$x + 825 = 1570$$
$$x + 825 - 825 = 1570 - 825$$
$$x = 745.$$

Therefore, the number of shares of $1 stock is 745.

EXAMPLE 5   The depth of a gear tooth is found by dividing the difference between the outside diameter and the bottom diameter by 2. Find the outside diameter if the bottom diameter is 5.5 inches and the depth of the gear tooth is .25 inches.

Solution   Let D = the outside diameter, D_1 = the bottom diameter, and h = the depth of a gear tooth.

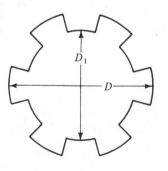

The formula, then, is:

$$h = \frac{D - D_1}{2}.$$

From the problem, $D_1 = 5.5$ inches and $h = 0.25$ inches. Substituting into the formula,

$$h = \frac{D - D_1}{2}$$
$$0.25 = \frac{D - 5.5}{2}$$
$$2(0.25) = 2\left(\frac{D - 5.5}{2}\right)$$
$$0.5 = D - 5.5$$
$$0.5 + 5.5 = D - 5.5 + 5.5$$
$$6 = D.$$

Therefore, the outside diameter is 6 inches.

We repeat here the steps for solving written problems:

1.   Read the problem.
2.   Express the unknowns with variables.
3.   Make a chart or diagram if helpful.
4.   Write an equation or formula.
5.   Solve the equation or formula.
6.   Answer the question.

Review Exercises

In Exercises 1–8 solve for x.

1. $x + a - b = c$ 2. $m = x - n + p$

3. $q = \dfrac{xw}{4 - c}$ 4. $\dfrac{xy}{a + b} = 8$

5. $\dfrac{4bx}{9m} = n$ 6. $d = \dfrac{3ax}{5y}$

7. $y = \dfrac{x + 3q}{7}$ 8. $\dfrac{x - 5a}{b} = c$

9.   Write 6 more than the product of x and 7 as an algebraic expression.

10.   Write 78 divided by the sum of m and n as an algebraic expression.

11.   Find the weight of a brass bar that has dimensions x feet by y feet by z feet. Brass weighs 510 pounds per cubic foot.

12. Find the weight of a bar of cast aluminum x millimeters by y millimeters by z millimeters. Cast aluminum weighs 0.0258 grams per cubic millimeter.

EXERCISES
10.12

In Exercises 1–12 write the necessary equation and find the number.

1. Twenty-three more than a number is thirty-seven.

2. A number less fifteen equals eight.

3. Six less than a number is fourteen.

4. Eight more than a number is twenty-one.

5. The sum of a number and twenty-six is fifty.

6. Thirteen less than a number is thirteen.

7. The quotient of four times a number and five is eight.

8. The quotient of eight more than a number and three is twelve.

9. The quotient of five less than a number and seven is eleven.

10. The sum of a number and twice the same number is fifty-four.

11. The sum of twice a number and five times the same number is ninety-one.

12. The sum of three numbers is thirty-six. The second number is twice the first, and the third number is three times the first.

Find the solutions in Exercises 13–29.

13. A rectangular field is bounded on one side by a river. The other side and two ends are to be fenced. Each end is 145 feet, and 670 feet of fencing are used. What is the length of the field?

14. After climbing to the top of a mountain, Michael descended to a point 450 feet lower than the peak. There he found a marker that read "5,290 feet above sea level." How many feet above sea level is the peak?

15. Find the price per foot of molding if the total cost of edging the top of a rectangular table 8 feet by 2 feet is $23.80.

16. The sum of three electric currents that come together at a point in a circuit is zero amperes. Find the current in the third circuit if the first two are 3.75 amperes and −2.5 amperes.

17. A student received 6 points for every correct answer and lost 2 points for every incorrect answer. Six questions were answered wrong, and the score was 72. How many questions were answered correctly?

18. How many pipes with radius one inch are needed to carry as much water as a pipe with a radius three inches?

19. How long does it take two machines working together to produce 117 items if the rates of the machines are 16 items per hour and 20 items per hour?

20. Two airplanes leave a terminal traveling in opposite directions. The average speeds of the planes are 400 mph and 550 mph. How soon will the planes be 3800 miles apart?

21. How long does it take two pipes together to drain a 2850-gallon tank if one pipe drains 200 gallons per hour and the other drains 250 gallons per hour?

22. A student received test scores of 65 and 82. What score must he get on a third test to have an average of 80 for the three tests?

23. A man traveled from town A to town B in four hours at 55 mph. He passed a historical marker 40 miles from town B. How far is the historical marker from town A?

24. Find the outside diameter of a gear if the bottom diameter is 15.8 inches and the depth of the gear tooth is 0.78 inches. (See Example 5.)

25. The current in a circuit with resistor R and a battery with voltage V and resistance r is the voltage divided by the sum of the resistances. Find the voltage of a battery with resistance 2.5 ohms if the current in a circuit with resistance 4.7 ohms is 5 amperes.

26. The pressure in pounds per square foot is the force in pounds divided by the area in square feet. What is the force on the bottom of a tank with area 54 square feet if the pressure is 600 lb/ft²?

27. The horsepower of an electric motor is the product of its voltage and current divided by 746. Find the current to the nearest tenth of an ampere of a motor with 10.3 horsepower at a voltage of 218.5.

28. The pitch diameter of a gear is equal to the product of the number of teeth and the outside diameter divided by the sum of two and the number of teeth. Find the outside diameter if the number of teeth is 54 and the pitch diameter is 13.5 inches.

29. In Exercise 28, find the outside diameter if the number of teeth is 18 and the pitch diameter is 7.65 inches.

Calculator Exercises

Use a calculator to solve Exercises 30–32.

30. An electric drill has 0.75 horsepower at 120 volts. Find the current to the nearest hundredth of an ampere. (See Exercise 27.)

31. What is the force to the nearest tenth of a pound on the bottom of a tank with a square bottom if one side of the bottom is 6 feet 9 inches and the pressure is 475 lb/ft²? (See Exercise 26.)

32. What is the force to the nearest tenth of a pound on the bottom of a tank with a circular bottom of diameter 10.5 feet? The pressure is 625 lb/ft². (See Exercise 26.)

MORE PROBLEM SOLVING

11.1
EQUATIONS

Equations are open sentences having symbols for variables in place of unknown numbers. For example, $3 + y = 10$ is an equation where y stands for the unknown number, and $x + 2y = 12$ is an equation containing two variables.

Writing equations that represent the "action" of a problem is probably the most often used problem-solving tactic.

In order to solve equations, one must have some knowledge of algebra as a prerequisite capability. Most people learn how to solve simple linear equations in junior high. One or two years of high school algebra are needed in order to be able to solve systems of equations and equations having a degree greater than one (such as $2x^2 - x + 4 = 0$).

Often when a problem contains information about two different variables, one of them can be written in terms of the other.

EXAMPLE 1   One number is three times another number. Their sum is 52. What are the numbers?

Solution   If x represents the least of the two numbers, then $3x$ represents the greatest. Since their sum is 52, the equation becomes:

$$x + 3x = 52$$

which simplifies to $4x = 52$, and solves to $x = 13$.

EXAMPLE 2   The sum of three consecutive odd numbers is 57. What are the three odd numbers?

Solution   Letting x be the least of the three *consecutive* odd numbers, $x + 2$ is the next, and $x + 2 + 2$ is the third. The equation becomes:

$$x + (x+2) + (x+4) = 57$$

which simplifies to $x + x + x + 2 + 4 = 57$, then to $3x + 6 = 57$. Transposing the 6 yields $3x = 51$, which solves to $x = 17$. Thus, the three numbers are 17, 19, and 21.

In the following problem the equation will be in quadratic form.

EXAMPLE 3   Three times the square of a number plus six times the number is 189. Find the number.

Solution   Using x^2 for the square of the number, the equation is

$$3x^2 + 6x = 189$$

in which all three terms can be divided by 3 to get

$$x^2 + 2x = 63$$
$$\text{which is } x(x + 2) = 63.$$

Since x times a number two greater is 63, x must be 7.

Solving two equations with two variables is dealt with in Section 11.4, and can often be useful when solving certain types of problems.

EXERCISES
11.1

(Remember that perimeter means "distance around" a figure.)

1. If one side of a triangle is one-fourth the perimeter, the second side is 7 units long,

and the third side is two-fifths the perimeter, how long is the perimeter?

2. Find a number such that 10 less than two-thirds the number is one-fourth the number.

3. Find the width of a rectangle if its width is one-sixth its length and its perimeter is 84 units.

Solutions   Do not look at these answers until you have tried to solve all the problems. If any one of your answers is incorrect, **forget the answer given here** and try again to solve the problem.

1. 20 units

2. 24

3. 6 units

11.2
TIME LINES

A time line is a number line indicating times, dates, years, and so on. Prerequisite capabilities needed for solving time line problems include ability to work with fractions and decimals, and being able to set up and solve simple linear equations.

EXAMPLE 1   A woman spent one-eighth of her life as a girl, one-eighth of her life as a young lady, was married for 7 years and then divorced, remained a divorcee for one-eleventh of her life before remarrying, became a widow in exactly 20 years, and died 31 years later. How old was she when she died?

Solution   Read the problem carefully several times, then draw a time line like the one below.

```
                a   total   of   x   years
0 ─────────────────────────────────────── Death
  x/8 │ x/8 │ 7 │ x/11 │    20    │    31
```

The number of years in each segment is indicated below the line.

You can set up an equation by adding all the segments and setting the expression equal to x,

$$x/8 + x/8 + 7 + x/11 + 20 + 31 = x$$

and can solve the equation by combining like terms in the left member first, to get

$$30/88x + 58 = x$$

then subtracting $30/88x$ from each member, to get

$$58 = 58/88x$$

then multiplying each member by 88, to get

$$88(58) = 58x,$$

and finally dividing each member by 58 to get $x = 88$. You can check that she was 88 years old when she died by calculating that she spent 11 years as a girl, 11 years as a young lady, 7 years married, 8 years as a divorcee, 20 years again married, and 31 years as a widow before death. These numbers add up to 88 and fit the conditions of the problem.

EXERCISES
11.2

1. Half the cost of a dress *plus* one-third the cost of a dress is $38.25 *less than* the cost of the dress. What is the cost of the dress?

2. A bottle of beverage plus its cork cost $10.50. The bottle of beverage cost $10 more than the cork. How much did the cork cost?

3. When an airliner exploded over the ocean, its "black box" fell to the ocean, through the water, and buried itself 50 feet deep in silt. Find the total distance from the airliner (in feet) the box traveled straight downward to its final destination if it fell 540/601 of that total distance through the air and 60/601 of that total distance through water.

Solutions   Do not look at these answers until you have tried to solve all three problems. If any one of your answers is incorrect, **forget the answer here** and try again to solve the problem.

1. $229.50
2. 25 cents
3. 30,050 ft.

11.3
DIAGRAMS

A variety of different types of diagrams can be used to make solving certain problems easier. One type helps set up equations for solving "mixture" problems.

EXAMPLE 1   Exactly how much pure antifreeze must be added to a radiator that is four-fifths of its 20-quart capacity full of a solution containing 10% pure antifreeze and 90% pure water in order to produce a final solution in the radiator that contains 20% pure antifreeze?

Solution   The following diagram shows the "action" of the problem and helps set up an equation to solve for the answer.

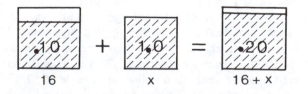

The far left figure in the diagram represents a radiator that is 4/5 full (note that 4/5 of 20 is 16)

of a 10% antifreeze solution (note that .10 represents 10%). The middle figure represents x quarts of pure antifreeze that is being added to the radiator (note that 1.0 represents 100%). The right figure in the diagram represents the radiator *after* the pure antifreeze is added. It is still not quite full; the .20 represents a 20% antifreeze solution, and the $16+x$ represents the total amount of liquid now in the radiator.

Since the amount of pure antifreeze represented in the two left figures must equal the amount of pure antifreeze represented in the right figure, the following equation can be written to represent the problem situation:

$$.10(16) + 1.0(x) = .20(16+x),$$

which simplifies to

$$1.6 + x = 3.2 + .2x.$$

Transposing yields $.8x = 1.6$; and dividing both members by .8 yields $x = 2$. Therefore, two quarts of pure antifreeze must be added to 16 quarts of a 10% antifreeze solution to obtain 18 quarts of a 20% antifreeze solution.

Diagrams representing the "action" of a problem, such as the one just illustrated, can help you set up the equation needed to solve the problem.

EXERCISES
11.3

1. How many quarts of alcohol must be added to 100 quarts of a 40% alcohol solution to obtain a 50% alcohol solution?

2. A musical production brought in $12,600 on the sale of 3,500 tickets. If the tickets sold for $2 and $4, how many of each type was sold?

3. A 9-liter radiator contains a 50% solution of antifreeze and distilled water. How much should be drained and replaced with pure antifreeze to obtain a solution containing 70% antifreeze?

Solutions   Do not look at these answers until you have tried to solve all the problems. If any one of your answers is incorrect, **forget the answer given here** and try again to solve the problem.

1. 20 quarts
2. 700 @ $2, 2800 @ $4
3. 3.6 liters

11.4
SUBSTITUTION

Substitution is a powerful tool for solving certain types of problems. It helps codify your analytical thinking and simplifies complicated expressions.

Prerequisite capabilities needed for utilizing substitution include the ability to recognize one set of symbols as representing another set, and a knowledge of elementary algebra.

EXAMPLE 1   Joan spent $50 at a variety store. She spent $45 on clothes and school supplies, and she spent $42 on clothes and snacks. How much did she spend on school supplies?

Solution   If you substitute the letter B for the amount she spent on school supplies, the letter A for the amount she spent on clothes, and the letter C for the amount she spent on snacks, you can represent her total expenditure as $A + B + C$. Since that was $50, you have

$$A + B + C = 50.$$

According to the information in the problem, $A + B = 45$. *Substituting* 45 for $A + B$ in the first equation you get

$$45 + C = 50$$

which solves to $C = 5$.

Also, according to given information, $A + C = 42$. *Substituting* 5 for C yields $A = 37$.

Lastly, *substituting* 37 for A and 5 for C in the first equation you get

$$37 + B + 5 = 50$$

which solves to $B = 8$. Since B stands for the amount she spent on school supplies, the answer to the problem question is $8.

In elementary algebra the solution to a set of two equations each containing the same two variables can be found using *substitution*. For example, the set

$$3x + 2y = 23$$
$$2x + y = 14$$

is such a system of equations. One way to solve this system is to solve one of the equations for one variable in terms of the other, then *substitute* the result into the other equation.

Observe that the second equation is easily solved for y to get $y = 14-2x$. *Substituting* $(14-2x)$ for y in the first equation yields

$$3x + 2(14-2x) = 23$$

which simplifies to $3x + 28-4x = 23$. Solving for x yields $x = 5$.

Substituting 5 for x in either equation allows you to solve for y, i.e., in the second equation,

$$2(5) + y = 14$$

which solves to $y = 4$. Therefore, the solution to the set of equations is $y = 4$ when $x = 5$.

EXAMPLE 2   A change machine changes dollar bills into quarters and nickels. If you receive 12 coins after inserting a dollar bill, how many of each type of coin did you receive?

Solution   Letting $x =$ number of quarters and $y =$ number of nickels,

$$x + y = 12$$
$$25x + 5y = 100$$

where the second equation represents the value of the coins in cents.

Solving the first equation for y yields $y = (12-x)$. *Substituting* that into the second equation results in

$$25x + 5(12-x) = 100$$

which solves to $x = 2$; and *substituting* 2 for x in the first equation yields $y = 10$. Thus, you receive 2 quarters and 10 nickels for your dollar bill.

EXERCISES
11.4

1. There were 49 boys out for three sports. Thirty-five went out for both football and track; and 27 went out for both basketball and track. How many of them went out for only football?

2. A change machine gives out 8 coins for a dollar bill, all nickels and quarters. How many nickels do you get when you insert a dollar bill?

3. A jeweler has two bars of gold alloy in stock, one 12 carat and the other 18 carat (24 carat is pure gold, 12 carat gold is 12/24 pure, 18 carat gold is 18/24 pure). How many grams of each alloy must be mixed to obtain 10 grams of 14 carat gold?

Solutions   Do not look at these answers until you have tried to solve all three problems. If any one of your answers is incorrect, **forget the answer here** and try again to solve the problem.

1. 22
2. 5
3. 3-1/3 g. of 18 carat; 6-2/3 g. of 12 carat

11.5
READING WITH COMPREHENSION

Read the following sentence, then quickly count the number of F's in it, and compare your result with the answer given after the sentence.

IF JEFF JONES FORGETS TO FORM HIS F'S CORRECTLY IT FOLLOWS THAT IF FOUR PEOPLE FEAR THAT JEFF IS A FAR POORER WRITER OF SCRIPT THAN IS FELT NECESSARY.

Once you have counted all fifteen F's (oops! If you didn't get 15, try again) please reread the sentence once more before answering the following question. *Do not look at the question before rereading the sentence.*

What might Jeff Jones do to cause people to fear that he is deficient?

If you answered, "Forgets to form his F's correctly," you are good at reading with comprehension.

However, for problem-solving purposes you need to be able to read with comprehension and to see exactly and precisely the details in what you read. It's all a matter of concentration—of not allowing your mind to wander from the subject at hand.

Utilizing these capabilities helps solve problems such as the one below, provided you have a basic knowledge of the four operations of arithmetic with whole numbers, and their relationships.

EXAMPLE 1   Tom multiplied a number by 2½ correctly and got 50 for his answer. However, he was supposed to divide the number by 2½ to get the correct answer to his problem. What was the correct answer to his problem?

Solution   If you reread this problem carefully, you will realize that Tom multiplied an unknown number by 2½ and got 50; which can be represented as $2½x = 50$. You can solve to find the value of the unknown number, which is 20. Tom should have divided the number by 2½ to get the correct answer. Thus, if you divide 20 (the number) by 2½ you get the correct answer, which is 8.

EXAMPLE 2   The divisor equals the multiplier. The quotient equals the product. The multiplicand equals one-half the dividend. Find the square of the divisor.

Solution   Rereading the problem several times helps you to realize that you are trying to find the *square of the divisor*. You know, of course, that the multiplicand is the number being multiplied, and the multiplier is the number you are multiplying the multiplicand by, and that the answer is the product. On the other hand, you

know that the dividend is the number being divided, the divisor is the number you are dividing by, and the quotient is the answer.

Using substitution, let d = the dividend, D = the divisor, and Q = the quotient. Also, be sure to realize that the multiplier (D) times the multiplicand ($d/2$) equals the product (Q), which comes from the information in the problem. Therefore,

D x $Q=d$, from the first part of the above paragraph, and $d/2$ x $D = Q$, from the second part.

Substituting $dD/2$ from the above line for Q in D x $Q = d$, you get

$$D(dD/2) = d.$$

Dividing both members by d you get $D(D/2) = 1$.

Multiplying both members by 2 yields $D(D) = 2$,

which provides the answer, $D^2 = 2$.

Most readers will agree that it takes a knack for reading with comprehension just to follow the above solution!

EXERCISES
11.5

1. "How did you do on the test, Brian?" his father asked. "Well, I got two more right than Matt did," replied Brian. "That doesn't tell me much," said his father. "How did you do percentage-wise?" "I got five percent more than Matt did." How many questions were there on the test?

2. There were ten people at the police station. Six of them were cops and four of them robbers. One Mr. Miller had arrested another Mr. Miller, and one Mr. Smith had arrested another Mr. Smith. However, the robber named Smith was not arrested by his own brother. Nobody remembers who arrested Kelly, but he could only have been arrested by a Miller or a Smith. Name the robbers.

3. Four men have to catch the six o'clock train. Bill's watch is ten minutes slow but he thinks it is five minutes slow. John's watch is ten minutes slow but he thinks it is ten minutes fast. Joe's watch is five minutes slow but he thinks it is ten minutes fast. Henry's watch is five minutes fast but he thinks it is ten minutes slow. Each of them leaves to just catch the train if his time is what he thinks it is. Who catches the train?

Solutions   Do not look at these answers until you have tried to solve all the problems. If any one of your answers is incorrect, **forget the answer given here** and try again to solve the problem.

1. 40
2. Miller, Smith, Smith, Kelly
3. Henry

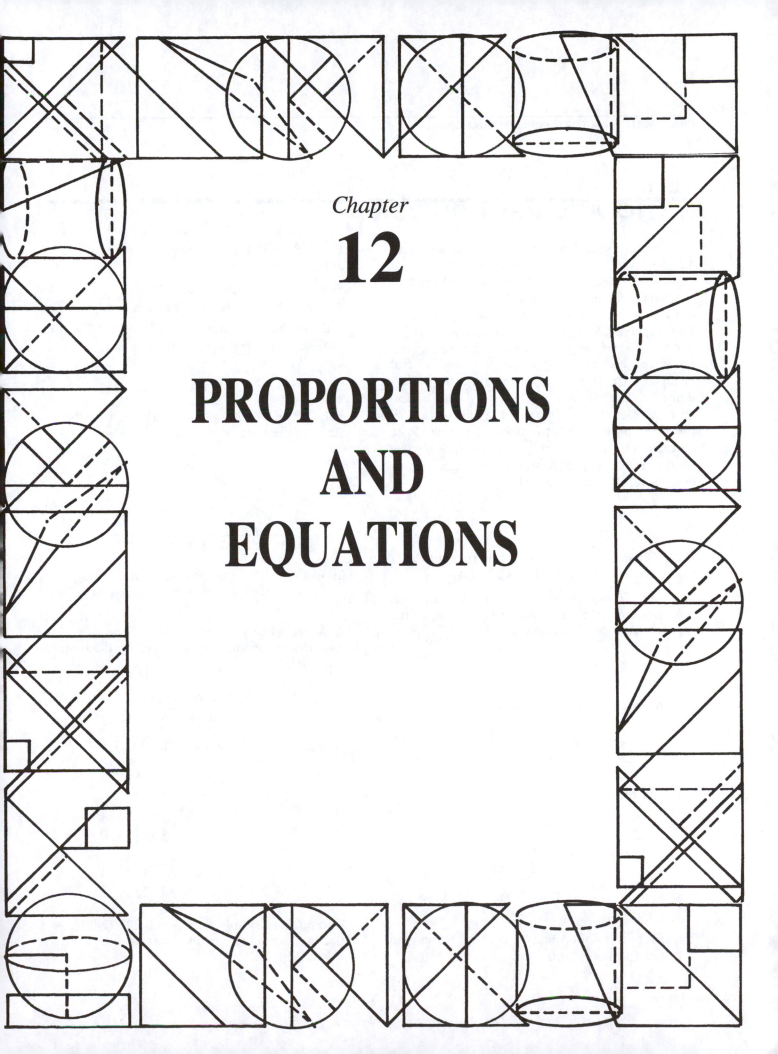

Chapter

12

PROPORTIONS AND EQUATIONS

12.1
RATIO AND PROPORTION

If a large gear has 63 teeth and a smaller gear has 36 teeth, then the ratio of the number of teeth of the large gear to the number of teeth of the smaller gear is $\frac{63}{36}$, which is $\frac{7}{4}$, or 1.75. A **ratio** is a comparison of numbers by division. *The ratio of one number to another is the quotient when the first is divided by the second.*

Rates are often expressed by ratios. For instance, the rate of speed is the ratio of distance to time. This is written $r = \frac{d}{t}$. An important number that is a ratio is the number π. It is the ratio of the circumference of any circle to its diameter. That is, $\pi = \frac{C}{d}$. The numerical value of this ratio (which is always the same) is the non-repeating decimal 3.14159265 . . . .

EXAMPLE 1 Find and simplify the ratio of 1 foot 4 inches to 6 feet 8 inches.

Solution A ratio of one quantity to another is the quotient of the first quantity divided by the second quantity. Thus, the ratio is:

$$\frac{1 \text{ ft } 4 \text{ in.}}{6 \text{ ft } 8 \text{ in.}}$$

This ratio can be simplified as follows:

$$\frac{1 \text{ ft } 4 \text{ in.}}{6 \text{ ft } 8 \text{ in.}} = \frac{1\frac{1}{3} \text{ ft}}{6\frac{2}{3} \text{ ft}}.$$

Then do the division:

$$1\frac{1}{3} \div 6\frac{2}{3} = \frac{4}{3} \div \frac{20}{3}$$
$$= \frac{4}{3} \cdot \frac{3}{20}$$
$$= \frac{1}{5}.$$

The ratio of 1 foot 4 inches to 6 feet 8 inches in simplest form is $\frac{1}{5}$, or 0.2.

EXAMPLE 2 Find and simplify the ratio of 300 miles to 5 hours.

Solution The ratio is the quotient of 300 miles divided by 5 hours. The ratio is $\frac{300 \text{ miles}}{5 \text{ hours}}$. The ratio of miles to hours is $\frac{300}{5}$, or 60. The simplified form of the ratio is 60 miles per hour.

EXAMPLE 3 The pitch of a roof may be defined as the ratio of the rise to the span.

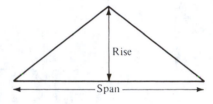

Find the pitch of a roof in simplest form if the span is 20 feet 3 inches and the rise is 4 feet 6 inches.

Solution The ratio of the rise to the span is the quotient when the rise is divided by the span. So the pitch is:

$$\frac{4 \text{ ft } 6 \text{ in.}}{20 \text{ ft } 3 \text{ in.}}$$

The ratio is simplified as follows:

$$\frac{4 \text{ ft } 6 \text{ in.}}{20 \text{ ft } 3 \text{ in.}} = \frac{4.5 \text{ ft}}{20\frac{1}{4} \text{ ft}}$$
$$= \frac{4.5 \text{ ft}}{20.25 \text{ ft}}$$
$$= \frac{4.5}{20.25}$$
$$= 0.22222 \ldots .$$

The pitch in simplest form is 0.22222 . . . . As a decimal rounded to the nearest hundredth it would be 0.22.

An equation such as $\frac{2}{5} = \frac{6}{15}$, which states that two ratios are equal, is called a **proportion**. The numbers 2 and 15 are called the **extremes**, and the numbers 5 and 6 are called the **means**. Note that $(2)(15) = 30$ and $(5)(6) = 30$. This is true for every proportion.

> In any proportion, the product of the means is equal to the product of the extremes.

EXAMPLE 4   Is $\frac{51}{85} = \frac{3}{5}$ a proportion?

Solution   The extremes are 51 and 5. The means are 85 and 3. If the product of the extremes equals the product of the means, then it is a proportion.

$$(51)(5) = 255$$
$$(85)(3) = 255$$

They *are* equal, so $\frac{51}{85} = \frac{3}{5}$ is a proportion.

EXAMPLE 5   Write at least four proportions with the numbers 7, 8, 21, and 24.

Solution   The four numbers will form a proportion whenever the product of the means is equal to the product of the extremes. For example,

$$\frac{7}{8} = \frac{21}{24}$$

is a proportion because $7(24) = 168$ and $8(21) = 168$;

$$\frac{7}{21} = \frac{8}{24}$$

is a proportion because $7(24) = 168$ and $21(8) = 168$;

$$\frac{24}{8} = \frac{21}{7}$$

is a proportion because $24(7) = 168$ and $8(21) = 168$; and

$$\frac{24}{21} = \frac{8}{7}$$

is a proportion because $24(7) = 168$ and $21(8) = 168$. On the other hand,

$$\frac{7}{8} = \frac{24}{21}$$

is *not* a proportion because $7(21) = 147$ and $8(24) = 192$; and

$$\frac{21}{8} = \frac{7}{24}$$

is *not* a proportion because $21(24) = 504$ and $8(7) = 56$.

Review Exercises

1. 52% of 60 = _____
2. 85% of 120 = _____

Solve for x in Exercises 3–6.

3. $7x + x = 80$
4. $90 = x - 10x$
5. $16 = \dfrac{2x}{3}$
6. $\dfrac{4x}{5} = 16$
7. Find the area of a circle with diameter 4 cm.
8. Find the area of a circle with radius 6 cm.
9. Find the circumference of a circle with radius 5 cm.
10. Find the circumference of a circle with diameter 6 cm.
11. Find the area of a triangle with base 5 m and altitude 4 m.
12. Find the area of a triangle with base 6 m and altitude 7 m.

EXERCISES
12.1

Write the expressions in Exercises 1–12 as ratios in simplest form.

1. 18 inches to 24 inches
2. 3 feet to 30 feet

3. 3.5 feet to 2 seconds
4. 60 miles to 1.5 hours
5. 24 centimeters to 9 centimeters
6. 75 kilometers to 10 hours
7. 5 meters to 2.5 seconds
8. 3.6 centimeters to 9 centimeters
9. 4 yards 2 feet to 5 yards 1 foot
10. 6 feet 3 inches to 3 feet 9 inches
11. 5 feet 10 inches to 4 feet 2 inches
12. 1 foot 9 inches to 2 feet 11 inches

In Exercises 13–32 decide which are proportions. Answer yes or no.

13. $\dfrac{2}{3} = \dfrac{8}{12}$

14. $\dfrac{24}{20} = \dfrac{18}{14}$

15. $\dfrac{7}{28} = \dfrac{18}{5}$

16. $\dfrac{2}{8} = \dfrac{5}{18}$

17. $\dfrac{6}{16} = \dfrac{3}{6}$

18. $\dfrac{24}{32} = \dfrac{3}{4}$

19. $\dfrac{13}{5} = \dfrac{18}{6}$

20. $\dfrac{16}{64} = \dfrac{3}{12}$

21. $\dfrac{9}{48} = \dfrac{3}{16}$

22. $\dfrac{14}{9} = \dfrac{20}{15}$

23. $\dfrac{\frac{2}{3}}{2\frac{1}{2}} = \dfrac{4}{15}$

24. $\dfrac{\frac{1}{2}}{2\frac{1}{2}} = \dfrac{1}{5}$

25. $\dfrac{5}{6} = \dfrac{2\frac{1}{2}}{3\frac{1}{3}}$

26. $\dfrac{3}{2} = \dfrac{\frac{1}{2}}{\frac{1}{3}}$

27. $\dfrac{\frac{1}{2}}{\frac{3}{8}} = \dfrac{4}{3}$

28. $\dfrac{\frac{1}{2}}{\frac{2}{3}} = \dfrac{12}{17}$

29. $\dfrac{13}{3.9} = \dfrac{0.8}{2.4}$

30. $\dfrac{24}{1.3} = \dfrac{0.48}{0.26}$

31. $\dfrac{2}{0.04} = \dfrac{45}{0.9}$

32. $\dfrac{0.8}{32} = \dfrac{0.05}{2}$

Exercises 33–36. For each proportion write three other proportions.

33. $\dfrac{2}{3} = \dfrac{6}{9}$

34. $\dfrac{3}{4} = \dfrac{15}{20}$

35. $\dfrac{18}{16} = \dfrac{9}{8}$

36. $\dfrac{30}{12} = \dfrac{5}{2}$

37. What is the ratio of the perimeter of a square with side 2 inches to the perimeter of a square with side 3 inches?

38. What is the ratio of the area of a square with side 2 inches to the area of a square with side 3 inches?

39. Find the pitch of a roof (ratio of rise to span) when the rise is 16 feet and the span is 56 feet.

40. The depth of a square screw thread is the ratio of 0.5 to the number of threads per inch. Find the depth of a square screw thread if the number of threads per inch is 10.

41. The depth of a Widget Company screw thread is the ratio of 0.6495 to the number of threads per inch. Find the depth of a Widget Company screw thread if the number of threads per inch is 15.

42. The factor of safety is the ratio of the ultimate stress of a bar to the actual unit stress that exists in a bar. Find the factor of safety of a copper casting under compression if the unit stress is 7500 pounds. The ultimate stress is 40,000 pounds.

43. The rear-axle ratio of a car is the ratio of the number of teeth on the ring gear to the number of teeth on the pinion gear. Find the rear-axle ratio to the nearest tenth if the number of teeth on the ring gear is 44 and the number of teeth on the pinion gear is 12.

44. Force in pounds is the ratio of work in foot-pounds to the distance in feet. Find the force when 160 foot-pounds of work is done from a distance of 25 feet.

45. The compression ratio of an engine is the ratio of the maximum space in the cylinder to the minimum space in the cylinder. Find the compression ratio if the maximum space is 357 cubic centimeters and the minimum space is 42 cubic centimeters.

46. Find the ratio of the circumference of a circle with radius 3 meters to that of a circle with radius 4 meters.

47. Find the ratio of the area of a circle with radius 3 meters to that of a circle with radius 4 meters.

Calculator Exercises

Use a calculator to determine the answers to Exercises 48–50.

48. The horsepower in an electrical circuit is the ratio of the number of watts to 746. Find the horsepower of 6000 watts to the nearest hundredth.

49. Find the rear-axle ratio to the nearest thousandth if the ring gear has 44 teeth and the pinion gear has 13 teeth. (See Exercise 43.)

50. Find the depth of a Widget Company screw thread to four significant digits when the number of threads per inch is 52. (See Exercise 41.)

12.2
SOLVING PROPORTIONS

In the last section you saw that, in a proportion, the product of the means is equal to the product of the extremes. This is useful in finding a missing part of a proportion, often called **solving a proportion.**

EXAMPLE 1 Find the missing value x in the proportion $\frac{5}{x} = \frac{6}{4}$.

Solution

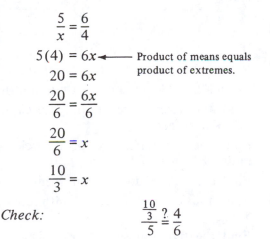

$$\frac{5}{x} = \frac{6}{4}$$

$5(4) = 6x$ ◄—— Product of means equals product of extremes.

$$20 = 6x$$

$$\frac{20}{6} = \frac{6x}{6}$$

$$\frac{20}{6} = x$$

$$\frac{10}{3} = x$$

Check:

$$\frac{\frac{10}{3}}{5} \overset{?}{=} \frac{4}{6}$$

$$\frac{10}{3}(6) \overset{?}{=} 5(4)$$

$$20 = 20 \text{ is true.}$$

The missing number of the proportion is $\frac{10}{3}$.

EXAMPLE 2 Solve the proportion $\frac{5}{7.5} = \frac{2y}{9}$.

Solution

$$\frac{5}{7.5} = \frac{2y}{9}$$

$(5)(9) = (7.5)(2y)$ ◄—— Product of means equals product of extremes.

$$45 = 15y$$

$$\frac{45}{15} = \frac{15y}{15}$$

$$3 = y$$

Check:

$$\frac{5}{7.5} \overset{?}{=} \frac{2(3)}{9}$$

$$5(9) \overset{?}{=} 7.5(2)(3)$$

$$45 = 45 \text{ is true.}$$

The missing number of the proportion is 3.

When you must set up a proportion it is helpful to recall that this can be done in several ways. For example, the proportion $\frac{5}{6} = \frac{15}{18}$ can also be written as:

$$\frac{5}{15} = \frac{6}{18},$$

$$\frac{6}{5} = \frac{18}{15},$$

and $\frac{15}{5} = \frac{18}{6}$.

EXAMPLE 3 A certain alloy is manufactured using a ratio of 2 pounds of metal A to 5 pounds of metal B. How much metal B is needed to mix with 2700 pounds of metal A?

Solution Let $x =$ the number of pounds of metal B. The ratio 2 pounds (A) to 5 pounds (B) is:

$$\frac{2 \text{ lb}}{5 \text{ lb}} \text{ or } \frac{2}{5}.$$

This problem involves two equal ratios, which is a proportion. In setting up the proportion, the amount of A must be in the numerator of the second ratio if the amount of A is in the numerator of the first ratio. We have

$$\frac{2}{5} = \frac{2700 \text{ lb}}{x}$$
$$2x = 5(2700 \text{ lb})$$
$$\frac{2x}{2} = \frac{5(2700) \text{ lb}}{2}$$
$$x = 6750 \text{ lb}.$$

Thus, 6750 pounds of metal B are needed.

The words "is proportional to" are often used in technical areas. For example, to say that the amount of money earned is proportional to the number of hours worked means that the ratio of any money earned to hours worked is equal to the ratio of any other amount of money earned to the corresponding number of hours worked.

EXAMPLE 4 Under certain conditions the weight on the end of a spring is proportional to the length that the spring stretches. A weight of 12 pounds stretches a spring 4 inches. What weight will stretch it 3 inches?

Solution To say that the weight is proportional to the length means that the ratio of any weight to the corresponding length is equal to the ratio of any other weight to the corresponding length. The given ratio is 12 pounds to 4 inches or:

$$\frac{12 \text{ lb}}{4 \text{ in.}}.$$

Let $x =$ the weight in pounds that will stretch the spring 3 inches. Then the second ratio is:

$$\frac{x}{3 \text{ in.}}.$$

The proportion is:

$$\frac{12 \text{ lb}}{4 \text{ in.}} = \frac{x}{3 \text{ in.}}$$
$$(12 \text{ lb})(3 \text{ in.}) = (4 \text{ in.})x$$
$$\frac{(12 \text{ lb})(3 \text{ in.})}{4 \text{ in.}} = \frac{(4 \text{ in.})x}{4 \text{ in.}}$$
$$\frac{(12 \text{ lb})(3 \text{ in.})}{4 \text{ in.}} = x$$
$$\frac{(12)(3) \text{ lb}}{4} = x$$
$$9 \text{ lb} = x$$

Therefore, the weight that will stretch the spring 3 inches is 9 pounds.

Review Exercises

1. $\sqrt[3]{27} =$ _____   2. $\sqrt[3]{8} =$ _____

3. If $x = -3$, then $x^2 - 2x + 3 =$ _____.

4. If $x = -2$, then $x^2 - 2x + 3 =$ _____.

5. Round 2.3473 to the nearest thousandth.

6. Round 2.3473 to the nearest hundredth.

7. Find the circumference of a circle with diameter 6 feet.

8. Find the circumference of a circle with radius 9 feet.

9. Find the area of a circle with radius 7 cm.

10. Find the area of a circle with diameter 8 cm.

11. Find the hypotenuse of a right triangle if the lengths of the legs are 4 cm and 7 cm (to the nearest hundredth cm).

12. Find the hypotenuse of a right triangle if the lengths of the legs are 8 feet and 3 feet (to the nearest hundredth of a foot).

EXERCISES
12.2

Solve the proportions in Exercises 1–22.

1. $\dfrac{x}{30} = \dfrac{4}{5}$

2. $\dfrac{5}{6} = \dfrac{y}{36}$

3. $\dfrac{1}{7} = \dfrac{-2}{b}$

4. $\dfrac{1}{11} = \dfrac{-3}{a}$

5. $\dfrac{2}{m} = \dfrac{3}{5}$

6. $\dfrac{3}{n} = \dfrac{4}{5}$

7. $\dfrac{x}{4} = \dfrac{10}{8}$

8. $\dfrac{7}{4} = \dfrac{b}{28}$

9. $\dfrac{9}{y} = \dfrac{18}{6}$

10. $\dfrac{16}{12} = \dfrac{4}{x}$

11. $\dfrac{2}{4} = \dfrac{x}{3}$

12. $\dfrac{p}{8} = \dfrac{12}{3}$

13. $\dfrac{3}{x} = \dfrac{9}{21}$

14. $\dfrac{7}{8} = \dfrac{y}{24}$

15. $\dfrac{x}{3.5} = \dfrac{6}{16}$

16. $\dfrac{5}{4.5} = \dfrac{x}{27}$

17. $\dfrac{\frac{1}{2}}{400} = \dfrac{q}{100}$

18. $\dfrac{150}{\frac{3}{4}} = \dfrac{s}{2}$

19. $\dfrac{3x}{2} = \dfrac{10}{8}$

20. $\dfrac{7}{4} = \dfrac{2a}{15}$

21. $\dfrac{4}{17} = \dfrac{5}{2y}$

22. $\dfrac{9}{4b} = \dfrac{1}{10}$

The numbers given in Exercises 23–27 are proportional. Write and solve a proportion to answer each question.

23. One hundred fifty tiles cost $31.50. How much will 420 tiles cost?

24. If there is a deduction of $23.80 from a paycheck of $85, what deduction will there be from a paycheck of $200?

25. If 6 castings cost $25.20, how much will 10 castings cost?

26. It requires 25 days to build a pipeline $1\frac{1}{2}$ miles long. How long will it take to build 4 miles of pipeline?

27. At $4.60 for 3 square feet, what will be the cost of a concrete walk that is $2\frac{1}{2}$ feet wide and 60 feet long?

The relationship between the number of teeth and the revolutions of two gears is the proportion

$$\frac{T}{t} = \frac{n}{N}$$

where T = the number of teeth in the driver gear, t = the number of teeth in the driven gear, N = the number of revolutions (or rpm) of the driver gear, and n = the number of revolutions (or rpm) of the driven gear. Answer Exercises 28–31.

28. Find n when $T = 32$, $N = 120$ rpm, and $t = 80$.

29. Find T when $t = 320$, $N = 420$ rpm, and $n = 210$ rpm.

30. Find t when $T = 60$, $N = 240$ rpm, and $n = 720$ rpm.

31. Find N when $T = 180$, $t = 270$, and $n = 126$ rpm.

The relationship between the diameters of two pulleys and their surface speeds (rpm) can be written as the proportion

$$\frac{D}{d} = \frac{s}{S}$$

where D = the diameter of the driver pulley, d = the diameter of the driven pulley, S = the speed of the driver pulley, and s = the speed of the driven pulley. Answer Exercises 32–35.

32. Find S when $d = 5$ inches, $D = 25$ inches, and $s = 340$ rpm.

33. Find D when $d = 9$ inches, $S = 270$ rpm, and $s = 360$ rpm.

34. Find d when $D = 15$ inches, $S = 450$ rpm, and $s = 750$ rpm.

35. Find s when $D = 24$ inches, $d = 18$ inches, and $S = 330$ rpm.

36. To find the gears to cut a required number of threads per inch the proportion

$$\frac{C}{N} = \frac{S}{L}$$

is used, where C = the lathe screw constant, N = the number of threads per inch, S = the number of teeth in the gear on spindle stud, and L = the number of teeth in the gear on lead screw. Find the number of teeth in the gear on lead screw if the lathe screw constant is 4, the number of threads per inch is 16, and the number of teeth in the gear on spindle stud is 24.

37. A board is to be cut into two pieces with lengths in the ratio of 3 to 5. Find the length of the long piece if the length of the short piece is 9 feet.

38. The dimensions of a room are in the ratio of 3 to 5. The room is longer than it is wide, and its width is 15 feet. How long is the room?

39. The commission that a salesperson gets is proportional to the amount of sales. Sales of $550 bring a commission of $45. How much of a commission will there be on $1200 worth of sales?

40. The dimensions of a photograph are proportional to any enlargement. A snapshot is 5 inches by 7 inches. What is the length of an enlargement if its width is 17.5 inches?

41. The voltage across part of an electric circuit is proportional to the current. If the voltage is 12 volts when the current is 3 amperes, find the current when the voltage is 15 volts.

42. In a vacuum, the speed of an object falling freely from rest is proportional to the length of time that it falls. If an object was falling at 64 feet per second 2 seconds after beginning its fall, how fast was it falling 8 seconds later?

Calculator Exercises

Use a calculator to determine the answers to the following exercises. Solve the proportion in each of Exercises 43–46.

43. $\dfrac{332}{x} = \dfrac{747}{675}$

44. $\dfrac{415}{664} = \dfrac{185}{x}$

45. $\dfrac{5.78}{86.7} = \dfrac{x}{4.23}$

46. $\dfrac{x}{0.368} = \dfrac{3.09}{55.2}$

47. If 15 cubic centimeters of a solution contains 2.8 grams of salt, how many grams (to the nearest hundredth) of salt would 47 cubic centimeters of the same solution contain?

48. The resistance of a wire in an electrical circuit is proportional to its length. If the resistance is 0.0028 ohms when the length is 47 feet, what is the resistance (with 5 significant digits) when the length is 130 feet?

12.3
SIMILAR TRIANGLES AND APPLICATIONS

Two triangles are **similar** if they have the same shape, but not necessarily the same size. *If two triangles are similar the measures of their corresponding angles are equal and the measures of their corresponding sides are proportional.* In Figure 12.1, triangle ABC is similar to triangle XYZ; corresponding angles are equal:

$$\angle A = \angle X, \angle B = \angle Y, \text{ and } \angle C = \angle Z;$$

and corresponding sides are proportional:

$$\frac{AB}{XY} = \frac{BC}{YZ} = \frac{AC}{XZ}.$$

FIGURE 12.1

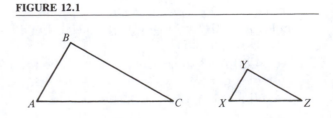

In Figure 12.2 triangle *MNP* is similar to triangle *MRS;* corresponding angles are equal:

$$\angle M = \angle M, \angle MRS = \angle N, \angle MSR = \angle P;$$

and corresponding sides are proportional:

$$\frac{MN}{MR} = \frac{NP}{RS} = \frac{MP}{MS}.$$

FIGURE 12.2

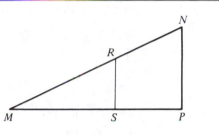

> To show that two triangles are similar we need only show that :
> 1. the corresponding angles are equal, or
> 2. the corresponding sides are proportional.

If one of these conditions is true, then the other condition is also true. If two pairs of corresponding angles are equal, then the third pair of corresponding angles are also equal. Therefore, we know two triangles are similar if two pairs of corresponding angles are equal. When it is known that triangles are similar, missing sides can be found by setting up proportions and solving for the unknown.

EXAMPLE 1 Find the missing sides *x* and *y* from the two similar triangles shown here.

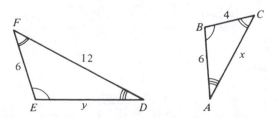

Solution The known ratio of corresponding sides is *BC* (opposite $\angle A$) to *EF* (opposite

$\angle D$), which is $\frac{4}{6}$ (using the small triangle as the first triangle, and using identical markings to indicate equal angles). These proportions are used to find the missing sides.

$$\frac{4}{6} = \frac{x}{12} \qquad \frac{4}{6} = \frac{6}{y}$$
$$(4)(12) = 6x \qquad 4y = (6)(6)$$
$$48 = 6x \qquad 4y = 36$$
$$\frac{48}{6} = \frac{6x}{6} \qquad \frac{4y}{4} = \frac{36}{4}$$
$$8 = x \qquad y = 9$$

The missing side in $\triangle ABC$ is 8 units. The missing side in $\triangle DEF$ is 9 units.

EXAMPLE 2 Given $\triangle ABC$ and $\triangle ADE,$ where $AB = 6,$ $AD = 15,$ and $DE = 20,$ decide if the triangles are similar, and if they are, find *BC.*

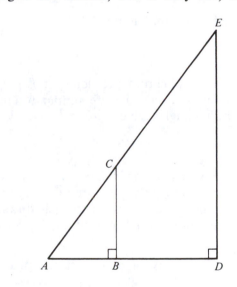

Solution Angle *B* and $\angle D$ are both right angles (see figure), so they are congruent. Angle *A* is in both triangles, and an angle is congruent to itself. Angle *C* and $\angle E$ are congruent because, if two angles of one triangle are congruent to two angles of another triangle, then the third angles are also congruent. Therefore, the triangles are similar. Corresponding sides of similar triangles are proportional, so *BC* is found as follows:

AB (opposite $\angle C$) corresponds to AD (opposite $\angle E$);

BC (opposite $\angle A$) corresponds to ED (opposite $\angle A$).

The proportion of corresponding sides is:

$$\frac{AB}{AD} = \frac{BC}{ED}$$

or

$$\frac{6}{15} = \frac{BC}{20}.$$

Solving for BC,

$$\frac{6}{15} = \frac{BC}{20}$$
$$15(BC) = 6(20)$$
$$\frac{15(BC)}{15} = \frac{6(20)}{15}$$
$$BC = 8.$$

The triangles are similar and $BC = 8$.

In many application problems, it is helpful to remember that triangles are similar whenever corresponding angles are congruent.

EXAMPLE 3 When the shadow of a 6-foot tall man is 4 feet long, the shadow of a flagpole is 12 feet long. If the tips of the shadows are at the same point on the ground, how tall is the pole?

Solution First draw a sketch of the problem.

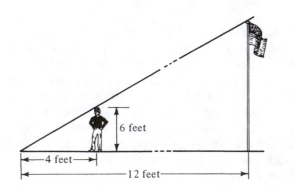

One triangle is formed with the man and his shadow. The other triangle is formed with the pole and its shadow. The two triangles are similar because the corresponding angles are congruent. Since the triangles are similar, corresponding sides are proportional. The known ratio of corresponding sides is

$$\frac{4 \text{ ft}}{12 \text{ ft}}.$$

Let x be the height of the flagpole. The proportion, then, is:

$$\frac{4 \text{ ft}}{12 \text{ ft}} = \frac{6 \text{ ft}}{x}$$
$$\frac{4}{12} = \frac{6 \text{ ft}}{x}$$
$$4x = (12)(6 \text{ ft})$$
$$\frac{4x}{4} = \frac{(12)(6 \text{ ft})}{4}$$
$$x = 18 \text{ ft}.$$

The height of the flagpole is 18 feet.

EXAMPLE 4 The taper of a piece of metal can be determined using the rules regarding similar triangles. What is the taper t over a length of 7 inches on a metal gauge if the taper over a length of 12 inches is 3.4 inches?

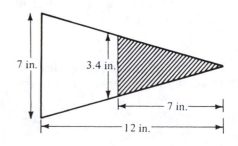

Solution

$$\frac{t}{3.4 \text{ in.}} = \frac{7 \text{ in.}}{12 \text{ in.}}$$
$$(12)(t) = (7)(3.4)$$
$$t \doteq 1.98 \text{ in.}$$

The ratio of the areas of two similar triangles is equal to the ratio of the squares of any pair of corresponding sides of the triangles.

EXAMPLE 5 Triangle ABC is similar to triangle XYZ. Side $AC = 4$ cm, $XZ = 6$ cm, and the area of triangle XYZ is 18 cm². What is the area of triangle ABC?

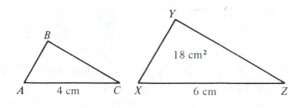

Solution

$$\frac{\text{area of triangle } ABC}{\text{area of triangle } XYZ} = \frac{(AC)^2}{(XZ)^2}$$

$$\frac{\text{area of triangle } ABC}{18 \text{ cm}^2} = \frac{(4 \text{ cm})^2}{(6 \text{ cm})^2}$$

$$(36)(\text{area of triangle } ABC) = (16)(18)$$

$$\text{area of triangle } ABC = 8 \text{ cm}^2$$

EXAMPLE 6 Right triangle LMN is similar to right triangle QRS. Side $LN = 5$ cm, $RQ = 7$ cm, and $QS = 12$ cm. If side LN corresponds to side QS what is the area of triangle LMN?

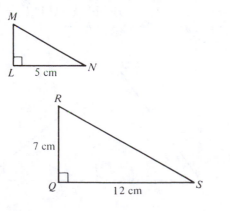

Solution

$$\text{Area of triangle } QRS = \frac{1}{2}(7 \text{ cm})(12 \text{ cm})$$

$$= 42 \text{ cm}^2$$

$$\frac{\text{area of triangle } LMN}{42 \text{ cm}^2} = \frac{(5 \text{ cm})^2}{(12 \text{ cm})^2}$$

$$(144)(\text{area of triangle } LMN) = (25)(42)$$

$$\text{area of triangle } LMN \doteq 7.29 \text{ cm}^2$$

Review Exercises

1. What is the distance from the origin of a set of coordinate axes to the point $P(-5, 9)$?

2. What is the distance from the origin of a set of coordinate axes to the point $Q(12, -10)$?

3. How many significant digits are there in the number 0.03800?

4. How many significant digits are there in the number 804,000?

5. What is the precision of a measure of 16.47 m?

6. What is the precision of a measure of 0.0590 in.?

7. Given $A = \frac{1}{2}h(b_1 + b_2)$, solve for h if $A = 84$, $b_1 = 5$, and $b_2 = 9$.

8. Given $A = \frac{1}{2}h(b_1 + b_2)$, solve for h if $A = 135$, $b_1 = 10$, and $b_2 = 18$.

9. Simplify: $16 - (-4) - (3 - 8)$.

10. Simplify: $-(5 - 7) + (-4) - (-12)$.

11. Given $j = \frac{3y}{k + p}$, solve for y.

12. Given $d = \frac{x + q}{4}$, solve for q.

EXERCISES

12.3

1. Triangle ABC is similar to triangle EFG. What are the lengths of sides AB and BC?

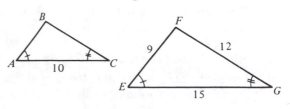

2. Triangle *MNP* is similar to triangle *XYZ*. What are the lengths of sides *MP* and *PN*?

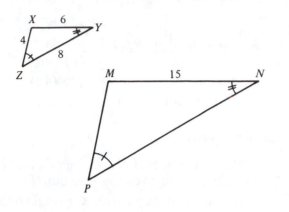

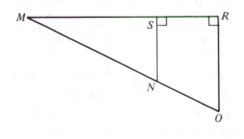

Use the figure below and the information in each of Exercises 3–6 to find:

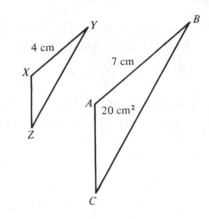

3. *NS* if *MS* = 3, *MR* = 4, and *OR* = 5.

4. *MR* if *MS* = 5, *MN* = 7, and *MO* = 9.

5. *MS* if *NS* = 4.3, *MR* = 10, and *OR* = 8.2.

6. *MO* if *MN* = 9.1, *NS* = 5.3, and *OR* = 7.7.

7. The sides of a triangle are 3.6 cm, 7.2 cm, and 9.8 cm. The shortest side of a similar triangle is 2.7 cm. Find the lengths of the other sides of the second triangle.

8. The sides of a right triangle are 9 m, 12 m, and 15 m. The hypotenuse of a similar triangle is 2.5 m. Find the lengths of the other sides of the second triangle.

9. Triangle *HJK* is similar to triangle *LMN*. Side *HJ* corresponds to side *LM*. If the area

of triangle *HJK* is 12 in.², what is the area of triangle *LMN*?

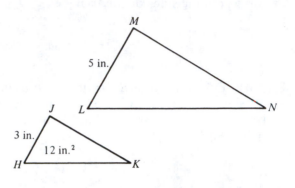

10. Triangle *ABC* is similar to triangle *XYZ*. Side *AB* corresponds to side *XY*. If the area of triangle *ABC* is 20 cm², what is the area of triangle *XYZ*?

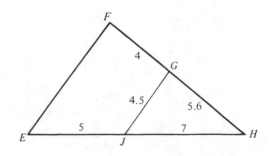

11. Triangle *EFH* is similar to triangle *GHJ*. If the area of triangle *GHJ* is 15 m², what is the area of triangle *EFH*?

12. Triangle *ABC* is similar to triangle *ADE*. If the area of triangle *ABC* is 23 yd², what is the area of triangle *ADE*?

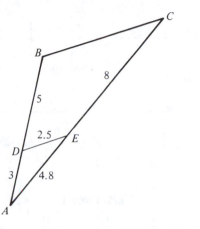

13. Triangle *EFG* is similar to triangle *EHK*. What are the lengths of sides *EF* and *FG*?

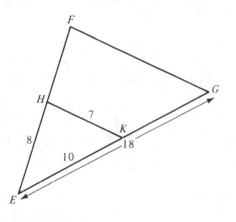

14. Triangle *ADG* is similar to triangle *ABC*. What are the lengths of sides *AD* and *AG*?

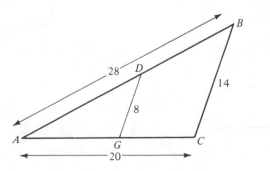

15. What is the taper on a metal pin 9.5 cm long if the taper is 2.7 mm per cm?

16. What is the taper on a metal gauge 7 inches long if the taper is $2\frac{3}{8}$ inches per foot?

17. A metal triangular gusset has sides 4.3 in., 2.7 in., and 3.4 in. If you make a larger gusset that is similar in shape, with the shortest side 5.3 in. long, what are the lengths of the other two sides?

18. A triangular concrete slab has sides of lengths 6.2 m, 8.3 m, and 10.7 m. The longest side of another triangular slab, similar in shape, is 5.9 m. What are the lengths of the other two sides of this slab?

19. Find the height of a building that makes a shadow of 40 feet on the ground at the same time that a 6-foot stick makes a shadow of 8 feet.

20. Find the width of the river *(AB)* if *AC* = 400 feet, *CE* = 80 feet, and *DE* = 70 feet. The line *DB* is a straight line.

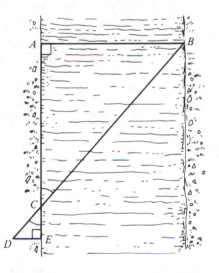

21. A man places a mirror on the ground 16 feet from a tree. He then walks backward until he just sees the top of the tree reflected in the mirror. If he walked back 4 feet from the mirror and his eye is $5\frac{1}{2}$ feet above the ground, find the height of the tree.

22. A telephone pole casts a shadow 20 feet long when a 3-foot stick casts a shadow

2 feet long. What is the height of the telephone pole?

23. Towns *A*, *B*, *C*, *D*, and *E* are located as shown in the sketch. (*DB* and *AE* are straight lines.) The distance from Town *B* to Town *D* is 35 miles; from Town *D* to Town *E* is 36 miles; and from Town *C* to Town *D* is 25 miles. Find the distance from Town *A* to Town *B*.

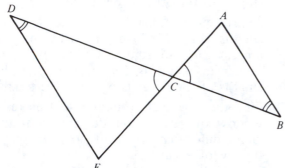

24. The shadow of a six-foot man is 5 feet at the same time that the shadow of a rocket standing vertically at its launch pad is 125 feet. Find the height of the rocket.

25. How high is a light pole, if it casts a shadow of 42 feet when a man 5 feet 9 inches tall casts a shadow of 10 feet?

26. Use the figure to find the widest part of the lake.

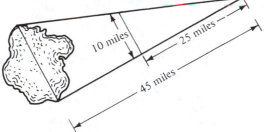

27. What is the minimum diameter of round stock if a square rod, 4.3 cm on a side, can be milled from the round stock?

28. What is the height of a triangular cut in a metal block if the cut is 34 mm wide and the length of the slant height is 25 mm?

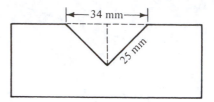

Calculator Exercises

Give the answers to Exercises 29–32 to two decimal places.

29. Triangle *ABC* is similar to triangle *EFG*. What are the lengths of sides *EF* and *FG*?

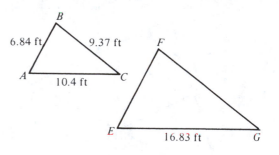

30. Triangle *MNS* is similar to triangle *TRS*. What are the lengths of sides *TS* and *NS*?

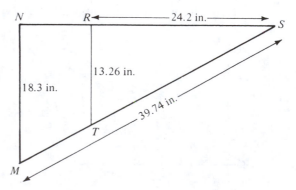

31. Triangle *EFG* is similar to triangle *HJK*. What is the area of triangle *EFG*?

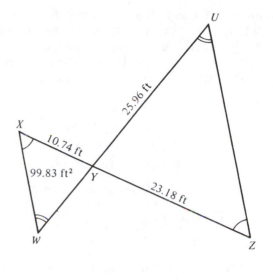

32. Triangle *WXY* is similar to triangle *UZY*. What is the area of triangle *UZY*?

12.4
DIRECT VARIATION

To convert a given number of feet to inches you can multiply the number of feet by 12. This can be represented as:

$$y = 12x$$

if *x* represents the number of feet and *y* represents the number of inches.

An equation of this form is said to express a **direct variation.** We say that *the number of inches varies directly as the number of feet; or the number of inches is (directly) proportional to the number of feet.*

Direct variation expresses the fact that a ratio of two quantities remains **constant.** In the example above, this constant is 12 inches/foot.

The ratio is, for example, $\frac{12 \text{ inches}}{1 \text{ foot}}$. The expression $\frac{12 \text{ inches}}{1 \text{ foot}}$, or 12 inches/foot, is called the **constant of variation** or the **constant of proportionality.**

The expression "*y* varies directly as *x*" means that $y = kx$. The constant of variation is *k*.

Here *x* and *y* are variables. Although *k* is a letter, its value will remain the same for a given situation. Letters used in this way are called **letter constants,** or just **constants.**

EXAMPLE 1 A law of physics states that the pressure *P* of a gas varies directly as, or is directly proportional to, the absolute temperature *T*. The constant of variation is *R*. Write this statement as a formula.

Solution To say that *y* varies directly as *x* means that $y = kx$. Pressure *P* varying directly as absolute temperature *T* then means $P = kT$. Since *R* is the constant of variation, the formula is $P = RT$. If *R* is the *constant* of variation, this means that, for different temperatures and pressures of the *same* gas, *R* will stay the *same*. It is possible that *R* will have a different value for different kinds of gases.

EXAMPLE 2 The weight of an object on the moon varies directly as its weight on earth. An astronaut who weighs 180 pounds on the earth weighs 30 pounds on the moon. Write the direct variation formula.

Solution To say that y varies directly as x means that $y = kx$. The weight on the moon (m) varying directly as the weight on earth (e) then means $m = ke$. To find the constant of variation we substitute into the formula and solve for k.

$$m = ke$$
$$30 \text{ lb} = k(180 \text{ lb})$$
$$\frac{30 \text{ lb}}{180 \text{ lb}} = \frac{k(180 \text{ lb})}{180 \text{ lb}}$$
$$\frac{30}{180} = k$$
$$\frac{1}{6} = k$$

The constant of variation is $\frac{1}{6}$, so the direct variation formula is $m = \frac{1}{6}e$.

EXAMPLE 3 If y varies directly as x, and $x = 12$ when $y = 54$, find the value of y when $x = 8$.

Solution The expression "y varies directly as x" means $y = kx$. To find k, substitute known values of x and y into the formula and solve for k:

$$y = kx$$
$$54 = k(12)$$
$$\frac{54}{12} = \frac{k(12)}{12}$$
$$\frac{9}{2} = k.$$

This direct variation formula, then, is $y = \frac{9}{2}x$. To find the value of y when $x = 8$, substitute and solve for y:

$$y = \frac{9}{2}x$$
$$y = \frac{9}{2}(8)$$
$$y = 36.$$

The required value is $y = 36$.

Sometimes one variable varies directly as a power of another variable. For example, "y varying directly as the square of x" means $y = kx^2$. The procedures all remain the same, but the formula will be different.

EXAMPLE 4 In a vacuum, the distance d a freely falling object falls varies directly as the square of the time t the object falls. If an object falls 256 feet in 4 seconds, how far does it fall in 5 seconds?

Solution To say that the distance (d) varies directly as the square of the time (t) means that $d = kt^2$. To find k:

$$d = kt^2$$
$$256 \text{ ft} = k(4 \text{ sec})^2$$
$$256 \text{ ft} = k(16 \text{ sec}^2)$$
$$\frac{256 \text{ ft}}{16 \text{ sec}^2} = \frac{k(16 \text{ sec}^2)}{16 \text{ sec}^2}$$
$$16 \text{ ft/sec}^2 = k.$$

Thus, the constant of variation k is 16 feet/second^2. The direct variation formula is:

$$d = \left(\frac{16 \text{ ft}}{\text{sec}^2}\right)t^2$$

or just

$$d = 16t^2.$$

To find d when $t = 5$ seconds:

$$d = \left(\frac{16 \text{ ft}}{\text{sec}^2}\right)(5 \text{ sec})^2$$
$$d = \left(\frac{16 \text{ ft}}{\text{sec}^2}\right)(25 \text{ sec}^2)$$
$$d = 16(25) \text{ ft}$$
$$d = 400 \text{ ft}.$$

Therefore, the answer to the original question is 400 feet.

One variable can also vary directly as the product of other variables. For example, "p

varies as the product of q and r" means $p = kqr$, where k is the constant of variation.

Review Exercises

1. Simplify: $(-4)(-8) + (-3)^2$.

2. Simplify: $2^3 - (-6)(-9)$.

3. If $ax + x = b$, solve for x.

4. If $e = x + dx$, solve for x.

5. If $3k = \dfrac{x + 2}{h}$, solve for x.

6. If $\dfrac{5m}{n + q} = t$, solve for m.

7. If $A = p + prt$, and $A = \$4720$, $r = 6\%$, and $t = 3$ years, solve for p.

8. If $A = p + prt$, and $A = \$4350$, $r = 9\%$, and $t = 5$ years, solve for p.

9. Find the perimeter of a rectangle in feet and inches if the length is 6 feet 5 inches and the width is 2 feet 7 inches.

10. Find the perimeter of a rectangle in feet and inches if the length is $8\frac{1}{2}$ feet and the width is $4\frac{3}{4}$ feet.

11. Triangle QRS is similar to triangle EFG. What are the lengths of sides QR and EG?

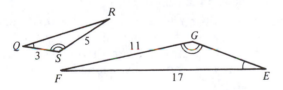

12. Triangle HJK is similar to triangle LMN. What are the lengths of sides MN and JH?

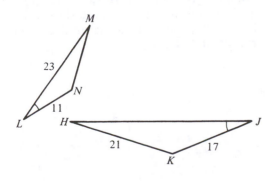

EXERCISES
12.4

Write each given statement in Exercises 1–6 as a direct variation formula. Use k as the constant of direct variation.

1. m varies directly as n.

2. p varies directly as q.

3. y varies directly as the square of x.

4. s varies directly as the cube of t.

5. v varies directly as the product of s and t.

6. y varies directly as the product of m and n.

In Exercises 7–18 write each statement as a direct variation formula. Use k as the constant of direct variation if it is not given.

7. At a constant rate of 24 knots, the distance d in nautical miles varies directly as the time t in hours.

8. The perimeter P of a square varies directly as the length s of a side. The constant of variation is 4.

9. Within a certain range, the amount that a spring stretches, s, varies directly as the weight of the object, w, hung on the spring.

10. The reaction distance is the distance a car travels between the time the driver gets a signal to stop and the time he puts his foot on the brake. The reaction distance d varies directly as the car's speed s.

11. At \$6 per hour, the gross wages w vary directly as the number of hours worked h.

12. The estimated cost c of a building varies directly as the number of square feet of floor space f.

13. The weight w of a pipe in pounds varies directly as its length ℓ in feet.

14. The circumference C of a circle varies directly as its diameter d. The constant of variation is π.

15. The distance d a car travels after the brakes are applied varies directly as the square of the car's speed s.

16. The distance d traveled by a falling object,

starting from rest, varies directly as the square of the time t it traveled.

17. The volume of a sphere V varies directly as the cube of the radius r. The constant of variation is $\frac{4}{3}\pi$.

18. The horsepower h required to propel a motorboat varies directly as the cube of the speed s of the boat.

Find the constant of direct variation in Exercises 19–24.

19. s varies directly as t and $s = 14$ when $t = 7$

20. w varies directly as x and $w = 15$ when $x = 45$

21. p varies directly as the cube of d and $p = 24$ when $d = 2$

22. y varies directly as the square of x and $y = 30$ when $x = 3$

23. p varies directly as the product of s and t, and $p = 45$ when $s = 4$ and $t = 5$

24. v varies directly as the product of w and z, and $v = 36$ when $w = 9$ and $z = 6$

In Exercises 25–32 write a direct variation formula and then find:

25. y when $x = 8$, if y varies directly as x and $y = 4$ when $x = 6$.

26. s when $t = 4$, if s varies directly as t and $s = 25$ when $t = 5$.

27. p when $q = 21$, if q varies directly as p and $p = 10$ when $q = 15$.

28. r when $s = 18$, if s varies directly as r and $r = 12$ when $s = 9$.

29. y when $x = 5$, if y varies directly as the square of x and $x = 3$ when $y = 36$.

30. p when $q = 6$, if p varies directly as the square of q and $p = 8$ when $q = 4$.

31. w when $x = 6$, if w varies directly as the cube of x and $w = 9$ when $x = 3$.

32. t when $s = 3$, if t varies directly as the cube of s and $t = 128$ when $s = 4$.

33. The pressure beneath the surface of a liquid varies directly as the depth of the liquid. If the pressure is 120 pounds per square inch at a depth of 30 feet below the surface, what is the pressure at 350 feet below the surface?

34. The weight of an iron ingot varies directly as the volume of the ingot. The weight is 2500 pounds when the volume is 12 cubic feet. What is the volume when the weight is 3000 pounds?

35. The weight of a steel beam with uniform cross-sectional area varies directly as the length. If a 10-foot beam weighs 450 pounds, how much does an 8-foot beam weigh?

36. The velocity of an object falling under the influence of gravity varies directly as the time the object falls. The velocity of an object after falling 1 second is 32 feet/second. Find the velocity of an object after falling 3.5 seconds.

37. The volume of a sphere varies directly as the cube of the radius. If a sphere with radius 6 centimeters has volume 288π cm³, what is the volume of a sphere with radius 3 centimeters?

38. The volume of a cube varies directly as the cube of the length of an edge. If the volume is 8 cubic centimeters when an edge is 2 centimeters, find the volume when an edge is 5 centimeters.

39. The power needed to propel a boat varies directly as the cube of the speed of the boat. If 3 horsepower drives a boat at 6 mph, what power is needed to run it at 15 mph?

40. The amount of material needed to mold a solid ball varies directly as the cube of the radius of the ball. If 64 ounces of material are needed for a ball of radius 2 inches, how many ounces of material are needed for a ball of radius $\frac{3}{4}$ inch?

Calculator Exercises

Use a calculator to determine the answers to Exercises 41–46. Write a direct variation formula and then find:

41. y when $x = 150.4$, if y varies directly as x and $y = 75.2$ when $x = 37.6$.

42. r when $s = 55.62$, if r varies directly as s and $r = 259.56$ when $s = 111.24$.

43. y when $x = 15$, if y varies directly as the square of x and $y = 874.8$ when $x = 6$.

44. q when $p = 24$, if q varies directly as the square of p and $q = 44{,}100$ when $p = 35$.

45. y when $x = 3.5$, if y varies directly as the cube of x and $y = 16.9344$ when $x = 1.2$.

46. s when $t = 26$, if s varies directly as the cube of t and $s = 205{,}770$ when $t = 95$.

12.5
SOLVING EQUATIONS (III)

This section is devoted to applying the method for solving equations which we learned earlier to new situations.

> Remember that you may:
>
> 1. add any number to both sides of an equation,
> 2. subtract any number from both sides,
> 3. multiply both sides by any number, or
> 4. divide both sides by any number except zero.

EXAMPLE 1   Solve $4x + 5 = 29$ for x.

Solution

$$4x + 5 = 29$$
$$4x + 5 - 5 = 29 - 5 \quad \longleftarrow \text{Subtract 5 from both sides.}$$
$$4x = 24$$
$$\frac{4x}{4} = \frac{24}{4} \quad \longleftarrow \text{Divide both sides by 4.}$$
$$x = 6$$

Check:
$$4x + 5 \overset{?}{=} 29$$
$$4(6) + 5 \overset{?}{=} 29$$
$$24 + 5 \overset{?}{=} 29$$
$$29 = 29 \text{ is true.}$$

The solution of the equation is 6.

The objective in solving any equation is always the same—to add to, subtract from, multiply, or divide both sides of the equation to get the variable alone on one side of the equal sign.

EXAMPLE 2   Solve $\frac{4x}{5} - 6 = 12$ for x.

Solution

$$\frac{4x}{5} - 6 + 6 = 12 + 6 \quad \longleftarrow \text{Add 6 to both sides.}$$
$$\frac{4x}{5} = 18$$
$$\frac{5}{4}\left(\frac{4x}{5}\right) = \frac{5}{4}(18) \quad \longleftarrow \text{Multiply both sides by } \tfrac{5}{4}.$$
$$x = \frac{5}{4}(18)$$
$$x = \frac{45}{2} \quad \text{or} \quad x = 22.5$$

Check:
$$\frac{4x}{5} - 6 = 12$$

$$\frac{4(22.5)}{5} - 6 \overset{?}{=} 12$$

$$\frac{90}{5} - 6 \overset{?}{=} 12$$

$$18 - 6 \overset{?}{=} 12$$

$$12 = 12 \text{ is true.}$$

The solution is 22.5.

Sometimes an equation must be solved where the variable is on both sides of the equation. To solve such equations, terms must first be added to or subtracted from both sides of the equation to get the *variable terms on one side and number terms on the other side* of the equal sign.

EXAMPLE 3   Solve $4x + 10 = 6x - 8$ for x.

Solution   We must first get the variable terms on one side and the number terms on the other side of the equal sign.

$$4x + 10 = 6x - 8$$
$$4x - 4x + 10 = 6x - 4x - 8 \quad \text{— Subtract } 4x \text{ from both sides.}$$
$$10 = 2x - 8$$
$$10 + 8 = 2x - 8 + 8 \quad \text{— Add 8 to both sides.}$$
$$18 = 2x$$
$$\frac{18}{2} = \frac{2x}{2}$$
$$9 = x$$

Note that subtracting $4x$ from $4x$ gives zero. That is, $4x - 4x = 0$.

Sometimes one side of the equation must be simplified before the variable can be isolated on one side of the equation.

EXAMPLE 4   Solve $15 + x = 0.4(50 + x)$.

Solution   Before solving, the right-hand side of the equation must be simplified. That is, the quantity $50 + x$ must be multiplied by 0.4.

$$15 + x = 0.4(50 + x) \quad \text{— Multiply using } a(b+c) = ab + ac.$$
$$15 + x = 0.4(50) + 0.4x$$
$$15 + x = 20 + 0.4x$$
$$15 + x - 0.4x = 20 + 0.4x - 0.4x$$
$$15 + 1x - 0.4x = 20 \quad \text{— } 0.4x - 0.4x = 0 \text{ and } x = 1x.$$
$$15 + 0.6x = 20$$
$$0.6x = 5$$
$$\frac{0.6x}{0.6} = \frac{5}{0.6}$$
$$x = 8\frac{1}{3}$$

The solution of the equation is $8\frac{1}{3}$ or $\frac{25}{3}$. The solution is checked by substituting this value into the original equation.

A troublesome situation is what to do with, for example, $-x = 6$. If the opposite of a number is 6, then the number is -6. That is, if $-x = 6$, then $x = -6$. A situation like this is shown in the next example.

EXAMPLE 5   Solve $3x = 4x + 9$ for x.

Solution

$$3x - 4x = 4x - 4x + 9$$
$$-1x = 9$$
$$-x = 9$$
$$x = -9$$

Similarly, if, for example, $-x = -7$, then $x = 7$.

Review Exercises

Simplify in Exercises 1–4.

1. $(18.45)(0.084)$ 2. $(0.076)(8.05)$

3. $5(-3) - (-8) - (5 - 8)$

4. $-(-9) - (3 - 7) - 2(3)$

5. Give your answer to the nearest hundredth: $467.8 \div 38$.

6. Give your answer to the nearest thousandth: $6.76 \div 20.8$.

7. Solve $\frac{7}{15} = \frac{x}{48}$ for x.

8. Solve $\frac{x}{32} = \frac{18}{5}$ for x.

9. Solve $ay + b = x - d$ for x when $a = 4$, $y = -3$, $b = 6$, and $d = -4$.

10. Solve $y - 3d = m + rs$ for y when $m = 6$, $r = -5$, $s = 2$, and $d = -1$.

11. Triangle RST is similar to triangle RMN. What are the lengths of segments RM and NT?

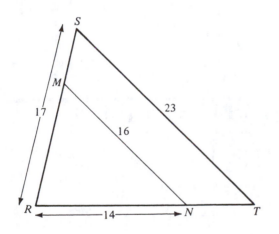

12. Triangle XYZ is similar to triangle UVZ. What are the lengths of segments YZ and XV?

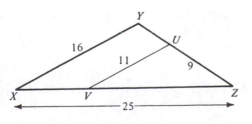

EXERCISES
12.5

Solve for x in Exercises 1–70.

1. $3x - 5 = 1$
2. $4x + 6 = 26$
3. $8 = 2x + 6$
4. $15 = 3x - 9$
5. $5x - 10 = 5$
6. $4x - 2 = 30$
7. $13 = 1 + 12x$
8. $6 = 6 + 6x$
9. $2x + 2 = 2$
10. $7x + 3 = 10$
11. $38 = 9x - 7$
12. $50 = 6x - 10$
13. $\frac{x}{6} + 4 = 12$
14. $\frac{x}{2} + 5 = 11$
15. $10 = \frac{x}{3} - 8$
16. $12 = \frac{x}{8} - 2$
17. $22 + \frac{x}{9} = 56$
18. $4 + \frac{x}{4} = 14$
19. $\frac{x}{7} - 3 = 4$
20. $\frac{x}{13} - 8 = 2$
21. $5 = 5 + \frac{x}{6}$
22. $4 = 4 + \frac{x}{9}$
23. $-6x + 4 = 16$
24. $-5x + 7 = 17$
25. $4 - 3x = 0$
26. $5 - 8x = 0$
27. $15 = 9 - 3x$
28. $12 = 20 - 2x$
29. $11 = 4 - 5x$
30. $9 = 3 - 2x$
31. $4x - x + 2 = 20$
32. $7x - 2x + 1 = 31$
33. $5x - 2 = 19 - 2x$
34. $4 - 3x = 14 - 8x$
35. $6x - 2 = 2x + 6$
36. $8x + 3 = 5x - 2$
37. $3x + 8 = x$
38. $5x + 9 = 2x$
39. $5x = 11 - 2x$
40. $6x = 26 - 3x$
41. $4x + 5 = 3x$
42. $7x - 8 = 6x$
43. $9x = 8x - 6$
44. $12x = 11x + 7$
45. $5x - 4 = 4x$
46. $9x - 12 = 8x$
47. $3x - x - 12x = 30$
48. $x + 6x - 10x = -20$
49. $8x - 4x = -9$
50. $2x + 12 = 7x + 2$
51. $2(x - 1) = x + 4$
52. $5(x + 1) = x - 7$
53. $2(x + 1) = 3(x - 3)$
54. $-2(x + 1) = 3(x + 6)$
55. $2(x - 5) = 24$
56. $20 = 5(2x + 1)$
57. $-5x - 5 = -6(2x + 5)$

58. $2(2x + 7) = 21$

59. $2(x + 1) + 2(x + 6) = 38$

60. $2(3x + 4) + 2(x + 5) = 60$

61. $0.4(x + 8) = 0.6x + 3$

62. $0.1x + 9 = 0.3(x - 4)$

63. $0.85(x - 4) = 0.6x + 8$

64. $0.6x + 6 = 0.75(x + 2)$

65. $\dfrac{3x - 7}{2} = 4$ 66. $\dfrac{5x - 10}{8} = 0$

67. $\dfrac{x + 3}{2} = 6 - x$ 68. $5 + x = \dfrac{2 + x}{4}$

69. $2x(x - 6) = 2x^2 - 48$

70. $2x^2 + 40 = 2x(x + 5)$

Calculator Exercises

Using a calculator, solve for x in Exercises 71–76. Give your answers to four decimal places.

71. $5.36x + 13.87 = 403.28$

72. $7.02x - 90.37 = 0.052$

73. $3.25 + 16.94x = 4.72x + 27.35$

74. $75.2 - 8.58x = 6.57x - 12.48$

75. $4.3(2.97x + 3.8) = -5.9(0.37 - 11.43x)$

76. $\dfrac{17.9x - 23.84}{6.5} = 12.49 - 2.34x$

12.6
SOLVING FORMULAS USING NUMBERS (III)

This section further expands the ideas and skills of Sections 10.3 and 10.9. We will solve formulas for a particular variable when the values for all but one letter are given.

EXAMPLE 1 Solve $\frac{ax}{b} + c = d$ for x when $a = -2$, $b = 5$, $c = 6$, and $d = 4$.

Solution First, substitute the given values for the variables:

$$\frac{ax}{b} + c = d$$

$$\frac{-2x}{5} + 6 = 4.$$

Then solve for the unknown:

$$\frac{-2x}{5} + 6 - 6 = 4 - 6 \longleftarrow \text{Subtract 6 from both sides.}$$

$$\frac{-2x}{5} = -2$$

$$\left(\frac{-2x}{5}\right)5 = (-2)5 \longleftarrow \text{Multiply both sides by 5.}$$

$$-2x = -10$$

$$\frac{-2x}{-2} = \frac{-10}{-2} \longleftarrow \text{Divide both sides by } -2.$$

$$x = 5$$

> **To find the unknown quantity in a formula:**
>
> 1. substitute the known values into the formula,
> 2. simplify if necessary, and
> 3. solve the equation for the unknown (variable).

EXAMPLE 2 Use the formula $h = vt - 16t^2$ to find v when $h = 80$ feet and $t = 4$ seconds.

Solution

$$h = vt - 16t^2$$
$$80 = v(4) - 16(4^2)$$
$$80 = v(4) - 16(16)$$
$$80 = v(4) - 256$$
$$336 = v(4)$$
$$\frac{336}{4} = \frac{v(4)}{4}$$
$$84 = v$$

The 16 in the formula is 16 feet/second², so the units of v are feet/second. Therefore, $v = 84$ feet/second.

EXAMPLE 3 Use the formula $A = p + prt$ to find t when $A = \$1450$, $p = \$1000$, and $r = 7\frac{1}{2}\%$ per year.

Solution Change $7\frac{1}{2}\%$ to a decimal (0.075) and then substitute given values for the variables.

$$A = p + prt$$
$$1450 = 1000 + (1000)(0.075)t$$

Then solve for t:

$$1450 = 1000 + (75)t$$
$$450 = 75t$$
$$\frac{450}{75} = \frac{75t}{75}$$
$$6 = t.$$

Since $r = 7\frac{1}{2}\%$ per *year*, t is 6 years.

The next example shows that sometimes one side of an equation can be simplified quite a bit before the equation is solved.

EXAMPLE 4 Solve $V = IR_1 + \frac{P}{I} + IR_2$ for P (in watts) when $V = 18$ volts, $I = 3$ amperes, $R_1 = 1.5$ ohms, and $R_2 = 3.5$ ohms.

Solution

$$V = IR_1 + \frac{P}{I} + IR_2$$
$$18 = 3(1.5) + \frac{P}{3} + 3(3.5)$$
$$18 = 4.5 + \frac{P}{3} + 10.5$$
$$18 = \frac{P}{3} + 4.5 + 10.5$$
$$18 = \frac{P}{3} + 15$$
$$18 - 15 = \frac{P}{3} + 15 - 15 \longleftarrow \text{Subtract 15 from both sides.}$$
$$3 = \frac{P}{3}$$
$$(3)3 = \left(\frac{P}{3}\right)3 \longleftarrow \text{Multiply both sides by 3.}$$
$$9 = P$$

Therefore, P is 9 watts.

The next example shows one side of an equation being simplified using $a(b + c) = ab + ac$ (the distributive property) before the equation is solved. That is, a number must be multiplied by a quantity before the equation can be solved.

EXAMPLE 5 Use the formula $A = \frac{1}{2}h(b_1 + b_2)$ for the area of a trapezoid to find b_1 when $A = 56$ square feet, $h = 8$ feet, and $b_2 = 5.5$ feet.

Solution

$$A = \frac{1}{2}h(b_1 + b_2)$$
$$56 = \frac{1}{2}(8)(b_1 + 5.5)$$
$$56 = 4(b_1 + 5.5)$$
$$56 = 4b_1 + 4(5.5) \longleftarrow \text{Multiply using } a(b + c) = ab + ac.$$
$$56 = 4b_1 + 22$$
$$34 = 4b_1$$
$$\frac{34}{4} = \frac{4b_1}{4}$$
$$8.5 = b_1$$

The unit of b_1 is feet, so b_1 is 8.5 feet, or 8 feet 6 inches.

Review Exercises

Find the value in Exercises 1–6.

1. (1000)(0.0576) 2. (428.5)(100)
3. $4.32 + 17 - 8.6$ 4. $15 + 2.9 - 0.65$
5. 48% of 695 6. 65% of 284
7. Fourteen is what percent of 20?
8. Thirty-nine is what percent of 260?
9. "Add on" 1.5% of 15 to 15.
10. "Add on" 2.6% of 34 to 34.
11. Simplify: $5x - 3y - 2x$.
12. Simplify: $x + 6y - 9x$.

EXERCISES
12.6

Given the formula $\frac{ax}{b} + c = d$ and the known quantities in Exercises 1–4, solve for the unknown.

1. $a = 5, b = 2, c = 4, d = -8$: Solve for x.
2. $a = 8, b = -3, c = 9, d = 4$: Solve for x.
3. $a = 11, b = 4, c = -1, d = 5$: Solve for x.
4. $a = -3, b = 8, c = 6, d = 9$: Solve for x.

Given the known quantities in Exercises 5–8 and the formula $h = vt - 16t^2$, where h = height, v = velocity, and t = time, solve for the unknown.

5. $h = 144$ feet, $t = 3$ seconds: Solve for v.
6. $h = 296$ feet, $t = 2$ seconds: Solve for v.
7. $h = 100$ feet, $t = 2.5$ seconds: Solve for v.
8. $h = 224$ feet, $t = 3.5$ seconds: Solve for v.

Given the formula $y = mx + b$ and the known quantities, solve for the unknown in Exercises 9–12.

9. $m = \frac{2}{3}, b = 8, y = 4$: Solve for x.
10. $m = -\frac{3}{4}, b = 5, y = 6$: Solve for x.
11. $x = 2, y = 3, b = \frac{1}{2}$: Solve for m.
12. $x = -5, y = 7, b = -8$: Solve for m.

Given the known quantities in Exercises 13–16 and the formula $A = p + prt$, where p = principal, r = rate of interest, t = time, and A = accumulated money, solve for the unknown.

13. $A = \$350, p = \$250, r = 10\%$ per year: Solve for t.
14. $A = \$2604, p = \$2400, r = 8.5\%$ per year: Solve for t.
15. $A = \$575, p = \$500, t = 2.5$ years: Solve for r.
16. $A = \$3330, p = \$3000, t = 2$ years: Solve for r.

Given the known quantities in Exercises 17–20 and the formula $V = IR_1 + \frac{P}{I} + IR_2$, where V = voltage, I = current, P = power (in watts), R_1 = resistance, and R_2 = resistance, solve for the unknown.

17. $V = 120$ volts, $I = 5$ amperes, $R_1 = 4.8$ ohms, $R_2 = 6.2$ ohms: Solve for P.
18. $V = 120$ volts, $I = 8.2$ amperes, $R_1 = 6$ ohms, $R_2 = 4$ ohms: Solve for P.
19. $V = 270$ volts, $I = 12$ amperes, $R_1 = 8$ ohms, $P = 1152$ watts: Solve for R_2.
20. $V = 4000$ volts, $I = 30$ amperes, $R_2 = 76.7$ ohms, $P = 4800$ watts: Solve for R_1.

Given the known quantities in Exercises 21–24 and the formula $A = \frac{1}{2}h(b_1 + b_2)$, where A = area of a trapezoid, h = height, b_1 = length of a base, and b_2 = length of a base, solve for the unknown. Answer in feet and/or inches.

21. $A = 39$ square inches, $h = 6$ inches, $b_1 = 5$ inches: Solve for b_2.
22. $A = 84$ square feet, $h = 12$ feet, $b_2 = 10$ feet: Solve for b_1.
23. $A = 42$ square feet, $h = 8$ feet, $b_2 = 6$ feet: Solve for b_1.
24. $A = 30$ square feet, $h = 9$ feet, $b_1 = 4$ feet: Solve for b_2.

Given the known quantities in Exercises 25–28 and the formula $P = 2\ell + 2w$, where P = perimeter of a rectangle, ℓ = length, and w = width, solve for the unknown. Answer in feet and inches.

25. $P = 50$ feet, $\ell = 8$ feet: Solve for w.
26. $P = 21$ inches, $w = 6$ inches: Solve for ℓ.
27. $P = 36$ feet, $\ell = 14$ feet: Solve for w.
28. $P = 21$ feet 4 inches, $w = 6$ feet: Solve for ℓ.

Given the known quantities in Exercises 29–32 and the formula $W = 1.5D + 0.125$, where W = the distance between the parallel sides of a bolt head, and D = diameter of a bolt, solve for the unknown.

29. $W = 3.125$ inches: Solve for D.
30. $W = 0.875$ inches: Solve for D.
31. $W = 2.375$ inches: Solve for D.
32. $W = 2.75$ inches: Solve for D.

33. Given $y = \dfrac{2}{3}x + 8$, solve for x when $y = 4$.

34. Given $y = -\dfrac{3x}{4} + 5$, solve for x when $y = 6$.

35. Given $S = L + 2B$, solve for B when $S = 26$ square meters and $L = 16$ square meters.

36. Given $F = k(R - L)$, solve for R when $k = 45$, $L = 0.4$, and $F = 9$.

37. Given $v_2 = v_1 + at$, solve for a when $v_2 = 24$ feet/second, $v_1 = 8$ feet/second, and $t = 2$ seconds.

38. Given $v_2 = v_1 + at$, solve for a when $v_2 = 18$ feet/second, $v_1 = 30$ feet/second, and $t = 3$ seconds.

39. Given $F = \dfrac{9}{5}C + 32$, solve for C when $F = 59°$.

40. Given $P = n(p - c)$, solve for p when $P = \$300$, $n = 25$, and $c = \$25$.

41. Given $v = 60 - 32t$, solve for t when $v = 12$ feet/second.

42. Given $D = D_1 + 2s$, solve for s when $D = 8$ inches and $D_1 = 6.5$ inches.

43. Given $N = DP - 2$, solve for P when $N = 82$ and $D = 12$ inches.

Calculator Exercises

Use a calculator to solve for the unknown in Exercises 44–49.

44. Given $h = vt - 16t^2$, where $t = 4.3$ seconds and $h = 217.85$ feet, solve for v.

45. Given $A = p + prt$, where $p = \$2875$, $r = 7.25\%$, and $A = \$3032.27$, solve for t.

46. Given $A = \dfrac{1}{2}h(b_1 + b_2)$, where $A = 1676.57$ cm², $h = 27.85$ cm, and $b_2 = 84.95$ cm, solve for b_1.

47. Given $P = n(p - c)$, where $p = \$387.57$, $n = 142$, and $P = \$47,369.78$, solve for c.

48. Given $A = p(1 + rt)$, where $A = \$456.34$, $r = 8.27\%$, and $t = 1.39$ years, solve for p.

49. Given $\dfrac{ax}{b} + c = d$, where $a = 3.51$, $b = 7.29$, $c = 14.03$, and $d = 6.48$, solve for x.

12.7
SOLVING FORMULAS USING LETTERS (III)

The formulas you solved in Sections 10.4 and 10.10 are formulas found in science and technology. You have used the following skills in solving equations and formulas:

1. Simplify the expression on each side of the equal sign if possible.
2. To get the variable alone on one side of the equal sign, use the proper combination of:
   (a) adding the same number or letter to each side
   (b) subtracting the same number or letter from each side
   (c) multiplying each side by the same number or letter
   (d) dividing each side by the same number or letter (except zero).

This section discusses other examples of combining these skills.

EXAMPLE 1   Solve $N = DP - 2$ for P.

Solution

$$N = DP - 2$$
$$N + 2 = DP - 2 + 2 \quad \text{Add 2 to both sides.}$$
$$N + 2 = DP$$
$$\frac{N + 2}{D} = \frac{DP}{D} \quad \text{Divide both sides by } D.$$
$$\frac{N + 2}{D} = P$$

The fraction bar must be under the quantity $N + 2$ because the entire left side of the formula was divided by D.

EXAMPLE 2   Solve $P = 2\ell + 2w$ for w.

Solution

$$P = 2\ell + 2w$$
$$P - 2\ell = 2\ell - 2\ell + 2w \quad \text{Subtract } 2\ell \text{ from both sides.}$$
$$P - 2\ell = 2w$$
$$\frac{P - 2\ell}{2} = \frac{2w}{2} \quad \text{Divide both sides by 2.}$$
$$\frac{P - 2\ell}{2} = w$$

WATCH OUT! Note that the 2's may not be cancelled in the expression $\frac{P-2\ell}{2}$, since there is subtraction. Fractions are reduced by cancelling only when the number in question is a factor of the numerator and denominator.

EXAMPLE 3   Solve $y = \frac{3}{4}x + 9$ for x.

Solution

$$y = \frac{3}{4}x + 9$$
$$y - 9 = \frac{3}{4}x$$
$$\frac{4}{3}(y - 9) = \frac{4}{3}\left(\frac{3}{4}x\right) \quad \text{Multiply both sides by } \frac{4}{3}.$$
$$\frac{4}{3}(y - 9) = x$$
$$\frac{4}{3}(y) - \frac{4}{3}(9) = x \quad \text{Simplify using } a(b - c) = ab - ac.$$
$$\frac{4}{3}y - 12 = x$$

Each part of the difference $y - 9$ is multiplied by $\frac{4}{3}$ so that the answer can be written without parentheses.

EXAMPLE 4   Solve $S = \frac{n}{2}(a + t)$ for t.

Solution

$$S = \frac{n}{2}(a + t)$$

$$S = \frac{n}{2}a + \frac{n}{2}t \quad \longleftarrow \quad \text{Simplify using } a(b+c) = ab + ac.$$

$$S - \frac{n}{2}a = \frac{n}{2}t$$

Before multiplying both sides by $\frac{2}{n}$, the left side can be simplified.

$$S - \frac{na}{2} = \frac{n}{2}t$$

$$S\left(\frac{2}{2}\right) - \frac{na}{2} = \frac{n}{2}t$$

$$\frac{2S}{2} - \frac{na}{2} = \frac{n}{a}t$$

$$\frac{2S - na}{2} = \frac{n}{2}t$$

$$\frac{2}{n}\left(\frac{2S - na}{2}\right) = \frac{2}{n}\left(\frac{n}{2}t\right) \quad \longleftarrow \quad \text{Multiply both sides by } \frac{2}{n}.$$

$$\frac{2}{n}\left(\frac{2S - na}{2}\right) = t$$

$$\frac{2S - na}{n} = t$$

This formula could also have been solved for t as follows:

$$S = \frac{n}{2}(a + t)$$

$$S = \frac{n}{2}a + \frac{n}{2}t$$

$$S - \frac{n}{2}a = \frac{n}{2}t$$

$$\frac{2}{n}\left(S - \frac{n}{2}a\right) = \frac{2}{n}\left(\frac{n}{2}t\right)$$

$$\frac{2}{n}\left(S - \frac{n}{2}a\right) = t$$

$$\frac{2}{n}(S) - \frac{2}{n}\left(\frac{n}{2}a\right) = t$$

$$\frac{2S}{n} - a = t$$

The two answers $\dfrac{2S - na}{n}$ and $\dfrac{2S}{n} - a$ are different forms of the same expression. To convince you of this, $\dfrac{2S}{n} - a$ can be changed to $\dfrac{2S - na}{n}$ as follows:

$$\frac{2S}{n} - a = \frac{2S}{n} - a\left(\frac{n}{n}\right)$$

$$= \frac{2S}{n} - \frac{na}{n}$$

$$= \frac{2S - na}{n}.$$

Unless a certain form is requested, either answer is acceptable.

The next example shows that, when the variable you are solving for appears on both sides, the first task is to add to or subtract from both sides to get all terms with that variable on one side of the equation.

EXAMPLE 5   Solve $Q_1 = P(Q_2 - Q_1)$ for Q_1.

Solution

$$Q_1 = P(Q_2 - Q_1)$$

$$Q_1 = PQ_2 - PQ_1$$

$$Q_1 + PQ_1 = PQ_2 \quad \longleftarrow \quad PQ_1 \text{ added to both sides}$$

$$1Q_1 + PQ_1 = PQ_2$$

$$(1 + P)Q_1 = PQ_2$$

$$\frac{(1 + P)Q_1}{1 + P} = \frac{PQ_2}{1 + P} \quad \longleftarrow \quad \text{Divide both sides by } 1 + P.$$

$$Q_1 = \frac{PQ_2}{1 + P}$$

Review Exercises

Find the value of the expressions in Exercises 1–4.

1. $\sqrt{81}$
2. $\sqrt{100}$
3. $\sqrt[3]{125}$
4. $\sqrt[3]{64}$
5. Solve for x, $\dfrac{3}{x} = \dfrac{19}{9}$
6. Solve for y, $\dfrac{7}{27} = \dfrac{2y}{9}$

7. Use the formula $\dfrac{D}{d} = \dfrac{s}{S}$ to find S when $d = 3$ in., $D = 14$ in. and $s = 280$ rpm.

8. Use the formula $\dfrac{T}{t} = \dfrac{n}{N}$ to find t when $T = 60$, $n = 80$ rpm and $N = 320$ rpm.

9. Solve for x, $2(3x - \frac{1}{2}) = 17$

10. Solve for x, $3(x + 1) + 2(x - 2) = 9$

11. Solve for x, $\dfrac{x}{3} - 3 = 7 - x$

12. Solve for x, $5 + x = \dfrac{2 + x}{4}$

EXERCISES

12.7

In Exercises 1–28 solve the equation for the indicated letter.

1. $ax + b = c$ for x
2. $cx - d = f$ for x
3. $g = m + by$ for y
4. $j = k - ay$ for y
5. $n(x + c) = j$ for x
6. $(y - x)p = q$ for y
7. $y = \dfrac{2}{3}x + 8$ for x
8. $y = \dfrac{5}{2}x - 10$ for x
9. $y = 8x + b$ for b
10. $y = -6x + b$ for b
11. $y - k = 2 - 3y$ for y
12. $2m + p = 4m + q$ for m
13. $5n - k = n + r$ for n
14. $s - x = 2t - 5x$ for x
15. $m + 2n - p = q$ for n
16. $4k + 5g + 2m = k$ for m
17. $3k + 2 = r$ for k
18. $5 = 2p - q$ for p
19. $q = 6(L - p)$ for L
20. $(R + k)3 = m$ for R
21. $\dfrac{k - m}{r} = q$ for k

22. $p = \dfrac{A - b}{c}$ for A

23. $p = Q(r - d)$ for Q

24. $S(c + m) = g$ for S

25. $\dfrac{2}{3}x + b = c$ for x

26. $d = f - \dfrac{3}{5}y$ for y

27. $ax + b = x - d$ for x

28. $y - 3d = m + ry$ for y

The formulas in Exercises 29–50 arise in various technical areas. Solve for the indicated letter.

29. $R = \dfrac{kL}{D^2}$ for L (electricity)

30. $C = \dfrac{H}{L^2}a^2$ for H (civil highway)

31. $F = A_2 - A_1 + P(V_2 - V_1)$ for A_2 (energy)

32. $r = \dfrac{R^2}{2h} + \dfrac{h}{2} + B$ for B

33. $D = D_1 + 2s$ for s (gears)

34. $W = 1.5D + 0.125$ for D (mechanics)

35. $F = \dfrac{9}{5}C + 32$ for C (temperature)

36. $A = p + prt$ for r (interest)

37. $R = aT + b$ for T (temperature)

38. $h = vt - 16t^2$ for v (velocity)

39. $S = \pi rh + \pi r^2$ for h (area)

40. $s = s_0 + v_0 t - 16t^2$ for v_0 (physics: motion)

41. $L = \pi(r_1 + r_2) + 2d$ for d (mechanics)

42. $S = 2\ell w + 2wh + 2\ell h$ for w (area)

43. $d = vt + \dfrac{1}{2}at^2$ for v (distance)

44. $V = IR_1 + IR_2$ for R_2 (electricity)

45. $V = I(R_1 + R_2)$ for R_1 (electricity)

46. $H = R(S + L)$ for L

47. $L = L_0(1 + at)$ for t (temperature expansion)

48. $L = \pi(r_1 + r_2) + 2d$ for r_1 (mechanics)

49. $R = \dfrac{kA(T_1 + T_2)}{d}$ for T_1 (heat)

50. $P = \dfrac{V_1(V_2 - V_1)}{g^j}$ for V_2 (jet engine power)

Calculator Exercises

Use a calculator to solve for the unknown in Exercises 51–56. Give the answers to four decimal places.

51. Given $cx - d = f$, solve for x when $c = 8.47$, $d = 3.94$, and $f = 18.23$.

52. Given $n(x + c) = j$, solve for x when $n = 2.03$, $c = 5.46$, and $j = 3.82$.

53. Given $y - k = 2 - 6.4y$, solve for y when $k = 13.85$.

54. Given $4.96m + 2.39n - 18.42 = -9.34m$, solve for m when $n = 6.35$.

55. Given $\dfrac{3.6k - 4.37m}{8.4r} = 17.5 + 15.39k$, solve for k when $m = -6.48$ and $r = 0.26$.

56. Given $(h + d)7.49 = s - 5.3d$, solve for d when $h = -14.38$ and $s = -85.91$.

12.8
WRITTEN PROBLEMS (III)

The problem-solving skills used here are the same as those in Sections 10.6 and 10.12. To review that procedure:

1. Read the problem.
2. Represent the unknown with variables.
3. Make a chart or diagram if helpful.
4. Write an equation or formula.
5. Solve the equation or formula.
6. Answer the question.

EXAMPLE 1 Five less than twice a number is four. Find the number.

Solution Let x be the number. The equation, then, is:

$$2x - 5 = 4.$$

Next we solve the equation:

$$2x - 5 + 5 = 4 + 5$$
$$2x = 9$$
$$\frac{2x}{2} = \frac{9}{2}$$
$$x = \frac{9}{2}$$

The number is $\frac{9}{2}$, or 4.5.

Check: $2\left(\dfrac{9}{2}\right) - 5 = 4$ is true.

EXAMPLE 2 The length of a rectangle is 5 feet longer than three times the width. The perimeter is 20 feet. Find the dimensions of the rectangle.

Solution Let x be the width in feet. The length is 5 feet longer than three times x, or $5 + 3x$. The formula for the perimeter of a rectangle is:

$$P = 2\ell + 2w.$$

Substituting into the formula and solving the equation:

$$P = 2\ell + 2w$$
$$20 = 2(5 + 3x) + 2x$$
$$20 = 2(5) + 2(3x) + 2x$$
$$20 = 10 + 6x + 2x$$
$$20 = 10 + 8x$$
$$10 = 8x$$
$$\frac{10}{8} = \frac{8x}{8}$$
$$\frac{5}{4} = x.$$

The width is $\frac{5}{4}$, or 1.25 feet. The length is:

$$5 + 3x = 5 + 3(1.25)$$
$$= 5 + 3.75$$
$$= 8.75.$$

Thus, the dimensions of the rectangle are 1.25 feet by 8.75 feet. In feet and inches this is 1 foot 3 inches by 8 feet 9 inches.

EXAMPLE 3   On a trip between two towns, the train averages 10 miles per hour more than the bus. If the train makes the trip in 5 hours and the bus takes 6 hours, how far apart are the towns?

Solution   Let x be the rate of the bus. Then $x + 10$ is the rate of the train. Since $d = rt$, the distance between the towns for the bus is $6x$ miles, and for the train it is $5(x + 10)$ miles.

Town $\bullet$ —————Bus $6x$ miles————— $\bullet$ Town
————Train $5(x + 10)$ miles————

The sketch makes it clear that the distance traveled by the bus and train are equal, since they are traveling between the same two towns. Thus the equation is:

$$6x = 5(x + 10).$$

Solving the equation:

$$6x = 5(x) + 5(10)$$
$$6x = 5x + 50$$
$$x = 50.$$

The rate of the bus, then, is 50 mph. The distance traveled is $6x$ or (6 hours)(50 mph), which is 300 miles. Therefore, the towns are 300 miles apart.

EXAMPLE 4   How many grams of a mixture containing 30% copper must be melted with 100 grams of a mixture containing 50% copper to obtain a mixture containing 35% copper?

Solution   Let x be the number of grams of 30% copper mixture. The total amounts of each mix-

ture are x grams of the 30% mixture, 100 grams of the 50% mixture, and $x + 100$ grams of the 35% mixture. The amount of the 35% mixture is $x + 100$ grams because the 30% and 50% mixtures are to be combined to obtain the 35% mixture. The amounts of pure copper in each mixture, then, are:

30% mixture: $(30\%)(x \text{ g}) = 0.3x \text{ g}$
50% mixture: $(50\%)(100 \text{ g}) = 50 \text{ g}$
35% mixture: $(35\%)(x + 100)\text{g} = 0.35(x + 100)\text{g}$

The equation can be written now, since the sum of the pure copper in the 30% and 50% mixtures will equal the amount of pure copper in the 35% mixture.

$$0.3x + 50 = 0.35(x + 100)$$
$$0.3x + 50 = 0.35x + 0.35(100)$$
$$0.3x + 50 = 0.35x + 35$$
$$0.3x + 15 = 0.35x$$
$$15 = 0.05x$$
$$\frac{15}{0.05} = \frac{0.05x}{0.05}$$
$$300 = x$$

Therefore, 300 grams of 30% copper must be added to obtain a 35% copper mixture.

EXAMPLE 5   The length of a rectangle is twice the width. If the width is increased by 6 feet, the area is increased by 96 square feet. Find the dimensions of the original rectangle.

Solution   Let x be the width of the original rectangle. Then $2x$ is the length of the original rectangle. A sketch is helpful here.

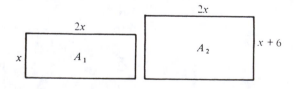

The area of the original rectangle, A_1, is $2x(x)$. The area of the new rectangle, A_2, is $2x(x + 6)$.

The equation is:

$$A_1 + 96 = A_2$$
or
$$2x(x) + 96 = 2x(x + 6).$$
Solving,
$$2x^2 + 96 = 2x(x) + 2x(6)$$
$$2x^2 + 96 = 2x^2 + 12x$$
$$2x^2 - 2x^2 + 96 = 2x^2 - 2x^2 + 12x$$
$$96 = 12x$$
$$8 = x.$$

The width of the original rectangle is 8 feet. The length is $2(x)$ or $2(8)$, which is 16 feet. Therefore, the dimensions of the original rectangle are 8 feet by 16 feet.

Review Exercises

1. Solve $cx - d = m$ for x.
2. Solve $y = px + q$ for x.
3. Solve $3x + 5 = 6x - 7$ for x.
4. Solve $8x - 9 = x + 5$ for x.
5. Solve $\dfrac{3x}{h - d} = 5$ for x.
6. Solve $c = \dfrac{5 + x}{8}$ for x.
7. $5.87 \div 100 = \underline{\hspace{1cm}}$
8. $3678 \div 1000 = \underline{\hspace{1cm}}$
9. Find the length of a leg of a right triangle to the nearest thousandth if the other leg is 15 cm and the hypotenuse is 18 cm.
10. Find the length of a leg of a right triangle to the nearest thousandth if the other leg is 13 cm and the hypotenuse is 16 cm.
11. Using the figure, find a.
12. Using the figure, find b.

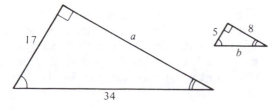

EXERCISES

12.8

In Exercises 1–8 write the necessary equation and find the number.

1. Eight more than three times a number is 29.
2. Twice a number increased by 12 is 44.
3. Four times a number decreased by 17 is 59.
4. Eleven less than five times a number is 49.
5. Five more than twice a number equals the sum of nine and the number.
6. Five less than four times the number is the sum of four and the number.
7. If $\frac{3}{8}$ of a number is subtracted from the number, the result is 45.
8. If a number is decreased by $\frac{4}{5}$ of the number, the result is 4.

In Exercises 9–14 write the necessary equation and find both numbers.

9. One number is 38 less than another number. Their sum is 146.
10. One number is 18 more than another. Their sum is 140.
11. One number is twice another. Their difference is 85.
12. One number is three times another. Their difference is 164.
13. One number is four less than another. The sum of the larger and three times the smaller is 216.
14. One number is five more than another. Six times the first, less the second, is 325.
15. The perimeter of a rectangle is 40 inches. The length is four inches longer than the width. Find the dimensions of the rectangle.
16. The perimeter of a rectangular lot is 160 feet. Its length is 20 feet more than its width. What are the dimensions of the lot?
17. A 32-inch board is divided into two pieces so that one piece is 6 inches shorter than the other. Find the length of each piece of board.

18. An 83-inch piece of metal stripping is cut into two pieces so that one piece is 7 inches longer than the other. How long is each piece of metal stripping?

19. The length of each of two equal sides of an isosceles triangle is 10 centimeters longer than the length of the third side. The perimeter is 68 centimeters. Find the lengths of the sides.

20. The length of the longest side of a triangle is twice that of the shortest side, and the third side is 8 feet longer than the shortest side. The perimeter is 44 feet. Find the lengths of the sides.

21. Two trains start from a city traveling in opposite directions. After 4 hours they are 320 miles apart, with one averaging 8 miles per hour more than the other. How fast is each train traveling?

22. Two people start in the same place and walk in opposite directions around a lake which has a 9-mile long shoreline. One person walks $\frac{1}{10}$ of a mile per hour faster than the other. If they meet in 2 hours, what is the average speed of each person?

23. A freight train leaves town X for town Y traveling at 30 mph. Four hours later a passenger train also leaves town X for town Y, traveling at 50 mph. How far from X does the passenger train pass the freight train?

24. A jet plane flying at 800 mph is to overtake another plane which started 3 hours earlier and is flying at 600 mph. How far from the starting point will the jet overtake the other plane?

25. How many ounces of a 20% alloy of silver must be added to 50 ounces of a 5% alloy to make a 15% alloy?

26. How many liters of a 20% salt solution must be added to 30 liters of a 50% salt solution to obtain a 40% salt solution?

27. How much pure silver must be added to 50 grams of a 70% silver alloy to make an 82% alloy?

28. How many ounces of acid must be added to 30 ounces of a 40% acid solution to make a 60% acid solution?

29. How many ounces of water must be added to 50 ounces of a 75% solution of acid to make a 30% solution?

30. How many liters of water should be added to one liter of pure acid to obtain a 25% acid solution?

31. If the two opposite sides of a square are decreased by 6 feet each, the area is decreased by 54 square feet. Find the length of a side of the square.

32. The length of a rectangle is three times its width. If the width is decreased by a foot, the area is decreased by 27 square feet. Find the dimensions of the original rectangle.

33. The length of a rectangle is 4 meters longer than the width. If the width were to be increased by 3 meters, the perimeter would be 38 meters. Find the dimensions of the original rectangle.

34. If the two opposite sides of a square were to be increased by 14 feet each, the perimeter would be 88 feet. Find the length of a side of the square.

Calculator Exercises

35. One number is 4.834 less than another. The sum of the larger and 6.2 times the smaller is 516.286. What are the two numbers?

36. One number is 13.24 greater than a second number. The sum of 3.5 times the greater number and 4.5 times the smaller number is 69.348. What are the two numbers?

37. The perimeter of a rectangle is 216.38 centimeters. The length of the rectangle is 5.75 centimeters longer than the width. What are the dimensions of the rectangle?

38. The perimeter of a triangle is 39.701 inches. The longest side is 12.4 inches longer than the shortest side, and the third side is 0.52 times as long as the longest side. What are the lengths of the three sides of the triangle?

12.9
SQUARE ROOTS IN FORMULAS (I)

The Pythagorean Theorem (Section 7.5) is a formula that uses square roots. In this section we look at other equations and formulas involving square roots. To review the definition of square roots:

If $c^2 = 25,$

then $c = \sqrt{25},$

or $c = 5.$

> In general, if $x^2 = b$, then x is a square root of b. (The number b is non-negative since any number squared is non-negative.)

If $x^2 = 49$, then $x = 7$ because $7^2 = 49$. It is also true that $x = -7$ since $(-7)^2 = 49$. Therefore, both 7 and -7 are square roots of 49. This is sometimes written $x = \pm 7$.

In solving equations and formulas in this section the first task will be to get the variable alone on one side. Then we will use the definition of square roots.

EXAMPLE 1 Solve $d = \frac{1}{2}gt^2$ for t when $g = 32$ feet/second^2, and $d = 144$ feet.

Solution

$$d = \frac{1}{2}gt^2$$

$$144 = \frac{1}{2}(32)t^2 \quad \longleftarrow \text{Substitute given values for variables.}$$

$$144 = 16t^2$$

$$\frac{144}{16} = \frac{16t^2}{16} \quad \longleftarrow \text{Divide both sides by 16.}$$

$$9 = t^2$$

Then t will equal a square root of 9. The square roots of 9 are 3 and -3. (This can be written $t = \pm 3$.) The units for t are seconds, so only the positive square root is a possible answer. Therefore, $t = 3$ seconds.

EXAMPLE 2 Find the length of a side of a square with area 92 square meters.

Solution The formula for the area of a square is:

$$A = s^2.$$

Substituting for A,

$$92 = s^2.$$

Solving for s,

$$\sqrt{92} = s.$$

Again, the only possible answer is the positive square root of 92, since s is a length. From a square root table, Table 1, at the back of the book, $\sqrt{92}$ is approximately 9.592, so $s \doteq 9.6$ meters.

EXAMPLE 3 Solve $E = mgh + \frac{1}{2}mv^2$ for v (velocity in meters/second) when $E = 10$ joules, $m = 0.05$ kilogram, $h = 10$ meters, and $g = 9.8$ meters/second.

Solution

$$E = mgh + \frac{1}{2}mv^2$$

$$10 = (0.05)(9.8)(10) + \frac{1}{2}(0.05)v^2$$

$$10 = 4.9 + 0.025v^2$$
$$10 - 4.9 = 4.9 - 4.9 + 0.025v^2$$
$$5.1 = 0.025v^2$$
$$\frac{5.1}{0.025} = v^2$$
$$204 = v^2$$
$$\pm\sqrt{204} = v$$
$$\pm 14.3 \doteq v$$

Without further knowledge it cannot be determined whether the velocity is positive or negative. Therefore, we cannot decide whether to use the positive or negative square root. So v is approximately ± 14.3 meters/second. The notations $\pm\sqrt{204}$ and ± 14.3 indicate that both the positive and negative numbers are to be considered.

Many formulas can be solved without knowing values for the variables. In the following cases, the solutions will contain a square root radical.

EXAMPLE 4   Solve $A = \pi r^2$ (area of a circle) for r.

Solution

$$A = \pi r^2$$
$$\frac{A}{\pi} = \frac{\pi r^2}{\pi}$$
$$\frac{A}{\pi} = r^2$$
$$\sqrt{\frac{A}{\pi}} = r$$

We need only the non-negative square root here, since physical units of length and area are always non-negative.

EXAMPLE 5   Solve $V = \frac{1}{3}\pi r^2 h$ (volume of a cone) for r.

Solution

$$V = \frac{1}{3}\pi r^2 h$$

$$3(V) = 3\left(\frac{1}{3}\pi r^2 h\right)$$

$$3V = \pi r^2 h$$

$$\frac{3V}{\pi h} = r^2$$

$$\sqrt{\frac{3V}{\pi h}} = r$$

Again, we need only the non-negative square root here, since units of length and volume are non-negative.

The following rule is the key to solving an equation or formula where the variable is squared.

If $x^2 = a$, then $x = \pm\sqrt{a}$.

Before looking at a formula which involves a cube root let us review the definition of a cube root.

If $x^3 = b$, then x is a cube root of b.

The cube root of b is written $\sqrt[3]{b}$. The sign of the cube root is the same as the sign of the number. For example, if $x^3 = 64$, then $x = \sqrt[3]{64}$ or 4 because $4^3 = 64$. The cube root of 64 is 4. This is written $\sqrt[3]{64} = 4$. In general, the key to solving an equation or formula where the variable is cubed is the following rule.

If $x^3 = a$, then $x = \sqrt[3]{a}$.

EXAMPLE 6 Solve $V = \frac{4}{3}\pi r^3$ (volume of a sphere) for r.

Solution The algebra is less complex if the formula is rewritten first. Multiplying by $\frac{4}{3}$ is the same as multiplying by 4 and then dividing by 3. So

$$V = \frac{4}{3}\pi r^3$$

can be rewritten as

$$V = \frac{4\pi r^3}{3}.$$

Solving for r,

$$3(V) = 3\left(\frac{4\pi r^3}{3}\right)$$

$$3V = 4\pi r^3$$

$$\frac{3V}{4\pi} = r^3$$

$$\sqrt[3]{\frac{3V}{4\pi}} = r \longleftarrow \begin{array}{l}\text{If } x^3 = a, \\ \text{then } x = \sqrt[3]{a}.\end{array}$$

Review Exercises

1. Solve $5x + 9 = 3 - x$ for x.
2. Solve $x - 11 = 3x + 1$ for x.
3. Solve $\frac{3x + 5}{2} = 7$ for x.
4. Solve $8 = \frac{2x - 6}{4}$ for x.
5. Given $P = 2\ell + 2w$, solve for w when $P = 96$ and $\ell = 15$.
6. Given $P = 2\ell + 2w$, solve for w when $P = 84$ and $\ell = 9$.
7. How many significant digits are there in the number 875,400?
8. How many significant digits are there in the number 0.07150?
9. Find the distance from the origin of a set of coordinate axes to the point $P(-3, 8)$.
10. Find the distance from the origin of a set of coordinate axes to the point $Q(-7, -4)$.

11. Find x.

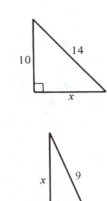

12. Find x.

EXERCISES

12.9

In Exercises 1–16 solve the equation for the unknown. Then use the tables of square roots and cube roots to find the solution of each equation to the nearest thousandth.

1. $x^2 = 8$
2. $x^2 = 14$
3. $93 = y^3$
4. $76 = y^3$
5. $2x^2 = 50$
6. $3x^2 = 48$
7. $5m^3 = 155$
8. $4n^3 = 172$
9. $11 = \frac{x^2}{3}$
10. $4 = \frac{x^2}{6}$
11. $8 = \frac{2x^3}{3}$
12. $15 = \frac{5x^3}{2}$
13. $49 = x^2 - 9$
14. $64 = y^2 + 36$
15. $3x^2 + 8 = 17$
16. $6 + 4x^2 = 38$

Solve for the indicated letter in Exercises 17–24.

17. $a = m^2$ for m
18. $y^2 = n$ for y
19. $p = 2q^2$ for q
20. $r = 5s^2$ for s
21. $\frac{n^2}{6} = m$ for n
22. $\frac{q^2}{r} = p$ for q
23. $h = \frac{3r^2}{m}$ for r
24. $t = \frac{ax^2}{2}$ for x

In Exercises 25–38 use the given information and solve for the indicated unknown. Use the tables of square roots and cube roots (Tables 1 and 2) to express each answer to the nearest thousandth.

25. $A = s^2$ for s when $A = 30$ square feet.

26. $A = s^2$ for s when $A = 96$ square meters

27. $A = \pi r^2$ for r when $A = 32\pi$ square centimeters

28. $A = \pi r^2$ for r when $A = 78\pi$ square inches

29. $P = I^2 R$ for I (in amperes) when $P = 312$ watts and $R = 52$ ohms

30. $P = I^2 R$ for I (in amperes) when $P = 476$ watts and $R = 28$ ohms

31. $V = \frac{1}{3}\pi r^2 h$ for r when $V = 84\pi$ cubic feet and $h = 6$ feet

32. $V = \pi r^2 h$ for r when $V = 256\pi$ cubic meters and $h = 8$ meters

33. $H = 0.4d^2 n$ for d (inches) when $H = 64$ horsepower and $n = 8$

34. $H = 0.4d^2 n$ for d (inches) when $H = 24$ horsepower and $n = 6$

35. $V = \frac{4}{3}\pi r^3$ for r when $V = 48\pi$ cubic feet

36. $V = \frac{4}{3}\pi r^3$ for r when $V = 128\pi$ cubic meters

37. $E = mgh + \frac{1}{2}mv^2$ for v (velocity in feet/second) when $E = 410$ joules, $m = 0.5$ pound, $h = 0.20$ feet, and $g = 32$ feet/second

38. $E = mgh + \frac{1}{2}mv^2$ for v (velocity in meters/second) when $E = 14$ joules, $m = 0.1$ kilogram, $h = 10$ meters, and $g = 9.8$ meters/second

Use the correct area or volume formula to find the requested information in Exercises 39–44.

39. The side of a square in feet and inches to the nearest $\frac{1}{8}$ inch if the area of the square is 85 square feet.

40. The side of a square in feet and inches to the nearest $\frac{1}{16}$ inch if the area of the square is 20 square feet.

41. The radius of a circle in feet and inches to the nearest $\frac{1}{4}$ inch if the area of the circle is 66 square feet. (Use $\frac{22}{7}$ for π.)

42. The radius of a circle in feet and inches

to the nearest $\frac{1}{2}$ inch if the area of the circle is 132 square feet. (Use $\frac{22}{7}$ for π.)

43. The edge of a cube in feet and inches to the nearest inch if the volume of the cube is 90 cubic feet.

44. The radius of a sphere in feet and inches to the nearest inch if the volume of the sphere is 176 cubic feet. (Use $\frac{22}{7}$ for π.)

In Exercises 45–52 solve the formula for the indicated letter.

45. $P = I^2 R$ for I (power)

46. $d = 16t^2$ for t (distance)

47. $S = 4\pi r^2$ for r (surface area)

48. $V = b^2 h$ for b (volume)

49. $V = \pi r^2 h$ for r (volume)

50. $d = \frac{1}{2}gt^2$ for t (distance)

51. $H = 0.4d^2 n$ for d (horsepower)

52. $T^2 = 4\pi^2 \frac{L}{g}$ for T (time)

Calculator Exercises

Use a calculator to solve for the indicated letter in Exercises 53–56. Use 3.1416 for π. (The calculator should have a square root key to do these exercises.)

53. Given $H = 0.057 I^2 R t$, solve for I when $H = 670$, $t = 42$, and $R = 8$.

54. Given $H = 0.4D^2 N$, solve for D when $H = 42.04$ and $N = 8$.

55. Given $P = I^2 R$, solve for I when $P = 10328$ and $R = 20.4$.

56. Given $V = \frac{1}{3}\pi r^2 h$, solve for r when $V = 2828$ and $h = 14.6$.

12.10
SQUARE ROOTS IN FORMULAS (II)

In the formulas of the last section, square roots were used but did not appear in the formulas themselves. Now we look at formulas where a square root radical is in the formula itself.

EXAMPLE 1   A circular mil is the area enclosed by a circle with a diameter of one mil. (A mil is $\frac{1}{1000}$ or 0.001 of an inch.) The formula for the area of a circle in circular mils is $A = D^2$, where D represents the length of the diameter in mils. If we know the area of the circle then a formula for the length of the diameter in mils is:

$$D = \sqrt{A}.$$

Use this formula to compute the diameter D of a circle with an area A of 74 circular mils.

Solution $D = \sqrt{A}$

$D = \sqrt{74}$

$D \doteq 8.602$ (from Table 1)

The length of the diameter is approximately 8.6 mils.

EXAMPLE 2   A formula for finding the approximate time in seconds that it takes an object to fall under the influence of gravity is:

$$t = \sqrt{\frac{d}{16}}.$$

Find t when $d = 576$ feet.

Solution $t = \sqrt{\frac{d}{16}}$

$t = \sqrt{\frac{576}{16}}$ ←——— Substituting

$t = \sqrt{36}$

$t = 6$

The time is 6 seconds.

EXAMPLE 3   The formula for the period of a pendulum is:

$$T = 2\pi \sqrt{\frac{L}{g}}$$

where T = the period of the pendulum (in seconds), L = the length of the pendulum, and g = the acceleration due to gravity. Find T when $L = 8$ feet and $g = 32$ feet/second^2.

Solution $T = 2\pi\sqrt{\frac{L}{g}}$

$T = 2\pi\sqrt{\frac{8}{32}}$ ←——— Substituting

$T = 2\pi\sqrt{\frac{1}{4}}$

$T = 2\pi\left(\frac{1}{2}\right)$ ←——— (Since ($\frac{1}{2}$)^2 = $\frac{1}{4}$)

$T = 2\left(\frac{1}{2}\right)\pi$

$T = \pi$

Since T is in seconds, T is π seconds, or approximately 3.14 seconds.

To solve a formula for a variable that is under the square root radical, as in the examples above:

---
1. **Multiply or divide both sides by a number or letter to get the radical alone on one side of the equation.**
2. **Square both sides. ($(\sqrt{a})^2 = a$ for any non-negative number a.)**
3. **Simplify and solve the equation.**

EXAMPLE 4   Solve $T = 2\pi\sqrt{\frac{L}{g}}$ for L.

Solution

$$T = 2\pi\sqrt{\frac{L}{g}}$$

$$\frac{T}{2\pi} = \sqrt{\frac{L}{g}}$$

$$\left(\frac{T}{2\pi}\right)^2 = \left(\sqrt{\frac{L}{g}}\right)^2 \longleftarrow \text{Square both sides.}$$

$$\left(\frac{T}{2\pi}\right)\left(\frac{T}{2\pi}\right) = \frac{L}{g}$$

$$\frac{T^2}{4\pi^2} = \frac{L}{g}$$

$$\left(\frac{T^2}{4\pi^2}\right)g = \left(\frac{L}{g}\right)g$$

$$\frac{T^2 g}{4\pi^2} = L$$

EXAMPLE 5   Solve $f\sqrt{C} = \dfrac{1}{2\pi}$ for C.

Solution   First get the radical alone on one side of the equation.

$$f\sqrt{C} = \frac{1}{2\pi}$$

$$\frac{1}{f}\left(f\sqrt{C}\right) = \frac{1}{f}\left(\frac{1}{2\pi}\right)$$

$$\sqrt{C} = \frac{1}{2f\pi}$$

$$\left(\sqrt{C}\right)^2 = \left(\frac{1}{2f\pi}\right)^2 \longleftarrow \text{Square both sides.}$$

$$C = \left(\frac{1}{2f\pi}\right)\left(\frac{1}{2f\pi}\right)$$

$$C = \frac{1}{4f^2\pi^2}$$

Review Exercises

1. Solve $x - 6 = 5x + 6$ for x.
2. Solve $3 - x = 5 + x$ for x.
3. Solve $\dfrac{m + x}{p} = q$ for x.
4. Solve $k = \dfrac{x - r}{s}$ for x.

5. Find y when $x = 10$, where y is directly proportional to x and $y = 18$ when $x = 8$.
6. Find y when $x = 36$, where y is directly proportional to x and $y = 10$ when $x = 25$.
7. Solve $\dfrac{3x^2}{4} = 9$ for x.
8. Solve $15 = \dfrac{5x^2}{9}$ for x.
9. Given $V = \pi r^2 h$, solve for r when $V = 280\pi$ and $h = 7$.
10. Given $V = \pi r^2 h$, solve for r when $V = 186\pi$ and $h = 6$.
11. Find x.

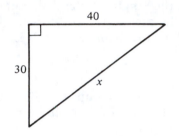

12. Find x.

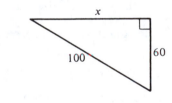

EXERCISES
12.10

Use your calculator or the table of square roots to solve for the unknown in Exercises 1-12. Round to the nearest hundredth. (Use a three-place approximation from the table and then round final answer.)

1. $x = \sqrt{a}$, $a = 9$
2. $y = \sqrt{b}$, $b = 16$
3. $m = \sqrt{k}$, $k = 28$
4. $n = \sqrt{g}$, $g = 32$
5. $w = 3\sqrt{u}$, $u = 68$
6. $z = 5\sqrt{m}$, $m = 54$
7. $m = \sqrt{\dfrac{q}{6}}$, $q = 150$
8. $n = \sqrt{\dfrac{c}{12}}$, $c = 120$

9. $x = \sqrt{\dfrac{d}{15}}$, $d = 300$ 10. $y = \sqrt{\dfrac{p}{8}}$, $p = 512$

11. $k = 3\sqrt{\dfrac{r}{6}}$, $r = 90$ 12. $j = 2\sqrt{\dfrac{s}{10}}$, $s = 240$

In Exercises 13–16 solve $T = 2\pi\sqrt{\dfrac{L}{g}}$ for T to the nearest tenth when:

13. $g = 9.8$ meters/second^2 and $L = 9.8$ meters.
14. $L = 2.45$ meters and $g = 9.8$ meters/second^2.
15. $g = 32$ feet/second^2 and $L = 8$ feet.
16. $g = 32$ feet/second^2 and $L = 32$ feet.

The formula $s = \sqrt{30fd}$ is used to estimate the speed s (in mph) that a car was traveling when d, the length of a skid mark, and f, a constant, are known. Find s to the nearest tenth when:

17. $f = 0.8$ and $d = 54$ feet.
18. $f = 0.4$ and $d = 8$ feet.

The formula $v = \sqrt{64h}$ gives the velocity v of the flow of a liquid from a container, where h is the distance of the opening in the container below the surface of the liquid. Find v (feet/second) when:

19. $h = 4$ feet.   20. $h = 1$ foot.
21. $h = 9$ feet.   22. $h = 2.25$ feet.
23. $h = 3$ inches.   24. $h = 1$ foot 9 inches.

Solve for the indicated letter in Exercises 25-36.

25. $v = \sqrt{\dfrac{E}{d}}$ for E (physics: motion)

26. $T = 2\pi\sqrt{\dfrac{m}{k}}$ for m (period of a spring)

27. $2\sqrt{LC} = \dfrac{1}{f}$ for L (electricity)

28. $2\sqrt{LC} = \dfrac{1}{f}$ for C (electricity)

29. $v = \sqrt{64h}$ for h (velocity)
30. $v = \sqrt{Rg}$ for R (velocity)

31. $t = \sqrt{\dfrac{d^3}{216}}$ for d (weather)

32. $s = \sqrt{30fd}$ for f (distance)

33. $s = \sqrt{30fd}$ for d (distance)

34. $r = \sqrt{\dfrac{A}{\pi}}$ for A (length)

35. $M\sqrt{1 + 4\pi^2f^2t^2} = 1$ for f (instrumentation)
36. $M\sqrt{1 + 4\pi^2f^2t^2} = 1$ for t (instrumentation)

Calculator Exercises

Use a calculator to determine each of the following.

The period T of the motion of a mass m hanging from a spring is given by the formula $T = 2\pi\sqrt{\dfrac{m}{k}}$, where k is the spring constant. Use 3.1416 for π. Find T with four significant digits when:

37. $m = 36$ and $k = 12$.
38. $m = 8$ and $k = 6$.
39. A formula from instrumentation for heat and heat transfer is:

$$\frac{X}{\theta_0} = \frac{1}{\sqrt{1 + (2\pi fT)^2}}$$

where X = indicated amplitude, θ_0 = actual amplitude, f = frequency, and T = time constant. Determine the amplitude ratio (that is, find $\dfrac{X}{\theta_0}$ with 3 significant digits) value for a thermometer with $T = 160$ seconds and $f = \dfrac{1}{300}$ cps. Use 3.1416 for π.

12.11
SETTING ASIDE UNTIL LATER

The important idea in this section is that you can utilize the power of your subconscious mind to help you solve problems. That part of your mind is on duty 24 hours per day and is perhaps your best problem solving tool!

Years ago when author George Henderson was in the midst of a concentrated problem-solving effort, he developed a technique for utilizing his subconscious to help solve seemingly impossible problems. The technique included the following eight steps:

1. Read and reread the problem until you fully understand the information given and the question asked.

2. Write the problem down on a piece of paper.

3. Make an attempt to solve the problem.

4. Write on the sheet of paper a condensed summary of your attempted strategy.

5. Place the paper in a desk drawer and wait until the next day to try again.

6. Look over what you did the previous day, then try another way to solve the problem.

7. If you are not successful on your second try, write a summary of your attempted strategy on the sheet of paper and put it away for another night.

8. Repeat steps 6 and 7 until you solve the problem.

In the author's case, this technique worked for all but two of the thousands of problems he applied it to. His subconscious mind apparently kept analyzing the problem while the sheet of paper was in the drawer, and he began the next day with new insight into the problem.

The following problem took one night's subconscious effort to get the answer, but took many years to obtain a general solution that would apply to every such situation.

EXAMPLE 1 How many perfect shuffles will it take to rearrange a new deck of 54 cards back

to its original order? (A perfect shuffle: split the deck into two equal halves, top half and bottom half. During the shuffle, cards are alternated from the halves, with the bottom card of the top half becoming the new bottom card, the bottom card from the bottom half becoming next-to-bottom, and so on.)

Solution Applying two of the tactics dealt with earlier in this book, "Act it Out" and "Gather Data and Look for Patterns," he began with a two-card deck and recorded his observations. Then he increased the number of cards in the deck. The results of this tactic were as follows:

Number of cards in the deck	Number of perfect shuffles needed
2	2
4	4
6	3
8	6
10	10
12	12
14	4
16	8
18	18
20	6
22	11
24	20
26	18
28	28
30	5
32	10

At first glance the right-hand column yielded no mathematical pattern. He noticed, however, that 2^1, 2^2, 2^3, 2^4, and 2^5 cards per deck took 2, 4, 6, 8, and 10 perfect shuffles, respectively. Later he was able to prove that 2^n cards take $2n$ shuffles.

There was a pattern embedded in the data, but it could not be used to solve the original problem, a 54-card deck. After summarizing his attempt and setting it aside for a day, the author became convinced that "Making a Model" would be a more effective technique than "Acting it Out." The model constructed was for an 8-card deck:

Original order of cards	Shuffle number one	two	three	four	five	six
1	5	7	8	4	2	1
2	1	5	7	8	4	2
3	6	3	6	3	6	3
4	2	1	5	7	8	4
5	7	8	4	2	1	5
6	3	6	3	6	3	6
7	8	4	2	1	5	7
8	4	2	1	5	7	8

The solid line represented the division between top and bottom equal halves of the deck. For each shuffle, the card above this line went to the bottom of the deck, and the bottom card became the next-to-bottom card; the card below the line became the top card after the shuffle, and so on, following the pattern of a perfect shuffle. *The sixth perfect shuffle put the cards back into their original order!*

Using this type of model, the author discovered that after ten perfect shuffles, a deck of 54 cards was in the reverse order from the beginning. In other words, the original top card was now the bottom card; the original bottom card was now the top card, and so on. The problem was *half solved*.

His conclusion then was that it would take 20 perfect shuffles to put a 54-card deck back into its original order. Satisfied that he had found the answer, he went on to other problems. Years later, when he finally found a general solution to the perfect shuffle problem, he became absolutely convinced of the correctness of his conclusion.

The problems that follow are all of the same type as the one he solved, and you can make models similar to the one he used. The three problems are in order of increasing difficulty.

EXERCISES

12.11

1. How many perfect shuffles does it take to put 64 cards in their original order?

2. How many perfect shuffles does it take to put 38 cards in their original order?

3. How many perfect shuffles does it take to put 48 cards in their original order?

Solutions   Do not look at these answers until you have tried to solve all the problems. If any one of your answers is incorrect, **forget the answer given here** and try again to solve the problem.

1. 12
2. 12
3. 21

EXTRA PROBLEMS
CHAPTER 12

1. Write the expressions as ratios in the simplest form.
   a. Resistance of 60 ohms to a resistance of 2.4 kilo ohms.
   b. A primary winding of 350 turns to a secondary winding of 17.5 turns.
   c. A shaft diameter of 0.625" to a bearing length of 1 7/8".
   d. A rise of 0.91 m in a roof layout to a span of 5.46 m.
   e. A volume of 12.5 cm^3 to a compressed volume of 9.375 cm^3

2. In a two gear set the driver gear has 54 gear teeth and the driven gear has 132 gear teeth. What is the ratio of the driver gear to the driven gear?

3. A small pulley turns 385 rpm and is belted to a larger pulley that turns 70 rpm. What is the ratio of their speeds?

4. The density of kerosene is 0.82 g/cm^3. The density of gasoline is 0.68 g/cm^3. What is the ratio of the density of gasoline to the density of kerosene?

5. The percentage by volume of nitrogen in the atmosphere is 78.0% and the percentage of oxygen in the atmosphere is 21.0%. What is the ratio of the volume of oxygen to the volume of nitrogen in the atmosphere?

6. A power plant produces 4.5×10^4 joules per hour of electrical energy for every 1.5×10^5 joules per hour of energy input. What is the efficiency (ratio of output to input) of this power plant?

7. Which of the following expressions are true proportions?

    a. $\dfrac{14}{35} = \dfrac{15}{38}$ c. $\dfrac{132}{528} = \dfrac{42}{168}$

    b. $\dfrac{8.25}{10.50} = \dfrac{28.875}{36.75}$ d. $\dfrac{3.21}{12.84} = \dfrac{7.85}{31.4}$

8. A large gear with 90 teeth revolves at a speed of 250 rmp and drives a smaller gear with 54 teeth. What is the speed (rpm) of the smaller gear?

9. A motor driven pump discharges 560 liters of water in 2.5 minutes. How long does it take for the pump to discharge 14.56 kiloliters of water?

10. On a scale model of a building 3" = 10'. If the model is 2' 3" high, how high will the actual building be?

11. If a 12 volt solar cell can produce 36 ampere-hours of electricity in 15 hours, how many hours will it take to produce 348 ampere-hours?

12. The relationship between pressure and volume of a confined gas can be represented as a proportion:

    $$\frac{P1 \ (\text{original pressure})}{P2 \ (\text{new pressure})} = \frac{V2 \ (\text{new volume})}{V1 (\text{original volume})}$$

    Find V2 when P1 = 1.5 lb/sq in, P2 = 9 lb/sq in and V1 = 1.2 cu ft.

13. If your "expected return" for each dollar invested in a state lottery is 30 cents, how much would you have to invest in the lottery to win $10,000 (assuming you get exactly the expected value)?

14. The shadow of a building is 35 meters long. At the same time, a meter stick casts a shadow of 70 centimeters. How high is the building?

15. The plot of land illustrated in the figure is a triangle ABC with DE parallel to AC. Measurements are made as shown in the figure. The section of land between points D and E is heavily wooded and a direct measurement from D to E cannot be made. Use the relationships between corresponding parts of similar triangles to estimate the distance from point D to point E.

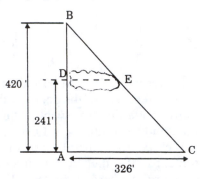

16. The triangles ABC and EDC are similar right triangles. What is the length of segment AC?

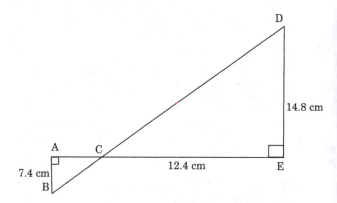

17. Triangle MNS is similar to triangle PQR. Side MS corresponds to side PR, where MS = 40 cm and PR = 70 cm. If the area of triangle MNS is 480 cm^2 what is the area of triangle PQR?

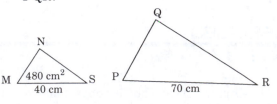

18. Hooke's law states that stress is directly proportional to strain for an elastic body. Applied to a spring this means that the force F on a spring varies directly as the distance d that the spring stretches. A force of 9.3 pounds stretches a spring 0.60 inch. Find the constant of variation for this force.

19. The cost of computer time on a certain high speeed computer varies directly as the amount of time used.
    a) If 10 minutes of computer time costs $500 find the constant of proportionality for this relationship.
    b) At the same rate, how much will 45 minutes of computer time cost?

20. The distance D of the horizon at sea varies directly as $\sqrt{h}$ where h is the height of the observer's eye. An observer, whose eye height is 15 m above sea level, can see 10 nautical miles to the horizon.
    a) What is the constant of proportionality for this relationship? (Round off your answer to two significant digits.)
    b) How much further will the observer be able to see if he/she climbs to the top of a mast so that his/her eye height is 23 m above sea level?

21. Under certain conditions the stopping distance of a car is directly proportional to the square of the speed. If a car traveling 20 mph stops in 35 feet:
    a) What is the constant of proportionality for the relationship between stopping distance and square of the speed?
    b) How many feet will it take to stop the car at a speed of 55 mph?

22. The volume of a cylinder varies directly as the square of the radius of the cylinder. The volume of a cylindrical tank is 6.4 dm^3. What will be the volume if the radius of the tank is tripled?

23. The change in resistance in an aluminum wire is directly proportional to the change in the temperature of the wire.
    a) What is the constant of proportionality if the resistance in the wire increases 0.042 ohms when the temperature of the wire increases from 10°C to 70°C?
    b) What is the decrease in resistance if the temperature decreases from 100°C to 0°C?

24. The velocity of a fluid being discharged from the bottom of a storage tank varies directly as the square root of the height of the fluid's surface. When the tank is full the height of the fluid is 11.56 meters and the velocity is 10.2 m/s.
    a) What is the constant of proportionality for this relationship?
    b) What will be the velocity of the discharged fluid when the tank is only half full?

Chapter

13

USING THE
QUADRATIC
FORMULA

13.1
PROPERTIES OF SQUARE ROOTS

The expression $\sqrt{100}$ can be written as $\sqrt{4 \times 25}$ such that each factor under the radical is a perfect square, and re-written as $\sqrt{4} \times \sqrt{25}$. Since $\sqrt{4} = 2$ and $\sqrt{25} = 5$, then $\sqrt{4} \times \sqrt{25} = 2 \times 5 = 10$.

However, $\sqrt{4}$ is also -2 and $\sqrt{25}$ is also -5. This makes it possible to write $\sqrt{4} \times \sqrt{25}$ as $\pm 2 \times \pm 5$, which, no matter what combination you use, is 10 or -10. (It should be noted that ± 2 means either +2 or -2.)

Based on the ideas in the two paragraphs above, it should be obvious that $\sqrt{100} = \pm 10$.

Similarly, $\sqrt{17} = \pm\sqrt{17}$, $\sqrt{3} = \pm\sqrt{3}$, and so on, always yielding two square roots for every positive number.

Factored form for square roots of expressions can be in different types of notation. For example, factors can be written under a single radical sign, like $\sqrt{3 \times 5}$. The radical sign can also be split, so $\sqrt{3 \times 5} = \sqrt{3} \times \sqrt{5}$. Another most used notation indicates all possible square roots of factors that are perfect squares, i.e., $\sqrt{16} = \sqrt{4 \times 4} = \sqrt{4} \times \sqrt{4} = \pm 2 \times \pm 2$; $\sqrt{8} = \sqrt{4 \times 2} = \sqrt{4} \times \sqrt{2} = \pm 2\sqrt{2}$; $i\sqrt{18} = i\sqrt{9 \times 2} = i\sqrt{9} \times \sqrt{2} = i \times \pm 3 \times \sqrt{2} = \pm 3i\sqrt{2}$, which is in final factored form. A radical in final factored form is written last, proceeded by the other factors listed with numbers before letters (variables); and variables are usually listed in alphabetical order.

A final example is $y\sqrt{ab^2} = y\sqrt{a}(\pm b)$, which should be written $\pm by\sqrt{a}$.

EXERCISES
13.1

1. $\sqrt{36} =$

2. $\sqrt{16} =$

3. $\sqrt{36} = \sqrt{6 \times (\ )} = \sqrt{(\ )} \times \sqrt{(\ )}$

4. $\sqrt{16} = \sqrt{4 \times (\ )} = \sqrt{(\ )}\sqrt{(\ )}$

5. $\sqrt{72} = \sqrt{(\ ) \times 36} = \sqrt{(\ )} \times \sqrt{(\ )} = (\ ) \times \pm (\ )\sqrt{(\ )} = \pm (\ )\sqrt{(\ )}$

6. $\sqrt{48} = \sqrt{3 \times 16} = \sqrt{(\ )} \times \sqrt{(\ )} = \pm(\ )\sqrt{3}$

7. Write $\sqrt{72}$ in factored form.

8. Write $\sqrt{48}$ in factored form.

9. Write $\sqrt{34}$ in factored form.

10. Write $\sqrt{26}$ in factored form.

11. Write $27\sqrt{27}$ in factored form.

12. Write $7\sqrt{36}$ in factored form.

13. Write $a\sqrt{xy}$ in factored form.

14. Write $q\sqrt{rs}$ in factored form.

15. Write $i\sqrt{39}$ in factored form.

16. Write $i\sqrt{35}$ in factored form.

17. Write $i\sqrt{36}$ in factored form.

18. Write $i\sqrt{49}$ in factored form.

19. Write $i\sqrt{4 \times 3 \times 5}$ in factored form.

20. Write $i\sqrt{4 \times 20 \times 6}$ in factored form.

13.2
IMAGINARY AND COMPLEX NUMBERS

Whenever you square a decimal number (real number), the answer is always positive; $3^2=9$, $(-3)^2=9$, $5^2=25$, $(2.5)^2=6.25$, etc.

More than a hundred years ago mathematicians wrestled with the idea of having a special number that could be squared to obtain a negative number. Consequently, they "made up," or invented, such a number. We call it the imaginary number and symbolize it using i; and define it using $i^2=-1$.

The invention of this new number, i, such that $i^2=-1$, opened up the development of new areas of mathematics, some of which helped explain electronic theory in mathematical terms. Electronics, of course, led to technological advancements in electricity and its uses, and to the development of computers and other electronic devices. It also led to a new set of numbers called complex numbers.

A **complex number** can be described or defined by combining real numbers with the imaginary number into the form **a + bi** where **a** and **b** are real numbers and **i²=-1**

It should be noted that if $i^2=-1$, then $\sqrt{-1} = i$. This makes it possible to write $\sqrt{-8}$ as $\sqrt{-1 \times 8}$ which is $\sqrt{-1}\sqrt{8}$, or $\sqrt{-1}\sqrt{4}\sqrt{2}$. In simplest form this is written $2i\sqrt{2}$.

EXERCISES
13.2

1. $i^2 =$
2. $2(i)^2 =$
3. $(2i)^2 =$
4. $(3i)^2 =$
5. $7(i^2) =$
6. $7(2i)^2 =$
7. $ai^2 =$
8. $a(2i)^2 =$
9. $(i\sqrt{2})^2 =$
10. $(i\sqrt{3})^2 =$
11. $3 + i^2 =$
12. $3 - i^2 =$
13. $2a + 2i^2 =$
14. $2a - 2i^2 =$
15. $i^2 - i^2 + 2i^2 =$
16. $2i^2 - i^2 + 3i^2 =$
17. $2i^2 + i =$
18. $-2i^2 + i =$
19. $-b + \sqrt{i^2} =$
20. $-b - \sqrt{i^2} =$

13.3
QUADRATIC EQUATIONS

A **quadratic equation** is an equation of the form **ax² + bx + c = 0.** X is the variable, and **a, b,** and **c** are constants (actual numbers). One example of a typical form of quadratic equation is $2x^2 - 3x + 5 = 0$, with x the variable, and **a** = 2, **b** = –3, and **c** = 5. Another is $-4y^2 + 3y - 8 = 0$, with y the variable, and **a** = –4, **b** = 3, and **c** = –8.

Variables can be expressions as well as single letters, i.e., x, y, z, p, q, r, (a+b), (c–d), etc. Using (c–d) as the variable, the form of the quadratic equation is $a(c-d)^2 + b(c-d) + e = 0$, with the standard form **c** equal to e; thus **a**= a, **b** = b, and **c** = e.

EXERCISES
13.3

List **a**, **b**, and **c** for the following quadratic equations.

1. $y^2 - 2y + 4 = 0$
2. $3y^2 + 4y - 5 = 0$
3. $2p^2 - 3p - 4 = 0$
4. $p^2 + 3p + 4 = 0$
5. $q^2 - q - 1 = 0$
6. $2q^2 + 2q - 2 = 0$
7. $5r^2 - 6 = 0$
8. $2r^2 + 4 = 0$
9. $4s^2 + 3 = 0$
10. $s^2 - 3 = 0$
11. $(a+b)^2 + (a+b) + 1 = 0$
12. $2(a+b)^2 - 3(a+b) + 5 = 0$
13. $(a-b)^2 - (a-b) + 6 = 0$
14. $3(a-b)^2 + 2(a-b) - 6 = 0$
15. $(a+c)^2 - 8 = 0$
16. $3(a+c)^2 + 8 = 0$
17. $3(a+b+c)^2 + (a+b+c) - 4 = 0$
18. $2(a+b+c)^2 - (a+b+c) - 5 = 0$
19. $(q+r)x^2 + vx - 14 = 0$
20. $(r-s)y^2 - (q+r)y + 7 = 0$

13.4
FINDING SOLUTIONS TO QUADRATIC EQUATIONS

Quadratic equations are extremely useful during the study of mathematics, and their solutions often need to be calculated. There are several techniques available for finding solutions to quadratic equations, but the one most useful in this course is using the **quadratic formula.**

The quadratic formula, indicating the two solutions to $\mathbf{ax^2 + bx + c = 0}$ can be written in the form

$$x = \frac{-b + \sqrt{b^2 - 4ac}}{2a} \quad \text{or} \quad \frac{-b - \sqrt{b^2 - 4ac}}{2a}$$

EXAMPLE 1   The solutions to $x^2 - 2x + 1 = 0$, using the quadratic formula where $a = 1$, $b = -2$, and $c = 1$:

$$x = \frac{2 + \sqrt{4 - 4(1)(1)}}{2(1)} \quad \text{or} \quad \frac{2 - \sqrt{4 - 4(1)(1)}}{2(1)}$$

simplifies to

$$x = \frac{2 + 0}{2} \quad \text{or} \quad x = \frac{2 - 0}{2} \quad \text{which is}$$

$$x = 1 \quad \text{or} \quad x = 1$$

When the same number is a solution for both roots, it means that there is only one solution to that particular quadratic equation. This happens every time that the number under the square root sign is 0.

EXAMPLE 2   Solve $x^2 - x - 6 = 0$. Note that $a = 1$, $b = -1$, $c = -6$.

Both parts of the quadratic formula can be indicated in one expression as

$$x = \frac{-b \pm \sqrt{b^2 - 4ac}}{2a}$$

substituting for a, b, and c, $x = \dfrac{1 \pm \sqrt{1 - (-24)}}{2}$

which yields $(1 \pm 5)/2$
and the solutions for x are 3 or –2.

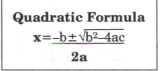

Quadratic Formula
$$x = \frac{-b \pm \sqrt{b^2 - 4ac}}{2a}$$

Note that $(b^2 - 4ac)$ is called the **discriminant** when a part of the quadratic formula. Its positive or negative value determines the number and types of solutions to the quadratic equation involved.

> If the discriminant is positive, then there are two real roots (roots that are real numbers). If the discriminant is negative, then there are two complex roots (roots that are complex numbers). If the discriminant is zero, then there is only one real root.

Approximate Versus Exact Answers

Square roots of nonperfect-squares cannot be written exactly as fractions or decimals. For example, the square root of 25 can be written *exactly* as ±5, but the square root of 31 cannot be written exactly.

If you use a calculator, you can compute square roots of positive numbers *approximately*, to the nearest tenth, hundredth, thousandth, etc.

Using a calculator, the square root of 3 can be written 1.7 to the nearest tenth, 1.73 to the nearest hundredth, and 1.732 to the nearest thousandth. The only way to represent the *exact* square root is to write $\pm\sqrt{3}$, which cannot be written exactly in decimal form.

When solving quadratic equations, we must make choices whether or not to write exact answers. For example, $\pm\sqrt{5}$ already is written in exact form; but it can be expressed in decimal form as 2.23 to the nearest hundredth, and 2.236 to the nearest thousandth.

For the most part, numbers used in technology are accepted in approximate form. Answers for exercises in this chapter usually are required in exact form.

EXAMPLE 3   Solve $2(a+b)^2+(a+b)-6 = 0$ for a if b = –1.

Solution

$$(a+b)=\frac{-1\pm\sqrt{1+48}}{4} \ \ =\frac{-1\pm7}{4} \ \ =3/2 \text{ or } -2$$

Since (a+b) = 3/2 or –2, and b = –1, it must be true that a = 5/2 or a = –1.

EXAMPLE 4   Solve $-3x^2 + 2x + 4 = 0$

Solution

$$x = \frac{-2\pm\sqrt{4-(-48)}}{-6}$$
$$= \frac{-2\pm\sqrt{52}}{-6} = \frac{-2\pm\sqrt{4}\sqrt{13}}{-6} = \frac{-2\pm2\sqrt{13}}{-6} = \frac{2(-1\pm\sqrt{13})}{-6}$$

which simplifies to $\frac{-1\pm\sqrt{13}}{-3}$ or $\frac{1\pm\sqrt{13}}{3}$

EXAMPLE 5   Solve $x^2 + x + 1 = 0$

Solution   a = 1, b = 1, c = 1

$$x = \frac{-1\pm\sqrt{1-4(1)(1)}}{2} = \frac{-1\pm\sqrt{-3}}{2} = \frac{-1\pm\sqrt{-1}\sqrt{3}}{2}$$
$$= \frac{-1\pm i\sqrt{3}}{2}$$

EXAMPLE 6
Solve $5x^2 -4x + 3 = 0$

Solution

a = 5, b = –4, c = 3

$$x = \frac{4\pm\sqrt{16-60}}{10} = \frac{4\pm\sqrt{(-1)(4)(11)}}{10} = \frac{4+2i\sqrt{11}}{10}$$
$$= \frac{2\pm i\sqrt{11}}{5}$$

EXAMPLE 7   Find the solutions to $3x^2 + 3x -5 = 0$ to the nearest hundredth (use a calculator).

Solution

a = 3, b = 3, c = –5

$$x = \frac{-3\pm\sqrt{9+60}}{6} = -1/2\pm1/6\sqrt{69}$$
$$= -.5\pm(.167)(8.31) = -.5 \pm 1.39 = 0.89 \text{ or } -1.89$$

EXAMPLE 8   Using a calculator, and expressing decimal numbers to the nearest hundredth, write the solutions to $2x^2 + 3x + 5 =0$ in complex number form.

Solution

a = 2, b = 3, c = 5

$$x = \frac{-3 \pm \sqrt{9 - 40}}{4} = \frac{-3 \pm i\sqrt{31}}{4}$$

$$= -.75 \pm 1.39i$$

$$= -.75 + 1.39i, \text{ or } -.75 - 1.39i$$

EXERCISES
13.4

Solve the following quadratic equations using the quadratic formula.

1. $x^2 + 6x + 9 = 0$
2. $x^2 - 8x + 16 = 0$
3. $x^2 - 12x - 15 = 0$
4. $x^2 + 7x - 10 = 0$
5. Solve for a if b = 1: $(a-b)^2 + 2(a-b) - 15 = 0$
6. Solve for b if a = -1: $2(a+b)^2 + 7(a+b) - 15 = 0$
7. $2a^2 + 8a - 10 = 0$
8. $-9a^2 - 3a - 6 = 0$
9. $3x^2 - 2x + 4 = 0$

10. $-2x^2 + 3x - 6 = 0$
11. $5x^2 - 3x + 7 = 0$
12. $3x^2 - 5x + 7 = 0$
13. $x^2 + 2x + 2 = 0$
14. $2x^2 + 6x + 5 = 0$
15. $2x^2 - 3x + 5 = 0$
16. $3x^2 + 2x + 3 = 0$
17. Express solutions to nearest hundredth:
$$5x^2 - 2x - 3 = 0$$
18. Express solutions to nearest hundredth:
$$3x^2 - 2x - 5 = 0$$
19. Express solutions in complex number form with all real numbers to the nearest hundredth:
$$3x^2 + 2x + 4 = 0$$
20. Express solutions in complex number form with all real numbers to the nearest hundredth:
$$5x^2 + 2x + 3 = 0$$

13.5
WRITTEN PROBLEMS

Being able to solve quadratic equations is useful when working challenging written problems. One worked out example is shown below, followed by a variety of application problems that require quadratic equations for solution.

EXAMPLE Crafty Morgan wants to create a metal picture frame that looks sort of three dimensional by making a tin rectangular box, painting it appropriately, and inserting a trimmed 4" x 6" photograph inside on the bottom. She has only one rectangular piece of tin that measures 8" by 10".

To construct the box, she has to cut congruent square pieces of tin out of the four corners of the original piece, fold up the edges, and solder the corners. What size squares (length of one edge), to the nearest hundredth, must she cut out to form a box with a bottom 20 square inches in area (so that a 4 x 6 picture can be trimmed to 3.58 x 5.58 to fit the bottom of the box)?

Solution Let the size of each cut out corner be x-inches by x-inches. That makes the bottom of the box **8 - 2x inches by 10 - 2x inches**. The area of the bottom can be found by multiplying its length by its width, which is **(8 - 2x)(10 - 2x)**, which is to be 20 square inches.

The following equation can thus be written: **(8 - 2x)(10 - 2x) = 20**, which simplifies to **$4x^2 - 36x + 80 = 20$**, which can be changed to **$x^2 - 9x + 15 = 0$**, a quadratic equation.

Since, in the quadratic formula, a = 1, b = -9, and c = 15, you can substitute and get

$$\frac{9 \pm \sqrt{21}}{2} = x$$

Using a calculator, **x = 6.79 or 2.21**. Since the longest dimension of the bottom is 10 - 2x, it is impossible for x to be 6.79 because twice that is more than 10. Therefore, the answer is 2.21 inches which makes the dimensions of the bottom 5.58 and 3.58, which has an area of 20 to the nearest one.

EXERCISES
13.5

1. Suppose Crafty Morgan wants to make a similar metal box out of a piece of tin 8.5 x 11 inches such that it has a bottom containing 24 square inches. What size squares must she cut out of the corners before folding and soldering (give answer to the nearest hundredth)?

2. Suppose Crafty Morgan wants to make another similar metal box out of a piece of tin 12 by 12 inches such that the bottom is 30 square inches. What size corners must she cut out of the corners before folding and soldering (give answer to the nearest hundredth)?

3. Find two consecutive positive even integers whose product is 168.

4. Find two positive numbers having a sum of 21 and a product of 104.

5. The sum of a number and its reciprocal is 5/2. Find the number.

6. If the reciprocal of a number is subtracted from the number, the difference is 8/3. Find the number.

7. If the length and width of a 4 cm by 2 cm rectangle are each increased by the same amount, the area of the new rectangle will be twice that of the old. What are the dimensions of the new rectangle (to the nearest hundredth)?

8. The width of a rectangle is 2 meters less than its length. Its area is 12 square meters. Find the width and length to the nearest hundredth.

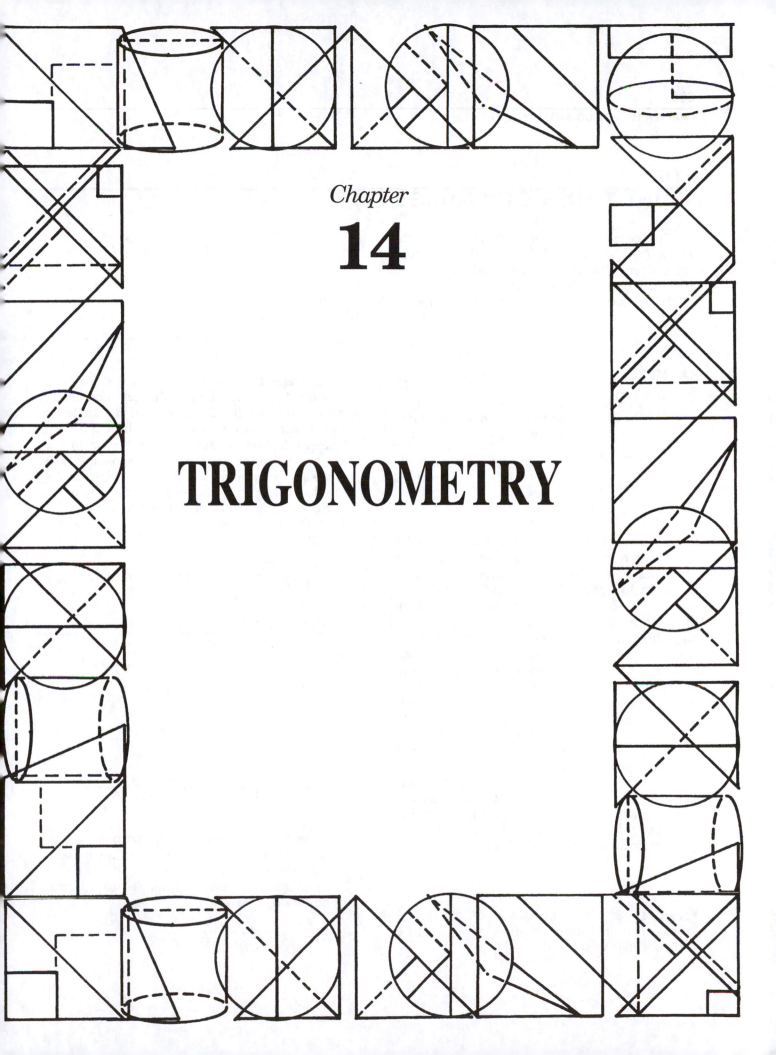

Chapter

14

TRIGONOMETRY

14.1
FUNCTIONS OF ANGLES (1)

The common unit of angle measure is the **degree** (symbolized by "°," such that 9° is read "nine degrees"). The degree is subdivided into 60 equal parts called **minutes** (symbolized by "′," such that 35′ is read "thirty-five minutes").

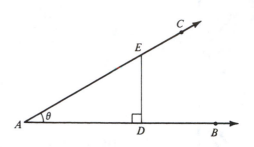

EXAMPLES

1 $\angle A = 27°43'$ Angle A has a measure of twenty-seven degrees, forty-three minutes.

2 $\angle W = 8°57'$ Angle W has a measure of eight degrees, fifty-seven minutes.

The sum of the measures of the three angles in any triangle is 180°. In a right triangle, one angle has a measure of 90°; therefore, each of the other two angles is less than 90°, since their sum is 90°. Because we will limit our work in this section to right triangles, we will consider only those angles with measures varying from 0° to 90°. A variable that is commonly used for angle measure is the Greek letter "theta," symbolized by "θ." Thus, in Sections 14.1 and 14.2, θ is a variable having a value in **degrees** of anywhere from 0° to 90°. For example, "$\angle A = \theta$" means that $\angle A$ has a measure of 90° or less.

If $\angle A = \theta$, for every value of θ except 0° and 90°, it is possible to construct a right triangle having $\angle A$ as one of its angles. The right triangle can be constructed by drawing a line segment perpendicular to one side of the angle such that it extends to the opposite side of the angle. (Perpendicular means "at a 90° angle to.")

EXAMPLE 3 The right triangle AED in the following figure is formed by drawing segment ED perpendicular to side AB.

For any $\angle A = \theta$, where $0° < \theta < 90°$, it is possible to construct an unlimited number of right triangles with $\angle A$ as one of the angles of each triangle (see Figure 14.1). All of the right triangles formed in this way are similar to one another. Since the lengths of corresponding sides of the similar triangles are proportional, we can see that the ratio of the length of the side opposite $\angle A$ to the length of the side adjacent to $\angle A$ (not the hypotenuse) is the same for *all* the triangles. That is,

$$\frac{ED}{AD} = \frac{GF}{AF} = \frac{IH}{AH} = \frac{KJ}{AJ} = \frac{ML}{AL}, \text{ and so on.}$$

FIGURE 14.1

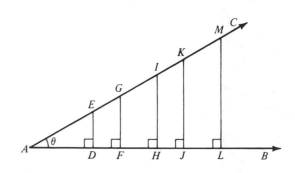

It is important to note that this ratio is independent of the length of the sides when the measure of the angle remains constant.

The ratio of the hypotenuse to the side opposite $\angle A$,

$$\frac{AE}{ED} = \frac{AG}{GF} = \frac{AI}{IH} = \cdots,$$

and the ratio of the hypotenuse to the side adjacent to $\angle A$,

$$\frac{AE}{AD} = \frac{AG}{AF} = \frac{AI}{AH} = \cdots,$$

also remain constant when the measure of $\angle A$ remains constant. Similar triangles and their properties can be used to solve problems such as the following.

EXAMPLE 4 Triangles EDC and EAB are similar right triangles. If $EB = 18$, $EC = 2$, and $DC = 3$, what is the length of AB?

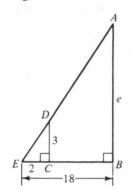

Solution Triangle EDC is similar to triangle EAB. Thus,

$$\frac{EC}{EB} = \frac{DC}{AB} \text{ or } \frac{2}{18} = \frac{3}{e}.$$

Solving the last proportion for e, we have $2e = 3 \cdot 18$ or $2e = 54$. Therefore, $e = 27$.

EXAMPLE 5 Triangles DCB and EAB are similar right triangles. If $CB = 100$, $AB = 25$, $AE = 18$, and $\angle ABE = \angle DBC$, what is the length of DC?

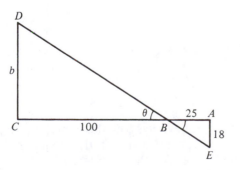

Solution Triangle DCB is similar to triangle EAB. Thus,

$$\frac{b}{18} = \frac{100}{25}.$$

Solving this proportion for b, we have $25b = 100 \cdot 18$ or $25b = 1800$. Therefore, $b = 72$.

In a right triangle ABC, with $\angle A = \theta$, where $0° < \theta < 90°$, we call side BC the side *opposite* $\angle A$, side AC the side *adjacent* to $\angle A$, and side AB the *hypotenuse* as shown in Figure 14.2.

FIGURE 14.2

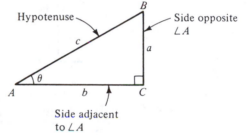

Using the ratios of the sides of the triangle, we define three functions of $\angle A$. These are called the *sine*, *cosine*, and *tangent* of the angle.

sine $\angle A = \dfrac{\text{length of side opposite } \angle A}{\text{length of hypotenuse}}$
cosine $\angle A = \dfrac{\text{length of side adjacent to } \angle A}{\text{length of hypotenuse}}$
tangent $\angle A = \dfrac{\text{length of side opposite } \angle A}{\text{length of side adjacent to } \angle A}$

The ratio defined as sine $\angle A$ is usually abbreviated as sin $\angle A$, sin A, or sin θ. Similarly, cosine $\angle A$ is abbreviated as cos $\angle A$, cos A, or cos θ, and tangent $\angle A$ is abbreviated as tan $\angle A$, tan A, or tan θ. We also use a, b, and c to represent the lengths of the sides of the triangle (see

Figure 14.3. Thus, we can abbreviate the three functions $\angle A$ as:

FIGURE 14.3

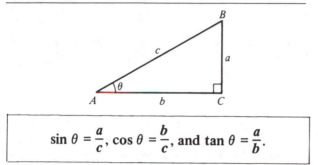

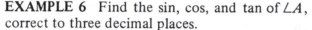

$$\sin \theta = \frac{a}{c}, \cos \theta = \frac{b}{c}, \text{ and } \tan \theta = \frac{a}{b}.$$

EXAMPLE 6 Find the sin, cos, and tan of $\angle A$, correct to three decimal places.

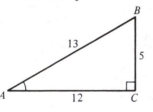

Solution

$$\sin A = \frac{a}{c} = \frac{5}{13} \doteq 0.385$$

$$\cos A = \frac{b}{c} = \frac{12}{13} \doteq 0.923$$

$$\tan A = \frac{a}{b} = \frac{5}{12} \doteq 0.417$$

EXAMPLE 7 Find the sin, cos, and tan of $\angle Q$, correct to two decimal places.

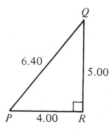

Solution

$$\sin Q = \frac{q}{r} = \frac{4.00}{6.40} \doteq 0.62$$

$$\cos Q = \frac{p}{r} = \frac{5.00}{6.40} \doteq 0.78$$

$$\tan Q = \frac{q}{p} = \frac{4.00}{5.00} = 0.80$$

To find approximations to the tangent, sine, or cosine of an angle you can use a scientific hand calculator or a table of trigonometric function values similar to Table 3, which is included in Appendix B at the back of your book. Since directions for using hand calculators vary, specific directions for using a calculator for estimating values of trigonometric ratios are not included here. Refer to the directions included with your calculator for the details of this process. A brief discussion of how to use Table 3 is included here. Additional directions for using the table are included in Appendix B.

A portion of Table 3 is reproduced in Table 14.1. From Table 14.1 it is clear that approximations of tan 31°, sin 31°, and cos 31° are 0.6009, 0.5150, and 0.8572 respectively. Likewise, tan 34° = 0.6745, cos 30° = 0.8660, sin 33° = 0.5446, and so on. The table can also be used to determine the measure of an angle if the tangent, sine, or cosine of the angle is known.

TABLE 14.1 VALUES OF TRIGONOMETRIC RATIOS

Angle	Tangent	Sine	Cosine
29°	0.5543	0.4848	0.8746
30°	0.5774	0.5000	0.8660
31°	0.6009	0.5150	0.8572
32°	0.6249	0.5299	0.8480
33°	0.6494	0.5446	0.8387
34°	0.6745	0.5592	0.8290

EXAMPLE 8 If $\cos \theta \doteq 0.8387$, we see from Table 14.1 that $\theta = 33°$. Likewise, if $\tan \theta \doteq 0.5774$, $\theta = 30°$.

Note that the values of the sine and tangent *increase* as the measure of the angle increases, but the value of the cosine *decreases* as the measure of the angle increases.

Review Exercises

In Exercises 1–4 find the length of the hypotenuse, correct to the nearest tenth.

1.

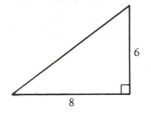

2.

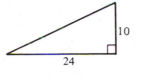

3.

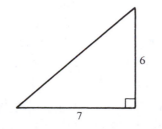

4.

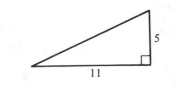

In Exercises 5–8 find the length of the leg, correct to the nearest tenth.

5.

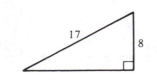

6.

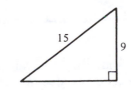

7.

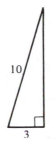

8.

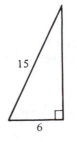

9. Find the length of the hypotenuse of a right triangle to the nearest tenth, given that the legs are 7 inches and 15 inches long.

10. One leg of a right triangle is 30 cm long, and the hypotenuse is 50 cm long. What is the length of the other leg?

11. Triangle ABC is a right triangle with $\angle C = 90°$. If $a = 12$ and $b = 10$, what is c (to the nearest hundredth)?

12. The hypotenuse of a right triangle is 8 cm long. If the two legs are the same length, what is the length of each leg?

EXERCISES
14.1

In Exercises 1–4 identify the values for sin, cos, and tan θ, correct to two decimal places.

1.
2.

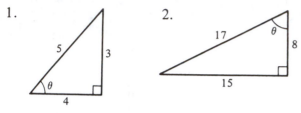

3.
4.

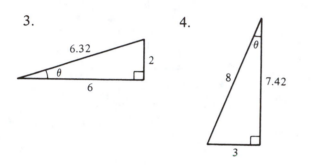

Use the figure below to identify the correct replacement for x in the expressions in Exercises 5–11.

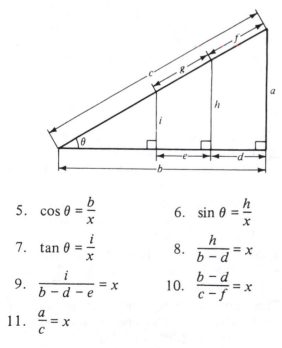

5. $\cos \theta = \dfrac{b}{x}$ 6. $\sin \theta = \dfrac{h}{x}$

7. $\tan \theta = \dfrac{i}{x}$ 8. $\dfrac{h}{b-d} = x$

9. $\dfrac{i}{b-d-e} = x$ 10. $\dfrac{b-d}{c-f} = x$

11. $\dfrac{a}{c} = x$

Using the figure below, identify the values in Exercises 12–16.

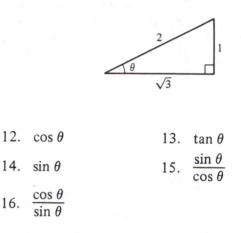

12. $\cos \theta$ 13. $\tan \theta$

14. $\sin \theta$ 15. $\dfrac{\sin \theta}{\cos \theta}$

16. $\dfrac{\cos \theta}{\sin \theta}$

Use your calculator or Table 3 at the back of the book to find approximations for each of the values in Exercises 17-26.

17. $\sin 85°$ 18. $\tan 80°$
19. $\cos 13°$ 20. $\tan 30°$
21. $\sin 3°$ 22. $\cos 87°$
23. $\sin 38°$ 24. $\cos 52°$
25. $\tan 45°$ 26. $\sin 12°$

Use your calculator or Table 3 at the back of the book to find the measure of θ in Exercises 27–36.

27. $\tan \theta \doteq 0.2679$ 28. $\sin \theta \doteq 0.7771$
29. $\cos \theta \doteq 0.2419$ 30. $\tan \theta \doteq 1.428$
31. $\cos \theta \doteq 0.9703$ 32. $\sin \theta \doteq 0.6293$
33. $\sin \theta \doteq 0.3090$ 34. $\tan \theta \doteq 0.4663$
35. $\cos \theta \doteq 0.8572$ 36. $\tan \theta \doteq 9.514$

37. The height of the building shown here in feet can be found by calculating 120(tan 25°). Find the height of the building to the nearest foot.

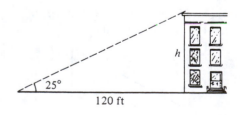

38. The width of the field shown here can be found by calculating 400(tan 40°). Find the width of the field to the nearest yard.

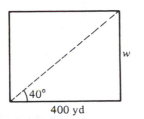

39. The distance r from a corner point P to the center of a hole drilled in the metal plate shown in the figure can be found by calculating the quotient $\dfrac{8.3}{\cos 52°}$. Find the length of r to the nearest tenth of an inch.

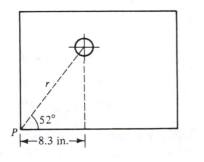

40. The width of a certain river in meters can be found by calculating 20(tan 38°). Find the width of the river to the nearest tenth of a meter.

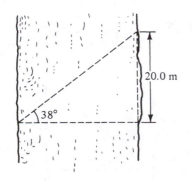

41. The coefficient of friction between an object and an inclined plane is the tangent of the angle that the plane makes with the horizontal (if the object moves down the plane with a constant speed). If an inclined plane makes an angle of 13° with the horizontal, find the coefficient of friction for an object moving down this plane.

42. If the coefficient of friction between an object and an inclined plane is 0.3249, at what angle is the plane inclined? (See Exercise 41.)

43. What is the size of angle θ, to the nearest degree, on a steel plate with the dimensions shown in the figure?

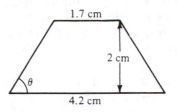

44. What is the size of angle θ, to the nearest degree, in a triangular wedge with the dimensions shown in the figure?

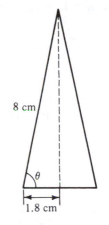

45. The work W done by a force F is defined as $W = Fd\cos\theta$, where F is the magnitude of the force, d is the distance through which it acts, and θ is the angle between the direction of the force and the direction of the motion. Given that a 35-pound force acts through a distance of 5 feet, and the angle between the force and the motion is 28°, how much work is done by the force?

46. The index of refraction of a medium is defined as $n = \dfrac{\sin i}{\sin r}$, where i is the angle between the perpendicular and a light ray in air, and r is the angle between the perpendicular and the light ray in the medium. Find the index of refraction for glass when $i = 60°$ and $r = 35°$.

14.2
SOLVING RIGHT TRIANGLES

Solving a triangle means finding the measures of the remaining parts of the triangle when the measures of several of the parts are given. (The **parts** of a triangle are its three angles and three sides.)

To solve a right triangle, the minimum information needed is the measure of one of its sides and the measure of one of its angles (other than the right angle). When the minimum information is given, the sine, cosine, and tangent functions, along with the Pythagorean Theorem, can be used to compute the measures of the remaining parts.

$$0.6249 \doteq \frac{a}{12}$$

$$a \doteq 12(0.6249)$$

$$a \doteq 7.499$$

$$\cos 32° = \frac{12}{c}$$

$$0.8480 \doteq \frac{12}{c}$$

$$c \doteq 12 \div (0.8480)$$

$$c \doteq 14.15$$

$$\angle B = 90° - 32° = 58°$$

Note: The Pythagorean Theorem could have been used to compute a or c after one of these values had been computed using a trigonometric function.

EXAMPLE 1 Solve the right triangle *ABC*. Round off each answer to four significant digits.

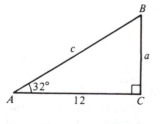

Solution

$$\tan 32° = \frac{a}{12}$$

EXAMPLE 2 Solve the right triangle *EFG*. Round off each answer to four significant digits.

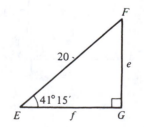

Solution

$$\sin 41°15' = \frac{e}{20}$$

$$0.6593 \doteq \frac{e}{20}$$

$$e \doteq 20(0.6593)$$

$$e \doteq 13.19$$

$$\cos 41°15' = \frac{f}{20}$$

$$0.7518 \doteq \frac{f}{20}$$

$$f \doteq 20(0.7518)$$

$$f \doteq 15.04$$

$$\angle F = 90° - 41°15' = 48°45'$$

EXAMPLE 3 Solve the right triangle *RST*.

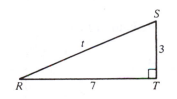

Solution

$$\tan R = \frac{3}{7}$$

$$\tan R \doteq 0.4286$$

$$\angle R \doteq 23°12'$$

$$t^2 = 7^2 + 3^2$$

$$t^2 = 49 + 9$$

$$t = \sqrt{58}$$

$$t \doteq 7.6$$

$$\angle S = 90° - 23°12' = 66°48'$$

An **angle of elevation** is the angle between the horizontal and the line of sight to some point at a higher elevation. An **angle of depression** is the angle between the horizontal and the line of sight to some point at a lower level.

Figure 14.4 illustrates angles of elevation and depression.

FIGURE 14.4

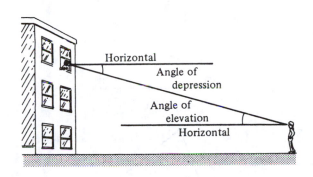

EXAMPLE 4 From the top of a 250-foot tower a forest ranger sees smoke coming from the ground at a point *x* feet from the tower. If the angle of depression from the observation point is 4°, how far away from the foot of the tower is the fire?

Solution Since the horizontal makes a right angle with the tower, the angle used in the triangle is 86°.

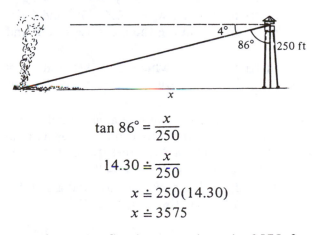

$$\tan 86° = \frac{x}{250}$$

$$14.30 \doteq \frac{x}{250}$$

$$x \doteq 250(14.30)$$

$$x \doteq 3575$$

Therefore, the fire is approximately 3575 feet from the foot of the tower.

Review Exercises

In Exercises 1–4 calculate the measures of the angles of the triangles which are not given.

1.

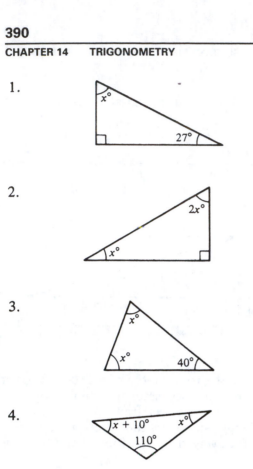

2.

3.

4.

5. In triangle *ABC*, where $\angle A = 13°$ and $\angle B = 34°$, what is the measure of $\angle C$?

6. In triangle *XYZ*, where $\angle X = 39°$ and $\angle Y = \angle Z + 18°$, what are the measures of $\angle Y$ and $\angle Z$?

7. In triangle *SRT*, where $\angle S = 90°$ and $\angle R = \angle T - 19°$, what are the measures of $\angle R$ and $\angle T$?

8. If a right triangle has one angle which is *not* a right angle and which has a measure one-third the sum of the measures of the other two angles, what is its measure?

9. Triangle *ABC* is similar to triangle *ADE*. What is the length of segment *DE*?

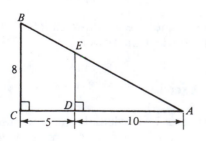

10. What is the value of *x*?

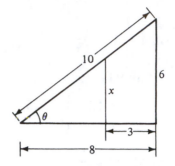

11. Determine the values for *x*, *y*, and *z*.

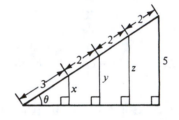

12. What is the value of *h*?

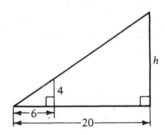

EXERCISES
14.2

Use your calculator or interpolation in Table 3 at the back of the book to find the values in Exercises 1–10.

1. $\tan 13°10'$	2. $\tan 84°40'$
3. $\cos 50°15'$	4. $\sin 6°24'$
5. $\tan 47°35'$	6. $\cos 29°40'$
7. $\tan 4°18'$	8. $\sin 49°56'$
9. $\tan 88°30'$	10. $\cos 85°32'$

Use your calculator or interpolation in Table 3 at the back of the book to find the measures of the angles in Exercises 11−20.

11. $\sin A \doteq 0.5620$ 12. $\tan A \doteq 0.4314$
13. $\cos A \doteq 0.5324$ 14. $\sin A \doteq 0.3007$
15. $\tan A \doteq 0.1075$ 16. $\cos A \doteq 0.9130$
17. $\sin A \doteq 0.9934$ 18. $\tan A \doteq 2.215$
19. $\cos A \doteq 0.5812$ 20. $\tan A \doteq 1.545$

Solve the right triangles in Exercises 21−26.

21.

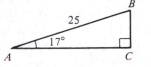

22.

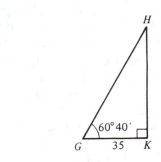

23.

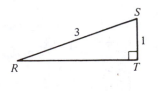

24.

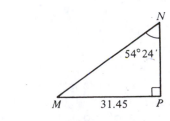

25.

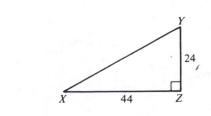

26.

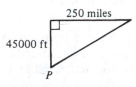

27. A tennis court is 78 feet by 36 feet. What is the length of its diagonal (distance from one corner to the opposite corner)?

28. What is the measure of $\angle P$?

29. What is the measure of $\angle A$?

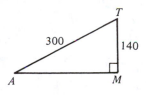

30. What is the length of side BD?

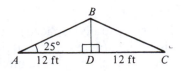

Given a right triangle ABC with $C = 90°$, find the missing parts for the triangles in Exercises 31−34.

31. $c = 16.4, \angle A = 24°50'$
32. $a = 24.6, b = 40.4$
33. $a = 25.3, \angle A = 38°40'$
34. $a = 3.81, c = 5.07$

35. From a point on the ground 200 feet from the foot of a building, the angle of elevation of the top is 30°. How high is the building?

36. What is the angle of elevation of the sun if an object 40 km high casts a shadow 60 km long?

37. A ladder 36 feet long reaches from the ground to a window 33 feet high. How far from the wall is its foot, and what angle does it make with the horizontal?

38. A boat is anchored near a bridge. From a point on the boat's deck 15 feet above the water, the angle of elevation of the top of the bridge is 39°, while the angle of depression of its image reflected in the water is 54°. Find the height of the bridge above the water and the distance of the boat from the bridge.

39. Ten bolt holes are equally spaced on the circumference of a circle. If the diameter of the circle is 20 inches, what is the distance d between the centers of two adjacent holes?

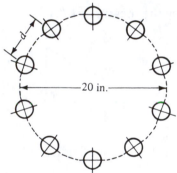

40. Nine holes (equally spaced) are to be drilled in a steel plate on the circumference of a circle which has a diameter of 12 inches. What will be the distance between the centers of any two adjacent holes?

41. Two points on the circumference of a wheel are 24 centimeters apart. The diameter of the wheel is 60 centimeters. What is the size of the angle formed by two radii to the two points?

42. The length of a roof rafter from the wall to the peak (without the overhang) is 25 feet, and the rise of the roof at the peak is 11 feet. What is the pitch of the roof (angle θ) to the nearest degree?

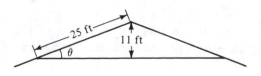

43. From an observation deck 85 feet above the water level a lookout sees a floating object at a point whose angle of depression from the observation deck is 2°. To the nearest 5 feet, how far is the object from the ship when it is first sighted by the lookout?

44. A pendulum 3 feet long swings through a total angle of 36°. Find the horizontal distance between the extreme positions of the pendulum.

45. A straight roadway from the foot to the top of a hill has a uniform incline of 28.5°. If the road is 1505 feet long, what is the height of the hill to the nearest foot?

46. A straight inclined railway is built from the bottom to the top of a hill with an elevation of 403 feet. The angle of elevation is a uniform 35.4°. What is the length of the inclined railway track to the nearest foot?

47. An engine has piston connecting rods 8.00 inches in length, and the maximum angle each rod makes with the line of center from the crankshaft to the piston is 18°. What is the length of the radius (r) of the crankshaft throw in this engine to the nearest hundredth of an inch?

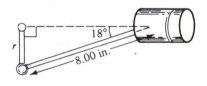

48. A wedge, 12.3 cm in length, has a taper of 16°, as shown in the figure. What is the minimum width of stock material needed to make this wedge?

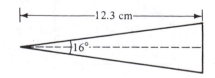

49. Find the angle (θ) of the taper for the piece of milled stock shown in the figure. Give the answer to the nearest 5' of a degree.

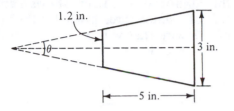

50. A guy wire 600 feet long is attached to a television antenna 430 feet above ground level. What angle does the wire make with the antenna, to the nearest 5' of a degree?

51. The angle of elevation from a fence line to the top of a building 240 feet high is 68°. To the nearest foot, how far out from the building is the fence line?

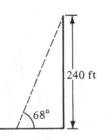

52. From a point A on the bank of a river, a sight is taken on a large rock directly across on the other bank of the river. Another sight is taken on the rock from a point B, 500 feet upstream from point A. The angle between the two bearings is 27°. To the nearest foot, how wide is the river between point A and the rock?

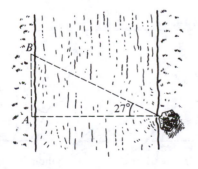

53. In alternating current circuits, the relationship between the resistance R (in ohms), the reactance X (in ohms), the impedance Z (in ohms), and the phase angle θ is represented by a right triangle. If the resistance is 530 ohms and the impedance is 720 ohms, what is the phase angle to the nearest 10' of a degree?

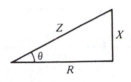

54. A hexagonal nut is to be constructed to fit a 1-inch bolt. The minimum wall thickness is to be 0.20 inch. What is the minimum diameter, to the nearest hundredth of an inch, of cylindrical stock material that can be used to make this nut?

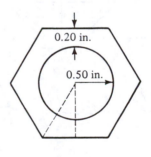

55. The figure shows a cross section of a gear track. Find the length x when $P = \frac{1}{2}$ in., $D = 0.26$ in., and $F = 0.18$ in.

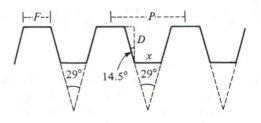

56. Using the figure in Exercise 55, find the length of x when $P = \frac{1}{4}$ in., $D = 0.16$ in., and $F = 0.04$ in.

57. The strongest joint a carpenter can use to fasten two boards together is called a dovetail joint. The mortise, or opening, of a dovetail is shown in the figure. What is the length of *x* in this mortise?

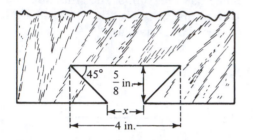

58. A tenon shaped like a wedge is designed to interlock with a mortise of the same size to form a dovetail joint. What is the length of *y* in the tenon shown in the figure?

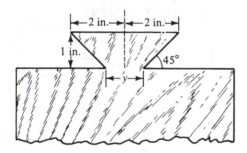

59. A machinist uses a method for gauging screw threads which involves finding a wire of proper size so that it will be flush with the top of the thread when placed in a thread groove. The figure shows a cross section of a screw thread. What is the diameter of the wire that will be flush with the top of the thread?

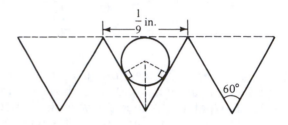

60. A design for a metal casting is shown in the figure. What is the measure of angle θ, to the nearest 5' of a degree?

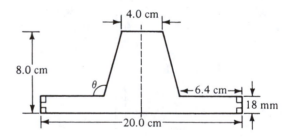

14.3
SOLVING OBLIQUE TRIANGLES

An **oblique triangle** is one that does not contain a right angle. This includes both acute and obtuse triangles. An **acute triangle** has three acute angles (angles with measures less than 90°). An **obtuse triangle** has one obtuse angle (an angle with a measure between 90° and 180°) and two acute angles. Figure 14.5 shows an example of an acute triangle and an obtuse triangle.

FIGURE 14.5

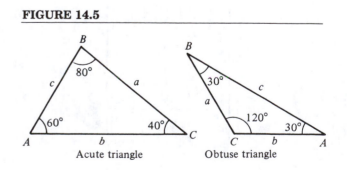

Acute triangle   Obtuse triangle

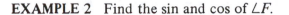

Since an oblique triangle may contain an obtuse angle, it is necessary to consider the functions of an obtuse angle. We shall let the Greek letter "alpha," symbolized by "α," be a variable that represents the measure of an obtuse angle. Thus, α is a variable having all values in degrees between 90 and 180. For any obtuse angle, the angle with measure $180° - \alpha$ is an acute angle. This is illustrated in Figure 14.6.

FIGURE 14.6

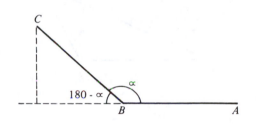

For any obtuse angle with measure α:

$\sin \alpha = \sin (180 - \alpha)$

$\cos \alpha = - \cos (180 - \alpha)$

$\tan \alpha = - \tan (180 - \alpha)$.

EXAMPLE 1   Find the sin, cos, and tan of $\angle Y$.

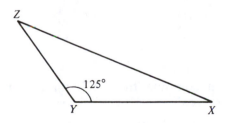

Solution

$$\sin 125° = \sin(180 - 125)°$$
$$= \sin 55°$$
$$\doteq 0.8192$$

$$\cos 125° = -\cos(180 - 125)°$$
$$= -\cos 55°$$
$$\doteq -0.5736$$

$$\tan 125° = -\tan(180 - 125)°$$
$$= -\tan 55°$$
$$\doteq -1.428$$

EXAMPLE 2   Find the sin and cos of $\angle F$.

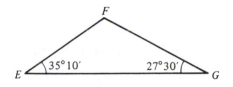

Solution

$$\angle F = 180° - (35°10' + 27°30')$$
$$= 180° - 62°40'$$
$$= 117°20'$$

$$\sin 117°20' = \sin(180° - 117°20')$$
$$= \sin 62°40'$$
$$\doteq 0.8964$$

$$\cos 117°20' = -\cos 62°40'$$
$$\doteq -0.4591$$

EXAMPLE 3   Find $\angle H$ if $\sin H = 0.5299$.

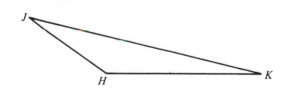

Solution   From Table 3 at the back of the book, we find that the angle whose sine is 0.5299 has a measure of 32°. However, $\angle H$ is an obtuse angle, so $\angle H = 180° - 32° = 148°$, and $\sin \angle H = \sin(180° - \angle H)$.

Special formulas are needed to solve an oblique triangle. One such formula, called the **Law of Sines**, is developed from the trigonometric ratios of a right triangle.

Let ABC be any oblique triangle, either acute or obtuse. From vertex B we draw the altitude h to the side AC (see Figure 14.7). In each triangle,

$$\sin A = \frac{h}{c} \text{ or } c(\sin A) = h,$$

and

$$\sin C = \frac{h}{a} \text{ or } a(\sin C) = h.$$

FIGURE 14.8

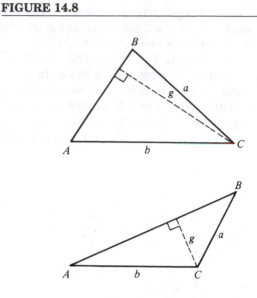

FIGURE 14.7

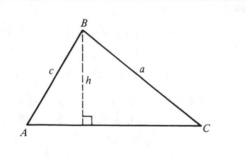

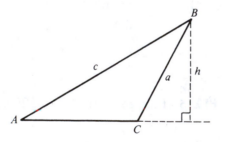

$$\sin A = \frac{g}{b} \text{ or } b(\sin A) = g,$$

and

$$\sin B = \frac{g}{a} \text{ or } a(\sin B) = g.$$

Thus,

$$b(\sin A) = a(\sin B),$$

and

$$\frac{\sin A}{a} = \frac{\sin B}{b}.$$

We can combine the last equation from each case to obtain the Law of Sines:

Since we have two expressions equal to h, we can say:

$$c(\sin A) = a(\sin C).$$

Dividing both sides of this equation by ac we have:

$$\frac{\sin A}{a} = \frac{\sin C}{c}.$$

If we now draw the altitude from vertex C to side AB we can show a similar relationship for $\sin A$ and $\sin B$ (see Figure 14.8). In each triangle,

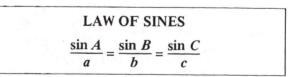

LAW OF SINES
$\dfrac{\sin A}{a} = \dfrac{\sin B}{b} = \dfrac{\sin C}{c}$

The Law of Sines can be used to solve an oblique triangle if: (1) two angles and a side are known, or (2) two sides and an angle opposite one of these sides is known.

EXAMPLE 4 If $\angle A = 65°$, $\angle B = 60°$, and the length of side AC is 10, solve the triangle.

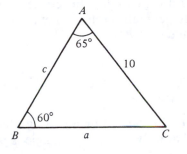

Solution

$$\angle C = 180° - (60° + 65°) = 180° - 125° = 55°$$

$$\frac{\sin B}{b} = \frac{\sin A}{a}$$

$$\frac{\sin 60°}{10} = \frac{\sin 65°}{a}$$

$$\frac{0.8660}{10} = \frac{0.9063}{a}$$

$$a(0.8660) = 10(0.9063)$$

$$a = \frac{9.063}{0.8660}$$

$$a \doteq 10.47$$

$$\frac{\sin C}{c} = \frac{\sin B}{b}$$

$$\frac{\sin 55°}{c} = \frac{\sin 60°}{10}$$

$$\frac{0.8192}{c} = \frac{0.8660}{10}$$

$$10(0.8192) = c(0.8660)$$

$$c = \frac{8.192}{0.8660}$$

$$c \doteq 9.46$$

EXAMPLE 5 Find the lengths of the two steel supports a and b as shown in the figure.

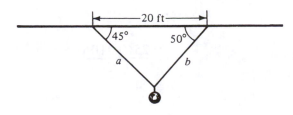

Solution The measure of the third angle is $180° - (45 + 50)° = 85°$.

$$\frac{\sin 50°}{a} = \frac{\sin 85°}{20}$$

$$a = \frac{20(\sin 50°)}{\sin 85°}$$

$$a = \frac{20(0.7660)}{0.9962}$$

$$a \doteq 15.4$$

$$\frac{\sin 45°}{b} = \frac{\sin 85°}{20}$$

$$b = \frac{20(\sin 45°)}{\sin 85°}$$

$$b = \frac{20(0.7071)}{0.9962}$$

$$b \doteq 14.2$$

The steel supports are approximately 15.4 feet and 14.2 feet.

EXAMPLE 6 If $\angle A = 25°$, $b = 20$, and $a = 11.2$, solve the triangle.

Solution The case where two sides and an angle opposite one of the sides are known is a somewhat ambiguous case since two triangles are possible to obtain from the given data. For this example the following two triangles are possible.

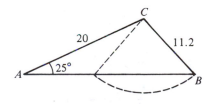

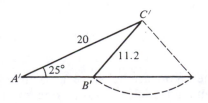

For triangle *ABC:*

$$\frac{\sin A}{a} = \frac{\sin B}{b}$$

$$\frac{\sin 25°}{11.2} = \frac{\sin B}{20}$$

$$\frac{0.4226}{11.2} = \frac{\sin B}{20}$$

$$\sin B = \frac{20(0.4226)}{11.2}$$

$$\sin B \doteq 0.7546$$

$$\angle B \doteq 49°$$

$$\angle C \doteq 180° - (25 + 49)° = 106°$$

$$\frac{\sin A}{a} = \frac{\sin C}{c}$$

$$\frac{\sin 25°}{11.2} = \frac{\sin 106°}{c}$$

$$(\sin 106° = \sin(180 - 106)°)$$

$$\frac{0.4226}{11.2} = \frac{0.9613}{c}$$

$$c = \frac{11.2(0.9613)}{0.4226}$$

$$c \doteq 25.5$$

For triangle *A'B'C'* the computation for sin *B'* is the same as for sin *B* in triangle *ABC*. Since ∠*B'* is obtuse,

$$\angle B' \doteq (180° - 49°) = 131°.$$

$$\angle C' \doteq 180° - (25 + 131)° = 24°$$

$$\frac{\sin A'}{a} = \frac{\sin C'}{c}$$

$$\frac{\sin 25°}{11.2} = \frac{\sin 24°}{c}$$

$$\frac{0.4226}{11.2} = \frac{0.4067}{c}$$

$$c = \frac{11.2(0.4067)}{0.4226}$$

$$c \doteq 10.8$$

When you are to solve a triangle given two sides and an angle opposite one of the sides, a scale drawing using the given data will show you if more than one triangle is possible. If two triangles are possible, the context of the problem will generally tell you which triangle is a reasonable solution.

EXAMPLE 7 If $b = 20$, $c = 25$, and $\angle C = 67°$, solve the triangle.

Solution A scale drawing using the given data shows that only one such triangle is possible.

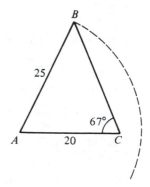

$$\frac{\sin B}{b} = \frac{\sin C}{c}$$

$$\frac{\sin B}{20} = \frac{\sin 67°}{25}$$

$$\frac{\sin B}{20} = \frac{0.9205}{25}$$

$$\sin B = \frac{20(0.9205)}{25}$$

$$\sin B \doteq 0.7364$$

$$\angle B \doteq 47°24'$$

$$\angle A \doteq 180° - (67° + 47°24') = 65°36'$$

$$\frac{\sin A}{a} = \frac{\sin C}{c}$$

$$\frac{\sin 65°36'}{a} = \frac{\sin 67°}{25}$$

$$\frac{0.9106}{a} = \frac{0.9205}{25}$$

$$a = \frac{25(0.9106)}{0.9205}$$

$$a \doteq 24.73$$

EXAMPLE 8 The crank and connecting rod of an engine like the one in the drawing, are 30 cm and 102 cm long, respectively. What angle does the crank make with the horizontal when the angle made by the connecting rod is 15°?

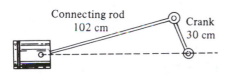

Solution There are two solutions possible for this example. In case 1 the required angle is an acute angle, and in case 2 the required angle is an obtuse angle. The computation is the same for both cases.

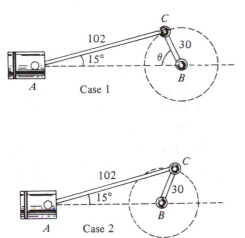

Case 1

$$\frac{\sin 15°}{30} = \frac{\sin \theta}{102}$$

$$\frac{0.2588}{30} = \frac{\sin \theta}{102}$$

$$\sin \theta = \frac{102(0.2588)}{30}$$

$$\sin \theta \doteq 0.8799$$

$$\theta \doteq 61°38'$$

Case 2

$$\frac{\sin 15°}{30} = \frac{\sin \alpha}{102}$$

$$\frac{0.2588}{30} = \frac{\sin \alpha}{102}$$

$$\sin \alpha = \frac{102(0.2588)}{30}$$

$$\sin \alpha \doteq 0.8799$$

Since α is an obtuse angle,

$$\alpha \doteq 180° - 61°38' = 118°22'.$$

When given the measures of two sides of a triangle and the angle between the two sides, or the measures of the three sides of a triangle, it is impossible to use the Law of Sines as a first step in solving the triangle. When data for either of these two situations is known, the triangle can be solved by using the **Law of Cosines**.

The Law of Cosines gives the relationship between three sides and one of the angles of any oblique triangle (see Figure 14.9). Therefore, if you are given any three of these parts, you can use one of the formulas of the Law of Cosines to compute the remaining part.

FIGURE 14.9

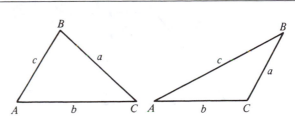

The development of the Law of Cosines is beyond the scope of this book. The formulas which make up the Law of Cosines are:

LAW OF COSINES
$a^2 = b^2 + c^2 - 2bc\,(\cos A)$
$b^2 = a^2 + c^2 - 2ac\,(\cos B)$
$c^2 = a^2 + b^2 - 2ab\,(\cos C)$

EXAMPLE 9   In triangle ABC, $\angle A = 40°$, $b = 6$, and $c = 7$. What is the length of side BC?

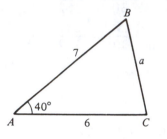

Solution

$$a^2 = b^2 + c^2 - 2bc(\cos A)$$
$$a^2 = 6^2 + 7^2 - 2(6)(7)(\cos 40°)$$
$$a^2 = 36 + 49 - 84(0.7660)$$
$$a^2 = 20.656$$
$$a = \sqrt{20.656}$$
$$a \doteq 4.5$$

After a is known, the Law of Sines can be used to compute the measures of the other parts of the triangle.

EXAMPLE 10   A metal frame is constructed in the form of a trapezoid as shown. Find the length of the diagonal brace.

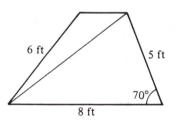

Solution   Let x be the length of the diagonal brace. Then,

$$x^2 = 8^2 + 5^2 - 2(8)(5)(\cos 70°)$$
$$x^2 = 64 + 25 - 80(0.342)$$
$$x^2 = 61.64$$
$$x = \sqrt{61.64}$$
$$x \doteq 7.85.$$

The diagonal brace is approximately 7.85 feet.

EXAMPLE 11   Find the angles of the triangle.

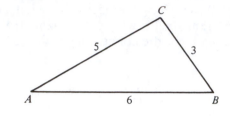

Solution

$$a^2 = b^2 + c^2 - 2bc(\cos A)$$
$$3^2 = 5^2 + 6^2 - 2(5)(6)(\cos A)$$
$$9 = 25 + 36 - 60(\cos A)$$
$$9 = 61 - 60(\cos A)$$
$$-52 = -60(\cos A)$$
$$\frac{-52}{-60} = \cos A$$
$$0.8667 = \cos A$$
$$\angle A \doteq 29°55'$$

$$b^2 = a^2 + c^2 - 2ac(\cos B)$$
$$5^2 = 3^2 + 6^2 - 2(3)(6)(\cos B)$$
$$25 = 9 + 36 - 36(\cos B)$$
$$25 = 45 - 36(\cos B)$$
$$-20 = -36(\cos B)$$
$$\frac{-20}{-36} = \cos B$$
$$0.5556 = \cos B$$
$$\angle B \doteq 56°15'$$

$$\angle C \doteq 180° - (29°55' + 56°15') = 93°50'$$

EXAMPLE 12   Two metal stringers, XY, 60 cm long, and YZ, 80 cm long, are welded together forming an angle of 135°. How long is the metal brace XZ?

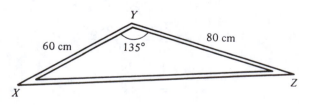

Solution   Using the Law of Cosines we have:

$$(XY)^2 = 60^2 + 80^2 - 2(60)(80)(\cos 135°).$$

You will recall that:

$$\cos 135° = -\cos(180 - 135)°$$
$$\cos 135° = -\cos 45°$$
$$\cos 135° \doteq -0.7071.$$

Thus,

$$(XZ)^2 = 60^2 + 80^2 - 2(60)(80)(-0.7071)$$
$$(XZ)^2 = 3600 + 6400 - (9600)(-0.7071)$$
$$(XZ)^2 = 10,000 + 6788.16$$
$$(XZ)^2 = 16,788.16$$
$$XZ = \sqrt{16,788.16}$$
$$XZ = 129.6.$$

The metal brace is approximately 129.6 cm long.

EXAMPLE 13 Three holes are drilled in a metal plate so that the distances between their centers are as shown in the figure. What is the measure of angle *ABC*?

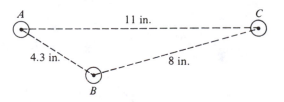

Solution

$$b^2 = a^2 + c^2 - 2ac(\cos B)$$
$$11^2 = 8^2 + 4.3^2 - 2(8)(4.3)(\cos B)$$
$$121 = 64 + 18.49 - 68.8(\cos B)$$
$$38.51 = -68.8(\cos B)$$
$$\frac{38.51}{-68.8} = \cos B$$
$$-0.5597 \doteq \cos B$$

Since cos *B* is a negative number, ∠*B* is an obtuse angle. If cos *B'* = 0.5597, then ∠*B'* = 55°58'. Thus,

$$\angle B \doteq 180° - 55°58' = 124°2'.$$

Review Exercises

Choose the correct answer in Exercises 1–6.

1. With reference to the following triangle, the ratio $\dfrac{\sqrt{5}}{2}$ is:

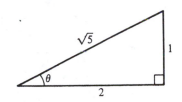

   (a)   tan θ
   (b)   sin θ
   (c)   cos θ
   (d)   none of the above

2. Selecting from the three given values, the best approximation of cos 38°24' is:
   (a)   0.7854
   (b)   0.7836
   (c)   0.7815

3. sin 33° = cos 57°
   (a)   True
   (b)   False

4. What is the approximate length of segment *DE*?

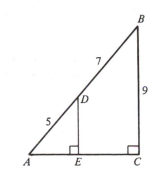

   (a)   2.33
   (b)   3.75
   (c)   6.43
   (d)   7.00

5. Triangle *RST* is similar to triangle *HJK*. What is the ratio of the corresponding sides?

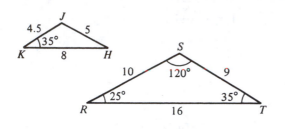

(a) $\dfrac{10}{4.5}$

(b) $\dfrac{10}{5}$

(c) $\dfrac{10}{8}$

6. In right triangle *PQR* the length of side *PR* is approximately 6.7.

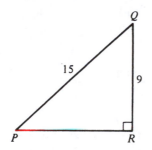

(a) True
(b) False

In Exercises 7-9 use your calculator or Table 3 at the back of the book to obtain a four-place approximation.

7. sin 68°40′

8. tan 11°18′

9. cos 75°50′

In Exercises 10-12 use your calculator or Table 3 at the back of the book to obtain the measure of θ, correct to the nearest 10′.

10. tan θ ≐ 4.225

11. cos θ ≐ 0.9093

12. sin θ ≐ 0.8273

EXERCISES 14.3

In Exercises 1–8 find a four-place approximation.

1. sin 115°   2. sin 130°

3. cos 95°   4. cos 147°

5. sin 164°20′   6. sin 126°45′

7. cos 108°30′   8. cos 117°20′

Solve the triangles in Exercises 9–12.

9.

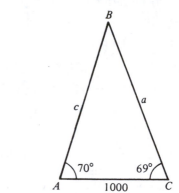

10.

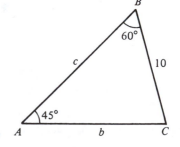

11.

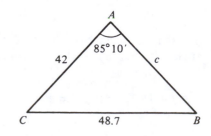

12.

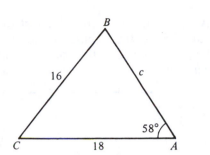

Solve the triangle *ABC* in Exercises 13–18.

13. ∠*B* = 30°, ∠*C* = 45°, *b* = 6
14. *a* = 10, ∠*B* = 45°, ∠*C* = 60°
15. *b* = 5, ∠*B* = 42°, ∠*C* = 28°
16. *a* = 4, ∠*C* = 60°, ∠*A* = 18°
17. *a* = 6425, ∠*B* = 73°29′, ∠*C* = 36°52′
18. ∠*A* = 37°56′, ∠*C* = 60°43′, *b* = 36.42

Find the third side of the triangle in each of Exercises 19–20.

19.

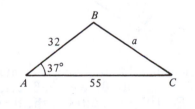

20.

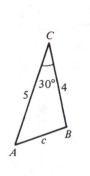

Find the measure of ∠*A* in the triangle in each of Exercises 21–22.

21.

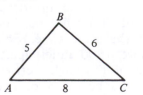

22.

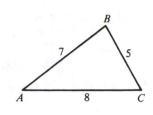

Solve the triangle in each of Exercises 23–26.

23.

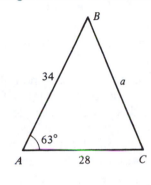

24.

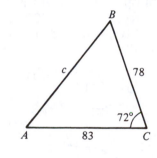

25.

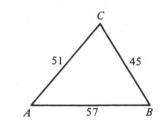

26.

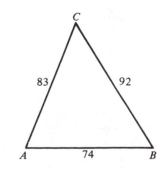

In Exercises 27–30 find the indicated parts of triangle *ABC*. Give the lengths of the sides to three significant digits, and the measures of the angles to the nearest minute of a degree.

27. Given $\angle B = 45°0'$, $a = 5.00$, and $c = 3.00$, find b.

28. Given $b = 12.00$, $c = 8.00$, and $\angle A = 44°0'$, find a.

29. Given $b = 54.0$, $c = 85.0$, and $\angle A = 42°20'$, find a and $\angle B$.

30. Given $b = 17.0$, $c = 12.0$, and $\angle A = 59°20'$, find a and $\angle C$.

31. A piece of wire 6 feet 9 inches long is bent into the shape of a triangle. If the lengths of two sides of the triangle are 2 feet 6 inches and 1 foot 3 inches, what are the measures, to the nearest 10′ of a degree, of the three angles of the triangle?

32. A surveyor at point *C* sights two points, *A* and *B*, on opposite sides of a lake. If *C* is 5000 feet from *A* and 7500 feet from *B*, and $\angle C$ is 110°, how wide is the lake to the nearest foot?

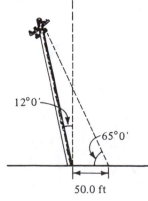

33. A telegraph pole casts a shadow 50.0 feet long when the angle of elevation of the sun is $65°0'$. The pole leans $12°0'$ from the vertical, directly toward the sun. What is the length of the pole to the nearest tenth of a foot?

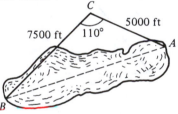

34. A metal frame is to be constructed in the form of a trapezoid, as shown. Find the length of each of the four sides to the nearest inch.

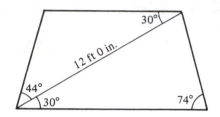

35. Find the lengths, to the nearest foot, of sides *x* and *y* of the four-sided piece of property represented by the figure.

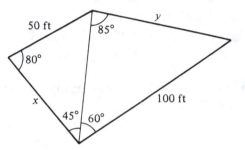

36. A triangular lot bounded by three streets has frontage of 500 feet, 250 feet, and 400 feet. Find the measures of the three angles, to the nearest 5′ of a degree, between the streets bounding the lot.

37. Find the length, to the nearest one-half inch, of the span of the roof shown in the figure. The length of the overhang is 2 feet.

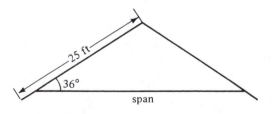

38. Two planes, one flying at an average of 450 miles per hour and the other flying at an average of 300 miles per hour, left an airport at the same time. Four hours later they were 1500 miles apart. What is the measure of the angle between their flight paths, to the nearest 5′ of a degree?

39. Three iron rods are welded together to form a triangular shape. If a 6.0-inch rod is welded to a 14.0-inch rod at an angle of 35°0', what is the length of the third rod, to the nearest $\frac{1}{4}$ of an inch?

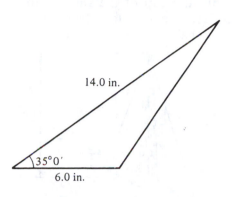

40. Find the slant height of the cliff shown in the figure if the angle of elevation at the base of the cliff is 85°, and another angle of elevation, measured at a point 40 feet from the base, is 58°. Give the answer to the nearest foot.

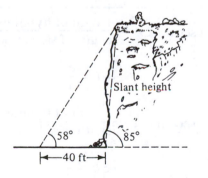

41. A guy wire, 60.0 feet long, is attached to a point 20.0 feet from the top of a pole. This wire forms an angle of 45°0' with the pole. How long, to the nearest tenth of a foot, is the guy wire attached to the top of the pole and anchored at the same point? What is the measure, to the nearest 10' of

a degree, of the angle between the pole and the guy wire attached at the top?

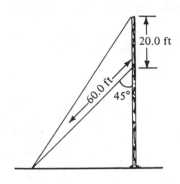

42. From a point on a horizontal plane, the angle of elevation to the top of a hill is 55°, and the angle of elevation to the top of a 60-foot tower standing at the top of the hill is 61°. To the nearest foot, what is the height of the hill?

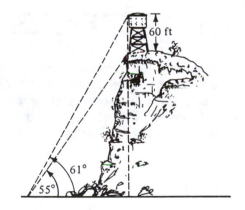

43. Two observers, 1500 feet apart on level ground, sight a weather balloon above and between them at the same time. The angles

of elevation for the two observers are 56°
and 76°. To the nearest foot, how high is
the balloon above ground level?

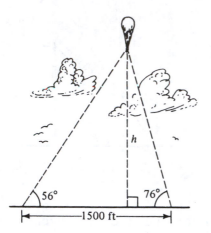

44. How long, to the nearest foot, is the guy
wire that stretches from the top of a
vertical antenna, 80 feet high, to an anchor
48 feet down a 28° slope from the bottom
of the antenna? How long is the guy wire
from the top of the antenna to an anchor
30 feet up the slope from the base of the
antenna?

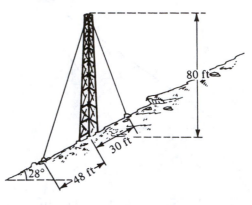

14.4
FUNCTIONS OF ANGLES (II)

In Section 14.1 the trigonometric functions sine,
cosine, and tangent were introduced. Three other
trigonometric functions—**cosecant, secant,** and
cotangent—can be defined in a similar way using
the ratios of the sides of a right triangle.

FIGURE 14.10

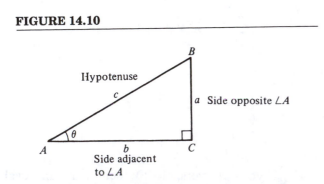

With reference to Figure 14.10, the following three
definitions can be stated:

$$\text{cosecant } \angle A = \frac{\text{length of hypotenuse}}{\text{length of side opposite}}$$

$$\csc \theta = \frac{c}{a}$$

$$\text{secant } \angle A = \frac{\text{length of hypotenuse}}{\text{length of side adjacent}}$$

$$\sec \theta = \frac{c}{b}$$

$$\text{cotangent } \angle A = \frac{\text{length of side adjacent}}{\text{length of side opposite}}$$

$$\cot \theta = \frac{b}{a}$$

Notice that the six trigonometric functions
can be arranged into three pairs of functions

which are reciprocals of each other. With reference to Figure 14.11, these three pairs are:

$$\sin \theta = \frac{a}{c} \text{ and } \csc \theta = \frac{c}{a}$$

so $\quad \sin \theta = \dfrac{1}{\csc \theta} \text{ and } \csc \theta = \dfrac{1}{\sin \theta}$

$$\cos \theta = \frac{b}{c} \text{ and } \sec \theta = \frac{c}{b}$$

so $\quad \cos \theta = \dfrac{1}{\sec \theta} \text{ and } \sec \theta = \dfrac{1}{\cos \theta}$

$$\tan \theta = \frac{a}{b} \text{ and } \cot \theta = \frac{b}{a}$$

so $\quad \tan \theta = \dfrac{1}{\cot \theta} \text{ and } \cot \theta = \dfrac{1}{\tan \theta}.$

FIGURE 14.11

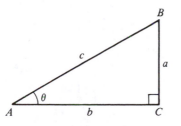

EXAMPLE 1   Find the csc, sec, and cot of angle *A*. Give each answer as a three-place decimal.

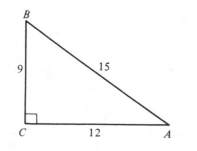

Solution

$$\csc \angle A = \frac{15}{9} \doteq 1.667$$

$$\sec \angle A = \frac{15}{12} \doteq 1.250$$

$$\cot \angle A = \frac{12}{9} \doteq 1.333$$

EXAMPLE 2   Find the csc, sec, and cot of $\angle R$. Give each answer as a fraction.

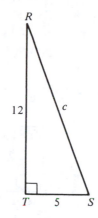

Solution   First we must compute the length of the hypotenuse.

$$c^2 = 5^2 + 12^2$$
$$c^2 = 169$$
$$c = \sqrt{169}$$
$$c = 13$$

Thus, $\csc \angle R = \frac{13}{5}$, $\sec \angle R = \frac{13}{12}$, and $\cot \angle R = \frac{12}{5}$.

EXAMPLE 3   Round off the answers for each of the following to four significant digits.

(a)   If $\cot \angle A = 1.428$, what is $\tan \angle A$?

(b)   If $\sin \angle B = 0.8192$, what is $\csc \angle B$?

(c)   If $\sec \angle A = 1.2207$, what is $\cos \angle A$?

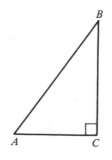

Solutions

(a)
$$\tan \angle A = \frac{1}{\cot \angle A}$$

$$= \frac{1}{1.428}$$

$$\doteq 0.7003$$

(b)
$$\csc \angle B = \frac{1}{\sin \angle B}$$

$$= \frac{1}{0.8192}$$

$$\doteq 1.221$$

(c)
$$\cos \angle A = \frac{1}{\sec \angle A}$$

$$= \frac{1}{1.2207}$$

$$\doteq 0.8192$$

The six trigonometric functions were defined for an angle in the first quadrant (from $0°$ to $90°$) of a coordinate system. With reference to Figure 14.12, the six trigonometric functions are:

$$\sin \theta = \frac{y}{r} \qquad \csc \theta = \frac{r}{y}$$

$$\cos \theta = \frac{x}{r} \qquad \sec \theta = \frac{r}{x}$$

$$\tan \theta = \frac{y}{x} \qquad \cot \theta = \frac{x}{y}$$

FIGURE 14.12

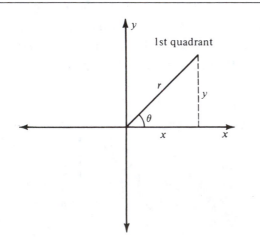

It is possible to develop formulas that will enable us to find the values of the six functions for angles in the other three quadrants of the plane: angles from $90°$ to $360°$. An angle in the second quadrant (that is, between $90°$ and $180°$) can be expressed as $180° - \theta$, where θ is less than $90°$ (see Figure 14.13). Thus, we can use Table 3 at the back of the book to find values of trigonometric functions of angles in the second quadrant.

FIGURE 14.13

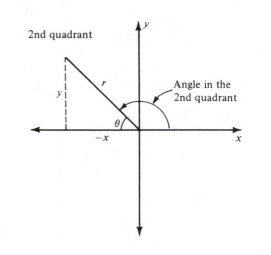

The functions of an angle in the second quadrant (between $90°$ and $180°$) can be expressed in terms of angle θ (see Figure 14.13) as follows:

$$\sin (180° - \theta) = \frac{y}{r} = \sin \theta$$

$$\cos (180° - \theta) = \frac{-x}{r} = -\cos \theta$$

$$\tan (180° - \theta) = \frac{y}{-x} = -\tan \theta$$

$$\csc (180° - \theta) = \frac{r}{y} = \csc \theta$$

$$\sec (180° - \theta) = \frac{r}{-x} = -\sec \theta$$

$$\cot (180° - \theta) = \frac{-x}{y} = -\cot \theta.$$

EXAMPLES

4   $\sin 130° = \sin (180 - 50)° = \sin 50° = 0.7660$

5   $\cos 113° = \cos (180 - 67)° = -\cos 67° = -0.3970$

6   $\tan 171° = \tan (180 - 9)° = -\tan 9° = -0.1584$

7   $\csc 95° = \csc (180 - 85)° = \csc 85° = 1.0038$

8   $\sec 148° = \sec (180 - 32)° = -\sec 32° = -1.179$

9   $\cot 100° = \cot (180 - 80)° = -\cot 80° = -0.1763$

The reciprocal relationships between the functions hold for angles in any quadrant. For example,

$$\sin 130° = \sin 50° = 0.7660,$$

$$\csc 130° = \csc 50° = 1.3055,$$

and

$$0.7660 = \frac{1}{1.3055}.$$

Thus, we can compute the values for csc, sec, and cot by computing the reciprocal values for the corresponding functions sin, cos, and tan, respectively.

An angle in the third quadrant (that is, between 180° and 270°) can be expressed as 180° + θ is less than 90°. (See Figure 14.14.)

FIGURE 14.14

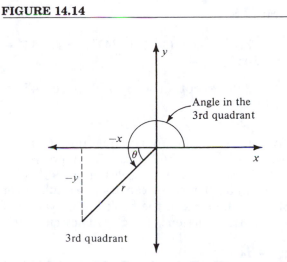

Angle in the 3rd quadrant

3rd quadrant

The functions of an angle in the third quadrant (between 180° and 270°) can be expressed in terms of angle θ (see Figure 14.14) as follows:

$$\sin (180° + \theta) = \frac{-y}{r} = -\sin \theta$$

$$\cos (180° + \theta) = \frac{-x}{r} = -\cos \theta$$

$$\tan (180° + \theta) = \frac{-y}{-x} = \tan \theta$$

$$\csc (180° + \theta) = \frac{r}{-y} = -\csc \theta$$

$$\sec (180° + \theta) = \frac{r}{-x} = -\sec \theta$$

$$\cot (180° + \theta) = \frac{-x}{-y} = \cot \theta.$$

EXAMPLES

10   $\sin 210° = \sin (180 + 30)° = -\sin 30° = -0.500$

11   $\cos 232° = \cos (180 + 52)° = -\cos 52° = -0.6157$

12   $\tan 257° = \tan (180 + 77)° = \tan 77° = 4.331$

13   $\csc 191° = \csc (180 + 11)° = -\csc 11° = -5.2411$

EXAMPLES

14   sec 214° = sec (180 + 34)° = −sec 34° ≐ −1.2063

15   cot 225° = cot (180 + 45)° = cot 45° = 1.0000

An angle in the fourth quadrant (that is, between 270° and 360°) can be expressed as 360° − θ, where angle θ is less than 90°. (See Figure 14.15.) Thus, we can use a calculator or Table 3 at the back of the book to find values of trigonometric functions of angles in any quadrant.

FIGURE 14.15

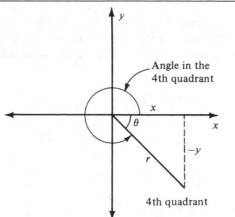

The functions of an angle in the fourth quadrant (between 270° and 360°) can be expressed in terms of θ (see Figure 14.15) as follows:

$$\sin (360° - \theta) = \frac{-y}{r} = -\sin \theta$$

$$\cos (360° - \theta) = \frac{x}{r} = \cos \theta$$

$$\tan (360° - \theta) = \frac{-y}{x} = -\tan \theta$$

$$\csc (360° - \theta) = \frac{r}{-y} = -\csc \theta$$

$$\sec (360° - \theta) = \frac{r}{x} = \sec \theta$$

$$\cot (360° - \theta) = \frac{x}{-y} = -\cot \theta$$

EXAMPLES

16   sin 340° = sin (360 − 20)° = −sin 20° = −0.3420

17   cos 289° = cos (360 − 71)° = cos 71° = 0.3256

18   tan 318° = tan (360 − 42)° = −tan 42° = −0.9004

19   csc 300° = csc (360 − 60)° = −csc 60° = −1.1547

20   sec 357° = sec (360 − 3)° = sec 3° = 1.0014

21   cot 294° = cot (360 − 66)° = −cot 66° = −0.4452

Figure 14.16 is a summary of the signs of the function values in each of the four quadrants of the coordinate plane. The functions of the angles 0°, 90°, 180°, and 270° can be defined as follows. If we consider a radius with a length of one unit which begins at the origin of a coordinate plane, then, for any angle of rotation, with the radius as one side of the angle, we can form a right triangle with legs of lengths x and y and hypotenuse of length r = 1. This is shown in Figure 14.17. This is similar to what was used to define the functions of the angles in all four quadrants in the previous discussion.

FIGURE 14.16

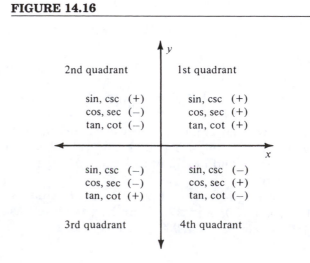

FIGURE 14.17

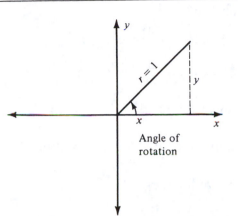

Angle of
rotation

The angles 90°, 180°, and 270° can be represented as angles of rotation in a similar manner, as shown in Figure 14.19. We can construct the following table of function values for these angles using the values for r, x, and y in Figure 14.19 (see Table 14.2).

FIGURE 14.19

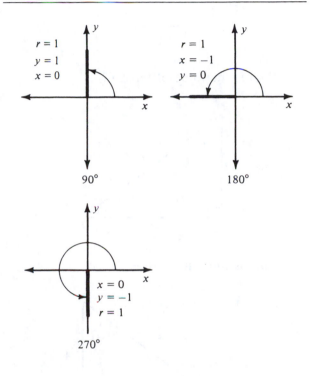

For an angle of 0° (see Figure 14.18) we can consider $r = 1$, $x = 1$, and $y = 0$. The functions of 0° are then defined as follows:

$$\sin 0° = \frac{y}{r} = \frac{0}{1} = 0$$

$$\cos 0° = \frac{x}{r} = \frac{1}{1} = 1$$

$$\tan 0° = \frac{y}{x} = \frac{0}{1} = 0$$

$$\csc 0° = \frac{r}{y} = \frac{1}{0} = \text{undefined}$$

$$\sec 0° = \frac{r}{x} = \frac{1}{1} = 1$$

$$\cot 0° = \frac{x}{y} = \frac{1}{0} = \text{undefined}$$

Recall that division by 0 is not defined.

FIGURE 14.18

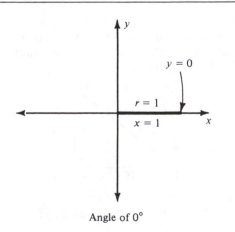

Angle of 0°

TABLE 14.2

	0°	90°	180°	270°
sin	0	1	0	−1
cos	1	0	−1	0
tan	0	undefined	0	undefined
csc	undefined	1	undefined	−1
sec	1	undefined	−1	undefined
cot	undefined	0	undefined	0

Review Exercises

1. What is the value of sin $\angle A$?

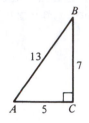

2. What is the value of tan $\angle T$?

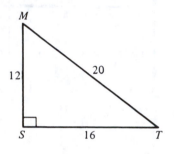

3. What is the value of cos $\angle H$?

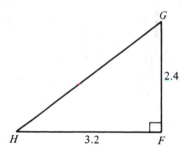

4. What is the value of sin $\angle ABC$?

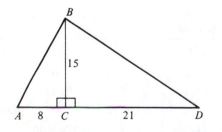

5. If tan θ = 2.301, what is the measure of angle θ?

6. If cos θ = 0.5983, what is the measure of angle θ?

7. Solve for h: $\dfrac{30}{h} = \dfrac{8}{21}$.

8. Solve for p: $\dfrac{4}{15} = \dfrac{17}{p}$.

9. Solve for x: $\dfrac{3(x+2)}{2} = x-4$

10. Solve for m: $2(5-2m) = 3(m+1)-7$

11. Solve for x: $\dfrac{3x}{y} = 5z$.

12. Solve for y: $\dfrac{3x}{y} = 5z$.

EXERCISES
14.4

1. Find the csc, sec, and cot of $\angle A$. Round off each answer to four significant digits.

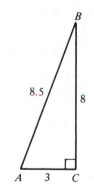

2. Find the csc, sec, and cot of $\angle B$. Round off each answer to four significant digits.

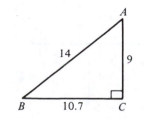

3. If tan θ = 1.325, what is cot θ?

4. If $\sec \theta = 1.2867$, what is $\cos \theta$?
5. If $\sin \theta = 0.4384$, what is $\csc \theta$?
6. If $\cot \theta = 2.356$, what is $\tan \theta$?
7. If $\cos \theta = 0.5736$, what is $\sec \theta$?
8. If $\csc \theta = 1.0864$, what is $\sin \theta$?

In Exercises 9–14 find the measure of angle θ in degrees and minutes. (*Hint:* Use the reciprocal functions and a calculator or Table 3 at the back of the book to determine the angle measures.)

9. $\csc \theta = 1.2522$ 10. $\cot \theta = 1.8041$
11. $\cot \theta = 0.854$ 12. $\sec \theta = 3.7425$
13. $\sec \theta = 1.1481$ 14. $\csc \theta = 1.0069$

In Exercises 15–30 find the values of the indicated trigonometric functions.

15. $\sin 149°$ 16. $\cos 108°$
17. $\tan 305°$ 18. $\sin 238°$
19. $\cos 196°$ 20. $\tan 154°$
21. $\sin 329°$ 22. $\sin 112°$
23. $\cos 298°$ 24. $\tan 250°$
25. $\cot 185°$ 26. $\sec 347°$
27. $\csc 284°$ 28. $\cot 99°$
29. $\sec 118°$ 30. $\csc 236°$

14.5
RADIAN MEASURE

In Section 5.2 a method was introduced for changing angle measures given in degrees, minutes, and seconds to decimal numerals in degrees only. A method for the reverse process was introduced as well. A brief review of these methods is presented in the following examples.

EXAMPLE 1 Change $15°42'$ to degrees.
Solution

$$42' = 42' \times \frac{1°}{60'} = \frac{42°}{60} = 0.7°$$

Thus, $15°42' = 15.7°$.

EXAMPLE 2 Change $40°25'38''$ to degrees.
Solution

$$25'38'' = \left(25' \times \frac{60''}{1'}\right) + 38'' = 1538''$$

$$1538'' = 1538'' \times \frac{1°}{3600''} = \frac{1538°}{3600} \doteq 0.427°$$

(to the nearest thousandth)

Thus, $40°25'38'' \doteq 40.427°$.

EXAMPLE 3 Change $29.85°$ to degrees, minutes, and seconds.

Solution

$$0.85° = 0.85° \times \frac{60'}{1°} = 51'$$

Thus, $29.85° = 29°51'0''$.

EXAMPLE 4 Change $8.372°$ to degrees, minutes, and seconds.

Solution

$$0.372° = 0.372° \times \frac{60'}{1°} = 22.32'$$

$$0.32' = 0.32' \times \frac{60''}{1'} = 19.2''$$

Thus, $8.372° = 8°22'19''$ (to the nearest second).

The degree measurement of angles is not always convenient for solving problems. Another unit of angle measure, which is better suited for some applications, is called the **radian** (see Figure 14.20). **A radian is the measure of an angle** with the following conditions: (1) the angle's vertex is at the center of a circle, and (2) the length of the intercepted arc is equal to the length of the radius of the circle. The abbreviation for radian is "rad."

FIGURE 14.20

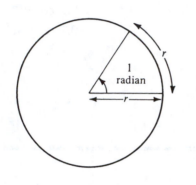

FIGURE 14.21

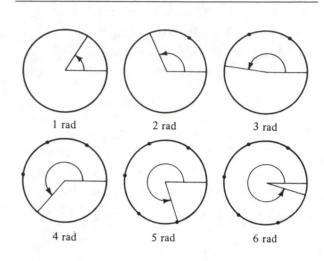

| 1 rad | 2 rad | 3 rad |
| 4 rad | 5 rad | 6 rad |

FIGURE 14.22

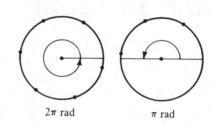

2π rad   π rad

The circumference of a circle is πd units, or $2\pi r$ units, so we can say the ratio of the circumference to the radius of a circle is 2π ($c = 2\pi r$, so $\frac{c}{r} = 2\pi$). Thus, for any circle, the radius may be marked off 2π (approximately 6.28) times along the circumference of the circle. Figure 14.21 illustrates the relationship of radians to angles of rotation. One complete rotation has a measure of 2π radians, and one-half of a rotation has a measure of π radians. (See Figure 14.22.) Since one complete rotation is also equal to 360°, the relation between degrees and radians may be written as $360° = 2\pi$ radians. From this equality we can show the following.

$$360° = 2\pi \text{ rad}$$

$$1° = \frac{2\pi}{360} \text{ rad} = \frac{\pi}{180} \text{ rad} \doteq 0.0175 \text{ rad}$$

and

$$2\pi \text{ rad} = 360°$$

$$1 \text{ rad} = \frac{360°}{2\pi} = \frac{180°}{\pi} \doteq 57.3°.$$

This can be summarized as follows:

1. To convert degrees to radians, multiply the number of degrees by $\frac{\pi}{180}$, or 0.0175.
2. To convert radians to degrees, multiply the number of radians by $\frac{180}{\pi}$, or 57.3.

EXAMPLE 5   Change 38° to radians.

Solution

$$38° = 38 \times 0.0175 \text{ rad} = 0.665 \text{ rad}$$

EXAMPLE 6   Change 247° to radians.

Solution

$$247° = 247 \times 0.0175 \text{ rad} = 4.3225 \text{ rad}$$

EXAMPLE 7   Change $130°48'30''$ to radians.

Solution

$$48'30'' = 2910'' = \frac{2910°}{3600} \doteq 0.808°$$

$$130.808° = 130.808 \times 0.0175 \text{ rad} \doteq 2.289 \text{ rad}$$

EXAMPLE 8   Change 4 radians to degrees.

Solution

$$4 \text{ rad} = 4 \times 57.3° = 229.2°$$

EXAMPLE 9   Change 2.73 radians to degrees, minutes, and seconds.

Solution

$$2.73 \text{ rad} = 2.73 \times 57.3° = 156.429°$$

$$0.429° = 0.429° \times \frac{60'}{1°} = 25.74'$$

$$0.74' = 0.74' \times \frac{60''}{1'} = 44.4''$$

Thus, $2.73 \text{ rad} \doteq 156°25'44''$ (to the nearest second).

For convenience and greater accuracy in complex computations radian measure is often expressed in terms of π. This is illustrated in the following examples.

EXAMPLES

10    $30° = 30 \times \frac{\pi}{180} \text{ rad} = \frac{30\pi}{180} \text{ rad} = \frac{\pi}{6} \text{ rad}$

11    $180° = 180 \times \frac{\pi}{180} \text{ rad} = \frac{180\pi}{180} \text{ rad} = \pi \text{rad}$

12    $72° = 72 \times \frac{\pi}{180} \text{ rad} = \frac{72\pi}{180} \text{ rad} = \frac{2\pi}{5} \text{ rad}$

13    $\frac{\pi}{15} \text{ rad} = \frac{\pi}{15} \times \frac{180°}{\pi} = \frac{180\pi°}{15\pi} = 12°$

14    $\frac{5\pi}{12} \text{ rad} = \frac{5\pi}{12} \times \frac{180°}{\pi} = \frac{5 \times 180\pi°}{12\pi} = 75°$

15    $\frac{2\pi}{7} \text{ rad} = \frac{2\pi}{7} \times \frac{180°}{\pi} = \frac{2 \times 180\pi°}{7\pi} \doteq 51.429°$

Radian measure is a ratio of arc length to radius length, so it is actually a number without dimension. Therefore, we often see radian measures written without a unit. For example,

$$30° = \frac{\pi}{6} \qquad\qquad 12° = \frac{\pi}{15}$$

$$180° = \pi \qquad\qquad 75° = \frac{5\pi}{12}$$

$$72° = \frac{2\pi}{5}$$

all represent radian measures. An angle measure expressed as a number without a unit is understood to mean the radian measure of the angle. Also, when radian measure is used in computations, it is treated as a number without a unit. Radian measure is useful in certain application problems. The following discussion presents two of these applications.

A **central angle** of a circle is an angle formed by two radii, with the center of the circle as its vertex. The length of the arc of a circle intercepted by a central angle is related to the measure of the angle in radians as shown by the following formula:

$$S = \theta r$$

where S is the length of the intercepted arc, θ is the measure of the central angle in radians, and r is the length of the radius of the circle. (See Figure 14.23.)

FIGURE 14.23

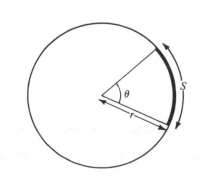

EXAMPLE 16 In a circle with a radius of 3 inches, a central angle has a measure of 2 radians. What is the length of the intercepted arc of the central angle?

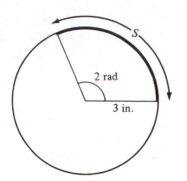

Solution

$$S = \theta r$$
$$S = 2 \text{ rad} \times 3 \text{ in.}$$
$$S = 6 \text{ in.}$$

The length of the arc is 6 inches.

EXAMPLE 17 If a circle has a radius of 8 centimeters, what is the length of the arc intercepted by a central angle of 140°?

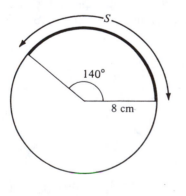

Solution   First, change 140° to radians.

$$140° = 140 \times 0.0175 \text{ rad} = 2.45 \text{ rad}$$

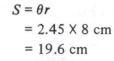

$$S = \theta r$$
$$= 2.45 \times 8 \text{ cm}$$
$$= 19.6 \text{ cm}$$

The length of the arc is 19.6 centimeters.

EXAMPLE 18 Find the measure of the central angle of a circle if the radius of the circle is 5 inches and the length of the intercepted arc is 18.35 inches.

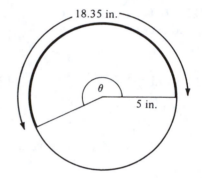

Solution

$$S = \theta r$$
$$\theta = \frac{S}{r} \quad \longleftarrow \text{ Dividing both sides of the equation by } r$$
$$\theta = \frac{18.35 \text{ in.}}{5 \text{ in.}}$$
$$\theta = 3.67$$

The measure of the central angle is 3.67 radians.

Another application of radian measure is with regard to the relationship between **linear velocity** (moving along a straight path) and **angular velocity** (moving around a circular path). Linear velocity is described by the formula:

$$v \text{ (velocity)} = \frac{S \text{ (distance)}}{t \text{ (time)}}.$$

EXAMPLE 19 What is the average velocity of an automobile traveling on a straightaway if, after 3 hours, it has traveled a distance of 120 miles? Give the answer as miles per hour and feet per minute.

Solution

$$v = \frac{S}{t}$$

$$v = \frac{120 \text{ mi}}{3 \text{ hr}} = 40 \text{ mi/hr}$$

$$40 \frac{\text{mi}}{\text{hr}} \times \frac{1 \text{ hr}}{60 \text{ min}} \times \frac{5280 \text{ ft}}{1 \text{ mi}} = \frac{40 \times 5280}{60} \frac{\text{ft}}{\text{min}}$$

$$= 3520 \text{ ft/min}$$

EXAMPLE 20 The average velocity of a runner traveling along a straightaway is 0.18 mi/min. How far will the runner travel in 10 minutes?

Solution

$$S = vt$$

$$S = 0.18 \frac{\text{mi}}{\text{min}} \times 10 \text{ min}$$

$$= 1.8 \text{ mi}$$

Angular velocity is described by the formula:

$$\omega \text{ (angular velocity)} = \frac{\theta \text{ (radians)}}{t \text{ (time)}}.$$

In this formula θ represents an angle of rotation, which could be more than one complete rotation. Angular velocity is usually expressed as radians per minute (rad/min) or radians per second (rad/sec).

From the previous discussion of arc length we know that $S = \theta r$. If we divide both sides of this equation by t we have:

$$\frac{S}{t} = \frac{\theta r}{t}.$$

Since $v = \frac{S}{t}$ (linear velocity) and $\omega = \frac{\theta}{t}$ (angular velocity), we can substitute in the equation above to obtain the formula:

v (linear velocity)

$= \omega$ (angular velocity) $\times r$ (radius).

The formula $v = \omega r$ relates linear velocity and angular velocity.

EXAMPLE 21 A flywheel with a radius of 20 cm is rotating with an angular velocity of 3 rad/sec. What is the linear velocity of a point on the rim of the flywheel?

Solution

$$v = \omega r$$

$$v = 3 \text{ rad/sec} \times 20 \text{ cm}$$

$$v = 60 \text{ cm/sec}$$

Thus, the linear velocity of the point is 60 cm/sec, which is equivalent to 23.6 in./sec, or 118 ft/min.

EXAMPLE 22 A pulley wheel with a radius of 5 inches is rotating with an angular velocity of 50 revolutions per minute. What is the linear velocity of a point on the outer rim of the pulley wheel?

Solution One complete revolution is equal to 2π radians.

$$50 \frac{\text{rev}}{\text{min}} = 50 \frac{\text{rev}}{\text{min}} \times 2\pi \frac{\text{rad}}{\text{rev}} = 100\pi \text{ rad/min}$$

$$v = \omega r$$

$$v = 100(3.14) \text{ rad/min} \times 5 \text{ in.}$$

$$v = 1570 \text{ in./min}$$

The linear velocity of a point on the outer rim of the pulley wheel is 1570 in./min, which is equivalent to 130.8 ft/min and 2.18 ft/sec.

Review Exercises

1. What is a four-place approximation of cos 125°?

2. What is a four-place approximation of sin 160°?

3. What is the length of side *AC*? Round off your answer to 3 significant digits.

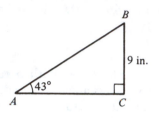

4. What is the length of side *BC*? Round off your answer to 3 significant digits.

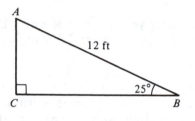

5. If sin *θ* = 0.3987, what is the measure of angle *θ*?

6. If tan *θ* = 1.063, what is the measure of angle *θ*?

7. How many significant digits are there in the number 8,350,000?

8. How many significant digits are there in the number 0.0368?

9. If *kr* = 3*t*, solve for *r*.

10. If $8mx = \frac{5}{3}$, solve for *x*.

11. Find the circumference of circle *O*. Round off the answer to 4 significant digits.

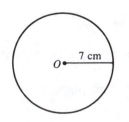

12. Find the circumference of circle *Q*. Round off the answer to 4 significant digits.

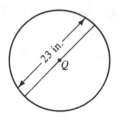

EXERCISES
14.5

In Exercises 1–8 convert the angle measures to radians. Round off each answer to three significant digits.

1. 83°   2. 112°
3. 193.7°   4. 46.5°
5. 15°40′   6. 163°15′
7. 73°28′17″   8. 88°35′48″

In Exercises 9–16 convert the angle measures to degrees. Round off each answer to four significant digits.

9. 2 rad   10. 3.6 rad

11. $\frac{\pi}{2}$ rad   12. $\frac{\pi}{6}$ rad

13. $\frac{5\pi}{7}$ rad   14. $\frac{8\pi}{9}$ rad

15. 0.036 rad   16. 0.154 rad

In Exercises 17–24 change each angle measure to radians and express the answers in terms of *π*.

17. 270°   18. 225°
19. 200°   20. 80°
21. 96°   22. 315°
23. 150°   24. 54°

In Exercises 25–30 change each angle measure to degrees, minutes, and seconds.

25. 3 rad   26. 5 rad

27. $\frac{3\pi}{11}$ rad

28. $\frac{2\pi}{7}$ rad

29. 4.8 rad

30. 1.3 rad

Complete the table for Exercises 31–38.

	Radius of the circle	Measure of a central angle	Length of the intercepted arc
31.	6 in.	4 rad	————
32.	12 cm	3.5 rad	————
33.	15 mm	85°	————
34.	7.3 in.	140°	————
35.	20 cm	————	54 cm
36.	8 in.	————	29.4 in.
37.	————	$\frac{\pi}{4}$ rad	7 cm
38.	————	$\frac{6\pi}{5}$ rad	17 in.

39. A pendulum 18.7 inches long oscillates through an angle of 37°40'. Find the length of the arc described by the end of the pendulum as it swings from one extreme position to the other.

40. A curve on a highway is laid out on an arc of a circle with a radius of 623.4 yards. How long is the arc that subtends a central angle of 29°30'?

41. If we assume the earth to be a sphere with a radius of 3959 miles, then the length of an arc at the equator that subtends an angle of 1 minute at the center of the earth is known as the nautical mile. What is the length of the nautical mile to the nearest foot? (The length of a land mile is 5280 feet.)

42. A protractor is to be constructed so that each degree is represented by an arc of $\frac{4}{64}$ inch. What will be the length of the radius of the protractor to the nearest $\frac{1}{64}$ inch?

43. A flywheel turns at a uniform rate of 3000 revolutions per minute. What is the angular velocity of the flywheel in radians per second?

44. A propeller blade turns at a uniform rate of 2700 revolutions per minute. What is the angular velocity of the propeller in radians per second?

45. A flywheel with a radius of 8 inches has an angular velocity of 30 rad/sec. What is the linear velocity of a point on the rim of the flywheel?

46. A wheel with a radius of 18 cm is rotating with an angular velocity of 1000 rad/min. What is the linear velocity of a point on the outer rim of the wheel?

47. The angular velocity of a grindstone is 1500 rev/min. The diameter of the grindstone is 24 cm. What is the linear velocity, in ft/sec, of a point on the rim of the grindstone?

48. A flywheel is rotating at 3600 rev/min. The radius of the flywheel is 10 inches. What is the linear velocity, in ft/min, of a point on the rim of the flywheel?

49. An automobile tire with a diameter of 30 inches rotates at 25 rev/sec. What is the linear velocity of a point on the outer rim of the tire, expressed in terms of ft/sec?

50. A centifuge with a radius of 10 cm rotates at 15,000 rev/min. What is the linear velocity of a point on the rim of the centrifuge, expressed in terms of cm/sec?

51. What is the linear velocity of a point on the edge of an 8-in., 45-rpm record in in./sec?

52. A pulley 18 inches in diameter is driven by a belt at the rate of 90 revolutions per minute. What is the linear velocity of the belt in feet per second?

53. A pulley is driven by a belt which moves 40 feet per second. If the pulley makes 360 revolutions per minute, find the diameter of the pulley.

54. What is the diameter of a wheel if a point on the rim is traveling 40 miles per hour and the wheel turns through 2180 radians per minute?

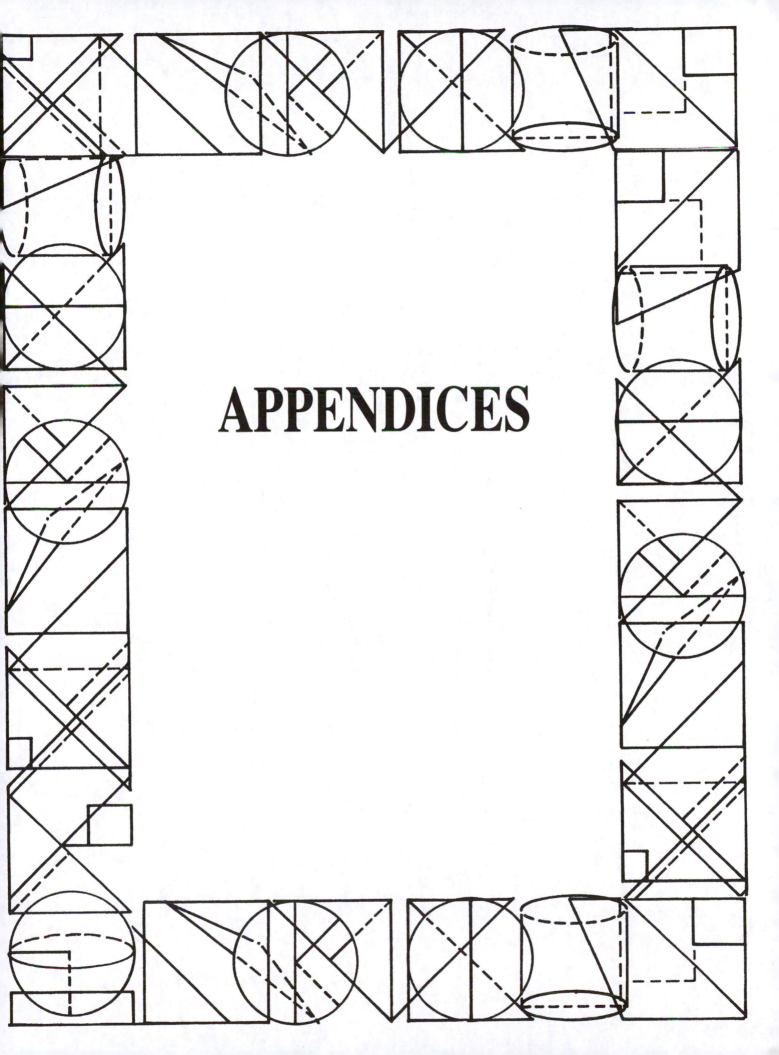

APPENDICES

Appendix

A

Using Interpolation to Estimate Square Roots and Cube Roots

For numbers less than 100, and for other selected numbers, square roots and cube roots can be determined directly from Tables 1 and 2. Examples 1-4 illustrate how the tables can be used in this manner.

EXAMPLE 1 What is the square root of each of the numbers 3, 22, and 53?

Solution To find the square root of a whole number from 1 to 100, first locate the number in the column labeled N. The square root of the number is listed to the right of the number in the column labeled $\sqrt{N}$ (second column to the right of the number).

$$\sqrt{3} \doteq 1.732$$
$$\sqrt{22} \doteq 4.690$$
$$\sqrt{53} \doteq 7.280$$

Note that most of the numbers in the column labeled $\sqrt{N}$ are approximations of the square roots rounded off to the the nearest thousandth. For example, 1.732 is an approximation of the square root of 3, since 1.732 X 1.732 = 2.999824.

EXAMPLE 2 What is the square root of each of the numbers 441, 2601, and 5184?

Solution The square root of selected numbers greater than 100 may be read directly from the table. For a number greater than 100 search the column labeled N^2 (square of number) for the given number. If the number is in this column, the square root of the number is immediately to the left of the number in the column labeled N.

$$\sqrt{441} = 21 \qquad \text{(since } 21^2 = 441\text{)}$$
$$\sqrt{2601} = 51 \qquad \text{(since } 51^2 = 2601\text{)}$$
$$\sqrt{5184} = 72 \qquad \text{(since } 72^2 = 5184\text{)}$$

EXAMPLE 3 What is the cube root of each of the numbers 7, 49, and 88?

Solution Use the process described in the solution of Example 1, but use Table 2.

$$\sqrt[3]{7} \doteq 1.913$$
$$\sqrt[3]{49} \doteq 3.659$$
$$\sqrt[3]{88} \doteq 4.448$$

These answers are estimates so there will be a small error factor for each answer. For example, $(3.659)^3 = 48.9877$.

EXAMPLE 4 What is the cube root of each of the numbers 343, 9261, and 262,144?

Solution Use the process described in the solution of Example 2, but use Table 2.

$$\sqrt[3]{343} = 7 \text{ (since } 7^3 = 343)$$

$$\sqrt[3]{9261} = 21 \text{ (since } 21^3 = 9261)$$

$$\sqrt[3]{262,144} = 64 \text{ (since } 64^3 = 262,144)$$

The tables also can be used to estimate a square root or cube root of a number greater than 100 that is not listed in the column labeled N^2 or the column labeled N^3. The process that is used is called *interpolation*. Interpolation is a means of estimating a missing functional value by taking a weighted average of known functional values at neighboring points.

The following examples illustrate how to use Table 1 and Table 2 to interpolate square roots and cube roots respectively.

EXAMPLE 5 What is the square root of 930?

Solution The number 930 is not listed in the table, but we can find numbers in the table that are *close* to 930. The table shows that $\sqrt{900} = 30$, and $\sqrt{961} = 31$, so we know that $\sqrt{930}$ is a number between 30 and 31.

We can estimate the value of $\sqrt{930}$ as follows: $930 - 900 = 30$, and $961 - 900 = 61$, so the value of $\sqrt{930}$ is $\frac{30}{61}$, or about $\frac{1}{2}$ of the way between 30 and 31. Thus, $\sqrt{930}$ is approximately 30.5.

$$\left.\begin{array}{l} 30 = \sqrt{900} \\ \ ? = \sqrt{930} \\ 31 = \sqrt{961} \end{array}\right\} \begin{array}{l} 30 \\ \ \end{array} \right\} 61$$

If we check this estimate we find that $(30.5)^2 = 930.25$. When we use the method of interpolation we can expect that there will be a small error factor in our result.

EXAMPLE 6 What is the square root of 1320?

Solution The number 1320 is not listed in the square root table, but we can find numbers in the table that are close to 1320. From the table we see that $\sqrt{1320}$ is between 36 and 37. By estimation we find that $\sqrt{1320}$ is $\frac{24}{73}$, or about $\frac{1}{3}$ of the way between 36 and 37. Thus, $\sqrt{1320}$ is approximately 36.3.

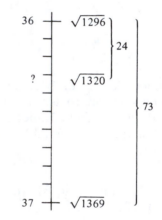

EXAMPLE 7 What is the square root of 4679?

Solution The number 4679 is not listed in the square root table, but it is between the numbers 4624 and 4761, which are in the table. The value of $\sqrt{4679}$ is $\frac{55}{137}$, or about 0.4 of the way between 68 and 69. Thus, $\sqrt{4679}$ is approximately 68.4.

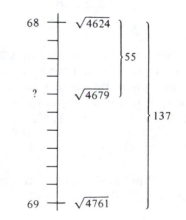

EXAMPLE 8   What is the cube root of 1550?

Solution   The number 1550 is not listed in the second column of Table 2, but we can find numbers that are close to 1550. The cube root table shows that $\sqrt[3]{1331} = 11$ and $\sqrt[3]{1728} = 12$ so we know that $\sqrt[3]{1550}$ is between 11 and 12.

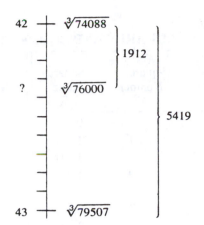

We estimate that $\sqrt[3]{1550}$ is $\dfrac{219}{397}$ or about $\dfrac{11}{20}$ of the way between 11 and 12.

Thus, $\sqrt[3]{1550}$ is approximately 11.55.

If we check this result we find that $(11.55)^3 = 1541$. This is an error factor of 0.06% which is tolerable.

EXAMPLE 9   What is the cube root of 76,000?

Solution   The number 76,000 is not listed in Table 2. Using numbers close to 76,000 we find that $\sqrt[3]{76000}$ is between 42 and 43. Using the interpolation method we estimate that $\sqrt[3]{76,000}$ is $\dfrac{1912}{5419}$ of the way between these numbers or approximately 42.35.

EXAMPLE 10   What is the cube root of 410,000?

Solution   The number 410,000 is not in Table 2. From the table we find

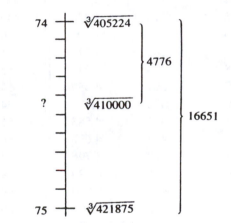

We estimate the cube root of 410,000 to be $\dfrac{4776}{16651}$ of the way between 74 and 75 or $\sqrt[3]{410000} \doteq 74.29$.

TABLE 1 SQUARES AND SQUARE ROOTS

Number N	Square of Number N^2	Square Root $\sqrt{N}$	Number N	Square of Number N^2	Square Root $\sqrt{N}$	Number N	Square of Number N^2	Square Root $\sqrt{N}$
0	0	0.000	34	1156	5.831	68	4624	8.246
1	1	1.000	35	1225	5.916	69	4761	8.307
2	4	1.414	36	1296	6.000	70	4900	8.367
3	9	1.732	37	1369	6.083	71	5041	8.426
4	16	2.000	38	1444	6.164	72	5184	8.485
5	25	2.236	39	1521	6.245	73	5329	8.544
6	36	2.449	40	1600	6.325	74	5476	8.602
7	49	2.646	41	1681	6.403	75	5625	8.660
8	64	2.828	42	1764	6.481	76	5776	8.718
9	81	3.000	43	1849	6.557	77	5929	8.775
10	100	3.162	44	1936	6.633	78	6084	8.832
11	121	3.317	45	2025	6.708	79	6241	8.888
12	144	3.464	46	2116	6.782	80	6400	8.944
13	169	3.606	47	2209	6.856	81	6561	9.000
14	196	3.742	48	2304	6.928	82	6724	9.055
15	225	3.873	49	2401	7.000	83	6889	9.110
16	256	4.000	50	2500	7.071	84	7056	9.165
17	289	4.123	51	2601	7.141	85	7225	9.220
18	324	4.243	52	2704	7.211	86	7396	9.274
19	361	4.359	53	2809	7.280	87	7569	9.327
20	400	4.472	54	2916	7.348	88	7744	9.381
21	441	4.583	55	3025	7.416	89	7921	9.434
22	484	4.690	56	3136	7.483	90	8100	9.487
23	529	4.796	57	3249	7.550	91	8281	9.539
24	576	4.899	58	3364	7.616	92	8464	9.592
25	625	5.000	59	3481	7.681	93	8649	9.644
26	676	5.099	60	3600	7.746	94	8836	9.695
27	729	5.196	61	3721	7.810	95	9025	9.747
28	784	5.292	62	3844	7.874	96	9216	9.798
29	841	5.385	63	3969	7.937	97	9409	9.849
30	900	5.477	64	4096	8.000	98	9604	9.899
31	961	5.568	65	4225	8.062	99	9801	9.950
32	1024	5.657	66	4356	8.124	100	10000	10.000
33	1089	5.745	67	4489	8.185			

TABLE 2 CUBES AND CUBE ROOTS

Number N	Cube of Number N^3	Cube Root $\sqrt[3]{N}$	Number N	Cube of Number N^3	Cube Root $\sqrt[3]{N}$	Number N	Cube of Number N^3	Cube Root $\sqrt[3]{N}$
0	0	0.000	34	39 304	3.240	68	314 432	4.082
1	1	1.000	35	42 875	3.271	69	328 509	4.102
2	8	1.260	36	46 656	3.302	70	343 000	4.121
3	27	1.442	37	50 653	3.332	71	357 911	4.141
4	64	1.587	38	54 872	3.362	72	373 248	4.160
5	125	1.710	39	59 319	3.391	73	389 017	4.179
6	216	1.817	40	64 000	3.420	74	405 224	4.198
7	343	1.913	41	68 921	3.448	75	421 875	4.217
8	512	2.000	42	74 088	3.476	76	438 976	4.236
9	729	2.080	43	79 507	3.503	77	456 533	4.254
10	1000	2.154	44	85 184	3.530	78	474 552	4.273
11	1331	2.224	45	91 125	3.557	79	493 039	4.291
12	1728	2.289	46	97 336	3.583	80	512 000	4.309
13	2197	2.351	47	103 823	3.609	81	531 441	4.327
14	2744	2.410	48	110 592	3.634	82	551 368	4.344
15	3375	2.466	49	117 649	3.659	83	571 787	4.362
16	4096	2.520	50	125 000	3.684	84	592 704	4.380
17	4913	2.571	51	132 651	3.708	85	614 125	4.397
18	5832	2.621	52	140 608	3.733	86	636 056	4.414
19	6859	2.668	53	148 877	3.756	87	658 503	4.431
20	8000	2.714	54	157 464	3.780	88	681 472	4.448
21	9261	2.759	55	166 375	3.803	89	704 969	4.465
22	10648	2.802	56	175 616	3.826	90	729 000	4.481
23	12167	2.844	57	185 193	3.849	91	753 571	4.498
24	13824	2.884	58	195 112	3.871	92	778 688	4.514
25	15625	2.924	59	205 379	3.893	93	804 357	4.531
26	17576	2.962	60	216 000	3.915	94	830 584	4.547
27	19683	3.000	61	226 981	3.936	95	857 375	4.563
28	21952	3.037	62	238 328	3.958	96	884 736	4.579
29	24389	3.072	63	250 047	3.979	97	912 673	4.595
30	27000	3.107	64	262 144	4.000	98	941 192	4.610
31	29791	3.141	65	274 625	4.021	99	970 299	4.626
32	32768	3.175	66	287 496	4.041	100	1 000 000	4.642
33	35937	3.208	67	300 763	4.062			

Appendix

B

Using Interpolation to Estimate Values of Trigonometric Ratios

The values of the trigonometric ratios tangent, sine, and cosine for angles whose measures are whole number degree measures from 1° to 89° can be determined directly from Table 3. This was discussed in Section 14.1. If the measure of an angle lies between two whole number degree measures then interpolation can be used to find the value of a trigonometric ratio. In a table of trigonometric ratios, interpolation is done by proportional comparison. For example, interpolation can be used to find tan 29° 20′, given that:

$$\tan 29° \doteq 0.5543$$

and

$$\tan 30° \doteq 0.5774.$$

Since tan 29°20′ is between tan 29° and tan 30°, the desired result is between 0.5543 and 0.5774. And, since 29°20′ is one-third of the way between 29° and 30°, we estimate that tan 29°20′ is one-third of the way between 0.5543 and 0.5774.

EXAMPLE 1 Find the tangent of 32°30′.

Solution

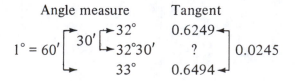

The difference between 32° and 33° is 1°, or 60′. The difference between 32° and 32°30′ is 30′, or $\frac{30}{60}$ of a degree. The difference between the tangents of 32° and 33° is 0.0245. Thus, the tangent of 32°30′ is

$$0.6249 + \frac{30}{60}(0.0245) = 0.6249 + 0.0122,$$

or approximately 0.6371.

EXAMPLE 2 Find the cosine of 54°20′.

Solution

	Angle measure	Cosine	
60′ [20′ [	54°	0.5878	] 0.0142
	54°20′	?	
	55°	0.5736	

Note: The value of the cosine *decreases* as the measure of the angle *increases*. Thus,

$$\cos 54°20′ \doteq 0.5878 - \frac{20}{60}(0.0142)$$
$$= 0.5878 - 0.0047$$
$$= 0.5831.$$

EXAMPLE 3 If sin $A \doteq 0.6587$, what is the measure of $\angle A$?

Solution

Angle measure   Sine

$$60' \begin{bmatrix} x' \begin{bmatrix} 41° & 0.6561 \\ ? & 0.6587 \end{bmatrix} 0.0026 \\ 42° & 0.6691 \end{bmatrix} 0.0130$$

The number 0.6587 is $\frac{0.0026}{0.0130}$ or $\frac{26}{130}$ of the way from 0.6561 to 0.6691. Thus, the measure of $\angle A$ is $\frac{26}{130}$ of the way from 41° to 42°. This can be expressed by the proportion $\frac{x}{60} = \frac{26}{130}$, or $\frac{x}{60} = \frac{1}{5}$. Solving this proportion for x, we find $5x = 60 \cdot 1$, or $x = 12$. Therefore, the measure of $\angle A$ is approximately 41°12′.

EXAMPLE 4 If $\cos B \doteq 0.3947$, what is the measure of angle B?

Solution

Angle measure   Cosine

$$60' \begin{bmatrix} x' \begin{bmatrix} 66° & 0.4067 \\ ? & 0.3947 \end{bmatrix} 0.0120 \\ 67° & 0.3907 \end{bmatrix} 0.0160$$

$$\frac{x}{60} = \frac{120}{160} \text{ or } \frac{x}{60} = \frac{3}{4}$$

Thus, $4x = 180$, or $x = 45$. The measure of $\angle B$ is approximately 66°45′.

TABLE 3   VALUES OF TRIGONOMETRIC RATIOS

Angle Measure in Degrees	Tangent	Sine	Cosine	Angle Measure in Degrees	Tangent	Sine	Cosine
1	0.0175	0.0175	0.9998	46	1.036	0.7193	0.6947
2	0.0349	0.0349	0.9994	47	1.072	0.7314	0.6820
3	0.0523	0.0523	0.9986	48	1.111	0.7431	0.6691
4	0.0699	0.0698	0.9976	49	1.150	0.7547	0.6561
5	0.0875	0.0872	0.9962	50	1.192	0.7660	0.6428
6	0.1051	0.1045	0.9945	51	1.235	0.7771	0.6293
7	0.1228	0.1219	0.9925	52	1.280	0.7880	0.6157
8	0.1405	0.1392	0.9903	53	1.327	0.7986	0.6018
9	0.1584	0.1564	0.9877	54	1.376	0.8090	0.5878
10	0.1763	0.1736	0.9848	55	1.428	0.8192	0.5736
11	0.1944	0.1908	0.9816	56	1.483	0.8290	0.5592
12	0.2126	0.2079	0.9781	57	1.540	0.8387	0.5446
13	0.2309	0.2250	0.9744	58	1.600	0.8480	0.5299
14	0.2493	0.2419	0.9703	59	1.665	0.8572	0.5150
15	0.2679	0.2588	0.9659	60	1.732	0.8660	0.5000
16	0.2867	0.2756	0.9613	61	1.804	0.8746	0.4848
17	0.3057	0.2924	0.9563	62	1.881	0.8829	0.4695
18	0.3249	0.3090	0.9511	63	1.963	0.8910	0.4540
19	0.3443	0.3256	0.9455	64	2.050	0.8988	0.4384
20	0.3640	0.3420	0.9397	65	2.145	0.9063	0.4226
21	0.3839	0.3584	0.9336	66	2.246	0.9135	0.4067
22	0.4040	0.3746	0.9272	67	2.356	0.9205	0.3907
23	0.4245	0.3907	0.9205	68	2.475	0.9272	0.3746
24	0.4452	0.4067	0.9135	69	2.605	0.9336	0.3584
25	0.4663	0.4226	0.9063	70	2.747	0.9397	0.3420
26	0.4877	0.4384	0.8988	71	2.904	0.9455	0.3256
27	0.5095	0.4540	0.8910	72	3.078	0.9511	0.3090
28	0.5317	0.4695	0.8829	73	3.271	0.9563	0.2924
29	0.5543	0.4848	0.8746	74	3.487	0.9613	0.2756
30	0.5774	0.5000	0.8660	75	3.732	0.9659	0.2588
31	0.6009	0.5150	0.8572	76	4.011	0.9703	0.2419
32	0.6249	0.5299	0.8480	77	4.331	0.9744	0.2250
33	0.6494	0.5446	0.8387	78	4.705	0.9781	0.2079
34	0.6745	0.5592	0.8290	79	5.145	0.9816	0.1908
35	0.7002	0.5736	0.8192	80	5.671	0.9848	0.1736
36	0.7265	0.5878	0.8090	81	6.314	0.9877	0.1564
37	0.7536	0.6018	0.7986	82	7.115	0.9903	0.1392
38	0.7813	0.6157	0.7880	83	8.144	0.9925	0.1219
39	0.8098	0.6293	0.7771	84	9.514	0.9945	0.1045
40	0.8391	0.6428	0.7660	85	11.43	0.9962	0.0872
41	0.8693	0.6561	0.7547	86	14.30	0.9976	0.0698
42	0.9004	0.6691	0.7431	87	19.08	0.9986	0.0523
43	0.9325	0.6820	0.7314	88	28.64	0.9994	0.0349
44	0.9657	0.6947	0.7193	89	57.29	0.9998	0.0175
45	1.000	0.7071	0.7071				

Appendix
C
Additional Challenging Problems to Solve

The following pages contain more problems to try. Each problem is coded according to mathematical background needed and level of difficulty. The following code is used:

Mathematical background needed:

M   middle/junior high school

A   eighth or ninth grade elementary algebra

AG   elementary algebra and elementary geometry

AM   advanced math, advanced algebra and/or pre-calculus mathematics

Level of difficulty of the problem:

E   relatively easy

A   average

D   difficult

ED   extremely difficult

The code will appear at the end of the problem in a form like this: (*M-E*)

meaning middle/junior high school mathematics required and that the problem is an easy one.

Browse through the problems, choosing ones that appeal to you. Eventually you may want to try them all.

(Answers can be found at the end of this appendix.)

1. Use four 7's, any operation symbols (+, −, x, and ÷), and parentheses to write an expression for the whole number 1. (*M-E*)

2. Fill in the missing digits in this multiplication:

   (*M-E*)

3. Place decimal points in each numeral to the left of the equal sign to make the sentence true. (*M-E*)

   1 3 2 + 6 0 5 + 5 3 4 = 2.459

4. In Jack's bank there are 7 ten-dollar bills, and six times as many five-dollar bills as ten-dollar bills. What is the total amount in his bank? (*M-E*)

5. Using only pennies, nickels, and dimes, list 21 coins that add up to one dollar. (*M-E*)

6. A number has four digits. It is less than 2,000. Three of its digits are 0, 5, and 5. It is an even number. What is the number? (*M-E*)

7. In a set of automobiles, 12 have engines under 300 h.p., 14 are blue, 13 have belted tires, 7 are blue and have engines under 300 h.p., 5 have belted tires and engines under 300 h.p., 6 are blue and have belted tires, 4 are blue and have engines under 300 h.p. and have belted tires. How many automobiles are there in the set? (*M-A*)

8. In how many different ways can a boy combine 3 pairs of shoes, 5 shirts, 4 pairs of trousers, 2 sweaters, and 6 pairs of socks into an outfit consisting of 1 pair of shoes, 1 shirt, 1 pair of trousers, 1 sweater, and 1 pair of socks? (*M-A*)

9. Fill in the missing digits:

$$\begin{array}{r} \_\,6\,\_ \\ +5\,\_\,2 \\ \hline \_\,3\,\_\,5 \end{array}$$

(M-A)

10. Mary and Betsy started a pen pal club. Once a month, each member sent a letter to every other member of the club. Also, each month one new member was added to the club. How many letters were mailed by club members during the month in which the sixth member joined the club, assuming that the sixth member sent letters that month? (M-A)

11. The RCA Record Company just released a new album. The first week they only received two orders. Then sales began to increase. The second week they received 27 orders; the third week, 57; the fourth week, 92. If this pattern continues, how many orders will they receive the sixth week? (M-A)

12. How many 2's are used to write all the numbers from 99 through 199? (M-A)

13. If a person starts a rumor by telling two other people within an hour; and each of the three tell two new people within the next hour; and this process continues with each person knowing the rumor telling two new people each hour; how many hours will it take for 19,683 people to know the rumor? (M-A)

14. Moving in a vertical, horizontal, or zig-zag to right and/or down fashion, how many different ways can you read the word "SCHOOL" in the following diagram?(M-A)

S
SC
SCH
SCHO
SCHOO
SCHOOL

15. A number is greater than 99,999; and it is less than 101,000. The hundreds digit is a 6. Two of its digits are 5 and 4. It is an odd number. What is the number? (M-A)

16. The principal wanted to declare a snow day. He called three teachers on Wednesday. He asked each of these three teachers to call three other teachers on Thursday. On Friday, each teacher who received a call on Thursday called three different teachers. If they all call the three teachers they have been asked to call, and the pattern continues, how many teachers will receive a call on Sunday for the snow day on Monday? (M-A)

17. Mr. and Mrs. Tubby and Mr. and Mrs. Smiles rode in a car, two in the front seat, two in the back seat. Mrs. Tubby's stepson-in-law drove. Percy sat by Laura's step-daughter. Felix sat by Grace's sister-in-law. Grace sat by Mrs. Tubby's son-in-law. Laura sat by Mr. Smiles' brother-in-law. Mr. Smiles' sister sat behind Percy. State the first and last name of each of the following:

The driver
Mr. Tubby's son-in-law
Mr. Smiles' brother-in-law
Mr. Tubby's daughter
Mr. Smiles' sister (M-D)

18. A group of people were on a hayride. On the way to their destination 10 wagons broke down, and the people on them transferred to the remaining wagons, one to each wagon. On the way home, 15 more wagons broke down, and the people on them transferred to the remaining wagons, two to each wagon. What unique number of people and corresponding unique number of wagons began the hayride? (M-ED)

19. A little girl said, "I am five years old, Grandpa; how old are you?" Grandpa replied, "Your father is eleven times as old as you, and I am as many years old as he will be when you are one-third of my age." How old is Grandpa? (A-A)

20. In writing the numbers from 1 through 99, how many times is the numeral 9 used? (M-A)

21.

8	1	6
3	5	7
4	9	2

The numbers 1-9 above form a *magic square* because the sums of each *row*, each *column*, and each *diagonal* are all equal. Make a *magic square* using 8, 10, 12, 14, 16, 18, 20, 22, and 24. What is the sum of each row, column, and diagonal? (M-D)

22. Half an hour ago it was twice as long after noon as it is from now until midnight. What time is it now? *(M-D)*

23. One of five brothers had broken a window. John said, "It was Henry or Thomas." Henry said, "Neither Ernest nor I did it." Thomas said, "You are both lying." David said, "No, one of them is speaking the truth, but not the other." Ernest said, "David, that is not true." Three of the brothers always tell the truth, but the other two cannot be trusted. Who broke the window? *(M-D)*

24. A square is divided into three congruent rectangles by three line segments connecting two of the parallel sides. The perimeter of one of the three rectangles is 16 meters. What is the perimeter of the square in centimeters? *(M-D)*

25. If your teacher walked the first one-fourth of a race; jogged for a while; then ran the last three-eighths of the race; what fraction of the race did she jog? *(M-D)*

26. In Mr. Geno's clock shop, two cuckoo clocks were brought in for repairs. One clock has the cuckoo coming out every six minutes, while the other one has the cuckoo coming out every eight minutes. Both cuckoos come out at exactly 12:00 noon. What time will it be when they both come out together again, the first time this happens after 12:00 noon? *(M-D)*

27. Draw a pentagon (5-sided polygon) and, using the numbers 1, 2, 3, 4, 5, 6, 7, 8, 9, and 10, place a numeral at each corner of the pentagon and a numeral midway along each side in such a way that the sum of the numbers represented along each side (including both corners) will be exactly 14. *(M-D)*

28. You sell a bicycle for $21, losing 12-1/2 percent of what you paid for it. What should you have sold it for to gain 12-1/2 percent of what you paid for the bicycle? *(M-D)*

29. Six boys fill six notebooks in six weeks, and four girls fill four notebooks in four weeks. How many notebooks will a class of 12 boys and 12 girls fill in 12 weeks? *(M-D)*

30. Four children set up vegetable stalls in the local farmer's market. Their names were Chris, Kim, Tracy, and Robin. Each was selling a different vegetable. Each had a different hair color. Each was wearing a different colored blouse. Chris has black hair. Kim sold carrots. Tracy wore an orange blouse. Chris wore a red blouse. Robin has red hair. The person who sold green peppers had a blue blouse. The person with brown hair was wearing a green blouse. Tracy did not sell brussels sprouts, but one of the others did. The person with blonde hair sold tomatoes. From the above information answer the following three questions: (1) Who was selling brussels sprouts? (2) What vegetable was Tracy selling? (3) What color is Kim's hair? *(M-D)*

31. Each letter represents a different digit. Convert the following pattern to numbers. *(M-D)*

NOEL
+ NOEL
BELLS

32. Ida was engaged to Miss Arizona's brother. Mary and Miss Maryland were at opposite ends of the line, facing the judges. Ann was at the judges' right, next to Miss Minnesota. Neither Maud nor Vera represented Iowa. Miss Vermont was between Katie and Miss Arizona. Miss Kansas was between Ida and Miss Minnesota. Vera was not next to the girl at the judges' left. Miss America was the only girl whose name began with the same letter as the name of her state. Who was Miss America? *(M-D)*

33. Larson, James, Murphy, and Smith are four men whose occupations are butcher, banker, grocer, and policeman, but not necessarily in that order. Larson and James are neighbors and take turns driving each other to work. James makes more money than Murphy. Larson beats Smith regularly at bowling. The butcher always walks to work. The policeman does not live near the banker. The only time the grocer had met the policeman was when the policeman arrested him for speeding. The policeman makes more money than the banker, and also makes more money than the grocer. What is each man's occupation? *(M-D)*

34. The prison warden decided to free some of her prisoners. Their cells were numbered from 1 to 25. Each cell had a lock that opened when you turned the key once, and locked when the key was turned again, and so on. One night when the prisoners were sleeping, she quietly turned all the locks once, opening them. Then she went back and turned every second lock. Then she went back and gave every third lock a turn. Then, every fourth, then every fifth, then every sixth, then every seventh, and so on up to every twenty-fifth. Of course, once she had turned every twelfth she only had to turn one lock each time. Which cells were open in the morning?(*M-D*)

35. Smith, Jones, and Robinson are the engineer, brakeman, and fireman on a train, but not necessarily in that order. Riding the train are three passengers with the same last names as the three crewmen, and each passenger is referred to as a Mister. Mr. Robinson lives in Los Angeles; the brakeman lives in Omaha; Mr. Jones forgot all the algebra he learned in high school; the passenger whose last name is the same as the brakeman's lives in Chicago; the brakeman and one of the passengers, a distinguished mathematical scientist, attend the same church; and Smith beat the fireman at billiards. Which men are engineer, brakeman, and fireman respectively? (*M-D*)

36. On a clock face there are 30 degree-units between each two of the twelve numbers; there are 360 degree-units in a circle; the minute hand moves six degree-units each minute; and the hour hand moves one-half a degree-unit each minute. Exactly how many degree-units are there in the least (smallest) angle between the hands of a clock at 6:45? (*M-ED*)

37. Each letter represents a different digit. Replace the letters by numbers to make the addition true.

$$\begin{array}{r} \text{SEND} \\ +\text{MORE} \\ \hline \text{MONEY} \end{array}$$ (*M-ED*)

38. Do the same as in problem 37.

$$\begin{array}{r} \text{PLEASE} \\ +\text{PARDON} \\ \hline \text{DELAYS} \end{array}$$ (*M-ED*)

39. A man spent one-sixth of his life as a child, one-twelfth of his life as a teenager, then one-seventh of it before getting married. Five years after he was married his son was born. The son died four years before his father did, and was half the age his father would be at his death. What whole number represents the father's age when he died? (*M-ED*)

40. The sum of the ages of Ann and Beth is 25. The sum of the ages of Ann and Carol is 20. The sum of the ages of Beth and Carol is 31. Who is oldest, and how old is she? (*A-D*)

41. A father's age is the same as the sum of the ages of his three children. In 9 years his age will be the sum of the ages of his two oldest children. Three years later, his age will be the sum of the ages of his oldest and youngest children. Three years after that, his age will be the sum of the ages of his two youngest children. What are the ages now of all four people? (*A-D*)

42. Draw an equilateral triangle. Connect each of the midpoints of the sides to the opposite vertices, using line segments. Connect all the midpoints of the sides using line segments. How many *different* triangles can be found in your final figure? (*AG-D*)

43. A boy is three-eighths of the way across a railroad bridge when he hears a train coming from behind him. The train is travelling 60 miles per hour. He runs to an end of the bridge, getting there exactly the same time the train does. It doesn't matter which end of the bridge he runs to, toward or away from the train. How fast can he run, assuming he ran full speed on the bridge? (*A-ED*)

44. A man and a woman can each paint one-half of an apartment in three hours. How many minutes will it take six women and eight men to paint ten such apartments if the women work continuously but the men each rest ten minutes after each one hour of work?(*AM-D*)

45. A restaurant owner spent $26 each on uniform pants and $25 each on uniform coats, buying at least one of each item. He spent exactly $1,000. How many more uniform pants than uniform coats did he buy if he bought a whole number of each? (*AM-D*)

Answers to Appendix C Problems

1. $(7 \div 7) + (7-7)$
2. 7, 2, 3, 2
3. 1.32 + .605 + .534
4. $280
5. 7D, 4N, 10P
6. 1550
7. 25
8. 720
9. 7, 3, 6, 1, 2
10. 30
11. 177
12. 20
13. 9
14. 32
15. 100,645
16. 243
17. Percy Smiles, Percy Smiles, Felix Tubby, Grace Smiles, Laura Tubby
18. 900 people, 100 wagons
19. 75
20. 20
21. 48
22. 8:10 p.m.
23. Thomas
24. 2400
25. 3/8
26. 12:24 p.m.
27. Clockwise from a corner, 4-8-2-7-5-6-3-10-1-9
28. $27
29. 60
30. Chris-tomatoes-brown
31. 9387
     9387
     ─────
     18774
32. Miss Kansas
33. Larson-banker; Murphy-butcher; James-grocer; Smith-policeman
34. Cells 1, 4, 9, 16, 25
35. Engineer-Smith, brakeman-Jones, fireman-Robinson
36. 67½
37. 9567
     1085
     ─────
     10652
38. 451681
     463927
     ──────
     915608
39. 84
40. Beth is 18
41. 36, 15, 12, 9
42. 47
43. 15 m.p.h.
44. 280 minutes
45. 11 more

ANSWERS TO ODD-NUMBERED EXERCISES

CHAPTER 1

Review Exercises, page 4

1. 7   3. 17   5. 6   7. 8   9. 14   11. 9

Exercises 1.1, page 4

1. seventy-six   3. ninety-three
5. three hundred fifty-four
7. two thousand four hundred five   9. 435
11. 394   13. 6203   15. hundreds
17. hundreds   19. ten thousands
21. seven hundred thousand sixty
23. eight million four hundred   25. 3,040,005
27. 7,005,304   29. 4017   31. 85   33. 55
35. 369   37. 459   39. 8097   41. 87
43. 896   45. 537   47. 477   49. 5999
51. 329   53. 22,419,598   55. 1452 board feet
57. 4222 ft²   59. $3256   61. 10,717 items
63. $28,193.00

Review Exercises, page 8

1. 8   3. 16   5. 7   7. 9   9. 110,419
11. 161,366

Exercises 1.2, page 8

$$\begin{array}{}
& \overset{5\ 14}{} & & \overset{3\ 10}{} & & \overset{0\ 19}{} & & \overset{2\ 11\ 18}{}
\end{array}$$
1. 6̸ 4̸   3. 4̸ 0̸   5. 1̸9̸   7. 3̸ 2̸ 8̸
   6 17 10   7 9 13   5 9 10
9. 7̸ 8̸ 0̸   11. 8̸ 0̸ 3̸   13. 6̸ 0̸ 0̸
   5 12 18 17   4 12 9 17   7 9 9 13
15. 6̸ 3̸ 9̸ 7̸   17. 3̸ 3̸ 0̸ 7̸   19. 8̸ 0̸ 0̸ 3̸
   5 9 9 9 10
21. 6̸ 0̸, 0̸ 0̸ 0̸   23. 42,211   25. 13,211
27. 35   29. 13   31. 10   33. 249   35. 517
37. 725   39. 363   41. 4898   43. 4658
45. 5545   47. 55,615   49. $2,004,500
51. 453 ft   53. 668 ft

Review Exercises, page 12

1. 63   3. 6372   5. 589
7. 504,861 refrigerators   9. (1) 780 miles,
   (2) 1089 miles, (3) 1988 miles,
   (4) 1089 miles, (5) 1936 miles
11. 868 kilowatt hours

Exercises 1.3, page 13

1. 84   3. 702   5. 1920   7. 1272
9. 4572   11. 5754   13. 29,391
15. 61,830   17. 2,562,111   19. 7,059,584
21. 18,150,462   23. 20,808 hours
25. $4625.00   27. 28,120 shingles
29. 12,388 turns   31. 235 total lumens
33. $888,404   35. 12   37. 87   39. 16
41. 14   43. 299   45. 58

Review Exercises, page 16

1. 63   3. 30   5. 162   7. 2506   9. 1288
11. 13,186

Exercises 1.4, page 16

1. 2 × 3   3. 2 × 7   5. 3 × 5   7. 5 × 5
9. 7 × 7   11. 2 × 2 × 3   13. 2 × 3 × 3
15. 3 × 3 × 3   17. 2 × 2 × 2 × 3 × 3
19. 2 × 2 × 2 × 5   21. 2 × 3 × 5   23. 2 × 2 × 7
25. 2 × 2 × 2 × 2 × 3   27. 2 × 5 × 7
29. 3 × 3 × 7   31. 2 × 2 × 2 × 3 × 3
33. 3 × 3 × 3 × 3   35. 6   37. 4   39. 8
41. 1   43. 5   45. 1   47. 9

Review Exercises, page 20

1. 63   3. 240   5. 1890   7. 13,500
9. 608,000   11. 34,821,664

Exercises 1.5, page 20

1. 21   3. 21   5. 26   7. 11   9. 14
11. 50   13. 11   15. 200, remainder 2
17. 85 yd² per day   19. 3-lb pully   21. 25 pieces
23. 23 ft   25. 45 full lengths   27. 108 ft
29. 27 hr   31. 15,920 mi/hr

Review Exercises, page 23

1. 7662   3. 1534   5. 28,236
7. 181, remainder 5   9. 2 × 2 × 2 × 3 × 3 × 5
11. 2 × 5 × 5 × 11

Exercises 1.6, page 23

1. 3   3. 4   5. 343   7. 1296   9. 256
11. 13,824   13. 48   15. 3   17. 7   19. 11
21. 14   23. 8   25. 5   27. 7   29. 169
31. 324   33. 3481   35. 729   37. 2197
39. 3375   41. 80   43. 28   45. 205   47. 9
49. 184   51. 576 ft   53. 343 cubic units

CHAPTER 2

Review Exercises, page 27

1. 88   3. 58,869   5. 41
7. 33, remainder 16   9. 11   11. 169

Exercises 2.1, page 27

1. 7   3. 5 of 9   5. $\frac{4}{4}, \frac{7}{7}, \frac{10}{10}$ 7. $\frac{1}{48}$ 9. $\frac{13}{31}$

11. $A = \frac{2}{5}, B = \frac{7}{5}$ 13. $\frac{65}{100}$

15. $\frac{3}{2}, \frac{4}{3}, \frac{9}{7}, \frac{105}{99}, \frac{15}{4}, \frac{2}{2}$

17. $\frac{2}{3}, \frac{4}{3}, \frac{3}{4}; \frac{15}{4}, \frac{3}{2}, \frac{2}{2}; \frac{1}{2}, \frac{9}{7}, \frac{4}{7}$ 19. $\frac{49}{835}$ 21. $\frac{12}{27}$

23. $\frac{53}{132}$ 25. $\frac{428}{750}$

Review Exercises, page 32

1. 8   3. $\frac{5}{9}$ 5. $\frac{5}{5}, \frac{7}{7}$, and $\frac{13}{13}$

7. $630 = 2 \times 3 \times 3 \times 5 \times 7$ 9. 21   11. 4

Exercises 2.2, page 33

1. $\frac{2}{4}, \frac{3}{6}, \frac{4}{8}, \frac{5}{10}, \frac{6}{12}, \frac{7}{14}, \cdots$

3. $\frac{6}{10}, \frac{9}{15}, \frac{12}{20}, \frac{15}{25}, \frac{18}{30}, \cdots$

5. $\frac{16}{10}, \frac{24}{15}, \frac{32}{20}, \frac{40}{25}, \frac{48}{30}, \cdots$ 7. $\frac{6}{9}, \frac{4}{6}, \frac{2}{3}$

9. $\frac{16}{28}, \frac{8}{14}, \frac{4}{7}$ 11. $\frac{20}{12}, \frac{10}{6}, \frac{5}{3}$ 13. gcf = 12

15. gcf = 21   17. gcf = 16   19. gcf = 3; $\frac{1}{3}$

21. gcf = 3; $\frac{6}{5}$ 23. gcf = 9; $\frac{3}{4}$ 25. gcf = 8; $\frac{5}{8}$

27. gcf = 4; $\frac{7}{10}$ 29. gcf = 7; $\frac{7}{9}$ 31. gcf = 24; $\frac{5}{3}$

33. gcf = 60; $\frac{14}{5}$ 35. $\frac{60}{156} = \frac{2 \times 2 \times 3 \times 5}{2 \times 2 \times 3 \times 13} = \frac{5}{13}$

37. $\frac{36}{132} = \frac{2 \times 2 \times 3 \times 3}{2 \times 2 \times 3 \times 11} = \frac{3}{11}$

39. $\frac{210}{98} = \frac{2 \times 3 \times 5 \times 7}{2 \times 7 \times 7} = \frac{15}{7}$

41. $\frac{300}{210} = \frac{2 \times 2 \times 3 \times 5 \times 5}{2 \times 3 \times 5 \times 7} = \frac{10}{7}$

43. $\frac{546}{910} = \frac{2 \times 3 \times 91}{2 \times 5 \times 91} = \frac{3}{5}$

Review Exercises, page 38

1. 67,928   3. 21   5. 13,860   7. 10
9. 236   11. $\frac{15}{49}$

Exercises 2.3, page 38

1. $\frac{60}{91}$ 3. $\frac{33}{112}$ 5. $\frac{15}{8}$ 7. 39   9. $\frac{36}{7}$

11. 15   13. $5\frac{2}{3}$ 15. $\frac{199}{13}$ 17. $7\frac{1}{3}$ 19. $\frac{22}{3}$

21. $\frac{221}{437}$ 23. $1\frac{144}{247}$ 25. $\frac{527}{35} = 15\frac{2}{35}$ 27. $4\frac{1}{120}$

29. $\frac{656}{189} = 3\frac{89}{189}$ 31. $\frac{425}{504}$ 33. $\frac{27}{64}$ 35. $\frac{9}{64}$

37. $\frac{8}{15}$ 39. $\frac{5}{7}$ 41. $\frac{15}{56}$ 43. $\frac{77}{5} = 15\frac{2}{5}$

45. 31   47. $\frac{1}{3}$ 49. $\frac{77}{3} = 25\frac{2}{3}$

Review Exercises, page 42

1. 23   3. 61,936   5. 217   7. $\frac{39}{68}$ 9. 10
11. $4\frac{5}{13}$

Exercises 2.4, page 43

1. $\frac{8}{12}, \frac{5}{12}$ 3. $\frac{3}{6}, \frac{1}{6}$ 5. $\frac{49}{315}, \frac{72}{315}$

7. $\frac{8}{24}, \frac{18}{24}, \frac{21}{24}$ 9. $\frac{9}{30}, \frac{24}{30}, \frac{5}{30}$ 11. $\frac{105}{360}, \frac{60}{360}, \frac{96}{360}$

13. $\frac{4}{5}$ 15. $\frac{2}{3}$ 17. $\frac{11}{8} = 1\frac{3}{8}$ 19. $\frac{7}{6} = 1\frac{1}{6}$

21. $8\frac{7}{20}$ 23. $3\frac{1}{6}$ 25. $\frac{65}{72}$ 27. $\frac{91}{40} = 2\frac{11}{40}$

29. $33\frac{3}{16}$ 31. $15\frac{19}{24}$ 33. $\frac{9}{10}$

35. $\frac{13}{15}$ of the building   37. $\frac{59}{60}$ of the gallon

39. $4\frac{37}{120}$ hr   41. $158\frac{7}{12}$ ft

Review Exercises, page 46

1. $\frac{15}{8} = 1\frac{7}{8}$ 3. 26   5. $\frac{36}{5} = 7\frac{1}{5}$ 7. 20

9. $\frac{253}{175} = 1\frac{78}{175}$ 11. $\frac{6}{9}, \frac{6}{9}$

Exercises 2.5, page 46

1. $\dfrac{20}{63}$ 3. $\dfrac{19}{40}$ 5. $\dfrac{11}{4} = 2\dfrac{3}{4}$ 7. $\dfrac{29}{15} = 1\dfrac{14}{15}$

9. $3\dfrac{41}{90}$ 11. $\dfrac{3}{2} = 1\dfrac{1}{2}$ 13. $\dfrac{7}{72}$ 15. 1

17. $\dfrac{41}{15} = 2\dfrac{11}{15}$ 19. $\dfrac{11}{3} = 3\dfrac{2}{3}$ 21. $\dfrac{65}{18} = 3\dfrac{11}{18}$

23. $5\dfrac{1}{4}$ ft   25. $\dfrac{21}{40}$ in.   27. $7\dfrac{11}{12}$ in.

29. $510\dfrac{1}{4}$ ft

Review Exercises, page 49

1. 23,732   3. 3608   5. $\dfrac{1}{32}$ 7. $\dfrac{4}{315}$

9. $\dfrac{19}{68}$ 11. $\dfrac{89}{40} = 2\dfrac{9}{40}$

Exercises 2.6, page 49

1. $\dfrac{35}{24} = 1\dfrac{11}{24}$ 3. 10   5. $\dfrac{55}{6} = 9\dfrac{1}{6}$

7. $\dfrac{19}{2} = 9\dfrac{1}{2}$ 9. $\dfrac{245}{187} = 1\dfrac{58}{187}$ 11. $\dfrac{27}{32}$

13. $\dfrac{2890}{1288} = 2\dfrac{157}{644}$ 15. $\dfrac{13}{3} = 4\dfrac{1}{3}$ 17. $\dfrac{7}{18}$

19. $\dfrac{57}{16} = 3\dfrac{9}{16}$ 21. $73\dfrac{1}{2}$ volts   23. $9\dfrac{3}{4}$ lawns

25. $17\dfrac{1}{2}$ bushels   27. $\dfrac{25}{56}$ lb per part

29. $54\dfrac{5}{8}$ in.

Review Exercises, page 53

1. $\dfrac{21}{40}$ 3. $\dfrac{35}{24} = 1\dfrac{11}{24}$ 5. $\dfrac{24}{35}$ 7. $\dfrac{55}{6} = 9\dfrac{1}{6}$

9. $\dfrac{15}{2} = 7\dfrac{1}{2}$ 11. $\dfrac{5}{24}$

Exercises 2.7, page 53

1. $\dfrac{2}{3} > \dfrac{8}{27}$ 3. $\dfrac{13}{15} > \dfrac{12}{14}$ 5. $\dfrac{3}{4} > \dfrac{7}{24}$

7. $\dfrac{14}{16} > \dfrac{13}{15}$ 9. $4 \times 8 > 9 \times 3$

11. $12 \times 20 = 15 \times 16$ 13. $5 \times 6 > 9 \times 3$

15. $7 \times 27 = 9 \times 21$ 17. $\dfrac{279}{999} < \dfrac{288}{1008}$

19. $\dfrac{5034}{5604} > \dfrac{5031}{5601}$ 21. $7\dfrac{9}{16} < 7\dfrac{10}{17}$

23.
```
 ├───────┼───┼┼───────┼───────────→
 0       1   33       2           1
         4   87       3
```

25. $\dfrac{2}{6}, \dfrac{3}{7}, \dfrac{4}{8}, \dfrac{5}{9}$ 27. 11 oranges for 92¢

Review Exercises, page 56

1. $\dfrac{39}{4} = 9\dfrac{3}{4}$ 3. $\dfrac{21}{16} = 1\dfrac{5}{16}$ 5. $\dfrac{13}{32}$ 7. $\dfrac{236}{13}$

9. $\dfrac{6}{7}$ 11. $\dfrac{4}{13}$

Exercises 2.8, page 57

1. $\dfrac{1 \text{ ft}}{12 \text{ in.}}$ 3. $\dfrac{1 \text{ m}}{1000 \text{ mm}}$ 5. $\dfrac{10 \text{ cm}}{1 \text{ dm}}$

7. $\dfrac{1 \text{ rd}}{5\frac{1}{2} \text{ yd}}$ 9. $\dfrac{4 \text{ qt}}{1 \text{ gal}}$ 11. $\dfrac{\frac{1}{2} \text{ pt.}}{1 \text{ cup}}$ 13. $\dfrac{5}{12}$ ft

15. 18 ft   17. 750 cm   19. $57\dfrac{1}{2}$ dm

21. $379\dfrac{1}{2}$ ft   23. 34 cups   25. $8\dfrac{3}{4}$ gal

27. $4\dfrac{18}{33}$ rd   29. $13\dfrac{8}{9}$ yd   31. $5\dfrac{5}{99}$ rd

33. $46\dfrac{7}{8}$ gal   35. 53 ft   37. $38\dfrac{5}{12}$ ft

39. 7 ft 4 in.   41. 8 ft 0 in.   43. 53 ft 5 in.

45. $5\dfrac{1}{2}$ kilowatts

CHAPTER 3

Review Exercises, page 61

1. $\dfrac{14}{76}$ 3. $\dfrac{76}{95}$ 5. $17\dfrac{67}{72}$ 7. $\dfrac{34}{63}$ 9. 4

11. $\dfrac{4}{8}, \dfrac{5}{9}, \dfrac{6}{10}, \dfrac{7}{11}$

Exercises 3.1, page 61

1. $(4 \cdot 100) + (7 \cdot 10) + (8 \cdot 1) + \left(3 \cdot \dfrac{1}{10}\right) +$
$\left(9 \cdot \dfrac{1}{100}\right) + \left(4 \cdot \dfrac{1}{1000}\right)$

3. $(2 \cdot 10) + (0 \cdot 1) + \left(1 \cdot \dfrac{1}{10}\right) + \left(4 \cdot \dfrac{1}{100}\right)$

5. $(1 \cdot 1000) + (0 \cdot 100) + (0 \cdot 10) + (0.1) +$
$\left(0 \cdot \dfrac{1}{10}\right) + \left(0 \cdot \dfrac{1}{100}\right) + \left(3 \cdot \dfrac{1}{1000}\right)$

7. $(3 \cdot 100) + (9 \cdot 10) + (4 \cdot 1) + \left(7 \cdot \dfrac{1}{10}\right) +$
    $\left(8 \cdot \dfrac{1}{100}\right) + \left(4 \cdot \dfrac{1}{1000}\right)$

9. $(5 \cdot 100) + (0 \cdot 10) + (9 \cdot 1) + \left(7 \cdot \dfrac{1}{10}\right) +$
    $\left(0 \cdot \dfrac{1}{100}\right) + \left(8 \cdot \dfrac{1}{1000}\right)$

11. four hundred seventy-eight and three hundred ninety-four thousandths
13. twenty and fourteen hundredths
15. one thousand and three thousandths
17. three hundred ninety-four and seven hundred eighty-four thousandths
19. five hundred nine and seven hundred eight thousandths   21. 2460   23. 30   25. 3590
27. 30   29. 5.39   31. 54.56   33. 54.57
35. 1.50   37. 3.55   39. 6.28   41. 55.56
43. 2.56   45. 1.40   47. 4.55   49. 950,000
51. 700   53. 340,000   55. 900

Review Exercises, page 63

1. 22,015   3. 39,778   5. 400.015   7. $\dfrac{3}{56}$

9. $\dfrac{1}{36}$ 11. $\dfrac{43}{432}$

Exercises 3.2, page 63

1. 15.9   3. 4856.7221   5. 16.81
7. 1310.994   9. 11.22   11. 24.65
13. 720.571   15. 47.4545   17. 7.14 in.
19. $67.57   21. 31.64 in.   23. 44.57 horsepower

Review Exercises, page 66

1. $\dfrac{248}{45} = 5\dfrac{23}{45}$ 3. $\dfrac{31}{48}$ 5. $9\dfrac{16}{81}$ 7. 3640.024
9. 34.510   11. 350

Exercises 3.3, page 66

1. 243.5   3. 5.79   5. 0.492   7. 1324.5
9. 349.45   11. 49.204   13. 0.0056
15. 3.421   17. 0.00043   19. 4.9325
21. 0.0000004   23. 97.908   25. 0.001849
27. 7189.057   29. 0.2390976
31. 15.76477728   33. 38,000 in.
35. $4118.81   37. 127,488.15 mi.
39. 240,443.6472 m²

Review Exercises, page 70

1. $\dfrac{64}{125}$ 3. $\dfrac{59}{864}$ 5. 18.49
7. 0.000000064   9. 1.006009
11. 0.06048

Exercises 3.4, page 70

1. 0.0   3. 4.162   5. 4.98   7. 0.00
9. 0.875   11. $0.3\overline{3}$ 13. 0.3125   15. 5.75
17. 0.345   19. 4.3375   21. 155.02
23. $0.0006\overline{6}$ 25. 0.00425   27. 0.028
29. 19.00045   31. 0.310   33. 2.67 m
35. $0.6\overline{6}$ 37. $0.250\overline{0}$ 39. $0.428571\overline{428571}$

Review Exercises, page 73

1. $49.16\overline{6}$ 3. 0.35   5. $0.925\overline{925}$ 7. 8.995
9. 81.309   11. $\dfrac{2}{5} < \dfrac{14}{34} < \dfrac{3}{7} < \dfrac{5}{11}$

Exercises 3.5, page 74

1. $\dfrac{9}{20}$ 3. $\dfrac{1}{40}$ 5. $3\dfrac{3}{8}$ 7. $9\dfrac{11}{2000}$ 9. $\dfrac{9}{16}$
11. $14\dfrac{321}{400}$ 13. $\dfrac{1}{3}$ 15. $\dfrac{1}{6}$ 17. $\dfrac{34}{99}$ 19. $\dfrac{124}{999}$
21. $2\dfrac{7}{110}$ 23. $\dfrac{11}{32}$ in.   25. $2\dfrac{5}{12}$ ft

Review Exercises, page 76

1. 27.68   3. 13.187   5. 0.826   7. 296.36649
9. 7.12   11. 0.21

Exercises 3.6, page 76

1. $\dfrac{44}{100} = 0.44$ 3. $\dfrac{241}{100} = 2.41$ 5. $\dfrac{1142}{100} = 11.42$

7. $\dfrac{\frac{1}{4}}{100} = 0.0025$ 9. $\dfrac{6.4}{100} = 0.064$

11. $\dfrac{66\frac{2}{3}}{100} = 0.6\overline{6}$ 13. $\dfrac{128.8}{100} = 1.288$ 15. 14.56
17. 23.8   19. 8.96   21. 0.276   23. 0.02
25. 2.547   27. 10.0024   29. 41.4
31. 18.0128   33. 27.36   35. 21.75
37. 43.18   39. $37.46

Review Exercises, page 78

1. 4.0809   3. 0.1852065   5. $\dfrac{9}{16}$ 7. 0.0125
9. 61.75   11. 612.14

Exercises 3.7, page 78

1. $3.62   3. $522.94   5. $4750.49
7. $15,247.50   9. $153.30   11. $2058.28

Review Exercises, page 80

1. 0.08   3. 2.36   5. 0.216   7. 0.08$\overline{3}$
9. 5.98   11. 5.304

Exercises 3.8, page 80

1. $47.07   3. $6.23   5. $6.64   7. $1674.00
9. $37.91   11. $91,536.75

Review Exercises, page 82

1. $\frac{13}{8} = 1\frac{5}{8}$ 3. $\frac{27}{64}$ 5. 0.001024   7. 78.037
9. 172.14   11. 4.375

Exercises 3.9, page 83

1. 45%   3. 5.5%   5. 475%   7. 0.05%
9. 380%   11. 40%   13. 16.6$\overline{6}$%
15. 233.3$\overline{3}$%   17. 537.5%   19. 501.83
21. 248.125   23. 999.86   25. $352.08
27. $66.66   29. $60.52
31. $8.87 (19% of $7.45 + $7.45 = $8.87)
33. $7.70 per hour   35. $8.12 per hour

Review Exercises, page 85

1. 12,852   3. $\frac{29}{48}$ 5. $\frac{3}{2}$
7. 0.035472   9. 74.863673   11. 99 in.

Exercises 3.10, page 85

1. 0.72 ft   3. 1.2 ft   5. 113.4 in.
7. 15.3 yd   9. 39.18 ft   11. 7.58 ft
13. 90.67 yd   15. 1.07 yd   17. 16 ft 1.5 in.
19. 50.67 ft   21. 33 yd 1 ft 0 in.
23. 24 ft 8.4 in.   25. $\frac{4}{8}$ in.   27. $3\frac{13}{16}$ in.
29. 98.12 in.

Extra Problems, page 86

1. $24.22   3. $225   5. $603.50   7. $16.80
9. $74   11. $117.87
13. $283.75, bank; $253.13, credit union
15. $54.19   17. $357.30   19. $132.20

CHAPTER 4

Answers for the problems are included in the chapter.

CHAPTER 5

Review Exercises, page 99

1. 0.00924   3. 10.222   5. 25   7. 46.17
9. 244.67   11. 43.52

Exercises 5.1, page 100

1. infinite number, one   3. ten
5. *AB, BC, CD, DE, EA, AC, AD*
7. *MN, NU, VN, MU*   9. ∠*LM,* ∠*LP,* ∠*LS,* ∠*LR*
11. ∠*HEJ*   13. ∠*WQN,* ∠*WQM,* ∠*QWZ*
15. ∠*WYZ,* ∠*ZYX,* ∠*XYW*
17. ∠*DAB,* ∠*B,* ∠*BCD,* ∠*D,* ∠*DAC,* ∠*CAB,* ∠*BCA,* ∠*ACD*
19. ∠*EAB,* ∠*B,* ∠*BCD,* ∠*D,* ∠*E,* ∠*EAC,* ∠*CAB,* ∠*BCA,* ∠*ACD*
21. *BF* ∥ *CDE, ABC* ∥ *FD, BF* ⊥ *AFE, CE* ⊥ *EFA*
23. *AF* ∥ *ED, AF* ∥ *BC, ED* ∥ *BC, AB* ∥ *DC, AF* ⊥ *EF, FE* ⊥ *ED*   25. quadrilateral
27.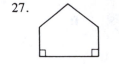
29. false   31. true   33. true   35. true
37. true   39. true
41. quadrilateral   43. yes

Calculator Exercises, page 102

45. $119.95   47. 5335.9084   49. 0.71498104

Review Exercises, page 108

1. ∠*MRN*   3. 0.4375   5. $\frac{19}{12} = 1\frac{7}{12}$
7. $\frac{77}{24} = 3\frac{5}{24}$ 9. $44\frac{3}{8}$ in.   11. $33\frac{10}{16}$ in.

Exercises 5.2, page 109

1. 40°   3. 125°   5. 540°   7. 720°
9. 25.650°   11. 4.273°   13. 135.142°
15. 42°36'0"   17. 89°10'48"   19. 120°21'0"
21. 71°, 161°   23. 35°23', 125°23'
25. 60°53'26", 150°53'26"   27. 60°

29. 66°5′, acute triangle   31. yes, no
33. yes, no   35. *LFGJ*   37. *MT, NT*
39.

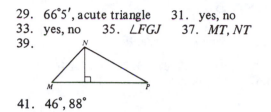

41. 46°, 88°

Calculator Exercises, page 110

43. 2251.3566   45. 692,470.49
47. 0.43640867

Review Exercises, page 118

1. $\frac{7}{24}$ 3. 6.14   5. 8.07   7. *SR*
9. 4°22′12″   11. quadrilateral

Exercises 5.3, page 118

1. 11 ft 5 in.   3. 6 yd 6 in.   5. 12 ft
7. 36 in.   9. 84.6 units   11. 62 ft
13. 8.667 ft   15. 11 ft   17. 2 ft $7\frac{6}{8}$ in.
19. 78.5 in.   21. 2.77 ft   23. 30.70 in.
25. 15.26 in.   27. 26.25 units   29. 16.28 units
31. 8 ft, 13 ft, 13 ft   33. 95.54 in.   35. 51.50 in.
37. 29.78 in.   39. 29.89 units

Calculator Exercises, page 122

41. 26.301   43. 497.391   45. 1127.028 ft

Review Exercises, page 130

1. 137.154   3. $\frac{27}{20} = 1\frac{7}{20}$ 5. $\frac{32}{27} = 1\frac{5}{27}$
7. 426.646   9. equilateral   11. 37.8 in.

Exercises 5.4, page 130

1. about 10 square units
3. about 18 square units   5. 240 ft²
7. $38\frac{1}{3}$ yd² or 345 ft²   9. 380.25 ft²
11. 45.901 ft²   13. 28.125 in.²
15. $\frac{9}{16}$ yd² or 0.5625 yd²   17. 76 in.²
19. 24 yd²   21. 2799.36 in.²
23. $28\frac{4}{9}$ yd² or 28.44 yd²   25. 217.5 in.²
27. 992.25 in.²   29. 180 in.² or 1.25 ft²

31. 6.61 yd² or $6\frac{11}{18}$ yd²   33. 819.06 in.²
35. $1792\frac{21}{32}$ sq. units   37. 6.75 units
39. 94.5 sq. units   41. 83 ft²   43. 503.5 in.²
45. 585.5 ft²

Calculator Exercises, page 134

47. 90.046 yd²   49. 1.256 yd²   51. 372.104 ft²

Review Exercises, page 139

1. 5.486   3. 234.09   5. $\frac{21}{80}$ 7. 21.98 in.
9. 27.567°   11. 1.63 yd²

Exercises 5.6, page 140

1. 102.02 in.²   3. 514.46 ft²   5. 1.008 ft²
7. 10.59 in.²   9. 25.905 in.²
11. 4 times as much   13. about 5520 pounds
15. 8.373 in.²   17. 763.19 yd²
19. 951.94 in.², 20.67%   21. 420.96 ft²
23. 47.944 in.²   25. 35.375 in.²   27. 1.112 in.²

Calculator Exercises, page 142

29. about 9.425 in.² or 41.094%
31. 49.646514 ft²   33. 52.159827 in.²

Review Exercises, page 148

1. 171   3. 4.42 ft   5. $\frac{11}{16}$ in.   7. 300.052
9. 67.5 ft²   11. octagon

Exercises 5.7, page 148

1. 504 ft³   3. 25.926 yd³   5. 16.74 lb
7. 15.12 lb   9. 163.28 in.³   11. 2.33 yd³
13. 215.345 in.³   15. 523.33 in.³
17. about 17,268 gal   19. 7.327 yd³   21. 90 ft³
23. 2.411 in.³   25. 2.361 ft³ or 4080 in.³
27. 4.612 in.³

Calculator Exercises, page 150

29. 2.0436293 lb   31. 6013.7663 lb
33. 0.37 in.³

Review Exercises, page 155

1. 47.885 in.   3. 0.385 in.²   5. 24.75 ft²
7. $\frac{27}{56}$ 9. 29.4   11. 0.4375

Exercises 5.8, page 155

1. 66 in.^{2} 3. 188.4 ft^2 5. 21.23 ft^2
7. 278.98 in.^{2} 9. 239.14 ft^2 11. 161.14 ft^2
13. 376.02 in.^{2} 15. about 5 gal   17. 12.52 ft^2
19. 53.589 ft^2 21. $51.91   23. 252 in.^{2}

Calculator Exercises, page 157

25. 94.7174 in.^{2} 27. about 158.765 in.^{2}
29. 25.912 in.^{2}

Extra Problems, page 158

1. a. tangent to a circle   e. chord of a circle
   b. sector of a circle   f. center of a circle
   c. diameter of a circle   g. radius of a circle
   d. point of tangency   h. arc of a circle
3. a. $\overline{FG}$ d. $\overline{EF}$ or $\overline{HG}$
   b. none   e. none
   c. $\overline{DEH}$ f. $\overline{AB}$ or $\overline{EF}$
5. a. $\overleftrightarrow{RS}$ or $\overleftrightarrow{RQ}$ or $\overleftrightarrow{SQ}$ (all names for the same line)
   b. $\overrightarrow{ST}$, $\overrightarrow{SQ}$, $\overrightarrow{SR}$, $\overrightarrow{RS}$ or $\overrightarrow{RQ}$, $\overrightarrow{QR}$ or $\overrightarrow{QS}$
7. a. triangle ABC
   b. triangle BDC and triangle ABC
   c. triangle BCD and triangle CDE
   d. triangle ABC
   e. triangle ABE and triangle BCE
9. 7 feet   11. 196.25 feet per minute
13. 0.33 square inches   15. $9179
17. 3.4 square feet   19. 48.32 ounces
21. 615.44 square inches

CHAPTER 6

Review Exercises, page 165

1. 4780   3. 0.0063   5. 75.402   7. 7.84
9. 40   11. 8

Exercises 6.1, page 165

1. 400   3. 0.053   5. 4930   7. 59
9. 0.017   11. mm   13. m   15. 73 cm
17. 84 cm, 8.4 dm   19. 44 cm   21. 37.68 cm
23. 5.024 mm   25. 4.2 m, 420 cm
27. 0.9112 m, 91.12 cm   29. 1.1 m, 110 cm
31. 0.6 m, 60 cm

Calculator Exercises, page 167

33. 11.62392 m   35. 52.944 cm
37. 4.9393805 m

Review Exercises, page 171

1. 168   3. 0.13   5. 62.5%   7. 1.728
9. 13.608   11. 9

Exercises 6.2, page 172

1. 510 cm^2 3. 28,500 cm^2 5. 452.16 cm^2
7. 125 cm^3 9. 1356.48 cm^3 11. 225 cm^2
13. 52 cm^2, 0.0832 m^2 15. 2327.71 cm^2
17. 3.375 hectares   19. 512 cm^2, 0.768 ℓ
21. 4.239 m^2, 0.954 kℓ   23. 75.36 cm^3

Calculator Exercises, page 174

25. 611.36321 m^2
27. 201.0624 dm^2, 268.0832 ℓ
29. 207.124 dm^2, 137.6715 ℓ

Review Exercises, page 176

1. 100   3. 3   5. 140
7. $\dfrac{33}{10} = 3\dfrac{3}{10}$ 9. 0.2495   11. $\dfrac{28}{5} = 5.6$

Exercises 6.3, page 177

1. 4420 g   3. 226.08 kg   5. 4919.752 g
7. 10.997 kg   9. 15.594 kg
11. 539.2 kg under the limit   13. 23.216 kg
15. 1.167 hours or 70 minutes   17. 57.5 °C
19. 27 mm per hour, 10.34 hours   21. 26.3 °C
23. 160 g of salt   25. 22.96 °C

Calculator Exercises, page 179

27. 14,855.4 mg   29. 287.62 kg, 0.3 t

Review Exercises, page 182

1. 1   3. 78.5 cm^2 5. 1.7   7. 4100
9. 100   11. 20

Exercises 6.4, page 182

1. 8 N   3. 2.5 N   5. 150 kg   7. 30 ohms
9. 333.3 watts   11. 256.41 volts
13. 4000 pascals   15. 0.02 ohms   17. 6 volts

Calculator Exercises, page 183

19. 114.745 W   21. 104.66067 V
23. 5996.9148 N

Extra Problems, page 183

1. 35.6 cm
3. a. 540   e. 25
   b. 300   f. 0.0873, 873
   c. 18,000   g. 450,000, 4,500
   d. 0.435   h. 90, 0.009
5. 345 mm   7. 300 cm, 3 m   9. 19.8 ha
11. a) surface area=180,864 cm^2
    b) volume=7,234.56ℓ
13. a) 365.496 m^2
    b) 548.244 kℓ
15. 2.232378 metric tons   17. 6.1866 kg   19. 36°C

CHAPTER 7

Review Exercises, page 189

1. 380 cm   3. 50 W   5. 3ℓ   7. 37 °C
9. 82 cm   11. 50.24 m^2

Exercises 7.1, page 190

1. approximate   3. exact   5. exact
7. exact   9. approximate   11. 35,000
13. 503,100   15. 5.33   17. 0.0028   19. 81
21. 0.04400   23. two   25. five   27. five
29. three   31. one   33. four   35. two
37. 2.5 × 10^4 39. 3.0840 × 10^6
41. 9.3005 × 10^1 43. 2.17 × 10^{-2} 45. 9 × 10^{-4}
47. 3.002 × 10^5 49. 4.0 × 10^{-1}
51. three, 7.08 × 10^{-3} 53. four, 3.650 × 10^2
55. four, 3.785 × 10^{-3} 57. four, 2.560 × 10^1

Calculator Exercises, page 191

59. 141,800   61. 5.067   63. 1048

Review Exercises, page 192

1. 287.44   3. five   5. 2.5 × 10^6
7. 0.0135 dam   9. dam   11. 8%

Exercises 7.2, page 193

1. 27.9, 27.9   3. both same accuracy, 0.043
5. 24.05, 0.734   7. 50.0, 0.05
9. 10.004, 10.004   11. 1.94 × 10^3, 4.1 × 10^{-1}

13. 18,000 and 2.4 × 10^3, 0.009
15. nearest inch, $\frac{1}{2}$ inch   17. nearest 100 g, 50 g
19. nearest one-hundredth of a milliliter, 0.005 mℓ
21. nearest one-thousandth of a centimeter, 0.0005 cm
23. nearest one-hundredth of a millimeter, 0.005 mm
25. nearest hundred gallons, 50 gal
27. 19 ± 0.5 in.   29. 18.3 ± 0.05 mm
31. 24 ± 0.5 mm   33. 399.5 ft – 400.5 ft
35. $\frac{1}{16}$ inch   37. 0.7180 cm – 0.7200 cm
39. 240.10 mℓ, 18.75 mℓ and 240.10 mℓ
41. nearest one-hundredth of a square centimeter

Calculator Exercises, page 193

43. 0.156   45. 0.0198   47. 1.31

Review Exercises, page 195

1. nearest ten inches   3. 0.5 m   5. 4940
7. 0.406 cm   9. 72   11. 36.04

Exercises 7.3, page 195

1. 491.9   3. 136.2   5. 10.68   7. 454.57
9. 6.8   11. 69.51   13. 340   15. 830,000
17. 30   19. 0.1800   21. 705 g   23. 0.18 cm
25. 9$\overline{1}$.5 cm   27. 5.59 lb   29. 1.08 watts

Calculator Exercises, page 196

31. 1,333,000   33. 11   35. 0.5715

Review Exercises, page 197

1. 420 m   3. 0.005 cm   5. four
7. 980.03 g   9. 61.2   11. 37.5%

Exercises 7.4, page 197

1. 0.020, 2.0%   3. 0.014, 1.4%
5. 0.06, 6%   7. 0.015, 1.5%
9. 0.000664, 0.0664%   11. 0.2, 20%
13. 0.00373, 0.373%   15. 0.02, 2%
17. 0.021, 2.1%   19. 0.00156, 0.156%

Calculator Exercises, page 198

21. 0.39161   23. 0.60847   25. 0.97959

Review Exercises, page 200

1. 0.379%   3. 382.7   5. 8.52 × 10^{-3}
7. 40.62 in.   9. 2 ft 8$\frac{11}{16}$ in.   11. $10.59

Exercises 7.5, page 200

1. $c = 10.0$ ft   3. $c = 17.0$ in.   5. $c = 7.62$ cm
7. $c = 9.22$ ft   9. $b = 20.0$ in.   11. $a = 30.0$ cm
13. $b = 8.66$ m   15. $a = 27.5$ cm
17. $b = 11.6$ mm   19. 7.6 ft   21. 7.8 ft
23. 23 cm   25. 15 dm   27. 30 ft

Calculator Exercises, page 202

29. 17.1   31. 16.3   33. 18.8

Extra Problems, page 203

1. a. Numbers obtained by counting or by definition.
   b. All numbers that are not exact numbers.
3. a. two   e. six
   b. five   f. three
   c. four   g. seven
   d. six   h. three

5. **Most Accurate Number**   **Most Precise Number**
   a. 83.510×10^3 0.0042
   b. 10,004.9 2.95×10^{-2}
   c. 0.063000   0.063000
   d. 358,000 8.5×10^{-3}
   e. 15.0072   0.00749
   f. 5.2810×10^5 25.6×10^{-5}
7. a. 0.192 cm   b. 0.096 cm
9. a. 242.4"   e. 34 square feet
   b. 807.85 cm   f. 0.013 grams
   c. 2260 mm^2 g. 130 miles
   d. 403 mℓ h. 51.0 cm
11. 2660 cubic inches   13. 8.84 g   15. $\ell \doteq 18.8$ ft.
17. brace is 22 inches   19. Run $\doteq 99.2$ inches
21. area $\doteq 294$ cm^2
23. No, the measure from point A to point B should be 7 feet 6 inches to be square.

CHAPTER 8

Review Exercises, page 210

1. 520 cm   3. 13.143%   5. 12.512
7. 706.5 cm^2 9. 240 in.^{3} 11. 1580 cm^2

Exercises 8.1, page 211

1. 130   3. 90
5. three and three-fourths figures   7. $12\frac{1}{2}$¢

9. $10.60   11. 250 million barrels
13. 175 million barrels   15. 30%   17. 38%
19. 63%   21. 6%
23.

Production of Cement in the U.S.

1997	☐☐☐☐☐☐☐☐☐ ⬒
1995	☐☐☐☐☐☐☐
1993	☐☐☐☐☐
1991	☐☐☐☐☐☐▷
1989	☐☐☐☐☐ ⬒
1987	☐☐☐☐ ⬒

☐ = 10 million tons of cement

Production in tons!

25.

Number of People Employed in a Manufacturing Company

Administration
Skilled workers
Unskilled Workers
Maintenance and Service
Sales

= 25 workers

Number of People Employed

27.

Item	Cost	Percent of total cost	Part of 360°
Land	$ 7,000	$11\frac{2}{3}$%	42°
House	$40,000	$66\frac{2}{3}$%	240°
Landscaping	$ 3,000	5%	18°
Furnishings	$10,000	$16\frac{2}{3}$%	60°
Totals	$60,000	100%	360°

Distribution of Costs for a New Home

House $40,000
Land $7000
Landscaping $3000
Furnishings $10,000

29.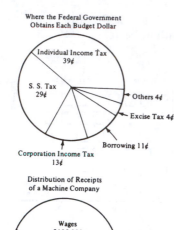

Where the Federal Government Obtains Each Budget Dollar

31.

Distribution of Receipts of a Machine Company

Calculator Exercises, page 214

33. 5682.7   35. 0.00624   37. 21.991%

Review Exercises, page 218

1. 40 cm^2 3. $\frac{21}{40}$ 5. 1.47108   7. 5
9. 0.012   11. 8.717°

Exercises 8.2, page 219

1. September, 5300   3. December
5. about 3000 units   7. $780   9. 22¢
11. 2 °C   13. 5 °C   15. 4.5 °C
17. 8:45 A.M.   19. 50 hp   21. 40 hp
23. 1700 rpm   25. true   27. false
29. false   31. true

33.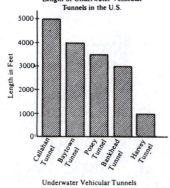

Length of Underwater Vehicular Tunnels in the U.S.

35.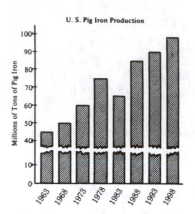

U. S. Pig Iron Production

37.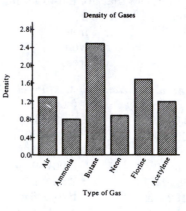

Density of Gases

39.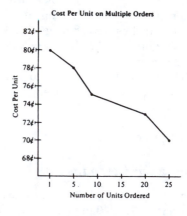

Cost Per Unit on Multiple Orders

41.

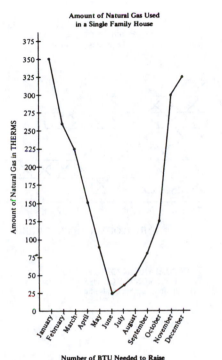

Amount of Natural Gas Used in a Single Family House

43.

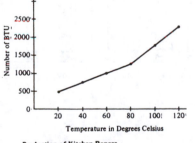

Number of BTU Needed to Raise the Temperature of a Metal

45.

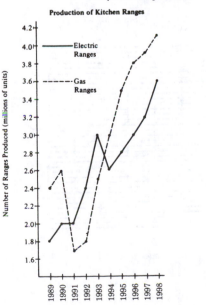

Production of Kitchen Ranges

Calculator Exercises, page 223

47. 0.02087   **49.** $225.03   **51.** 0.004747

Review Exercises, page 225

1. 42.474   **3.** 80.45 km/hr   **5.** $\frac{147}{38} = 3\frac{33}{38}$

7. $\frac{7}{5} = 1\frac{2}{5}$ **9.** 35°23′   **11.** $\frac{15}{2} = 7\frac{1}{2}$

Exercises 8.3, page 226

1. $A(3,3), B(-3,3), C(-3,-3), D(3,-3)$
3. $A(3,3), B(1,1), C(-2,-2), D(-4,-4)$
5. $A(3,4), B(3,2), C(3,0), D(3,-3)$
7. $A(0,4), B(-3,3), C(-4,0), D(0,0)$
9.

(graph with points: $A(5, 6)$, $D(-\frac{1}{2}, 4)$, $G(4, 0)$, $C(3, -1.5)$, $H(-2\frac{1}{2}, -2\frac{1}{2})$, $F(0, -3)$, $E(7, -3.5)$, $B(-2, -4)$)

Calculator Exercises, page 227

11. 14.7969   **13.** 1485.4167   **15.** 12.4175

Review Exercises, page 232

1. 10.2 ft   **3.** 0.0588%   **5.** 1.3
7. 0.085 m²   **9.** $\frac{22}{7} = 1\frac{7}{15}$ **11.** $\frac{7}{5} = 1\frac{2}{5}$

Exercises 8.4, page 232

1. 13.0   **3.** 17.0   **5.** 8.94   **7.** 10.6
9. 13.6   **11.** 15.3   **13.** 17.0   **15.** 4.0
17. 30.0   **19.** 12.1   **21.** 9.7   **23.** 12.4
25. 5.0   **27.** 12.8   **29.** 13.0   **31.** 15.6
33. 10.0   **35.** 8.5

Calculator Exercises, page 233

37. 10.0125   **39.** 17.2534   **41.** 9.6889

Review Exercises, page 237

1. A (3, 4); B(−2, −4)   3. 15   5. 4.0 in
7. approx. 68% efficient at 250 rpm
9. 6%   11. three, 5.20 X 10⁻³

Exercises 8.5, page 238

1. $\frac{3}{2}$ 3. $\frac{1}{9}$ 5. $-\frac{3}{2}$ 7. no slope   9. $-\frac{24}{5}$

11. $\frac{2}{1}$ 13. 0   15. $\frac{3}{2}$ 17. $-\frac{1}{2}$ 19. $-\frac{2}{3}$

21. neither   23. parallel   25. perpendicular
27. 3 ft   29. 36 ft   31. 7.2 ft

Calculator Exercises, page 240

33. 3096.9495   35. 105.75887   37. 37.541349

Extra Problems, page 244

1. a. 25,500   d. 10%
   b. ≐41.2%   e. ≐57.1%
   c. 6000   f. 50%

3.

COMMISSIONS EARNED FOR THE FIRST SIX MONTHS OF A YEAR

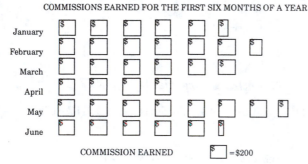

COMMISSION EARNED $\boxed{\$}$ = $200

5.

EXPENDITURES FOR RESEARCH AND DEVELOPMENT

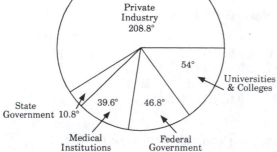

7.

UNIT PRODUCTION TOTALS
(Thousands of Units)

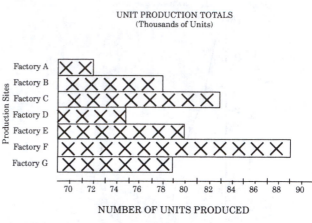

NUMBER OF UNITS PRODUCED

9.

AJAX COMPANY EXPORTS FOR THE
FIRST SIX MONTHS OF THE YEAR
(In hundreds of thousands of dollars)

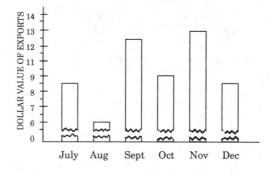

11.

CURRENT / RESISTANCE AT
A CONSTANT 150 WATTS

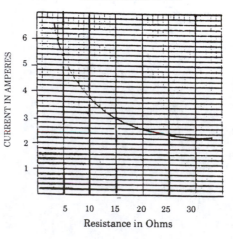

13.

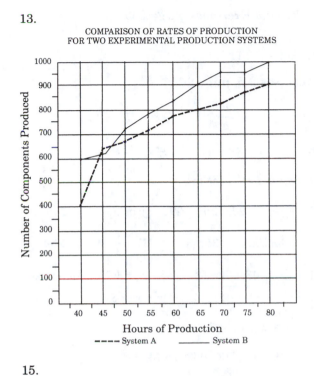

COMPARISON OF RATES OF PRODUCTION
FOR TWO EXPERIMENTAL PRODUCTION SYSTEMS

15.

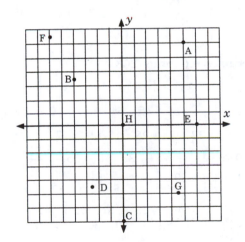

17. a. 10.0   d. 17.0
    b. $\doteq$12.5   e. $\doteq$6.02
    c. $\doteq$20.6   f. 5.60

19. PQ=SR=$\sqrt{13}$, the coordinates of point S are (2, 2).

21. a. neither   c. perpendicular
    b. parallel   d. neither

23. slope=$\frac{7}{100}$ 25. slope=$\frac{1}{8}$

27. The coordinates of point Y are (-4, 3).

CHAPTER 9

Review Exercises, page 253

1. 11.66   3. 14   5. 87   7. 34%
9. 25.2   11. 32 cm

Exercises 9.1, page 253

1.
3
5.

7. −72   9. 0   11. $-\dfrac{15}{16}$ 13. −31   15. −8
17. 6   19. −25   21. −10.7   23. 1.6
25. −4.06   27. −2.7   29. −10   31. −2
33. 6   35. −10   37. 21   39. −51   41. 8
43. −16   45. −282   47. $1\frac{1}{8}$ 49. 107,000
51. −9°   53. $91.20 loss   55. 1.6

Calculator Exercises, page 254

57. −55.61

Review Exercises, page 256

1. 40.58   3. $\dfrac{2}{3}$ or 0.6$\overline{6}$ 5. 0.065
7. 6.25   9. −1   11. 2.375

Exercises 9.2, page 256

1. −6   3. 0   5. 39   7. 6.94   9. 3
11. −18   13. −2   15. 18   17. −5   19. 7
21. −0.9   23. −3.26   25. −2.6   27. −10
29. −10   31. −4   33. −7   35. 1   37. 2
39. 8   41. −21   43. 21   45. −2   47. −31
49. −11

Calculator Exercises, page 256

51. −18.52   53. −770.73

ANSWERS TO ODD-NUMBERED EXERCISES

Review Exercises, page 259

1. 5.8   3. 18.93   5. $\dfrac{7}{12}$ 7. -16   9. -7
11. $8\frac{2}{4}$ in.

Exercises 9.3, page 260

1. $\dfrac{1}{4}$ 3. $\dfrac{8}{7}$ 5. -9   7. -65   9. -126
11. -2   13. $\dfrac{4}{5}$ 15. $-\dfrac{14}{5} = -2\dfrac{4}{5}$ 17. -8.1
19. 17   21. -4   23. 9   25. 1   27. -14
29. $-\dfrac{2}{3}$ 31. 6   33. -2.4   35. 80
37. -0.2   39. $-\dfrac{3}{5}$ 41. $-\dfrac{32}{3} = -10\dfrac{2}{3}$
43. Divide by 4 or multiply by $\dfrac{1}{4}$.
45. Divide by $-\dfrac{1}{2}$ or multiply by -2.
47. Divide by $-\dfrac{3}{8}$ or multiply by $-\dfrac{8}{3}$.
49. 0   51. 90   53. 72   55. -64   57. 3
59. -8   61. -10   63. 15   65. -6   67. 4

Calculator Exercises, page 261

69. 27.1625   71. -8

Review Exercises, page 262

1. 2.8   3. -5   5. 120   7. -9   9. 64 cm²
11. 67.52 cm²

Exercises 9.4, page 263

1. -26   3. -4   5. 13   7. 3   9. -10
11. 25   13. 16   15. -77   17. 52
19. -20   21. -22   23. -16   25. -1
27. -3   29. -2   31. -3
33. $-\dfrac{28}{5}$ or -5.6   35. 7   37. 12   39. 20
41. -45   43. -11   45. -14   47. 8.5
49. 55   51. $480   53. 1700 gal   55. 235 mi

Calculator Exercises, page 264

57. 6467.34   59. 563.67   61. 668.84
63. -0.065   65. 16.1   67. -98.375

Review Exercises, page 265

1. -4.5   3. -9   5. 12   7. 53.35°
9. 8950   11. 54

Exercises 9.5, page 266

1. $x + y$ 3. $\dfrac{c}{b}$ 5. $y - 8$ 7. $6 + x$
9. $\dfrac{p}{q}$ 11. $x - y$ 13. y^2 15. m^4
17. $x + 6$ 19. $3x + 8$ 21. $\dfrac{m+n}{4}$ 23. $\dfrac{13}{pq}$
25. $\dfrac{rs}{4+x}$ 27. $2p$ 29. $\dfrac{1}{2}y$ or $\dfrac{y}{2}$ 31. $24m$
33. $79m$ cents   35. $12b$ 37. $\dfrac{x}{3}$ 39. $27m$
41. $\dfrac{s}{27}$ 43. $2r + 2s$ 45. $x, x + 2$ 47. $x, 3x$
49. $x, 2x - 8$ 51. $2x + 2x + x$ or $5x$
53. perimeter: $2x + 2(3x)$ or $8x$
    area: $x(3x)$ or $3x^2$

Review Exercises, page 268

1. $\dfrac{3}{2}$ 3. $-\dfrac{1}{10}$ 5. 23   7. 0   9. $9\dfrac{10}{16}$ in.
11. 360 ft³

Exercises 9.6, page 268

1. 8   3. 36   5. -14   7. -3   9. $\dfrac{1}{8}$
11. 1   13. 96   15. 64   17. 10   19. 9
21. 7.8   23. 1   25. 18.72   27. $\dfrac{1}{2}$
29. 0.064   31. 5   33. 4   35. 1.44   37. 4
39. $\dfrac{1}{4}$ 41. $\dfrac{39}{32} = 1\dfrac{7}{32}$ 43. 2   45. 24
47. 151.25   49. 108   51. 6.25   53. 20
55. 48   57. 48.10   59. 30   61. 2
63. 55.5

Calculator Exercises, page 270

65. 15.56   67. 3.97

Review Exercises, page 272

1. -4.8   3. 310   5. 0.079   7. 0.48
9. 0.0275   11. 1.884 ft

Exercises 9.7, page 272

1. $20x$ 3. $24x$ 5. $18x$ 7. $21x$

9. $26.1x$ 11. x 13. x 15. $2x$ 17. x

19. $-2x$ 21. $\dfrac{4x}{3}$ 23. $\dfrac{5x}{8}$ 25. x 27. $\dfrac{x}{2}$

29. $\dfrac{2x}{3}$ 31. x 33. x 35. $\dfrac{x}{2}$ 37. $x+7$

39. $m-4$ 41. $2x+18$ 43. $4x-24$

45. $12x-8$ 47. $9x+22.5$ 49. $-2x-8$

51. $2dx+6d$ 53. $\dfrac{8m}{3}+6$ 55. R 57. v

59. h 61. $\dfrac{5S}{44}$ 63. f 65. $4(3y)=12y$

67. $\dfrac{120s}{144}=\dfrac{5s}{6}$ 69. $\dfrac{135x}{18}=7.5x$

Review Exercises, page 275

1. 2 3. -55 5. $-\dfrac{1}{2}$ 7. 30% 9. 3.6%

11. $385.6\ \text{ft}^2$

Exercises 9.8, page 275

1. like   3. like   5. unlike   7. unlike
9. unlike   11. like   13. $6a$ 15. $-4t$
17. $2x$ 19. $-y$ 21. $2x^2$ 23. 0 25. $1.9x$

27. $\dfrac{5}{4}t$ or $\dfrac{5t}{4}$ 29. $\dfrac{11}{2}q$ or $\dfrac{11q}{2}$ 31. $-\dfrac{1}{2}x$ or $\dfrac{-x}{2}$

33. $\dfrac{8x-3y}{12}$ 35. $\dfrac{4r+s}{8}$ 37. $\dfrac{2q+3p}{6}$

39. $\dfrac{3z+8w}{8}$ 41. $\dfrac{8a-5c}{8}$ 43. $(r+s)y$

45. $(3m-x)n$ 47. $(a+1)t$ 49. $(1-r)b$
51. $(2m-1)x$ 53. $9m$ 55. t 57. $-6s$
59. $7y+1$ 61. $12a+11b$ 63. $6s+5$ 65. 0
67. $8x$ 69. $24y$ 71. $2(2x)+2x=6x$
73. $3y$ 75. $4x+2y$
77. $d+2.89+d=2d+2.89$

Calculator Exercises, page 277

79. 106.506 81. 4.65230 83. 7.5282452

CHAPTER 10

Review Exercises, page 281

1. 1875 3. 1500.12 5. 68 7. x 9. x
11. -32

Exercises 10.1, page 281

1. 90 mi   3. 162.5 km   5. 80 mi
7. 11.25 km   9. 10 in.^{2} 11. 2 ft^2
13. 9920 cm^2 or 0.992 m^2 15. 9 ft
17. 16 ft $4\dfrac{6}{8}$ in.   19. 6 ft 4 in.   21. 52 volts
23. $1162.50   25. 225 mi   27. 3.6 m^2
29. $14\dfrac{2}{9}$ yd^3 31. 3 ft^2 100 in.^{2}
33. 4 yd^2 4 ft^2 35. 126 joules
37. 18 threads per inch   39. 25.12 cm^3
41. 64 cm^3 43. 256 ft   45. 500 lb
47. 1800 rpm

Calculator Exercises, page 283

49. 775.74 51. 546.44 53. 1085.25

Review Exercises, page 285

1. $15x$ 3. $-20x$ 5. $-2x$ 7. $-m+3$
9. 5 11. 0.405

Exercises 10.2, page 286

1. 5 3. 2 5. 6 7. $\dfrac{15}{2}=7\dfrac{1}{2}$ 9. 0

11. $\dfrac{2}{5}$ 13. 6.3 15. -1 17. -6 19. 9

21. 36 23. $-\dfrac{21}{2}=-10\dfrac{1}{2}$ 25. 112 27. $-\dfrac{3}{10}$

29. 8 31. 6.25 33. 25 35. 6 37. 4

39. -4 41. 2 43. 1 45. $\dfrac{4}{3}=1\dfrac{1}{3}$

47. $\dfrac{12}{5}=2\dfrac{2}{5}$ 49. 8

Calculator Exercises, page 286

51. 3 53. 1.5 55. -25 57. 6.8

Review Exercises, page 288

1. -3.66 3. 5 5. 11 7. 20% 9. 1.08
11. $(1+b)x$

Exercises 10.3, page 289

1. 3 hr   3. 7 hr   5. 60 mph   7. 2 in.
9. 6 m   11. 3 cm   13. 4 yr   15. $2600
17. 6%   19. 9 amperes   21. 3.5 ft
23. 16 ft^3 25. 140 rpm   27. 1320 lb
29. 3.8 ft   31. 14 in.   33. 12 lb

35. $\dfrac{1}{16} = 0.0625$ 37. $\dfrac{35}{9} = 3\dfrac{8}{9} = 3.89$

39. $2000   41. $3\dfrac{1}{2}$ ft

Calculator Exercises, page 290

43. 2.387 in.   45. 8   47. 18.904

Review Exercises, page 292

1. 1   3. −15.14   5. 2.65   7. 14
9. 0.12   11. 200

Exercises 10.4, page 292

1. $\dfrac{c}{a} = b$ 3. $gi = h$ 5. $\dfrac{bg}{f} = x$ 7. $\dfrac{pq}{3} = y$

9. $\dfrac{a}{3b} = x$ 11. $\dfrac{cb}{md} = y$ 13. $w = \dfrac{A}{\ell}$

15. $\dfrac{W}{f} = s$ 17. $\dfrac{I}{pt} = r$ 19. $\dfrac{E}{c^2} = m$

21. $\dfrac{H}{I^2} = R$ 23. $f\ell = v$ 25. $\dfrac{ds}{S} = D$

27. $\dfrac{2A}{b} = h$ 29. $\dfrac{3V}{4r^3} = \pi$ 31. $\dfrac{2d}{t^2} = g$

33. $\dfrac{2E}{v^2} = m$ 35. $\dfrac{d}{\pi r^2 s} = n$ 37. $\dfrac{550H}{62.4f} = d$

39. $fjdA = M$

Calculator Exercises, page 292

41. 61.9038   43. 6.3166   45. 1583.9984

Review Exercises, page 295

1. 5.43   3. 2.8   5. 17.64   7. −6490
9. −42.9   11. 11

Exercises 10.5, page 295

1. $R, B, 9.6$ 3. $P, R, 300$ 5. $P, B, 65\%$
7. $R, B, 18.75$ 9. $P, R, 4$ 11. $P, B, 265\%$
13. $R, B, 0.15$ 15. $P, R, 1200$ 17. $P, B, 0.5\%$
19. 50   21. 400   23. 8   25. 54 lb
27. 5%   29. 20   31. 66,300 watts   33. 80%
35. 20 watts   37. $3465   39. $1200
41. 621   43. 3.332 cm   45. $600
47. $34,500

Calculator Exercises, page 297

49. 2112.59   51. 29.2%   53. $8.28

Review Exercises, page 299

1. −24   3. $(a + 3)x$ 5. 28.8 in.²

---

7. 105 cm²   9. 58.0586 in.²   11. 144 ft

Exercises 10.6, page 299

1. $2x = 124; 62$ 3. $4x = 36; 9$ 5. $\dfrac{x}{4} = 11; 44$

7. $\dfrac{1}{4}x = 16; 64$ 9. $\dfrac{2}{5}x = 32; 80$

11. $\dfrac{4x}{5} = 20; 25$ 13. $8500   15. $120,000

17. 8 hp, 16 hp
19. A, 600 gal; B, 1200 gal; C, 1800 gal
21. 0.006 amperes, 0.012 amperes
23. 3 ft 9 in.   25. 28 m, 56 m   27. 53 mph
29. 7 hr

Review Exercises, page 303

1. 6   3. 5   5. 16   7. $-3x + 9$
9. 87.5%   11. $0.701\overline{6}$

Exercises 10.7, page 303

1. 56 ft/sec   3. −38 ft/sec   5. 3 m/s

7. 11 ft 6 in.   9. 16 ft $7\dfrac{3}{16}$ in.

11. 15 ft 8 in.   13. $7   15. $98.05
17. 1.5 ohms   19. 50 °F   21. $253.50
23. 0.661   25. 54 m/s   27. 1.25
29. 1.0625   31. 1.8125   33. 16 ft²
35. 65 ft² 72 in.²   37. 251 ft² 28.8 in.²
39. 8.2 amperes   41. 9   43. 14.75
45. 5 amperes   47. 0.125 in.   49. 36
51. 1 ohm

Calculator Exercises, page 305

53. 15.38%   55. 51 yd² 2 ft² 63 in.²

Review Exercises, page 307

1. −10   3. −27   5. $9x + 3$ 7. $9x - 6y$
9. 7.445   11. 8.6

Exercises 10.8, page 307

1. 31   3. 14   5. 8   7. 8   9. 15
11. 53   13. 0   15. 92   17. −4   19. 2
21. −5   23. 5.4   25. 8.2   27. 8.6   29. 10
31. −8   33. 22   35. $6\dfrac{3}{8}$ 37. −16   39. 9
41. −9   43. 18   45. 9   47. −4   49. 1.6
51. 14   53. 16   55. 35.2   57. 8.6

Calculator Exercises, page 308

59. 37.27   61. −3.976   63. 4.8   65. −6.17

Review Exercises, page 310

1. -2.1 3. 17.71 5. $-\dfrac{8}{9}$ 7. x

9. $6 + y$ 11. $m + 8$

Exercises 10.9, page 310

1. \$11.45   3. \$646.50   5. 3 m/s

7. 60 ft/sec   9. 2 ft 5 in.   11. 6 ft $2\frac{1}{2}$ in.

13. 4 ft 5 in.   15. 48   17. 60   19. 4 ft

21. 10 ft 3 in.   23. 7 ft 6 in.   25. 15

27. 16 m   29. 3.6 amperes   31. 27   33. 90

35. 15 volts   37. 6 volts   39. $2\frac{1}{2}$ in.

41. 6 watts   43. 16 ft^2 45. 225 per sec

Calculator Exercises, page 312

47. 49.3 watts   49. 2 in.

Review Exercises, page 313

1. 24.5   3. 0.52   5. 54.8   7. 855

9. 2   11. 2.67 ft

Exercises 10.10, page 314

1. $b - a = c$ 3. $e + c = x$ 5. $h - j = x$

7. $5(a + b) = z$ 9. $d(y - 4) = x$

11. $f - m - n = x$ 13. $k - a + c = x$

15. $g + b - f = x$ 17. $\dfrac{c(m + n)}{4} = y$

19. $\dfrac{d(m - 4)}{b} = a$ 21. $wk - 3 = r$ 23. $4d + q = x$

25. $a - bc = x$ 27. $p + kb = y$ 29. $\dfrac{12V}{\pi N} = D$

31. $\dfrac{RD^2}{k} = L$ 33. $v_2 - at = v_1$

35. $I - I_2 - I_3 = I_1$ 37. $\dfrac{F}{R - L} = k$

39. $S - 2B = L$ 41. $2A - b = a$ 43. $2S_0 + d = D$

45. $\dfrac{rMN(N + 1)}{24} = F$ 47. $\dfrac{L}{1 + at} = L_0$

Calculator Exercises, page 314

49. 8.35   51. 39.479167   53. 13.681592

Review Exercises, page 317

1. $\dfrac{m}{a} = x$ 3. 2   5. $\dfrac{5}{m + n} = x$ 7. $\dfrac{14q}{p} = x$

9. $35 - b = x$ 11. $\dfrac{9(r + s)}{3}$ or $3(r + s)$

Exercises 10.11, page 317

1. $w = fs$, where w = work, f = force, s = distance
3. $E = mc^2$, where E = energy, m = mass, c = speed
   of light
5. $I = \dfrac{V}{R + r}$, where I = current
7. $p = s - c$, where p = profit, s = selling price,
   c = cost   9. $A = s^2 + 9g^2$ 11. $C = 2mn$

13. $N = 18j$ 15. $d = 120 + 55t$ 17. $G = \dfrac{xy}{300}$

19. $N = 2a + 2b - 5.5$ 21. $N = 160j$

23. $P = 2b + 2d + 2c$ 25. $A = bc + \dfrac{1}{2}ac$

27. $A = 58x$

Calculator Exercises, page 318

29. $T = 0.0810255$ in.   31. 17.432606%

Review Exercises, page 320

1. $c - a + b = x$ 3. $\dfrac{q(4 - c)}{w} = x$ 5. $\dfrac{9mn}{4b} = x$

7. $7y - 3q = x$ 9. $7x + 6$ 11. $510xyz$ lb

Exercises 10.12, page 321

1. $x + 23 = 37; 14$ 3. $x - 6 = 14; 20$

5. $x + 26 = 50; 24$ 7. $\dfrac{4x}{5} = 8; 10$

9. $\dfrac{x - 5}{7} = 11; 82$ 11. $2x + 5x = 91; 13$

13. 380 ft   15. \$1.19 per ft   17. 14

19. 3 hr 15 min   21. 6 hr 20 min   23. 180 mi

25. 36 volts   27. 35.2 amperes   29. 8.5 cm

Calculator Exercises, page 322

31. 21,642.2 lb

CHAPTER 11

Answers for the problems are included in the chapter.

CHAPTER 12

Review Exercises, page 333

1. 31.2   3. 10   5. 24   7. 12.56 cm²
9. 31.4 cm   11. 10 m²

Exercises 12.1, page 333

1. $\frac{3}{4}$ 3. 1.75 ft/sec   5. $\frac{8}{3}$ 7. 2 m/s

9. $\frac{7}{8}$ or 0.875   11. $\frac{7}{5}$ or 1.4   13. yes

15. no   17. no   19. no   21. yes   23. yes
25. no   27. yes   29. no   31. yes

33. $\frac{2}{6}=\frac{3}{9},\frac{3}{2}=\frac{9}{6},\frac{9}{3}=\frac{6}{2}$ 35. $\frac{16}{18}=\frac{8}{9},\frac{9}{18}=\frac{8}{16},\frac{18}{16}=\frac{9}{8}$

37. $\frac{2}{3}$ 39. $\frac{2}{7}$ 41. 0.0433 in.   43. 3.7

45. 8.5   47. $\frac{9}{16}$

Calculator Exercises, page 335

49. 3.385

Review Exercises, page 336

1. 3   3. 18   5. 2.347   7. 18.84 ft
9. 153.86 cm²   11. 8.06 cm

Exercises 12.2, page 337

1. 24   3. −14   5. $\frac{10}{3}=3\frac{1}{3}$ 7. 5   9. 3

11. $\frac{3}{2}$ 13. 7   15. $\frac{21}{16}=1\frac{5}{16}$ 17. $\frac{1}{8}$ 19. $\frac{5}{6}$

21. 10.625 or $\frac{85}{8}$ 23. $88.20   25. $42.00

27. $230.00   29. 160   31. 189 rpm
33. 12 in.   35. 440 rpm   37. 15 ft
39. $98.18   41. 3.75 amperes

Calculator Exercises, page 338

43. 300   45. 0.282   47. 8.77

Review Exercises, page 341

1. $\sqrt{106} \doteq 10.296$ 3. 4   5. 0.01 m   7. 12
9. 25   11. $\frac{jk+jp}{3}=x$ or $\frac{j(k+p)}{3}=x$

Exercises 12.3, page 341

1. $AB=6, BC=8$ 3. $\frac{15}{4}=3\frac{3}{4}$ 5. 5

7. 5.4 cm, 7.35 cm   9. $33\frac{1}{3}$ in.²   11. 44.08 m²
13. $EF=14.4, FG=12.6$ 15. 25.65 mm
17. 6.674 in., 8.4407 in.   19. 30 ft   21. 22 ft
23. 14.4 mi   25. 24.15 ft   27. 6.0811 cm

Calculator Exercises, page 344

29. $EF \doteq 11.07$ ft, $FG \doteq 15.16$ ft   31. 12.26 cm²

Review Exercises, page 347

1. 41   3. $\frac{b}{a+1}=x$ 5. $3kh-2=x$
7. $4000   9. 18 ft 0 in.
11. $GE \doteq 6.6, QR \doteq 7.727$

Exercises 12.4, page 347

1. $m=kn$ 3. $y=kx^2$ 5. $v=kst$
7. $d=24t$ 9. $s=kw$ 11. $w=6h$

13. $w=k\ell$ 15. $d=ks^2$ 17. $V=\frac{4}{3}\pi r^3$

19. 2   21. 3   23. $\frac{9}{4}$ 25. $\frac{16}{3}=5\frac{1}{3}$ 27. 14

29. 100   31. 72   33. 1400 lb per in.²
35. 360 lb   37. 36π cm²   39. 46.875 hp

Calculator Exercises, page 349

41. $y=kx$; 300.8   43. $y=kx^2$; 5467.5
45. $y=kx^3$; 420.175

Review Exercises, page 350

1. 1.5498   3. −4   5. 12.31   7. 22.4
9. −10   11. $RM \doteq 11.826, NT \doteq 6.125$

Exercises 12.5, page 351

1. 2   3. 1   5. 3   7. 1   9. 0   11. 5
13. 48   15. 54   17. 306   19. 49   21. 0

23. −2   25. $\frac{4}{3}=1\frac{1}{3}$ 27. −2   29. $-\frac{7}{8}$

31. 6   33. 3   35. 2   37. −4

39. $\frac{11}{7}=1\frac{4}{7}$ 41. −5   43. −6   45. 4

47. −3   49. $-\frac{9}{4}=-2\frac{1}{4}$ 51. 6   53. 11

55. 17   57. $-\dfrac{25}{7} = -3\dfrac{4}{7}$ 59. 6   61. 1
63. 45.6   65. 5   67. 3   69. 4

Calculator Exercises, page 352

71. 72.6511   73. 1.9722   75. 0.3388

Review Exercises, page 354

1. 57.6   3. 12.72   5. 333.6   7. 70
9. 15.225   11. $3x - 3y$

Exercises 12.6, page 354

1. $-\dfrac{24}{5} = -4\dfrac{4}{5}$ 3. $\dfrac{24}{11} = 2\dfrac{2}{11}$ 5. 96 ft/sec

7. 80 ft/sec   9. -6 11. $\dfrac{5}{4} = 1\dfrac{1}{4}$ 13. 4 yr

15. 6%   17. 325 watts   19. 6.5 ohms
21. 8 in.   23. 4 ft 6 in.   25. 17 ft   27. 4 ft
29. 2 in.   31. 1.5 in.   33. -6 35. 5 m^2
37. 8 ft/sec^2 39. 15 °C   41. 1.5 sec
43. 7

Calculator Exercises, page 355

45. 0.7549985   47. $53.98   49. -15.680769

Review Exercises, page 357

1. 9   3. 5   5. $x = \dfrac{27}{19}$ 7. $S = 60$ rpm
9. $x = 3$ 11. $x = 7.5$

Exercises 12.7, page 358

1. $\dfrac{c - b}{a} = x$ 3. $\dfrac{g - m}{b} = y$

5. $\dfrac{j}{n} - c$ or $\dfrac{j - cn}{n} = x$ 7. $\dfrac{3}{2}y - 12 = x$

9. $y - 8x = b$ 11. $\dfrac{2 + k}{4} = y$ 13. $\dfrac{r + k}{4} = n$

15. $n = \dfrac{q + p - m}{2}$ 17. $\dfrac{r - 2}{3} = k$

19. $\dfrac{q + 6p}{6}$ or $\dfrac{q}{6} + p = L$ 21. $qr + m = k$

23. $\dfrac{p}{r - d} = Q$ 25. $\dfrac{3}{2}(c - b)$ or $\dfrac{3c - 3b}{2} = x$

27. $\dfrac{-b - d}{a + 1} = x$ or $x = \dfrac{b + d}{1 - a}$ 29. $\dfrac{RD^2}{k} = L$

31. $F + A_1 - P(V_2 - V_1) = A_2$ 33. $\dfrac{D - D_1}{2} = s$

35. $\dfrac{5}{9}(F - 32)$ or $\dfrac{5F - 160}{9} = C$ 37. $\dfrac{R - b}{a} = T$

39. $\dfrac{S - \pi r^2}{\pi r} = h$ 41. $\dfrac{L - \pi(r_1 + r_2)}{2} = d$

43. $\dfrac{d - \frac{1}{2}at^2}{t}$ or $\dfrac{d}{t} - \dfrac{at}{2} = v$ 45. $\dfrac{V - IR_2}{I} = R_1$

47. $\dfrac{L - L_0}{aL_0} = t$ 49. $\dfrac{Rd + kAT_2}{kA} = T_1$

Calculator Exercises, page 359

51. 2.6175   53. 2.1419   55. -0.3300

Review Exercises, page 361

1. $\dfrac{m + d}{c}$ 3. 4   5. $\dfrac{5(h - d)}{3}$ 7. 0.0587
9. 9.950   11. 27.2

Exercises 12.8, page 361

1. $8 + 3x = 29; 7$ 3. $4x - 17 = 59; 19$
5. $2x + 5 = x + 9; 4$ 7. $x - \dfrac{3}{8}x = 45; 72$
9. $x + (x - 38) = 146; 92$ and 54
11. $2x - x = 85; 85$ and 170
13. $x + 3(x - 4) = 216; 57$ and 53   15. 8 in., 12 in.
17. 19 in., 13 in.   19. 16 cm, 26 cm, 26 cm
21. 36 mph, 44 mph   23. 300 mi   25. 100 oz
27. 50 g   29. 75 oz   31. 9 ft   33. 6 m, 10 m

Calculator Exercises, page 362

35. 75.869, 71.035   37. 51.22, 56.97

Review Exercises, page 365

1. -1 3. 3   5. 33   7. 4   9. $\sqrt{73} \doteq 8.544$
11. 9.798

Exercises 12.9, page 365

1. $\pm\sqrt{8} \doteq \pm 2.828$ 3. $\sqrt[3]{93} \doteq 4.531$ 5. ± 5
7. $\sqrt[3]{31} \doteq 3.141$ 9. $\pm\sqrt{33} \doteq \pm 5.744$
11. $\sqrt[3]{12} \doteq 2.289$ 13. $\pm\sqrt{58} \doteq \pm 7.616$
15. $\pm\sqrt{3} \doteq \pm 1.732$ 17. $\pm\sqrt{a}$ 19. $\pm\sqrt{\dfrac{p}{2}}$

21. $\pm\sqrt{6m}$ 23. $\pm\sqrt{\dfrac{hm}{3}}$ 25. 5.477 ft^2
27. 5.657 cm^2 29. 2.449 amperes   31. 6.481 ft
33. 4.472 in.   35. 3.302 ft   37. 18.974 ft/sec
39. 9 ft 2$\dfrac{5}{8}$ in.   41. 4 ft 7 in.   43. 4 ft 6 in.

45. $\sqrt{\dfrac{P}{R}}$ 47. $\sqrt{\dfrac{5}{4\pi}}$ 49. $\sqrt{\dfrac{V}{h\pi}}$ 51. $\sqrt{\dfrac{H}{0.4n}}$

Calculator Exercises, page 366

53. 5.9146675   55. 22.500545

Review Exercises, page 368

1. -3 3. $pg - m$ 5. 22.5
7. $\pm\sqrt{12} \doteq \pm 3.464$ 9. 6.324   11. 50

Exercises 12.10, page 368

1. 3   3. 5.29   5. 24.74   7. 5   9. 4.47
11. 11.62   13. 6.3   15. 3.1   17. 36 ft
19. 16 ft/sec   21. 24 ft/sec   23. 4 ft/sec
25. $dv^2 = E$ 27. $\dfrac{1}{4Cf^2} = L$ 29. $\dfrac{v^2}{64} = h$
31. $\sqrt[3]{216t^2}$ or $6\sqrt[3]{t^2}$ 33. $\dfrac{s^2}{30f} = d$
35. $\sqrt{\dfrac{1 - M^2}{4M^2r^2t^2}}$ or $\dfrac{\sqrt{1 - M^2}}{2Mrt} = F$

Calculator Exercises, page 369

37. 10.88   39. 0.286

Extra Problems, page 371

1. a. $\dfrac{1}{40}$ b. $\dfrac{20}{1}$ c. $\dfrac{1}{3}$ d. $\dfrac{1}{6}$ e. $\dfrac{4}{3}$
3. $\dfrac{11}{2}$ 5. $\dfrac{7}{26}$
7. a. no   c. yes
   b. yes   d. yes
9. 65 minutes   11. 145 hours   13. $33,333.33
15. DE $\doteq$ 139 ft   17. 1470 cm^2
19. a. k=50,   b. $2250
21. a. $k = \dfrac{7}{80}$,   b. approximately 265 feet
23. a. k=0.0007   b. -0.07 ohms

CHAPTER 13

Exercises 13.1, page 375

1. ± 6 3. $\sqrt{6 \times 6} = \sqrt{6} \times \sqrt{6}$ 5. $\pm 6\sqrt{2}$
7. $\pm 6\sqrt{2}$ 9. $\sqrt{2} \times \sqrt{17}$ 11. $\pm 81\sqrt{3}$
13. $a\sqrt{x}\,\sqrt{y}$ 15. $i\sqrt{3}\,\sqrt{13}$ 17. $\pm 6\,i$
19. $\pm 2\,i\sqrt{3}\,\sqrt{5}$

Exercises 13.2, page 376

1. -1   3. -4   5. -7   7. -a   9. -2   11. 2
13. 2a-2   15. -2   17. -2+i   19. -b-i

Exercises 13.3, page 377

1. 1, -2, 4   3. 2, -3, -4   5. 1, -1, -1
7. 5, 0, -6   9. 4, 0, 3   11. 1, 1, 1
13. 1, -1, 6   15. 1, 0, -8   17. 3, 1, -4
19. (q+r), v, -14

Exercises 13.4, page 379

1. -3   3. $6 \pm \sqrt{51}$ 5. ± 4 7. 1 or -5
9. $(1 \pm i\sqrt{11})/3$ 11. $(3 \pm i\sqrt{131})/10$ 13. $-1 \pm i$
15. $(3 \pm i\sqrt{31})/4$ 17. 1 or -.60
19. -.33+1.11i or -.33-1.11i

Exercises 13.5, page 380

1. 2.35 in.   3. 12, 14   5. 2 or 1/2
7. 5.12 cm by 3.12 cm

CHAPTER 14

Review Exercises, page 385

1. 10   3. 9.2   5. 15   7. 9.5   9. 16.6 in.
11. 15.62

Exercises 14.1, page 386

1. $\sin\theta = 0.60$, $\cos\theta = 0.80$, $\tan\theta = 0.75$
3. $\sin\theta = 0.32$, $\cos\theta = 0.95$, $\tan\theta = 0.33$
5. c 7. $b - d - e$ 9. $\tan\theta$ 11. $\sin\theta$
13. $\dfrac{1}{\sqrt{3}} = \dfrac{\sqrt{3}}{3}$ 15. $\dfrac{1}{\sqrt{3}} = \dfrac{\sqrt{3}}{3}$ 17. 0.9962
19. 0.9744   21. 0.0523   23. 0.6157
25. 1.0000   27. 15°   29. 76°   31. 14°
33. 18°   35. 31°   37. 56 ft   39. 13.5 in.
41. 0.2309   43. 58°   45. 154.5 ft-lb

Review Exercises, page 389

1. $x = 63°$ 3. $x = 70°$ 5. $\angle C = 133°$
7. $\angle R = 54.5°$, $\angle T = 35.5°$ 9. $5\dfrac{1}{3}$
11. $x = 1\dfrac{2}{3}$, $y = 2\dfrac{7}{9}$, $z = 3\dfrac{8}{9}$

Exercises 14.2, page 390

1. 0.2340   3. 0.6394   5. 1.095
7. 0.0752   9. 42.96   11. 34°12'
13. 57°50'   15. 6°8'   17. 83°27'   19. 54°28'
21. $a \doteq 7.310, b \doteq 23.9075, \angle B \doteq 73°$
23. $\angle S \doteq 70°32', \angle R \doteq 19°28', s \doteq 2.83$
25. $\angle X \doteq 28°36', \angle Y \doteq 61°24', z \doteq 50.11$ 27. 85.9°
29. 27°49'   31. $a \doteq 6.89, b \doteq 14.9, \angle B = 65°10'$
33. $c \doteq 40.5, b \doteq 31.6, \angle B \doteq 51°20'$ 35. 115.5 ft
37. 14.4 ft, 66°27'   39. 6.18 in.   41. 47°10'
43. 2435 ft   45. 718 ft   47. 2.47 in.
49. 20°25'   51. 97 ft   53. 42°40'
55. 0.186 in.   57. 2.75 in.   59. 0.064 in.

Review Exercises, page 401

1. *D*   3. *A*   5. *B*   7. 0.9314   9. 0.2447
11. 24°40'

Exercises 14.3, page 402

1. 0.9063   3. −0.0872   5. 0.2700
7. −0.3173   9. $\angle B = 41°, a \doteq 1432.25, c \doteq 1422.95$
11. $\angle B \doteq 59°10', \angle C \doteq 35°40', c \doteq 28.50$
13. $\angle A = 105°, a \doteq 11.59, c \doteq 8.49$
15. $\angle A = 110°, a \doteq 7.02, c \doteq 3.51$
17. $\angle A \doteq 69°39', b \doteq 6570.29, c \doteq 4111.31$
19. $a \doteq 35.18$ 21. $\angle A \doteq 48°30'$
23. $\angle B \doteq 49°32', \angle C \doteq 67°28', a \doteq 32.80$
25. $\angle A \doteq 48°52', \angle B \doteq 58°35', \angle C \doteq 72°33'$
27. $b \doteq 3.58$ 29. $a \doteq 57.9, \angle B \doteq 38°53'$
31. 24°8', 100°58', and 54°54'   33. 201.4 ft
35. 58 ft, 87 ft   37. 37 ft $2\frac{1}{2}$ in.   39. $9\frac{3}{4}$ in.
41. 75.5 ft, 34°10'   43. 1624 ft

Review Exercises, page 412

1. $\frac{7}{13}$ or 0.5385   3. $\frac{3.2}{4}$ or 0.8   5. 66°30'
7. 78.75   9. $x = -14$ 11. $x = \frac{5yz}{3}$

Exercises 14.4, page 412

1. $\csc \angle A \doteq 1.062, \sec \angle A \doteq 2.833, \cot \angle A \doteq 0.3750$
3. $\cot \theta \doteq 0.7547$ 5. $\csc \theta \doteq 2.2810$
7. $\sec \theta \doteq 1.7434$ 9. 53°0'   11. 49°30'
13. 29°25'   15. $\sin 149° = \sin 31° \doteq 0.5150$
17. $\tan 305° = -\tan 55° \doteq -1.428$
19. $\cos 196° = -\cos 16° \doteq -0.9613$
21. $\sin 329° = -\sin 31° \doteq -0.5150$
23. $\cos 298° = \cos 62° \doteq 0.4695$

25. $\cot 185° = \cot 5° \doteq 11.4286$
27. $\csc 284° = -\csc 76° \doteq -1.0306$
29. $\sec 118° = -\sec 62° \doteq -2.1299$

Review Exercises, page 417

1. −0.5736   3. 9.65 in.   5. 23°30'   7. four
9. $r = \frac{3t}{k}$ 11. 43.96 cm

Exercises 14.5, page 418

1. 1.45 rad   3. 3.39 rad   5. 0.274 rad
7. 1.29 rad   9. 114.6°   11. 90°   13. 128.6°
15. 2.063°   17. $\frac{3\pi}{2}$ rad   19. $\frac{10\pi}{9}$ rad
21. $\frac{8\pi}{15}$ rad   23. $\frac{5\pi}{6}$ rad   25. 171°54'0"
27. 49°5'27"   29. 275°2'24"   31. 24 in.
33. 22.31 mm   35. 2.7 rad   37. 8.92 cm
39. 12.3 in.   41. 6097 ft   43. 314 rad/sec
45. 240 in./sec or 20 ft/sec or 1200 ft/min
47. 61.8 ft/sec   49. 196.25 ft/sec
51. 18.8 in./sec   53. 2.1 ft

APPENDIX C

Answers for the problems are included in the appendix.

INDEX

ISBN 1-57766-023-4

90000

9 781577 660231